Applied Mathematical Sciences

Volume 176

Editors

S.S. Antman
Department of Mathematics
and
Institute for Physical Science and Technology
University of Maryland
College Park, MD 20742-4015
USA
ssa@math.umd.edu

P. Holmes
Department of Mechanical and Aerospace Engineering
Princeton University
215 Fine Hall
Princeton, NJ 08544
USA
pholmes@math.princeton.edu

L. Sirovich
Laboratory of Applied Mathematics
Department of Biomathematical Sciences
Mount Sinai School of Medicine
New York, NY 10029-6574
USA
lsirovich@rockefeller.edu

K. Sreenivasan
Department of Physics
New York University
70 Washington Square South
New York City, NY 10012
USA
katepalli.sreenivasan@nyu.edu

Advisors
L. Greengard J. Keener
J. Keller R. Laubenbacher B.J. Matkowsky
A. Mielke C.S. Peskin A. Stevens A. Stuart

For further volumes:
www.springer.com/series/34

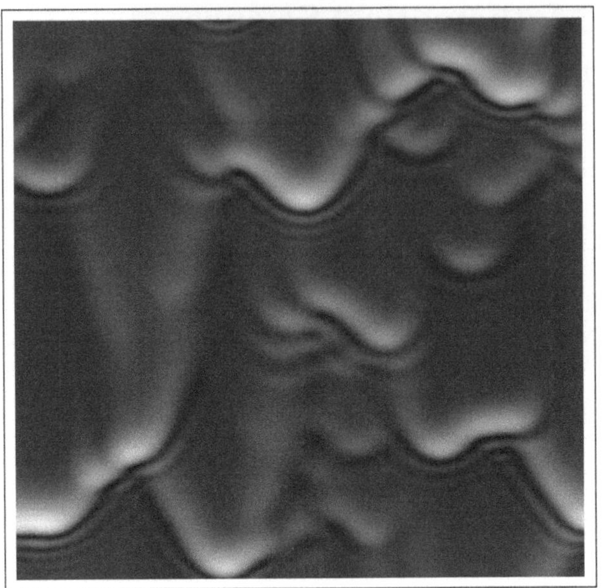

S. Kalliadasis · C. Ruyer-Quil · B. Scheid · M.G. Velarde

Falling Liquid Films

Springer

Prof. S. Kalliadasis
Department of Chemical Engineering
Imperial College London
London, UK
S.Kalliadasis@Imperial.ac.uk

Dr. C. Ruyer-Quil
Laboratoire FAST
Université Pierre et Marie Curie (UPMC)
Paris, France
Ruyer@Fast.u-psud.fr

Dr. B. Scheid
TIPs—Fluid Physics Unit
Université Libre de Bruxelles
Brussels, Belgium
BScheid@ulb.ac.be

Prof. M.G. Velarde
Instituto Pluridisciplinar
Universidad Complutense
Madrid, Spain
mgvelarde@pluri.ucm.es

Additional material to this book can be downloaded from http://extras.springer.com
Password: [978-1-84882-366-2]

ISSN 0066-5452   Applied Mathematical Sciences
ISBN 978-1-4471-2710-9   ISBN 978-1-84882-367-9 (eBook)
DOI 10.1007/978-1-84882-367-9
Springer London Dordrecht Heidelberg New York

British Library Cataloguing in Publication Data
A catalogue record for this book is available from the British Library

Mathematics Subject Classification: 00A69, 76-XX, 80-XX, 76Dxx, 76Exx, 76Mxx

Cover design: VTeX UAB, Lithuania

Printed on acid-free paper

Springer is part of Springer Science+Business Media (www.springer.com)

Preface

The wavy dynamics on a liquid film flow down an inclined plate is an everyday life phenomenon, easily observable on windows or on sloped pavements in the midst of a rainfall. It is a fascinating sight and so the design of many fountains includes falling liquid films to captivate and entertain passers-by. From the scientific point of view, such flows are part of the general class of free-boundary problems, which hold a strategic position both in pure and applied sciences. The occurrence of free-boundaries and interfaces, i.e., material or geometric frontiers between regimes with different physical properties not a priori prescribed, arises in disparate in nature, inherently nonlinear problems, from fluid and solid mechanics and combustion to financial mathematics, material science and glaciology. Not surprisingly therefore, the wavy dynamics of falling liquid films has attracted not only the attention of Sunday strollers but also of many researchers, and for several decades, to the point that literally hundreds, if not thousands of research papers have been devoted to this topic. Falling films have also been the subject of several books and monographs (see, e.g., [3, 44]) as well as reviews (see, e.g., [201]).

Such considerable interest, which continues up to date with several new developments, stems not only from the inner beauty of the phenomenon but also from its many technological applications, in particular in relation to chemical engineering processes. Typical examples include evaporators and related heat exchange processes with heat transport from a hot wall to a film and vapor condensation, absorbers/mass exchange processes with absorption of dilute gas and coating processes for which the hydrodynamic behavior of the initial liquid film can affect the quality of the final coated surface. The cooling of microelectronic equipment or the separation of multi-component mixtures in the chemical and food industries are often ensured by means of falling films. Falling films even represent the state-of-the-art technique in the sugar industry and constitute the basic components in sea-water desalination plants. Film heat exchangers are commonly used as condensers of cooling agents in cryogenic technology. In addition, films are also used as lubricant layers for the flow of crude oil in pipes and channels, or as the means of thermal protection of the combustion chamber walls in the design of rocket engines.

As far as the use of falling films for heat/mass transport applications is concerned, besides small thermal resistance and large contact area at small specific flow rates, another advantage is a drastic enhancement of heat/mass transport [58]. For example, Frisk & Davis [97] and Goren & Mani [106] have shown that heat/mass transport across a wavy film can increase by as much as 10–100% compared to flat films. Therefore liquid film flows play a central role in the development of efficient means for interfacial heat/mass transfer in engineering applications.

With regards to fundamental research efforts, it is not just the presence of a free boundary that contributes to the complexity of film flows, but also many other challenging aspects, e.g., heating effects and the way they influence the film flow, three-dimensional effects and chemical reactions, all with many different subtleties and peculiarities that have not been fully resolved as of yet. It is precisely for this reason that falling film flows are still the subject of active research, with several new developments as noted above.

In the recent past, the study of the transition from a state of order to one of disorder in spatially extended systems through low-dimensional dynamical models has been one of the many routes taken by physicists in their quest to understand the development of "spatio-temporal chaos" (or "low-dimensional turbulence") and even the onset of usual turbulence. A transition to spatio-temporal chaos also characterizes the dynamics of a falling liquid film. More specifically, one observes a well-organized cascade of bifurcations that leads from the flat film state (a "laminar" state) to a state of disorder/spatio-temporal chaos even at low Reynolds numbers. In the latter state, although the film surface appears to be random one can still identify robust "coherent structures," which continuously interact with each other. Such structures are described well with techniques from nonlinear dynamics and dynamical systems theory. The falling liquid film also shares many analogies and features with other open flow hydrodynamic systems, such as developing boundary layers.

However, unlike many open-flow hydrodynamic systems, the long wave nature of the instabilities observed on a falling film and the low-to-moderate values of the Reynolds number render the problem amenable to a thorough theoretical and numerical investigation within the framework of the long-wave theory. As the waves are long compared to the film thickness, or equivalently, deformations of the free surface are weak, the viscosity of the fluid ensures a great coherence of the flow across the film. These fortuitous characteristics inherent to the falling film problem, enable us to drastically reduce the complexity of the basic equations and to obtain systems of simplified model equations. The advantage of these models is to isolate the underlying physical mechanisms of the phenomena associated with the nonlinear wave evolution on a falling film and to simulate them extensively at a reduced analytical and numerical cost. Hence, a falling liquid film can serve as a canonical reference system for the study of the general problem of transition to spatio-temporal chaos and also for the study of other open-flow hydrodynamic systems.

The object of this research-oriented monograph is to summarize and report past and recent developments of the modeling of falling liquid films subjected or not to heat transfer. But because falling films are part of the general class of interfacial flows, we also outline the fundamentals of interfacial fluid mechanics. The conceptual framework, the underlying assumptions and the associated limits of applicability of the different methodologies are systematically given at each step of the derivations with the aim a ready-to-use text with easy access to mathematical models of different degrees of complexity. Details of the basic numerical methods we used, as well as an introduction to the software package AUTO-07P for continuation and bifurcation problems in ordinary differential equations [79], are provided in appendices and tutorials, with the purpose of offering easy access to the falling film area of research. These methods have other uses as well. For instance, the numerical solution of the Orr–Sommerfeld eigenvalue problem we offer is obviously useful not only for the falling film problem, but for numerous fluid flow problems as well, with or without interfaces. Gathering, ordering and giving a detailed overview and comprehensive, critical and pedagogical analysis of past and most up-to-date theoretical and wherever possible experimental advances on film flows have been demanding tasks in view of the numerous and vigorous efforts on the subject. Our sincere hope is that this monograph would be helpful to students and young scientists interested in the field and to scientists both in industry and academia already working with film flows or in general with interfacial fluid mechanics, hydrodynamic stability and nonlinear waves.

Before going any further, the reader will find useful Appendix A, where we render homage to two key scientists, P.L. Kapitza and C.G.M. Marangoni, who made pioneering contributions on falling liquid films and surface-tension-gradient phenomena. Their works have in turn inspired many applied mathematicians, physicists and engineers on falling liquid films and surface-tension-gradient phenomena, including ourselves.

London, UK	S. Kalliadasis
Paris, France	C. Ruyer-Quil
Brussels, Belgium	B. Scheid
Madrid, Spain	M.G. Velarde

Acknowledgements

This monograph originates from individual works and research collaboration of the co-authors for more than a decade.

The authors would like to acknowledge the following:

Paul Manneville for his deep insights, valuable comments and suggestions on falling films and for providing Figs. 8.3 and 8.4; Nikos Malamataris, Michalis Vlachogiannis and Vasilis Bontozoglou for the experimental image of falling film waves in Fig. 1.8 and the wave profile obtained by DNS shown in Fig. 6.4; Pierre Colinet, Eugene Demekhin, Oleg Kabov, Jean-Claude Legros, Alex Oron and Uwe Thiele for fruitful discussions on many aspects of modeling falling liquid films; Howard Stone for stimulating discussions on thin-film flows; Alberto Passerone and Giuseppe Loglio for correspondence concerning Marangoni's works, publications and biography.

Part of the research reported in this monograph was sponsored by the Franco-British Research Partnership Programme (Alliance), Engineering and Physical Sciences Research Council of the UK under Grants EP/F016492, EP/F009194, GR/S01023 and GR/M73767, European Union—Framework Programme 7 ITN 214919 (Multiflow), European Research Council Advanced Investigator Grant 247031, the F.R.S.-FNRS, and the Spanish Del Amo Foundation.

Contents

Chapter 1
Introduction

1.1 Brief Historical Perspective

The first systematic exploration of the flow characteristics of a liquid film flowing down an inclined plate ("plate" and "wall" are used indistinguishably in this monograph) in the region of moderate flow rates with and without forcing at the inlet was first provided by P.L. Kapitza [140, 141]. Kapitza did both theory and experiments. A short biography is given in Appendix A.1. As a physicist, his theory was built upon the intuitive simplest approach of balancing work or energy supply (due to the gravity field) with viscous dissipation. Curiously enough, even though his theoretical predictions apparently agreed with his and others' experiments, part of his analysis was in error.

Kapitza identified a dimensionless group, combining certain powers of the "kinematic surface tension" (as he called the ratio of the usual surface tension over the density of the liquid), kinematic viscosity (shear or dynamic viscosity divided by density) and gravity acceleration. This group, known today as the "Kapitza number" (to be defined in Chap. 2) is very useful in the study of wave growth and instability in falling liquid films. Kapitza predicted that if the Reynolds number (to be defined also in Chap. 2) was below a critical value which was a function of the Kapitza number, the liquid film flow would be uniformly laminar. The critical value "establishes well the moment of the transition from a laminar into an undulatory flow," which he also experimentally observed and described[1]; however, as said above, he was in part in error. We shall return to this point later in this chapter and in Chap. 4 when we define the transition between the two main regimes characterizing the waves on a falling liquid film.

Following Kapitza's pioneering theoretical and experimental work, many investigators have contributed to our understanding of the flow characteristics of falling

---

[1] The Kapitza number is actually "hidden" in equation (IX) of his first paper. In fact his prediction for the instability threshold is given in terms of the "reduced Reynolds number," to be defined in Chap. 4.

S. Kalliadasis et al., *Falling Liquid Films*, Applied Mathematical Sciences 176, DOI 10.1007/978-1-84882-367-9_1, © Springer-Verlag London Limited 2012

liquid films. It is not the purpose of this introductory chapter to summarize or item-ize in a list all significant works. Most of them will be discussed in the following chapters. However, there are two monographs, to our knowledge, that have dealt with the subject of the present monograph, which we ought to mention at this stage. One is the monograph by Alekseenko, Nakoryakov and Pokusaev [3] and the other is by Chang and Demekhin [44]. The former focuses on two-dimensional isother-mal films with some exposition to heat transfer and analyzes primarily experimental aspects, though some theoretical developments are also included. The latter restricts attention to mathematical and numerical results of isothermal falling liquid films and focuses primarily on two-dimensional effects with some short exposition to three-dimensional ones through a simple model equation. The present monograph aims to cover isothermal falling films (both two-dimensional and three-dimensional effects) and falling films in the presence of heat transfer/Marangoni effects (both two-dimensional and three-dimensional effects). This is accomplished through a balanced and detailed presentation of the state-of-the-art mathematical and numeri-cal methodologies used to describe the evolution of a falling liquid film in time and space, both isothermal and heated (and, whenever possible, the link between theory and experiments is illustrated). The concepts and tools required for the modeling of a falling liquid film are introduced in a way that the reader mastering them should be able to use them to analyze additional complexities and effects in film flows but also for other purposes/problems.

At this point it seems pertinent to discuss some of the phenomena to be studied at length in the subsequent chapters.

1.2 Basic Phenomena of a Falling Film

1.2.1 Surface Wave Instability

Experiments with a thin liquid film flowing down an inclined plate under isothermal conditions have shown the development of "long" wavelength deformations on its open (referred to as "free," as we shall justify in Chap. 2) surface, i.e., deformations much longer than the film thickness, as sketched in Fig. 1.1. "Short" waves have not been observed in experiments, at least not at smaller flow rates. These long waves seem to result from the instability of an initially uniform laminar flow (flat film base flow; it will be discussed in detail in Chap. 2). For a vertical geometry ($\beta = \pi/2$) wavy motions appear as soon as the film flows down the plate.

There are three related mechanisms influencing this *long-wave hydrodynamic instability* [256]: (i) One is due to the presence of gravity, more precisely its stream-wise component, which is a body force pushing the liquid to fall down to a minimum potential level; (ii) another is inertia, whose subtle role along with that of viscosity we shall carefully elucidate later in this monograph; (iii) the third one is the cross-stream component of gravity leading to hydrostatic pressure that tends to maintain equipotential levels and hence tends to prevent surface deformation. Needless to say,

Fig. 1.1 Sketch of a thin
liquid film of mean film
thickness $\bar{h}_N$ flowing down an
inclined plate of inclination
angle β with respect to the
horizontal direction; U is the
semiparabolic velocity profile
corresponding to the
"fully-developed" viscous
film flow

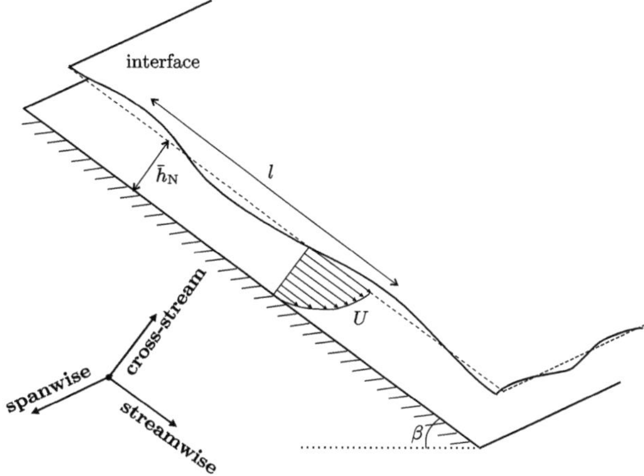

surface tension and, depending on the circumstances, surface tension gradients and
thermal diffusivity come into play. Let us now describe the above three mechanisms
in general physical terms:

i. Consider a perturbation to the flat liquid film flow in which the free surface is
deflected slightly upward over a lengthscale l that is much longer than the depth
$\bar{h}_N$ of the film (see Fig. 1.1). Because the height of the top surface varies slowly
in the streamwise direction, the velocity distribution at each streamwise location
will remain close to that of a fully developed viscous film flow characterized to
a good first approximation by a semiparabolic (half-Poiseuille) profile depicted
by U in Fig. 1.1. Indeed, by neglecting the hydrodynamic drag of the ambient
gas atmosphere, the theory predicts, that for low flow rates (or, equivalently,
low values of the Reynolds number), the velocity profile in the liquid film is
semiparabolic. It can also be shown that the net streamwise flow rate in the film
is positive and that it increases with the depth of the film. Thus, at the crests of
the deflection the streamwise flow rate is at a maximum, and it is at a minimum
at the troughs. The net result, as shown in Fig. 1.2, is that gravity draws fluid
toward the front face of a crest, deflecting it upward while at the same time it
draws fluid from the rear face, deflecting it downward. This first mechanism
produces a wavy downstream motion of the perturbation without growth and at
a phase speed higher than the velocity of any fluid particle in the undisturbed
film.

ii. Consider now at a particular instance in time a streamwise location that is at the
front face of a perturbation crest. Here, the surface height is increasing because
of the forward motion of the perturbation. The flow in the bulk of the film is
accelerating at this position because it is attempting to follow the fully developed viscous velocity profile dictated by the surface height increase. However,
inertia effects prevent the flow from accelerating fast enough to completely follow this velocity profile. The result is that the volume flux in the film is not as
large as one due to a fully developed film flow. At the rear face of the crest, the

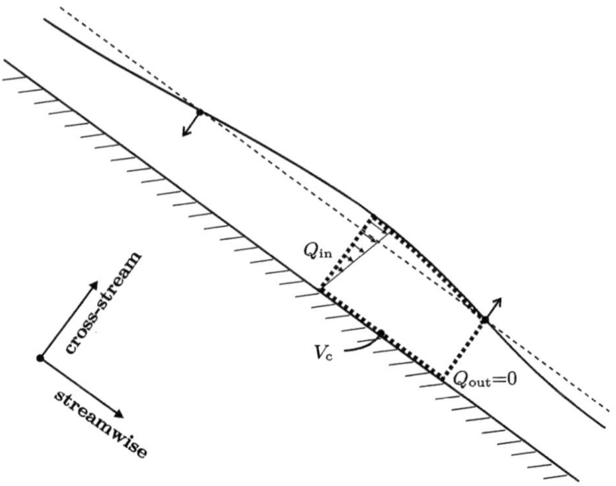

Fig. 1.2 Propagation
mechanism of a perturbation
(*solid line*) of the originally
undisturbed free surface
(*dashed line*). The control
volume V_c (*dotted box*)
experiences a net inflow Q_{in};
to conserve mass the interface
then must move upward.
Likewise, the interface at the
rear of the crest must move
downward
(upward/downward motions
are indicated by *arrows*)

Fig. 1.3 Film flow and
induced interfacial motion
produced by the effects of
inertia. The *dashed line* is the
originally undisturbed free
surface

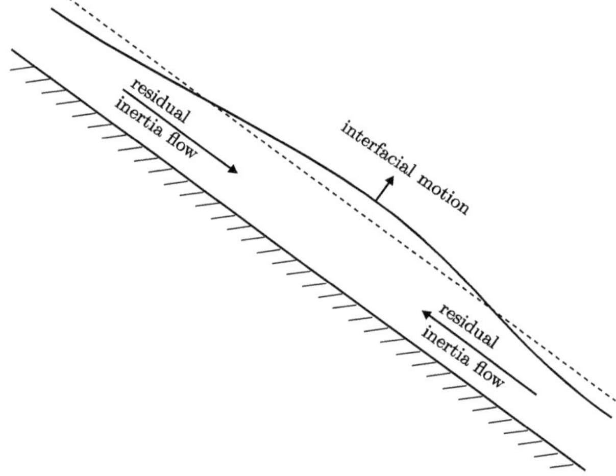

velocity is decreasing, but inertia effects similarly prevent the flow from decelerating too rapidly. Thus, the volume flux in the film is larger than that due to a fully developed film flow. The net effect of these two bulk fluxes results in an accumulation of fluid underneath the perturbation crest and an increase in the interfacial displacement, as shown in Fig. 1.3.

iii. Due to the cross-stream component of gravity, the perturbation also produces an increase in the hydrostatic pressure under the crest, proportional to the local depth of the film. This pressure tends to push fluid away from the crest and toward the troughs were the hydrostatic pressure is lower, resulting in a depletion of the fluid under the crest and a decrease in the depth of the film, as shown in Fig. 1.4. This stabilizing flow competes with the inertia accumulation of the fluid under the crest. If inertia is strong enough, the film is unstable and the perturbation grows. Hence, a film flowing a vertical wall is always unstable to

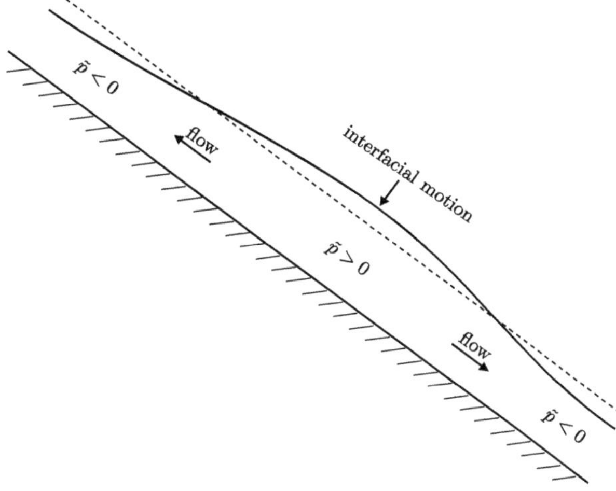

Fig. 1.4 The direction of the perturbation film flow and the induced interfacial motion when an increase of the hydrostatic pressure lies underneath a perturbation crest; $\tilde{p}$ is the change of hydrostatic pressure due to perturbations. The *dashed line* is the undisturbed free-surface position

free-surface perturbations since the effect of hydrostatic pressure is canceled out [304]. As far as surface tension is concerned, it does not need to come into play unless wavelengths reach the "capillary length," which is about a millimeter for, e.g., liquid water in standard conditions.

Noteworthy is that the intuitively appealing approach followed by Kapitza—that of determining the flow characteristics from a thermodynamic criterion in which energy dissipation due to viscosity is in balance with gravitational work—is naive and not sufficient in the case of a falling liquid film. Surely, the uniform laminar flow (half-Poiseuille flow) can be obtained from such a thermodynamic criterion. In general, the balance between viscous dissipation and energy supply leads to a family of steady solutions. The solution that actually occurs can then be determined by the minimization of the "viscous dissipation function" (defined in Appendix D.1, also referred to as "Rayleigh dissipation function") for given boundary conditions. But the uniform flow obtained from this minimization process in the case of a film on a plate is only observable for horizontal and inclined layers ($\beta \neq \pi/2$) and not when the plate is vertical ($\beta = \pi/2$). This point was the crux of the misunderstanding made by Kapitza: He thought of the wavy film dynamics as some kind of "equilibrium state" whose energy dissipation could be defined as a function of "state variables" (amplitude and wavenumber of the sinusoidal perturbations he considered).

But a sinusoidal perturbation to the flat film is a nonequilibrium state. In fact, for vertical layers, Benjamin [19] showed unequivocally that the previous result by Kapitza on the instability threshold was in error, when, in view of the apparent absence of waves on very "thin films," he concluded that, for the flow down a vertical plane, there exists a critical flow rate (or critical Reynolds number) calculated from the above thermodynamic criterion, below which the uniform laminar flow is entirely stable. Benjamin studied the stability of the uniform laminar flow for an arbitrary inclination angle $\beta \neq \pi/2$ and showed that although there is a range of low flow rates for which such base flow could be observed, this is not possible when the plate is vertical ($\beta = \pi/2$), in which case the flow is unstable for all flow rates so that a critical flow rate (or equivalently a critical Reynolds number) in the usual sense does not exist. In other words, for all finite values of the Reynolds number there is a class of sinusoidal perturbations which undergo unbounded amplification according to the linear theory.

Further, he showed that surface tension does not alter the general conclusion regarding the critical Reynolds number. In fact, a key point of the hydrodynamic instability mechanism leading to long-wavelength perturbations on the surface of the film, also referred to as the *Kapitza mode* or *H-mode*, is that the waves generated at the interface travel much faster than any fluid particles inside the film. Inertia plays a central role in the growth of the instability by introducing a shift between the vorticity field generated by the waves and the film surface displacement [19].

1.2.2 Flow Evolution Features

The instability onset as outlined in the previous section is just a small part of wave evolution on a falling film. According to observations, as well as predicted by theory, falling liquid films generally exhibit a cascade of symmetry-breaking bifurcations leading from the flat-film flow to a two-dimensional (streamwise dimension and height) periodic wave train, which eventually evolves into *solitary waves* and further to three-dimensional (streamwise, spanwise dimensions and height)[2] solitary waves and complex flows/wave patterns. Figure 1.5 illustrates experimentally this sequence of events. We note that since the pioneering experiments undertaken by Kapitza, many more experiments have been performed to describe in detail the flow evolution features in falling liquid films [1, 4, 25, 56, 57, 167–170, 187, 210, 211, 262, 264, 294, 295, 298]. A historical summary of experimental results can be found in [44]; here we focus on some of the most recent experimental results.

Let us look at the different stages of the long-wave (much longer than the liquid film thickness) evolution in more detail. The first stage of the instability is characterized by the linear growth of the H-mode. It has been established that such instability is of the *convective* type. In other words, the perturbation grows in the wave-moving frame of reference whereas at a fixed lab/plate point it decays. Equivalently, the perturbation is advected downstream by the flow while it is being fed by the noise upstream. We can say that the flow is kind of a *noise amplifier*. (N.B. As opposed to the convective instability, an *absolute instability* is one for which a perturbation seen at a fixed lab/plate point grows as time proceeds. These two instabilities will be discussed in detail in Chap. 7.)

---

[2]For a two-dimensional flow there is no spanwise dependence and no spanwise velocity component, while the surface elevation varies in the streamwise direction only. For a three-dimensional flow there is dependence in all three directions, while the surface elevation varies in both the streamwise and spanwise directions.

Fig. 1.5 Shadow image of waves naturally occurring on a film of water falling along a vertical plate at Reynolds number $Re = 33$. Reprinted with permission from Park and Nosoko, *AIChE*, 49(11):2715–2727, Wiley, 2003

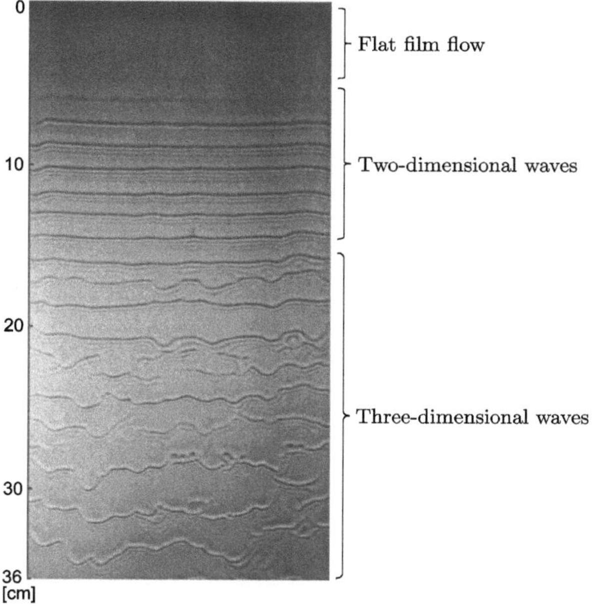

The first stage is followed by a second stage with linear saturation of this growth. Further downstream the wave loses stability, e.g., as a consequence of interaction between the wave harmonics, and the wave pattern is altered, embracing a wide spectrum. This process eventually yields highly asymmetric and nonlinear waves containing most of the liquid, and each of which consisting of a *hump* with a long flat tail behind. The front of the hump is steep and is preceded by small ripples (also denoted *radiation*) with a wavelength close to the originally linearly fastest growing wave. These highly asymmetric and nonlinear waves are what we already referred to as solitary waves and have a velocity that increases with their amplitude. Figure 1.6 shows quasi-two-dimensional solitary waves on a water film on the surface of a street during a rainy day.

As solitary waves propagate downward they become transversely modulated until they form horseshoe-like structures further downstream (not shown in Fig. 1.6; these events will be analyzed in detail in Chap. 8). At this stage of the evolution the interface is characterized by a turbulent-like dynamic in which horseshoe-like three-dimensional solitary waves continuously interact with each other.

The wave pattern in a falling film can be very sensitive to perturbations like time-dependent forcing at the inlet, as shown in Fig. 1.7. For sufficiently large frequencies, the waves are initially two-dimensional but they easily become three-dimensional, and exhibit rather irregular patterning, eventually leading to seemingly turbulent flow (at a forcing frequency of 100 Hz). Various distinct wave flow features can be identified as the flow rate is increased or the frequency of the forcing at the inlet is varied, as illustrated in Fig. 1.7. For instance, there is a first region that corresponds to growing infinitesimal perturbations considered as noise near the inlet that eventually yield downstream a two-dimensional wave of well-defined wavelength.

Fig. 1.6 Quasi-two-dimensional solitary pulses on a water film. They are disturbed by transverse perturbations resulting in solitary waves with curved fronts but each transverse cross-section of these waves resembles a two-dimensional solitary wave. The photograph was taken on a rainy day in Orsay by one of the authors: the film is falling down the sloped street Rue de la Colline in the neighborhood of Laboratoire FAST

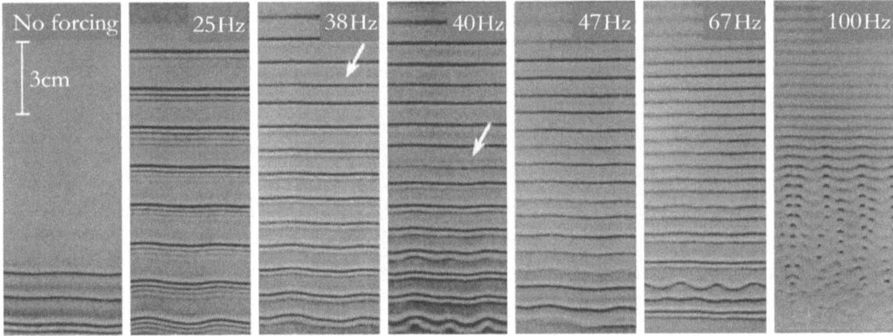

Fig. 1.7 Shadow images of waves in a film of water flowing down a vertical plate at $Re = 52$. With the exception of the *first panel on the left*, the inlet flow rate is periodically forced at the frequencies indicated in the *upper right corner of each panel* (in Hz). The *white arrows* indicate the appearance of double-peaked waves. Reprinted with permission from Nosoko and Miyara, *Phys. Fluids*, 16(4):1118–1126, American Institute of Physics, 2004

Without forcing (first panel from the left) the wave selected in the first stage corresponds to the fastest growing wave according to the linear stability analysis of the uniform laminar flow. Also, at sufficiently low-frequency of the inlet-forcing, solitary waves appear right after the linear waves exhibit maximum growth with no saturation of the latter (see, e.g., for 25 Hz in Fig. 1.7).

Noteworthy is that some experiments have been done at small inclination angles, so that the development of the free surface is more gradual and the waves remain two-dimensional for a longer distance, as in Fig. 1.6 or Fig. 1.8. Also, for small angles the film thickness is typically thicker, which allows for better resolution in the measurements. Examples of high quality measurements of two-dimensional solitary waves are shown in Fig. 1.9.

Fig. 1.8 Solitary pulses on the surface of a water film flowing down a plate forming an angle of 5° with the horizontal direction at $Re = 44$ and forced at a frequency of 2 Hz. The channel width is 45 cm. Photo courtesy of Profs. M. Vlachogiannis and V. Bontozoglou

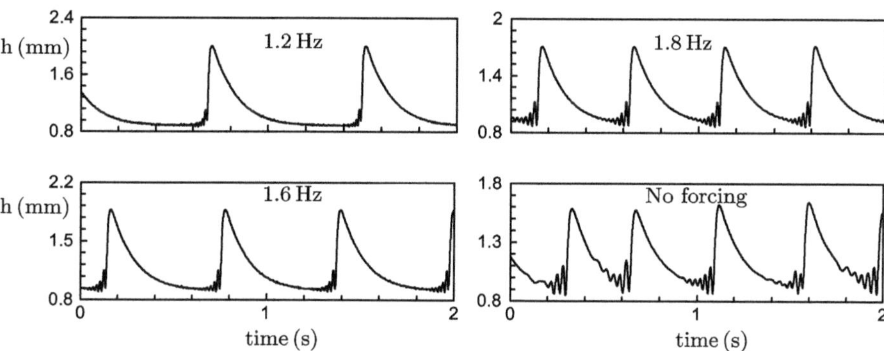

Fig. 1.9 Time traces of the film thickness measured on a film of a 5% by weight aqueous solution of polyalkylene glycol flowing down a plate forming an angle of 5° with the horizontal direction at $Re = 35$ and forced at certain frequencies. With permission from Springer Science+Business Media. Solitary waves on inclined films: their characteristics and the effects on wall shear stress, *Exp. Fluids*, 41, 2006, pp. 79–89, J. Tihon, K. Serifi, K. Argyiriadi, and V. Bontozoglou, Fig. 2, Springer, 2006

The rich phenomena of isothermal falling films will be analyzed in detail in Chaps. 7 and 8, where the two-dimensional and three-dimensional models developed in Chaps. 6 and 8 for the flow, respectively, will be compared to experimental data available in the literature [3, 167–170, 203].

1.2.3 Marangoni Effect

Surfactants or heat are known to significantly affect interfacial flows. The so-called *Marangoni effect*—named after C.G.M. Marangoni (see short biography in Appendix A.2)—is the appearance of flow or the modification of an existing flow due to surface tension nonuniformity, e.g., when the surface tension varies along an interface/free surface with temperature or concentration. This effect is referred to as the *thermocapillary* or *solutocapillary effect*, respectively; only the former will be considered in this monograph.

When a temperature gradient is applied across a horizontal fluid layer being heated from below (throughout this monograph we shall consider the heating from the plate only), the thermocapillary Marangoni effect can lead from a quiescent conducting state to motion. Two forms of motion may occur and the corresponding mechanisms have been classified by Goussis and Kelly [107] as the *P-mode* and the *S-mode*. The P-mode generally yields "steady convection rolls" or hexagonal or square cells, the size of which is of the same order of magnitude with the depth of the layer (convective motions in a wide variety of physical settings are discussed in [191]). These convective patterns are called after Bénard, who around 1900 provided their systematic experimental exploration. The instability leading to Bénard cells is known as the "Marangoni–Bénard instability" and was first studied by Pearson [206]. This instability may occur even with a nondeformable free surface (in fact, Pearson did not account for the deformability of the surface). On the other hand, the S-mode corresponds to significant long-scale deformations, whose horizontal size is much larger than the depth of the layer. This instability is referred to as the *long-wave Marangoni instability* and was clearly explained by K.A. Smith [255].[3] The P-mode will be analyzed in Chap. 3. However, since this monograph is devoted to thin films—with thicknesses much smaller than a millimeter—and their associated long-wave instabilities, the P-mode which is of short-wave type will be neglected in subsequent chapters. Therefore, for the remainder of the monograph, with the exception of Chap. 3 where we examine in detail the linear stability characteristics of heated falling films, when speaking about the Marangoni effect we refer to the long-wave instability, or S-mode.

Let us now examine in detail the mechanism of the surface-tension-driven long-wave thermocapillary instability, assuming that, as for most liquids, the surface tension decreases with increasing temperature. Consider a cross-section of a horizontal liquid layer heated from below at temperature T_w as sketched in Fig. 1.10; T_∞ is the temperature of the ambient gas. If a spontaneous infinitesimal deformation occurs at the interface at time t_1 as shown in the figure, the film temperature in the trough of a depression, T^+, will be larger than at the crest of an elevation, T^-, provided that $T_w > T_\infty$. Because surface tension decreases with temperature, a flow is induced along the interface moving liquid away from the hot spot. This flow acts so as to amplify the initial perturbation to that shown in time t_2. What mainly opposes the deformation is gravity through the hydrostatic pressure, which tends to maintain the flatness of the horizontal layer, i.e., leveled with an equipotential. For perturbations of sufficiently short wavelength, surface tension also acts and tends to suppress surface deformation, a consequence of a law found by Laplace. Therefore, the

---

[3] Actually, this mode was first obtained by Scriven and Sterling [246] but without gravity. Thus, they found that a horizontal liquid layer subjected to a vertical temperature difference is always unstable with respect to long-wave perturbations. Smith then added gravity and found the critical Marangoni number for the long-wave mode.

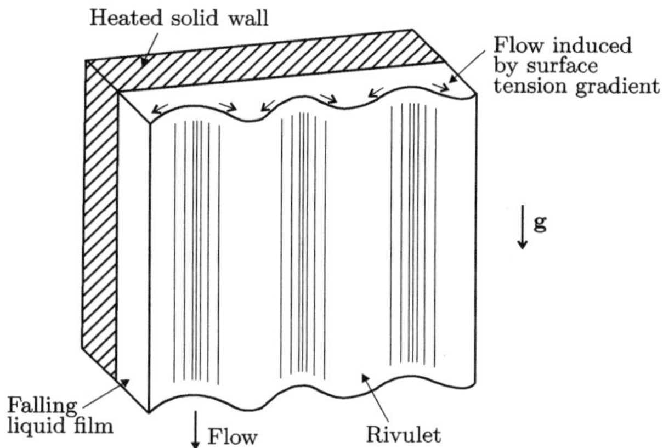

Fig. 1.10 Mechanism of the long-wave Marangoni instability (S-mode) for a film heated from below. T_w is the wall temperature and T_∞ is the temperature of the ambient gas phase above the film

Fig. 1.11 Convective rivulet pattern due to the S-mode at the surface of a falling liquid film heated from below. This pattern will be appropriately modified by the presence of hydrodynamic waves due to the H-mode

instability with respect to this mode should take the form of long-wavelength perturbations for which the hydrostatic pressure is a stabilizing force. This instability occurs when the balance between the Marangoni stress produced by the temperature gradient across the layer and the hydrostatic pressure turns in favor of the former. The instability can lead to formation of dry spots when at the troughs the liquid film becomes too thin. For a complete theory in this case we need to incorporate forces of nonhydrodynamic origin, i.e., long-range attractive intermolecular interactions (between the solid and the gas phases separated by the liquid). We shall not do this in the present monograph but we shall discuss the significance of including such forces in Chap. 9 when we analyze three-dimensional wave patterns in heated falling films.

The consequence of tilting the plate is that the liquid will flow down driven by gravity, while at the same time we have the long-wave Marangoni instability. This would lead to flow into downstream aligned rivulets as experimentally observed and drawn in Fig. 1.11. We shall return to this point in Chap. 9.

The coupling between the two instabilities, thermocapillary-driven motions and the surface wave instability of a falling liquid film, has been studied by several investigators. It has been shown from the linear stability analysis of a heated falling liquid film that the thermocapillary S-mode predominates at low enough flow rates—or equivalently for sufficiently thin films—where the Marangoni stress in the presence of surface deformation, generated by the destabilizing temperature gradient

Fig. 1.12 Coexistence between three-dimensional hydrodynamic waves flowing downstream and rivulets aligned with the flow induced by the thermocapillary effect for pure water at $Re = 22$ [133]. The wall is heated by a constant heat flux of $0.8\ \mathrm{W\,cm^{-2}}$. The heated section of the wall has a length of 150 mm and begins 120 mm from the inlet so that the hydrodynamic waves are already well developed before the Marangoni effect starts influencing their dynamics. Photo courtesy of Prof. O.A. Kabov

across the layer, overrules the stabilizing hydrostatic pressure [107]. On the other hand, the hydrodynamic H-mode prevails at high enough flow rates where the destabilizing inertia effects become dominant. However, for a wide range of parameter values, the S- and H-modes may coexist and reinforce each other [107, 128]. Figure 1.12 shows experimental evidence [133] of the coexistence between rivulet structures produced by transverse thermocapillary effect (as for Fig. 1.11) and three-dimensional hydrodynamic waves (that are also modified by thermocapillarity).

It has also been shown that when the S- and H-mode reinforce each other and the Marangoni effect is strong enough, we may have dry patch formations as sketched in Fig. 1.13. Again a complete theory of this phenomenon would require taking into account long-range attractive intermolecular interactions.

1.2.4 Inhomogeneous Heating

Experimental studies have also been performed on thin films falling along inhomogeneously heated vertical plates. Various convective patterns have been observed. Figure 1.14 illustrates the case of a locally heated film at rather low flow rates. At the upper edge of the heater, the temperature of the plate increases along the flow direction. Consequently, as the temperature of the fluid surface increases, surface tension decreases downstream. The concomitant surface tension gradient produces a Marangoni flow opposed to the gravitationally driven flow. This competition produces a horizontal bump of locally increased film thickness at the upper edge of the heater, which becomes unstable and develops rivulets with a well-defined wavelength in the spanwise direction. This problem has been analyzed theoretically in [138, 254].

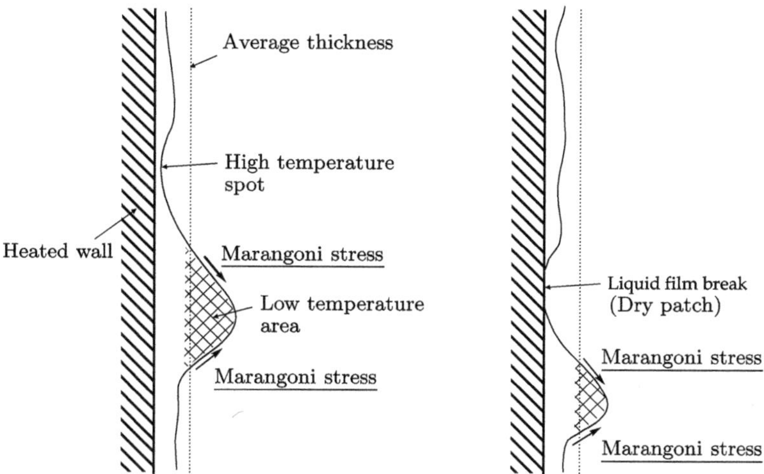

Fig. 1.13 The Marangoni effect reinforces the hydrodynamic wave instability and can lead to dry-patch formation [124]. *Left panel* shows the film thickness prior to the dry patch formation (*right panel*)

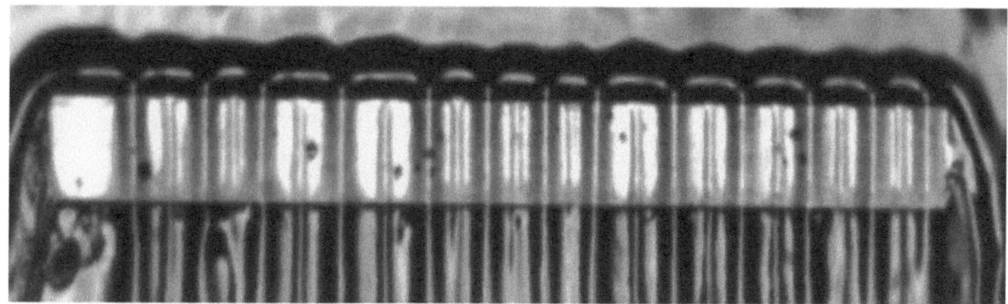

Fig. 1.14 Steady regular structure for a 10% ethyl-alcohol aqueous solution, $Re = 1$. The bright rectangular zone corresponds to the heater. The characteristic wavelength of the structure is 10 mm, while the film thickness before the heater is about 100 μm [134]. Photo courtesy of Prof. O.A. Kabov. Reprinted with permission of Gian Piero Celata

The phenomena shown in both Figs. 1.12 and 1.14 are key in this monograph. In both cases, the Marangoni effect (once again considered only when heating occurs from below) plays a crucial role in the convective/wavy pattern formation. In the homogeneous heating case of Fig. 1.12, the flow rate is sufficiently moderate such that hydrodynamic waves have relatively high amplitudes and become rapidly three-dimensional. In the local heating case of Fig. 1.14, the flow rate is low enough such that hydrodynamic waves have very small amplitudes while the thermocapillary effect is dominant and induces steady regular patterns.

1.3 Mathematical Modeling and Methodologies

Depending on the flow regime, different levels of reduction can be used to simplify the governing equations, namely Navier–Stokes and Fourier equations, and the wall and free-surface boundary conditions. As already noted, interfacial deformations (such as those in Figs. 1.12 and 1.14) are long compared to the film thickness. The cross-stream scale is then well-separated from the streamwise and spanwise scales in a way reminiscent of the separation of scales underlying the boundary layer theory in aerodynamics (e.g., [243]). The approximations that lead to the *boundary layer equations* there also apply for thin film flows where the pressure is mostly governed by gravity and surface tension. The disparity in scale between long-wavelength streamwise/spanwise deformations and film thickness is referred to as the *long-wave approximation* (or *lubrication approximation* for vanishing Reynolds numbers). This approximation enables us to perform a "gradient expansion" of the velocity and temperature fields and to subsequently obtain systems of equations of reduced dimensionality to model the dynamics of the flow.

In the region where inertia is not important, namely if the Reynolds number is small, the velocity and temperature fields can be considered slaved to the kinematics of the free surface and a single evolution equation for the film thickness h can be derived either from the full Navier–Stokes and Fourier equations and associated wall and free-surface boundary conditions or the boundary layer equations, as we shall see in Chap. 5. For isothermal films, this was first done by Benney [21] followed by several other authors (see the reviews by Oron, Davis and Bankoff [201] and Craster and Matar [61]). The resulting equation, frequently referred to as the *Benney equation* and denoted in the following discussions, as BE, has the form

$$\partial_t h + h^2 \partial_x h + \partial_x \left\{ \left(A h^6 - B h^3 \right) \partial_x h + C h^3 \partial_{xxx} h \right\} = 0, \tag{1.1}$$

where $h(x,t)$ denotes the location of the free surface of the liquid film and A, B and C are parameters that we shall define in Chap. 5. ∂_t and ∂_x denote time and space partial derivatives, respectively.

Benney's approach is exact close to criticality/instability threshold as far as critical, neutral conditions (critical Reynolds number, neutral curve) and interfacial quantities are concerned. However, as the Reynolds number increases, it breaks down rapidly at an $\mathcal{O}(1)$ value of the Reynolds number, leading to unacceptable finite-time blow-up behavior. This behavior is a sign of the intrinsic dynamics of the slaved modes (through the gradient expansion associated with the long-wave approximation) for higher Reynolds numbers. Ooshida realized that the gradient expansion leading to the BE is divergent and to overcome the unrealistic behavior of the BE he modified the expansion appropriately by using a "Padé approximants" regularization procedure for divergent asymptotic series [196]. However, although the equation he obtained does not suffer from the drawback of finite-time blow-up, it does fail to describe accurately the dynamics of the film at moderate Reynolds numbers (in the region ~ 10–50) since its solitary-wave solutions exhibit unrealistically small amplitudes and speeds.

There is also the so-called *integral boundary layer approximation* (IBL), which performs much better in the region of moderate Reynolds numbers. IBL is derived from the boundary layer equations and combines the assumption of a semi-parabolic velocity profile within the film with the Kármán–Pohlhausen averaging method of boundary layer theory in aerodynamics. This formulation was first used by Kapitza [140] to describe stationary waves and later on extended by Shkadov [248] to nonstationary two-dimensional flows and by Demekhin and Shkadov to three-dimensional ones [71]. We shall hence refer to this approach as the *Kapitza–Shkadov approximation*. In two dimensions, the approximation leads to two coupled evolution equations for film thickness h and streamwise flow rate q. Results for isothermal nonlinear waves far from criticality obtained from the Kapitza–Shkadov model are in quantitative agreement with the boundary layer [72] and full Navier–Stokes equations for moderate Reynolds numbers [70, 218, 232]. However, the Kapitza–Shkadov approach does not predict well neutral and critical conditions, except for large inclination angles. Indeed, it has been shown that for a vertical falling film, the Kapitza–Shkadov model gives the correct value for the critical Reynolds number, i.e., zero. For all other inclination angles, the model is off by 20%. This discrepancy originates in the velocity profile assumed in the Kapitza–Shkadov approximation [226]. Although this profile seems to be in agreement with the experiments by Alekseenko, Nakoryakov and Pokusaev [3] and hence does capture most of the physics, corrections to the profile known to exist at first order in the gradient expansion (a point to be discussed in detail in Chap. 6), are important for an accurate prediction of the linear instability threshold.

This shortcoming has been resolved by combining the gradient expansion with a "weighted residuals" technique using polynomials as test functions [227, 228]. The resulting models will be referred to hereinafter as *weighted residuals models*. In the simplest case, a *Galerkin* method, leads to a "first-order model" involving two coupled evolution equations for the film thickness h and the local flow rate q, much like the Kapitza–Shkadov model but with different numerical coefficients of the terms originating from inertia. The first-order model predicts the correct linear instability threshold for all inclination angles while at the same time it also predicts rather well the nonlinear flow features. This is a direct consequence of using a detailed representation of the velocity field accounting for its deviations from the semiparabolic profile. A "second-order model" has also been developed that takes into account the second-order viscous effects. It involves four equations by allowing the corrections to the semiparabolic velocity profile to evolve according to their own dynamics. Further, an approximation to this four-field model, a "simplified second-order model," has been proposed involving only two fields, which allows us to easily scrutinize the role played by the second-order viscous effects on wave profiles and stationary wave selection. The spatial evolution of the solutions of the simplified second-order model in the presence of noise or periodic forcing agrees rather well with both experiments and direct numerical simulations of the boundary layer equations. We shall be examining in detail these points in Chaps. 6 and 7.

A number of authors have considered the role of heat on the evolution of a falling liquid film. For the problem of a film flowing down along a uniformly heated wall,

Joo, Davis and Bankoff [128] included in addition to thermocapillary effects, evaporation and intermolecular forces. They used the long-wave approximation to obtain an equation for the evolution of the local film thickness. Without evaporation effects and intermolecular forces, which are beyond the scope of this monograph, their evolution equation will also be referred to hereinafter as BE. The equation has a structure similar to (1.1) but with some additional terms, of course, to account for the Marangoni effect. These authors compared the influence of the H and S modes on the shape of nonlinear waves by performing numerical experiments. They observed that both instability modes reinforce each other (see also our earlier discussion). They also noted that the H-mode is more sensitive to the local layer thickness—with the wave crests growing more rapidly compared to the troughs, which diminish in time—than the S-mode, for which the growth of the crests and troughs is similar.

The interaction between the S- and H-modes has also been studied for two-dimensional flows with moderate Reynolds numbers, i.e., outside the range of validity of the BE by appropriately modifying the Kapitza–Shkadov approximation to include the Marangoni effect. Therefore, as for the isothermal case, the Kapitza–Shkadov model for the heated falling film does not suffer any finite-time blow-up like the BE and solitary wave solutions have been obtained for higher Reynolds numbers than with the BE. Nevertheless, the model suffers from the same limitations with the Kapitza–Shkadov model for isothermal films, i.e., it does not predict accurately the behavior of the film close to the linear instability of the semiparabolic base flow when the plate is inclined. The limitations of the Kapitza–Shkadov approach for two-dimensional heated films have been corrected in previous studies [230, 240] by extending the weighted residuals approximations discussed above to include the Marangoni effect. This work will be reviewed in detail in Chap. 9, where we will also extend it to three dimensions.

Finally, we note that for small deviation amplitudes from the flat film, the models discussed above, e.g., BE and Kapitza–Shkadov, can be simplified substantially via a weakly nonlinear expansion, which leads to different weakly nonlinear evolution equations. This is done in Chap. 5 for the BE. A review of these far simpler and elegant equations, together with other prototypes occurring often in hydrodynamic stability and pattern formation in general, is given in Appendix C.5.

1.4 Structure and Contents of the Monograph

We begin with a foundational part in Chaps. 2, 3 and 4, which gives the governing equations and associated wall and free-surface boundary conditions, linear stability analysis and derivation of boundary layer equations. These equations are the basis for subsequent modeling. In Chaps. 5 and 6 we develop in detail the different methodologies and models used to analyze film flows. We also offer a critical assessment of their domain of validity and discuss their limitations. In Chaps. 7, 8 and 9 we use the models developed earlier and we present results on wave evolution on an isothermal and heated falling film. We critically compare the theory, numerics and experimental results (whenever possible).

More specifically:

In Chap. 2 we introduce the governing equations and their corresponding boundary conditions and we discuss at length possible scalings and the role of governing dimensionless groups and parameters entering the falling liquid film problem.

Chapter 3 is devoted to the Orr–Sommerfeld linear stability analysis of the base flow for isothermal and heated falling films. We consider transverse perturbations without any free-surface deformation and we clearly identify the S- and P-modes mentioned earlier. We then proceed to the study of evolution of streamwise perturbations. By performing a small wavenumber expansion for the H-mode also mentioned earlier, we obtain the neutral stability curve and identify the wavenumber corresponding to the maximum growth rate. Subsequently, the energy and vorticity balances associated with the onset and growth of wavy perturbations are studied, elucidating the role played by vorticity, a point not considered by Kapitza in his much simpler approach. A numerical scheme for the solution of the Orr–Sommerfeld eigenvalue problem is given in Appendix F.1.

In Chap. 4 we introduce the boundary layer equations, an approach which, as we already pointed out, is similar to that used in boundary layer theory in aerodynamics. The equations are scaled using a scaling proposed by Shkadov [248]. This scaling is inherent to the falling film problem in the region of moderate Reynolds numbers due to the separation of scales in the streamwise and cross-stream directions in this region that is in fact due to the strong effect of surface tension. It also makes apparent the balance among all forces, i.e., inertia, gravity, viscosity and surface tension, necessary to sustain strongly nonlinear waves. Using the boundary layer equations we identify two flow regimes. One occurs when the streamwise component of gravity is mainly balanced by the viscous drag with inertia playing little if any role, the "drag-gravity regime." The other is a "drag-inertia regime," where inertia balances viscous drag. The drag-gravity regime corresponds to low-Reynolds number flows whereas the drag-inertia flow regime corresponds to moderate values of the Reynolds number. The transition between the two regimes occurs at a "reduced Reynolds number" (obtained from the Shkadov scaling) of about one.

Chapter 5 develops the methodologies suitable for the study of flows at low values of the Reynolds number. These are based on what we have already referred to as a gradient expansion. We show that the rather complex free boundary problem describing the evolution of a falling liquid film can be reduced to the study of a single evolution equation for the free surface, what we already referred to as BE. The equation is also derived by including the Marangoni effect and we assess fully its region of validity, with and without the Marangoni effect. We subsequently derive both the Kuramoto–Sivashinsky (KS) and Kawahara equations defined in Appendix C.5 restricted to the evolution of small amplitude disturbances. We also discuss the paradigmatic role played by these model equations. Such models are typical cases of "driven–dissipative soliton-bearing equations" in a generalized sense ("driven–dissipative" refer to energy pumping-leak, respectively); indeed, numerical and experimental evidence shows that collisions between such *dissipative* solitary waves share common features with *solitons* in conservative systems, hence the coinage "dissipation solitons" (e.g., [55]). Dissipation expresses itself in the form of front (or rear) radiation of the clearly identifiable bump (or trough) of a wave. We also study the linear stability of the BE and we demonstrate that the resulting critical Reynolds number agrees fully with that found from the Orr–Sommerfeld analysis in Chap. 3. Moreover, the neutral curve and interfacial quantities obtained from the BE in the vicinity of criticality agree with those found from Orr–Sommerfeld. Hence the BE describes accurately the linear instability threshold (as far as critical/neutral conditions and interfacial quantities are concerned). However, far from criticality the BE is shown to blow up in finite time for both isothermal and heated films (this blowup behavior corresponds precisely to a reduced Reynolds number of about 1). Hence, the BE cannot describe correctly the drag-inertia regime and nonlinear waves far from criticality. An attempt to cure the deficiencies of the BE due to Ooshida [196] by employing a "Padé approximants" regularization scheme to the long-wave approximation is then discussed. A "continuation" scheme for the numerical construction of "traveling-wave" bifurcation diagrams is offered in Appendix F.2.

In Chap. 6 we consider methodologies for the study of flows at moderate Reynolds numbers. We review and reexamine in detail work done by ourselves and others with a view to explore further and deeply scrutinize the different methodologies/approaches. Our starting point is the derivation of the Kapitza–Shkadov model. This model forms the basis of a hierarchy of appropriately improved models based on weighted residuals approximations such as the Galerkin approach (after all, as we demonstrate in this chapter, the Kapitza–Shkadov model is a particular case of a simple weighted residuals modeling approach). We compare different possibilities, including the Galerkin, "collocation" and the "method of moments" and we show that the Galerkin approach performs best. We also present the "center manifold" approach of Roberts (e.g., [221]) and analyze its virtues and shortcomings.

Chapter 7 is devoted to an in-depth study of two-dimensional isothermal flows of the models developed in Chap. 6 and considers the virtues and shortcomings of the Kapitza–Shkadov model and the weighted residuals models developed in Chap. 6. For example, we examine the linear stability of these models and demonstrate that the Kapitza–Shkadov model predicts the critical Reynolds number with a 20% error as opposed to the weighted residuals models, which predict the correct value and fully resolve the instability onset. Subsequently we consider solitary waves in their moving frame by using elements from dynamical systems theory. We provide features of solitary waves amenable to experimental tests, such as speeds, amplitudes and shapes. Such waves are dissipative and are typically characterized by a primary hump followed by small-amplitude oscillations at its front (radiation) and a long tail at its back. The spatial evolution of two-dimensional waves is also analyzed when forcing at the inlet is introduced, as already done by Kapitza and subsequently by other experimentalists. This allows us direct comparisons with experiments by Gollub's group (e.g., [170]). Further, we discuss absolute and convective instabilities and offer new results dealing with "wave hierarchies" (i.e., "kinematic/dynamic waves"). A finite-differences scheme that can be used for the numerical solution of the two-dimensional model equations considered here (and for that matter, other nonlinear partial differential equations) is outlined in Appendix F.3.

In Chap. 8 we examine three-dimensional effects for isothermal films. This is the natural continuation of the analysis of preceding chapters. Theory, numerics and experiments are discussed, including the stability of two-dimensional periodic waves with respect to three-dimensional disturbances (using Floquet theory for periodic solutions). The subtleties of modulation effects on two-dimensional waves in particular are treated in depth. The role of noise at the inlet and its different effects compared to periodic forcing are also discussed (recall that the flow down the plate is a kind of noise amplifier). The results are contrasted with experiments by Gollub's group. The spectral representation of periodic solutions in Fourier space (including aliasing), crucial for obtaining the numerical results in this chapter, is outlined in Appendix F.4.

In Chap. 9, the weighted residuals approach introduced in Chap. 6 is extended to the problem of the influence of heating and the significant role played by the Marangoni effect on the dynamics and evolution of the falling liquid film. This is done at various levels of mathematical complexity and approximation. The linear stability is examined and fully nonlinear waves, i.e., solitary waves, are constructed. A wide variety of results is offered to both theoreticians/modelers and to experimentalists. The role of various parameters and constraints involved in the dynamics of heated films is discussed in detail. Particular attention is given to the study of three-dimensional patterns and "rivulets" arising from wave-wave interactions.

Finally, in Chap. 10 we offer suggestions for open problems for readers interested in pursuing research along the avenues explored in the previous chapters.

Chapter 2
Flow and Heat Transfer: Formulation

We present the full statement of the governing equations, boundary conditions and dimensionless groups for a film falling down a heated plate and exposed to ambient air. Due to the relatively low dynamic (shear) viscosity of the air the mechanical influence of the air motion on the liquid can be considered negligible, allowing us to simplify the boundary conditions on the open surface and thus consider it a "free" surface. We also ignore the heat transfer in the air by using an ad hoc boundary condition for the temperature on the free surface so that only the heat transfer process in the liquid is taken into account.

Our formulation considers in parallel an alternative with two possible thermal boundary conditions on the plate, namely a specified temperature (Dirichlet condition) and a heat flux distribution (Robin/mixed condition). For the latter to be realistic, heat losses between the wall and the surrounding medium must be included, which implies one additional empirical parameter in the equations, the heat transfer coefficient of the wall–air interface.

2.1 Governing Equations and Boundary Conditions

We consider the evolution of a viscous *thin film* flowing down a heated plate as illustrated in Fig. 2.1. The plate forms an angle β with the horizontal direction ($\beta = 0$, horizontal film; $\beta = \pi/2$, vertical film). For the isothermal case, the flow is then driven by the streamwise gravitational acceleration $g \sin \beta$. We introduce a Cartesian coordinate system (x, y, z) where x is the streamwise coordinate in the direction of flow, y is the outward-pointing coordinate normal to the plate and z is the spanwise coordinate. The plate is then located at $y = 0$ and the interface at $y = h(x, z, t)$, a single-valued function of x, z and time t. The main hypotheses are:

H1: The density ρ of the liquid is constant, or, equivalently the liquid remains incompressible. This assumption is valid for thin films ($\bar{h}_N < 1$ mm) where buoyancy can be neglected, i.e., the effect of thermal expansion upon density in the buoyancy force is negligible; see, e.g., [27, 58]. We examine this assumption

S. Kalliadasis et al., *Falling Liquid Films*, Applied Mathematical Sciences 176, DOI 10.1007/978-1-84882-367-9_2, © Springer-Verlag London Limited 2012

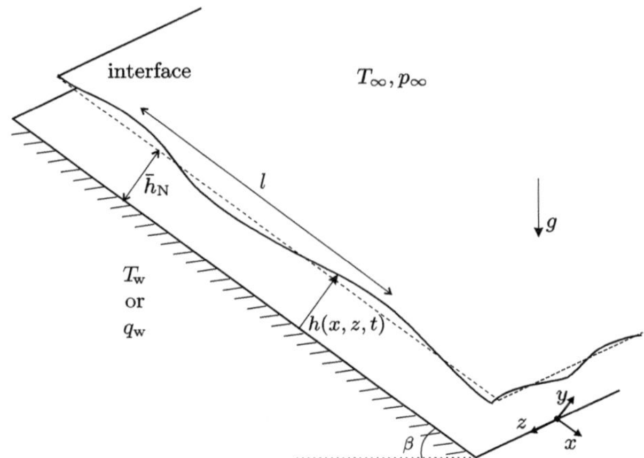

Fig. 2.1 Sketch of the geometry for a viscous thin film flowing down an inclined plate forming an angle β with the horizontal direction. The system is unbounded in the x and z directions. $h(x, z, t)$ is the local film thickness, $\bar{h}_N$ is the mean film thickness, typically < 1 mm, and g is gravitational acceleration. The surrounding gas phase is air maintained at temperature and pressure T_∞ and p_∞, respectively. The wall is heated either by a specified temperature distribution T_w or a given heat flux distribution q_w. l, a typical wavelength of the interfacial waves in the x direction, is much longer than $\bar{h}_N$

in more detail in Appendix D.1. We shall not, however, consider liquid thicknesses in the 100 nm range where intermolecular interactions become significant.

H2: The liquid is Newtonian and hence it obeys a linear stress-strain relationship whose proportionality coefficient is the dynamic (shear) viscosity μ. The kinematic viscosity is given by $\nu = \mu/\rho$.

H3: The plate is rigid and hence a no-slip (stick) and no-penetration condition for the velocity field applies on the plate (thus excluding the possibility of porous walls).

H4: The liquid is assumed to be nonvolatile so that in the range of temperatures we shall consider evaporation effects can be neglected.

H5: The air acts as a reservoir of infinite heat capacity and hence is maintained at the constant temperature T_∞. It is also maintained at the constant pressure p_∞. In addition, the air is assumed to be mechanically "passive" in the sense that the viscous stress from the air side is negligible compared to that from the liquid side (e.g., $\mu_{air}/\mu_{water} \approx 10^{-2}$). This "one-fluid" approach enables us to consider the momentum equation for the liquid without considering the momentum equation for the air. Note that despite the smallness of the dynamic viscosity ratio between air and liquid, the opposite is true for the kinematic viscosity, e.g., $\nu_{air}/\nu_{water} \approx 10$. This can make the one-fluid approach at times problematic when dealing with turbulent flows. However, a discussion of such flows is beyond the scope of this monograph.

H6: To obtain the constitutive equation for the surface tension σ, let us expand $\sigma(T_s)$, with T_s the interfacial temperature in a Taylor series at a reference temperature taken to be the temperature of the surrounding gas phase, T_∞. Taking the first two terms in the expansion leads to the linear decrease with temperature,

$$\sigma = \sigma_\infty - \gamma(T_s - T_\infty),\tag{2.1}$$

where σ_∞ is the surface tension at the gas temperature and $\gamma = -(d\sigma/dT_s)_{T_\infty}$ is positive for most liquids. An alternative for the reference temperature would be the interfacial temperature for a flat film. However, in this case we would have to assume that the interfacial temperature remains close to its flat film value while the ultimate aim of this monograph is to examine the nonlinear flow regimes where the film thickness departs significantly from its initial constant value. Further, unlike T_∞, the flat film temperature is not a control parameter in experiments. The above constitutive relation is further discussed in Appendix B.1.

H7: Any boundary between two phases, such as the liquid–gas or the liquid–solid interface, has typically a nonnegligible thermal resistance that leads to a difference in temperature across the boundary. This temperature difference is balanced by the heat flux normal to the boundary so that the following Robin/mixed condition applies there:

$$-\lambda\nabla T \cdot \mathbf{n} = \alpha(T - T_0),\tag{2.2}$$

where $\mathbf{n}$ is the outward-pointing (from the phase under consideration to the other side of the boundary) unit vector normal to the boundary, the dot is used to denote the dot product either of two vectors or of a tensor with a vector, λ is the thermal conductivity of the phase under consideration, T is the temperature at the boundary of the phase under consideration, T_0 is the temperature away from the boundary and α is the heat transfer coefficient that describes the rate of heat transport from the phase under consideration to the other phase across the boundary.

The main assumption here is that all the resistance to heat transfer (via conduction and convection) happens in a thin layer of the order λ/α in the immediate vicinity of the boundary, i.e., a significant temperature gradient exists over a small distance from the boundary to the other side. T_0 is then the temperature right outside this "thermal resistance layer." The larger the thickness of the layer the stronger the resistance to heat transfer. In general, the thickness of this layer is not known, although for high Reynolds number flows, e.g., in tube reactors, taking the thermal resistance layer to be the thermal boundary layer and T_0 to be the nearly constant temperature in the bulk outside the thermal boundary layer leads to a good approximation for the heat transfer process. Equation (2.2) is usually quoted as *Newton's law of cooling*. It is further discussed in Appendix B.2.

H8: The contribution of viscous dissipation in the heat equation is omitted. This is a reasonable assumption for thin liquid films and the thermal gradients we shall consider here (see Appendix D.1 for details).

H9: In addition to density (H1), all other fluid parameters, i.e., dynamic viscosity μ (and so kinematic viscosity ν), thermal conductivity λ and thermal diffusivity $\chi = \lambda/\rho c_p$ with c_p the constant pressure heat capacity, are not altered significantly by the action of the relatively small thermal gradients in the problem and are taken to be constant. Tacitly, this corresponds well to our situation of heating thin films only.

In the following we shall be making use of basic knowledge of fluid mechanics and interfacial phenomena as well as vector/tensor calculus. The reader should refer to some of the many textbooks available. For example, fundamental principles and derivations of the basic equations for fluid flow and heat transport are given in [26, 159, 160] while an interesting derivation of these equations using elements of nonequilibrium thermodynamics can be found in [69]. Reference [17] focuses on heat transfer while [158, 159] derive in detail the interfacial boundary conditions. References [12], Appendix A in [26] and [109] cover in detail vector/tensor calculus.

With hypotheses H1, H2, H8 and H9, the governing equations, namely continuity, momentum (Navier–Stokes) and energy (Fourier) equations, can be written as:

$$\nabla \cdot \mathbf{v} = 0, \tag{2.3}$$

$$\frac{D\mathbf{v}}{Dt} = -\rho^{-1}\nabla p + \nu\nabla^2\mathbf{v} + \mathbf{F}, \tag{2.4}$$

$$\frac{DT}{Dt} = \chi\nabla^2 T, \tag{2.5}$$

where $D/Dt \equiv \partial_t + \mathbf{v} \cdot \nabla$ is the "material derivative" (it is the "derivative following the motion," also called the "Lagrangian," "substantial" or "convective derivative"), $\nabla \equiv (\partial_x, \partial_y, \partial_z)$ is the gradient operator and the subscripts denote differentiation with respect to the corresponding variables. $\mathbf{v} = (u, v, w)$ is the fluid velocity vector with components u, v and w in the x, y and z directions, respectively. T and p are the temperature and total pressure of the fluid (including both dynamic and hydrostatic contributions), respectively, and $\mathbf{F} = (g\sin\beta, -g\cos\beta, 0)$ is the body force with g the gravitational acceleration.

Equations (2.3)–(2.5) are subject to the following boundary conditions:

At the Plate $y = 0$:

- The no-slip and no-penetration boundary condition (H3):

$$\mathbf{v} = \mathbf{0}. \tag{2.6}$$

- The wall heating, e.g., an electric heating device embedded in the wall, generates a temperature distribution in the film. We impose two types of boundary conditions for the temperature field in the film, namely a Dirichlet or *specified temperature* (ST) and a Robin/mixed or *heat flux* (HF). The ST thermal boundary condition is

$$T = T_{\mathrm{w}}, \tag{2.7}$$

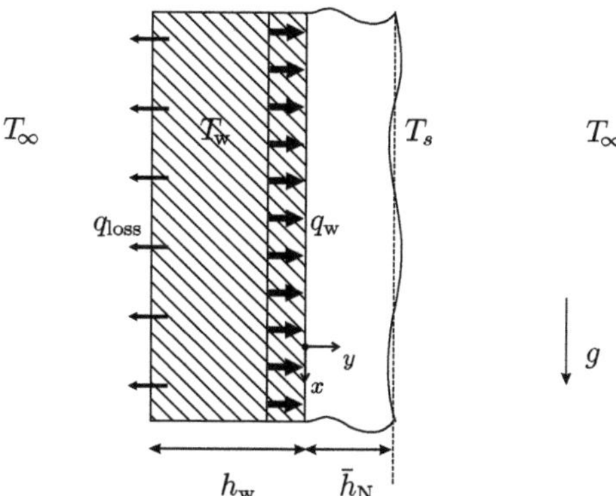

Fig. 2.2 Sketch of the heat fluxes at the wall for the HF case. q_w is the heat flux generated by the heater embedded in the wall, q_{loss} is the heat loss from the wall to the ambient gas phase and T_w, h_w, are the wall temperature and thickness, respectively. T_s denotes the interfacial temperature. Both q_w and q_{loss} contribute to the liquid temperature gradient at $y = 0$

i.e., the heater maintains the wall temperature at the value T_w ($> T_\infty$). The HF thermal boundary condition can be obtained by solving the energy equation in the wall. The derivation is given in Appendix C.1. The result is

$$\lambda \partial_y T = -q_w + \alpha_w (T - T_\infty), \tag{2.8}$$

where q_w is the heat flux generated by the heater and supplied by the plate to the liquid (see Fig. 2.2) and α_w is the heat transfer coefficient of the wall-air interface. The term $\alpha_w (T - T_\infty)$ is a measure of the heat losses to the gas phase in contact with the wall. Indeed, formally $q_{loss} = \lambda_w \partial_y T_w|_{y=-h_w}$. From (C.3a) in Appendix C.1 we find $\lambda_w \partial_y T_w\big|_{y=-h_w} = q_w + \lambda_w A$. But for $h_w \to 0$ as we did in Appendix C.1, $\lambda_w A = -q_w + \alpha_w (T - T_\infty)$ so that $q_{loss} = \alpha_w (T - T_\infty)$. Hence the mixed boundary condition in (2.8) expresses the simple physical fact that both the flux supplied by the plate to the liquid, q_w, and the heat losses, q_{loss}, to the gas phase in contact with the plate contribute to the temperature gradient at $y = 0$. In the particular case that this gradient vanishes, $q_w = \alpha_w (T - T_\infty)$, so that all the heat generated by the heater is lost to the gas phase in contact with the wall and we have a specified temperature boundary condition (Dirichlet condition). If, on the other hand, the wall is perfectly insulated from the air, i.e., $\alpha_w = 0$, we have a specified heat flux boundary condition (Newmann condition).

At the Free Surface $y = h(x, z, t)$:

• Provided H4, the *kinematic boundary condition* is obtained by differentiating $y - h(x, z, t) = 0$ with respect to t: $y - h$ is a scalar function which vanishes on the liquid interface so that its time derivative following any material point on the interface (which has velocity $\mathbf{v}$) also vanishes, that is,

$$\frac{D}{Dt}(y - h) = 0$$

at all points on the interface. Using the definition of the material derivative, this condition provides a relationship between the film thickness and the normal velocity component $v = Dy/Dt$ on the free surface:

$$v = \partial_t h + \mathbf{v} \cdot \nabla h. \tag{2.9}$$

It is a constraint on the material velocities in terms of the shape of the interface: A fluid particle on the free surface will remain there at all times and move with the velocity of the surface. We shall use extensively the kinematic boundary condition in this form throughout the monograph.

An alternative form combines the definition of the material derivative with that of a unit vector $\mathbf{n}$ normal to the surface that points into the surrounding gas as shown in Fig. 2.3,

$$\mathbf{n} = \frac{1}{n}(-\partial_x h, 1, -\partial_z h) \equiv \frac{1}{n}\nabla(y - h),$$

with $n = (1 + (\partial_x h)^2 + (\partial_z h)^2)^{1/2}$, i.e., the vector $(-\partial_x h, 1, -\partial_z h)$ is appropriately normalized so that its modulus is unity. Since, $v - \mathbf{v} \cdot \nabla h = \mathbf{v} \cdot \nabla(y - h)$, the kinematic boundary condition in (2.9) can be written as,

$$\frac{1}{n}\partial_t h = \mathbf{v} \cdot \mathbf{n},$$

where $\mathbf{v} \cdot \mathbf{n}$ is the component of $\mathbf{v}$ normal to the interface. In this form, the component of the velocity normal to the interface is balanced by the time variation of the interface; after all the interface changes for normal motions only (the tangential component causes motion on the interface but it does not change the material location of the interface).

- The *stress balance* or "momentum jump" on the free surface,

$$(\mathbf{T} - \mathbf{T}_\infty) \cdot \mathbf{n} = 2\sigma K(h)\mathbf{n} + \nabla_s \sigma, \tag{2.10}$$

where ∇_s is the surface gradient operator (see Appendix C.2) and $K(h)$ is the mean curvature of the free surface given by the average, K, of the two principal curvatures k_1, k_2, $K = (1/2)(k_1 + k_2) = (1/2)(1/R_1 + 1/R_2) = -(1/2)\nabla_s \cdot \mathbf{n}$ where R_1, R_2 are the principal radii of curvature. Hence:

$$K(h) = \frac{1}{2}\frac{\partial_{xx} h[1 + (\partial_z h)^2] + \partial_{zz} h[1 + (\partial_x h)^2] - 2\partial_x h \partial_z h \partial_{xz} h}{[1 + (\partial_x h)^2 + (\partial_z h)^2]^{3/2}}.$$

The quantity $\mathbf{T} = -p\mathbf{I} + \mathbf{P}$ is the stress tensor for the liquid with $\mathbf{I}$ the identity matrix, $\mathbf{P} = 2\mu\mathbf{E}$ (using H2) is the deviatoric stress tensor and $\mathbf{E} = (1/2)(\nabla\mathbf{v} + (\nabla\mathbf{v})^t)$ is the rate-of-strain tensor. For the gas $\mathbf{T}_\infty = -p_\infty\mathbf{I}$ (see H5). We then have:

$$(p_\infty - p)\mathbf{n} + \mathbf{P} \cdot \mathbf{n} = 2\sigma K(h)\mathbf{n} + \nabla_s \sigma. \tag{2.11}$$

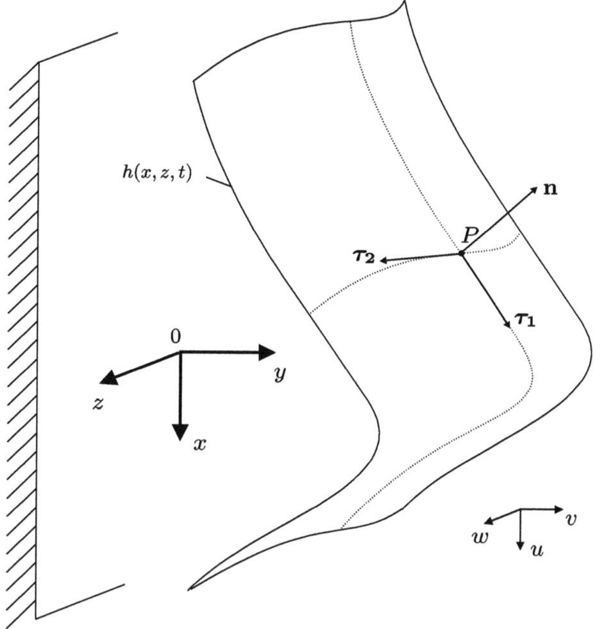

Fig. 2.3 Definition of normal **n** and tangential unit vectors $\boldsymbol{\tau}_{1,2}$ at point $P(x, y, z)$ on the liquid film free surface. $h(x, z, t)$ denotes the location of the free surface and u, v, w the streamwise, cross-stream and spanwise components of the velocity field, respectively

We now identify the physical meaning of the different terms in (2.10). The term $\mathbf{T} \cdot \mathbf{n}$ is the force per unit area acting on the liquid side of the interface and in the direction normal to the interface, the term $2\sigma K(h)\mathbf{n}$ is the Laplace surface tension term also acting in the normal direction, while the term $\nabla_{\!s}\sigma$ gives rise to a force tangent to the interface. The condition (2.10) then simply states that all forces acting on the interface must balance. Note that for rigid bodies, this condition is usually ignored as the stress in such materials is indeterminate and the condition does not provide a useful constraint on the stress.

To obtain the *normal stress balance* or *normal stress boundary condition*, we take the dot product of (2.11) with **n**. This gives:

$$p_\infty - p + (\mathbf{P} \cdot \mathbf{n}) \cdot \mathbf{n} = 2\sigma K(h). \tag{2.12}$$

Hence, in crossing the interface, the normal component of the total stress undergoes a "jump" equal to $\sigma(\nabla_{\!s} \cdot \mathbf{n})$. In the limiting case of no motion in the fluid, $p_\infty - p = -\sigma(\nabla_{\!s} \cdot \mathbf{n})$. Therefore, as Laplace noted, the pressure inside the convex region of a curved interface at equilibrium is larger than the outside by an amount that depends on both the curvature and σ. In the limiting case of a sphere, $R_1 = R_2 = -R$, the sphere's radius. Thus, for a spherical bubble or drop, $p = p_\infty + 2\sigma/R$ and the internal pressure exceeds the external one by $2\sigma/R$. For a planar geometry the factor of 2 does not appear.

Let us now define two unit vectors tangential to the free surface (see Fig. 2.3),

$$\boldsymbol{\tau}_1 = \frac{1}{\tau_1}(1, \partial_x h, 0) \quad \text{and} \quad \boldsymbol{\tau}_2 = \frac{1}{\tau_2}(0, \partial_z h, 1),$$

with $\tau_1 = (1 + (\partial_x h)^2)^{1/2}$ and $\tau_2 = (1 + (\partial_z h)^2)^{1/2}$, i.e., the vectors $(1, \partial_x h, 0)$ and $(0, \partial_z h, 1)$ are appropriately normalized so that their modulus is unity. The choice of these vectors is discussed in Appendix C.3. The projection of (2.11) on these vectors then gives the components of the *tangential stress balance* or *tangential stress boundary conditions*:

$$(\mathbf{P} \cdot \mathbf{n}) \cdot \boldsymbol{\tau}_i = \nabla_s \sigma \cdot \boldsymbol{\tau}_i, \quad i = 1, 2. \tag{2.13}$$

The evaluation of the right hand side is discussed in Appendix C.4. An important consequence of this condition is that systems with $\nabla_s \sigma \neq \mathbf{0}$, such as those considered here, must undergo motion even in the absence of an externally imposed flow: If $\nabla_s \sigma \neq \mathbf{0}$, then $\mathbf{P} \neq \mathbf{0}$ and thus, any mechanism which maintains $\nabla_s \sigma \neq \mathbf{0}$ will necessarily drive motion in the fluid or alter an existing one. This is the Marangoni effect defined in the Introduction. We refer to such motions as *Marangoni driven flows*.

- Newton's law of cooling (see H7),

$$-\lambda \nabla T \cdot \mathbf{n} = \alpha (T - T_\infty), \tag{2.14}$$

where α is the heat transfer coefficient that measures the rate of heat transport between the liquid film and the ambient air.

2.2 Dimensionless Equations, Scalings and Parameters

The system in (2.3)–(2.5) with corresponding plate and free-surface boundary conditions (2.6)–(2.8), (2.9) and (2.12)–(2.14) has a trivial or base solution corresponding to the plane-parallel base state with thickness $h = \bar{h}_N$,

$$U(y) = \frac{g \sin \beta}{2\nu} y(2\bar{h}_N - y), \tag{2.15a}$$

$$V(y) = 0, \tag{2.15b}$$

$$W(y) = 0, \tag{2.15c}$$

$$P(y) = p_\infty + \rho g \cos \beta (\bar{h}_N - y), \tag{2.15d}$$

$$\text{ST:} \quad \Theta(y) = T_w - \frac{\alpha(T_w - T_\infty)}{\lambda + \alpha \bar{h}_N} y, \tag{2.15e}$$

or

$$\text{HF:} \quad \Theta(y) = T_\infty + \frac{q_w(\lambda + \alpha(\bar{h}_N - y))}{\lambda(\alpha + \alpha_w) + \alpha \alpha_w \bar{h}_N}. \tag{2.15f}$$

The streamwise gravitational acceleration balances viscous forces giving rise to a semiparabolic velocity profile while heat propagates by pure conduction, giving rise to a linear temperature distribution. The semiparabolic velocity profile was first obtained by Nusselt [194] and we shall refer to the trivial solution in (2.15a)–(2.15f) as the *Nusselt flat film solution*. Note that the temperature distributions for the two thermal wall boundary conditions can also be written in the unified form,

$$\Theta(y) = T_\infty + \beta_T\big[\alpha(\bar{h}_N - y) + \lambda\big],$$

where

$$\text{ST:} \quad \beta_T = \frac{T_w - T_\infty}{\alpha\bar{h}_N + \lambda}$$

or

$$\text{HF:} \quad \beta_T = \frac{q_w}{\lambda(\alpha_w + \alpha) + \alpha_w\alpha\bar{h}_N}.$$

We now utilize the Nusselt flat film solution to introduce the nondimensionalization (details are given in Appendix D.1),

$$(x, y, z) \to \bar{h}_N(x, y, z), \qquad h \to \bar{h}_N h, \qquad t \to \frac{t_\nu l_\nu}{\bar{h}_N}t, \tag{2.16a}$$

$$(u, v, w) \to \frac{\bar{h}_N^2}{t_\nu l_\nu}(u, v, w), \qquad p \to p_\infty + \rho\frac{l_\nu\bar{h}_N}{t_\nu^2}p, \tag{2.16b}$$

$$\text{ST:} \quad T \to T_\infty + T\Delta T \tag{2.16c}$$

or

$$\text{HF:} \quad T \to T_\infty + T\Delta T_N, \tag{2.16d}$$

where the temperature scale ΔT is chosen as

$$\text{ST:} \quad \Delta T = T_w - T_\infty, \tag{2.16e}$$

or

$$\text{HF:} \quad \Delta T_N = \frac{q_w\bar{h}_N}{\lambda} = \Delta T h_N, \tag{2.16f}$$

and

$$l_\nu = \left(\frac{\nu^2}{g\sin\beta}\right)^{1/3} \quad \text{and} \quad t_\nu = \left(\frac{\nu}{(g\sin\beta)^2}\right)^{1/3}$$

are the *viscous-gravity* length and time scales built from the streamwise gravity acceleration and the kinematic viscosity. These scales make explicit the balance between gravity and viscous forces giving rise to the Nusselt flat film solution in (2.15a)–(2.15f) and are discussed in Appendix D.1. The subscript N in (2.16f) is used to denote that the corresponding temperature scale for HF is based on $\bar{h}_N$.

$\Delta T = q_{\mathrm{w}} l_\nu / \lambda$ is a temperature scale for HF based on l_ν and $h_{\mathrm{N}} = \bar{h}_{\mathrm{N}} / l_\nu$ is the dimensionless Nusselt flat film thickness based on l_ν. Note that the temperature scales for both ST and HF cases are natural control parameters in experiments. Note also that utilizing the streamwise gravitational acceleration in the scaling forbids us from taking the limit of an horizontal plane. As a matter of fact, for slightly inclined planes the dominant hydrodynamic mode is a shear mode associated with the semi-parabolic Nusselt profile [23, 94] and not the long-wave interfacial H-mode whose interaction with the long-wave thermocapillary S-mode is one of the key points in this monograph (see Introduction).

In terms of these dimensionless variables, the equations of motion and energy (2.3)–(2.5) become,

$$\partial_x u + \partial_y v + \partial_z w = 0, \tag{2.17}$$

$$3Re(\partial_t u + u\partial_x u + v\partial_y u + w\partial_z u) = -\partial_x p + \partial_{xx} u + \partial_{yy} u + \partial_{zz} u + 1, \tag{2.18}$$

$$3Re(\partial_t v + u\partial_x v + v\partial_y v + w\partial_z v) = -\partial_y p + \partial_{xx} v + \partial_{yy} v + \partial_{zz} v - Ct, \tag{2.19}$$

$$3Re(\partial_t w + u\partial_x w + v\partial_y w + w\partial_z w) = -\partial_z p + \partial_{xx} w + \partial_{yy} w + \partial_{zz} w, \tag{2.20}$$

$$3Pe(\partial_t T + u\partial_x T + v\partial_y T + w\partial_z T) = \partial_{xx} T + \partial_{yy} T + \partial_{zz} T, \tag{2.21}$$

where the dimensionless parameters Re and Pe will be defined shortly. These equations are subject to the dimensionless versions of the boundary conditions (2.6)–(2.9) and (2.12)–(2.14):

- At the plate $y = 0$:

$$u = v = w = 0, \tag{2.22}$$

$$\text{ST:} \quad T = 1 \tag{2.23a}$$

or

$$\text{HF:} \quad \partial_y T = -1 + B_{\mathrm{w}} T. \tag{2.23b}$$

- On the film surface $y = h(x, z, t)$:

$$v = \partial_t h + u\partial_x h + w\partial_z h, \tag{2.24}$$

$$p = \frac{2}{n^2}\big[(\partial_x h)^2 \partial_x u + (\partial_z h)^2 \partial_z w + \partial_x h \partial_z h(\partial_z u + \partial_x w) - \partial_x h(\partial_y u + \partial_x v)$$

$$- \partial_z h(\partial_z v + \partial_y w) + \partial_y v\big] - \frac{1}{n^3}(We - M\,T)\big[\partial_{xx} h\big(1 + (\partial_z h)^2\big)$$

$$+ \partial_{zz} h\big(1 + (\partial_x h)^2\big) - 2\partial_x h \partial_z h \partial_{xz} h\big], \tag{2.25}$$

$$0 = \frac{1}{n}\big[2\partial_x h(\partial_y v - \partial_x u) + \big(1 - (\partial_x h)^2\big)(\partial_y u + \partial_x v) - \partial_z h(\partial_z u + \partial_x w)$$

$$- \partial_x h \partial_z h(\partial_z v + \partial_y w)\big] + M(\partial_x T + \partial_x h \partial_y T), \tag{2.26}$$

$$0 = \frac{1}{n}\Big[2\partial_z h(\partial_y v - \partial_z w) + \big(1 - (\partial_z h)^2\big)(\partial_y w + \partial_z v) - \partial_x h(\partial_z u + \partial_x w)$$

$$- \partial_x h\partial_z h(\partial_y u + \partial_x v)\Big] + M(\partial_z T + \partial_z h\partial_y T), \tag{2.27}$$

$$BT = \frac{1}{n}(\partial_x h\partial_x T + \partial_z h\partial_z T - \partial_y T). \tag{2.28}$$

Let us now introduce the following dimensionless groups and parameters:

- The *inclination number*

$$Ct = \cot\beta, \tag{2.29}$$

  which compares the cross-stream component of the gravitational force to its streamwise component. It quantifies the contribution of the hydrostatic pressure that vanishes for a film falling down a vertical wall.
- The *Prandtl number*

$$Pr = \frac{\nu}{\chi}, \tag{2.30}$$

  which compares the momentum diffusivity to the thermal diffusivity.
- The *Kapitza number*

$$\Gamma = \frac{\sigma_\infty l_\nu}{\rho \nu^2} = \frac{\sigma_\infty}{\rho\,(g\sin\beta)^{1/3}\,\nu^{4/3}}, \tag{2.31}$$

  which compares the surface tension force $\sigma_\infty l_\nu$ to the force of inertia $\rho(u_\nu l_\nu)^2 = \rho\nu^2$, which is independent of the flow rate. The Kapitza number is a function of the liquid properties and β. For a vertical falling film the Kapitza number becomes $\Gamma_\perp = \sigma_\infty/\rho g^{1/3}\nu^{4/3}$, a *vertical Kapitza number*, and depends on the liquid properties only. It is thus fixed once the liquid is selected.
- The *Marangoni number*

$$Ma = \frac{\gamma\,\Delta T l_\nu}{\rho\nu^2} = \frac{\gamma\,\Delta T}{\rho(g\sin\beta)^{1/3}\nu^{4/3}} = \Gamma\frac{\gamma\,\Delta T}{\sigma_\infty}, \tag{2.32}$$

  which for both ST and HF compares the force induced by the surface tension gradient $\gamma\,\Delta T l_\nu$ to $\rho\nu^2$ (recall that for HF $\Delta T = q_{\mathrm{w}}l_\nu/\lambda$, but we use $\Delta T_{\mathrm{N}} = q_{\mathrm{w}}\bar{h}_{\mathrm{N}}/\lambda$ to nondimensionalize the temperature).
- The *free-surface Biot number*

$$Bi = \frac{\alpha\,l_\nu}{\lambda} = \frac{\alpha\,\nu^{2/3}}{\lambda\,(g\sin\beta)^{1/3}}, \tag{2.33}$$

  a dimensionless heat transfer coefficient describing the rate of heat transport from the liquid to the ambient gas.
- The *wall Biot number*

$$Bi_{\mathrm{w}} = \frac{\alpha_{\mathrm{w}}\,l_\nu}{\lambda} = \frac{\alpha_{\mathrm{w}}\,\nu^{2/3}}{\lambda\,(g\sin\beta)^{1/3}}, \tag{2.34}$$

a dimensionless heat transfer coefficient describing the rate of heat transport from the wall to the ambient gas. This dimensionless group appears only in the HF problem. Note that since the heat transfer coefficient of the liquid–gas interface is in general smaller to that of the solid–gas interface as pointed out in Appendix B.2, and Bi and Bi_w both scale with the thermal conductivity of the liquid, in general $Bi < Bi_w$. As also pointed out in Appendix. B.2 this situation can be reversed by insulating the solid, in which case $Bi > Bi_w$ (increasing the thermal conductivity of the liquid does increase Bi but it cannot really lead to $Bi > Bi_w$).

We now define the dimensionless groups in the system (2.17)–(2.28) and we write these groups in terms of h_N and the dimensionless groups in (2.30)–(2.34):

- The *Reynolds number*

$$Re = \frac{\bar{u}_N \bar{h}_N}{\nu} = \frac{\bar{q}_N}{\nu} = \frac{g \sin \beta \bar{h}_N^3}{3\nu^2} \tag{2.35}$$

compares inertia to viscous forces with $\bar{u}_N,$[1] the average velocity of the Nusselt flat film solution (2.15a), $\bar{u}_N = g \sin \beta \bar{h}_N^2/(3\nu)$, and $\bar{q}_N$, the specific volumetric flow rate (flow rate per unit width of wall in the transverse direction), defined as

$$\bar{q}_N = \int_0^{\bar{h}_N} U(y)\,dy = \frac{g \sin \beta \bar{h}_N^3}{3\nu}, \tag{2.36}$$

the control parameter that determines the Nusselt flat film thickness $\bar{h}_N$ in experiments. Hence, the Reynolds number in (2.35) is merely the dimensionless flow rate based on the viscous gravity scales, $q_N = \bar{u}_N \bar{h}_N/[(l_\nu/t_\nu)l_\nu] \equiv \bar{q}_N/\nu$. From (2.35) we can also directly relate the dimensionless Nusselt flat film thickness based on l_ν to the Reynolds number:

$$h_N = \frac{\bar{h}_N}{l_\nu} = (3Re)^{1/3}. \tag{2.37}$$

Clearly the definition of the Reynolds number can vary depending on the chosen velocity scale, i.e., one can use the average velocity $\bar{u}_N$, the velocity at the interface $3\bar{u}_N/2$ (see (2.15a)) or the speed of linear waves $3\bar{u}_N$ (see Chap. 3). In this monograph, we choose the definition (2.35).
- The *Péclet number*

$$Pe = PrRe \tag{2.38}$$

expresses the relative importance of convection and heat conduction. This number is also referred to as the "heat transport Péclet number" in combined heat-mass transport problems to distinguish it from the "mass transport Péclet number", $PrSc$, where $Sc = \nu/D$ is the "Schmidt number" and D is the molecular diffusivity.

---

[1]Bars are used to distinguish dimensional from dimensionless quantities unless the distinction is unnecessary.

– The *Weber number*

$$We = \frac{\sigma_\infty}{\rho g \bar{h}_N^2 \sin\beta} = \frac{\Gamma}{h_N^2} \qquad (2.39)$$

compares the surface tension pressure $\sigma_\infty/\bar{h}_N$ to the viscous normal stress generated by gravity at the film surface, $\mu \bar{u}_N/\bar{h}_N = \rho g \bar{h}_N \sin\beta$. For large We, the fluid behavior is mainly determined by surface tension (e.g., at small $\bar{h}_N$), while gravity dominates for small We. The Weber number allows us to access surface deformability due to the flow. High values of We mean that the viscous forces due to the flow fail to generate pressure capable of deforming the surface.

– The *film Marangoni number*

$$\text{ST:} \quad M = \frac{\gamma \Delta T}{\rho g \bar{h}_N^2 \sin\beta} = \frac{Ma}{h_N^2} \qquad (2.40\text{a})$$

or

$$\text{HF:} \quad M = \frac{\gamma \Delta T_N}{\rho g \bar{h}_N^2 \sin\beta} = \frac{Ma}{h_N} \qquad (2.40\text{b})$$

expresses the relative importance of the thermocapillary stress induced by the surface tension gradient, $\gamma \Delta T/\bar{h}_N$ for ST or $\gamma \Delta T_N/\bar{h}_N$ for HF, to the viscous normal stress generated by gravity at the film surface, $\mu \bar{u}_N/\bar{h}_N = \rho g \bar{h}_N \sin\beta$.

– The *free-surface* and *wall film Biot numbers*

$$B = \frac{\alpha \bar{h}_N}{\lambda} = Bi h_N \quad \text{and} \quad B_w = \frac{\alpha_w \bar{h}_N}{\lambda} = Bi_w h_N. \qquad (2.41)$$

As pointed out in Appendix D.1, since all dimensionless parameters in (2.35)–(2.41) depend on the Nusselt flat film solution, which is controlled by the flow rate, they vary when the flow rate is varied. Nevertheless, the parametrization in (2.35)–(2.41) demarcates clearly the dependence of the problem on the flow rate and the properties of the gas–liquid–solid system (physical properties of the gas–liquid system[2] and wall temperature/heat flux supplied by the heater) and inclination angle. As also pointed out earlier the nondimensionalization in (2.16a)–(2.16f) is based on the Nusselt flat film solution, here the term *Nusselt scaling* adopted in Appendix D.1. This is the most widely used scaling in the literature.

The definitions (2.39) and (2.40a), (2.40b) show that $M, We \to \infty$ as $h_N \to 0$, while $Re \to 0$ from (2.37). Hence, for very thin films, interfacial forces, i.e., Marangoni and capillary forces, dominate over inertia. On the other hand, with $h_N \to \infty$, $M, We \to 0$, so that interfacial forces are not important in the region of large film thicknesses and inertia forces dominate over interfacial ones. This point will be discussed further in Chap. 9 when we analyze the relative influence of inertia,

---

[2] As discussed in Appendix B.2, the liquid–gas heat transfer coefficient is practically independent of what is happening in the liquid and only dependent on the physical properties of the gas.

Marangoni and capillary forces on the bifurcation diagrams for single-hump solitary pulses.

Noteworthy is that in the literature one frequently encounters a "static Bond number" and a "dynamic Bond number." They can be written in terms of the conventional Weber number and film Marangoni number for the ST case defined above as: $Bo = \rho g \bar{h}_N^2 \sin\beta / \sigma_\infty \equiv We^{-1}$ and $Bo_d = \rho g \bar{h}_N^2 \sin\beta / (\gamma \Delta T) \equiv M^{-1}$. Strictly speaking, the static Bond number is appropriate for static problems, e.g., a motionless horizontal liquid layer or a drop at equilibrium. This group compares the role of gravity trying to make the free surface leveled to an equipotential, with surface tension trying to make the interface spherical. Setting $Bo = 1$ gives that $\bar{h}_N$ is equal to the capillary length, $l_\sigma = (\sigma_\infty / \rho g \sin\beta)^{1/2}$, which on Earth is about 2.5 mm for water at room temperature. Our case, however, is such that the flow is driven by $g \sin\beta$ and hence Bo as defined earlier is in fact related to the Kapitza number Γ through (2.39). But Γ is a direct consequence of $g \sin\beta$, and so flow; g here is a dynamic quantity "creating flow" and not a passive body force whose role is simply restricted to creating equipotential surfaces, like in static problems. Hence, due to the nature of our problem Bo appears as a flow-related parameter and compares viscous normal stresses with surface tension. This is the main reason we prefer to use the Weber number defined by (2.39) instead.

We close this section with comments on the different limits for the Reynolds and Péclet numbers. The equations of motion and energy in (2.18)–(2.21) describe the competition between two agents: inertia, which gives rise to the H-mode of instability for a falling film, and the long-wave thermocapillary S-mode (see Introduction). This competition is expressed by two dimensionless groups: the Reynolds number in (2.18)–(2.20) and the Péclet number in (2.21), which is a "thermal Reynolds number", as (2.18)–(2.20) and (2.21) are formally equivalent.

For $Pe \to 0$ (because $Pr \to 0$ like in liquid metals), the temperature field is slaved to the velocity field and we can drop the left hand side of (2.21). This then "freezes" the Marangoni mode, the time evolution is set by the velocity field and the system is driven by inertia (appropriately modified of course by the Marangoni effect). On the other hand, for $Re \to 0$ (while Pe remains finite), the velocity field is slaved to the temperature field and we can drop the left hand side of (2.18)–(2.20). This then "freezes" inertia, i.e., the H-mode, the time evolution is set by the temperature field and the system is driven by the Marangoni forces.

For $Pe \to 0$ and $Re \to \infty$, we expect the usual "Kolmogorov–Reynolds inertial turbulence", which is characterized by an energy transfer from long to short scales mostly dissipation-free (within the inertia interval; dissipation occurs only at the end of the cascade where viscosity kills the small eddies). In this case one can neglect the viscous terms in (2.18)–(2.20). Note that the "Tollmien–Schlichting instability" and transition to fully developed turbulence usually occurs for very large Re, in the region 1000–2000 [56]. This turbulent regime is beyond the scope of the monograph, which focuses on low and moderate Reynolds numbers (in the region 0–50).

On the other hand, for $Re \to 0$ and $Pe \to \infty$ (because $Pr \to \infty$, as with some silicone oils), one expects turbulence with strong dissipation [265] (the velocity scale is now set by the Marangoni effect which appears in the tangential boundary condition). In this case one can neglect the thermal boundary condition in (2.21). This limit should lead to an inverse cascade from small to large eddies (opposite to the one for Kolmogorov–Reynolds turbulence), e.g., for Bénard–Marangoni convection increasing the Péclet number increases the Bénard–Marangoni cells: dissipation increases and the cells grow to accommodate the large dissipation [291]. This dissipative turbulent regime is also beyond the scope of the monograph.

As a consequence of the above observations, different liquids should exhibit different behavior with respect to the instabilities considered here, depending on their Prandtl number. For falling films with liquids having low Prandtl numbers, e.g., liquid metals, the waves on the films should be controlled by the H-mode. On the other hand, for liquids having large Prandtl numbers, e.g., silicone oils, the waves should be controlled by the thermocapillary S-mode. The competition between the two modes will be discussed in detail in Chaps. 3 and 9.

2.3 On the Development of the Nusselt Flat Film Solution

For the isothermal case, the Nusselt flat film flow in (2.15a)–(2.15f) will in general develop very rapidly after the inlet at $x = 0$: once this happens the flow is "fully developed" (of course this flow will subsequently develop an instability). Clearly the location where the Nusselt flat film flow develops depends on the initial film thickness h_i provided by the manifold. From the integral analysis of the momentum boundary layer in the monograph by Alekseenko et al. [3], we find that for $h_i/\bar{h}_N = 3$, the distance x_h necessary for the film to reach the Nusselt flat film solution with an accuracy of 10^{-4}, is $x_h/\bar{h}_N \approx 1.2Re$ (which also seems to agree with experiments). For instance, a water film with $\bar{h}_N = 0.15$ mm and $Re = 11$ gives $x_h \sim 2$ mm.

For the nonisothermal case, simple scaling arguments can be used to show that the thickness δ_T of the thermal boundary grows proportionally with the Péclet number immediately downstream from the inlet at $x = 0$ where $\delta_T \ll \bar{h}_N$; at $y = \delta_T$, the terms $u\partial_x T$ and $\chi\partial_{yy} T$ of the energy equation in (2.5) scale as $u|_{y=\delta_T} T|_{y=\delta_T}/x$ and $\chi T|_{y=\delta_T}/\delta_T^2$, respectively. Balancing these two terms yields, $u|_{y=\delta_T}/x \sim \chi/\delta_T^2$. Note that from the continuity equation in (2.3), $u|_{y=\delta_T}/x \sim v|_{y=\delta_T}/\delta_T$ and hence the terms $u\partial_x T$ and $v\partial_y T$ of the energy equation balance automatically. But $u|_{y=\delta_T} \equiv U|_{y=\delta_T}$ where U is given in (2.15a)–(2.15f) or $u|_{y=\delta_T} = (g\sin\beta/2v)\delta_T(2\bar{h}_N - \delta_T) \sim g\sin\beta\bar{h}_N\delta_T/v \sim \bar{u}_N\delta_T/\bar{h}_N$ or $\delta_T \sim (\chi\bar{h}_N x/\bar{u}_N)^{1/3}$, i.e., $x/\bar{h}_N \sim (\delta_T/\bar{h}_N)^3 Pe$ with $\bar{u}_N$ the average velocity of the Nusselt flat film solution and Pe the Péclet number defined in (2.38).

This estimate shows that the development of the thermal boundary layer for moderate Péclet numbers (i.e., for liquids of moderate Prandtl number and flows of moderate Reynolds number) occurs close to the inlet. As an example, for a water film with $\bar{h}_N = 0.15$ mm, $Re = 11$ and $Pr = 7$, the location at which $\delta_T = \bar{h}_N/3$ is $x \approx 0.5$ mm. On the other hand, for a water film with $\bar{h}_N = 0.25$ mm, $Re = 50$ and $Pr = 7$, the location at which $\delta_T = \bar{h}_N/3$ is $x \approx 3.2$ mm. An accurate estimate of the length at which entrance effects associated with the inlet region are neglected would require a detailed integral analysis of the thermal boundary layer which is beyond the scope of this monograph.

Consequently, for liquids of moderate Prandtl number and flows at moderate Reynolds number, the semiparabolic velocity profile and linear temperature distribution can be assumed soon after the inlet so that the film reaches the state governed by (2.15a)–(2.15f), i.e., both flow and heat transfer are fully developed before they undergo any instability.

2.4 On the Two Wall Thermal Boundary Conditions: Retrieving ST from HF

For HF the temperature field has been nondimensionalized with $\Delta T_{\mathrm{N}} = q_{\mathrm{w}}\bar{h}_{\mathrm{N}}/\lambda$. An alternative scaling could have been $T^* = (T - T_\infty)/(q_{\mathrm{w}}/\alpha_{\mathrm{w}})$, which with $y \to \bar{h}_{\mathrm{N}}y$ would convert (2.8) to

$$\partial_y T = B_{\mathrm{w}}(-1 + T^*).\qquad(2.42)$$

In the limit $B_{\mathrm{w}} \to \infty$, (2.42) yields $T^* \to 1$, thus retrieving the boundary condition for ST (2.23a); but (2.23a) is obtained by scaling the temperature field with $\Delta T = T_{\mathrm{w}} - T_\infty$. This scaling must be related to that used to obtain (2.42). Converting $T^* = 1$ to dimensional variables and setting $T = T_{\mathrm{w}}$, yields $q_{\mathrm{w}} = \alpha_{\mathrm{w}}(T_{\mathrm{w}} - T_\infty)$: all the heat generated by the solid is now removed to the gas below so that the temperature T at $y = 0$ is kept constant at $T = T_{\mathrm{w}} = T_\infty + q_{\mathrm{w}}/\alpha_{\mathrm{w}}$. Note that although T refers to the temperature in the liquid, in the limit $B_{\mathrm{w}} \to \infty$, we lose the term $(1/B_{\mathrm{w}})\partial_y T$ in (2.42), hence the communication between the wall and the liquid, and so we must set $T = T_{\mathrm{w}}$. In other words, the wall is effectively removed from the problem and we are only concerned with the heat transfer between the liquid and the gas.

Taking the limit $B_{\mathrm{w}} \to \infty$ in (2.23b) yields $T \to 0$. It would then appear that we cannot retrieve the ST problem from (2.23b) in this limit. However, (2.23b) can be converted to (2.42) by using the transformation:

$$T = \frac{1}{B_{\mathrm{w}}}T^*.\qquad(2.43)$$

Thus, in the limit $B_{\mathrm{w}} \to \infty$, $T^* \to 1$ becomes $T \to 0$ and hence, the alternative form of the wall thermal boundary condition in (2.42) is equivalent to (2.23b). The advantage of (2.42) is that it makes the recovery of ST from HF in the limit $B_{\mathrm{w}} \to \infty$ transparent. On the other hand, the advantage of (2.23b) is that it makes the limit $B_{\mathrm{w}} \to 0$ more obvious, as in this limit we retrieve the HF case.

2.5 Role of the Biot Number

The role of the Biot number on the Nusselt flat film temperature distribution and how it influences the Marangoni effect is subtle. We discuss separately the ST and HF cases.

Let us nondimensionalize the temperature distributions in (2.15e) and (2.15f) with the gravity-viscous scaling, which then expresses these distributions in terms of h_N and Bi for ST, and h_N, Bi and Bi_w for HF:

$$\text{ST:}\quad \Theta(y) = \frac{1 + Bi(h_N - y)}{1 + Bih_N} \tag{2.44a}$$

or

$$\text{HF:}\quad \Theta(y) = \frac{1 + Bi(h_N - y)}{Bi + Bi_w(1 + Bih_N)}. \tag{2.44b}$$

ST

The temperature of the undeformed free surface is obtained from (2.44a) as

$$\Theta_s \equiv \Theta|_{y=h_N} = \frac{1}{1 + Bih_N}, \tag{2.45}$$

and consequently, the temperature gradient between the surface and the wall is

$$b_s \equiv \frac{\Theta|_{y=0} - \Theta|_{y=h_N}}{h_N} = \frac{Bi}{1 + Bih_N}. \tag{2.46}$$

Let us now consider the behavior of (2.45) and (2.46) in the limits of $Bi = 0$ and $1/Bi = 0$; the first limit corresponds to very poor heat transfer characteristics of the liquid–gas interface; the second one is not physical but it is mathematically useful.

- With $Bi = 0$, (2.45) shows that $\Theta_s = 1$. This means that the wall and the free surface have the same temperature. In fact, in this case the fluid temperature is uniform and equal to unity.
- In the limit $1/Bi = 0$, (2.45) shows that $\Theta_s = 0$ so that the free surface and the air have the same temperature.

In both cases, the temperature of the free surface is independent of the film thickness so that any perturbation of h does not affect the free-surface temperature distribution and the Marangoni instability (S-mode) does not occur. This can be made explicit by defining a film Marangoni number $M^\star$ based on the Nusselt flat film temperature difference between the wall and the free surface,

$$\Delta T_s \equiv (T_w - T_s) = b_s h_N(T_w - T_\infty = b_s h_N \Delta T, \tag{2.47}$$

$$M^\star \equiv \frac{\gamma\,\Delta T_s}{\rho g \bar{h}_N^2 \sin\beta} = \frac{MaBi}{h_N(1 + Bih_N)} = \frac{BM}{1 + B}, \tag{2.48}$$

referred to hereinafter as the *modified film Marangoni number*, and where the product $MaBi$ appears explicitly through (2.46). Therefore, $M^\star \to 0$ if $Bi \to 0$ so that there is no thermocapillary effect in this limit. Nevertheless, it appears that in the case of a small Biot number, $Bi \ll 1$, which is frequently the case for liquid films in contact with gases, the base state temperature gradient can be assumed to be independent of the film thickness, $b_s \approx Bi$. In this limit, the base state temperature gradient is uniquely defined by the heat transfer coefficient α and the thermal conductivity λ.

HF

In this case, the temperature of the undeformed free surface is obtained from (2.44b),

$$\Theta_s \equiv \Theta|_{y=h_N} = \frac{1}{Bi + Bi_w(1 + Bih_N)}, \tag{2.49}$$

and the temperature gradient between the free surface and the wall now reads:

$$b_s \equiv \frac{\Theta|_{y=0} - \Theta|_{y=h_N}}{h_N} = \frac{Bi}{Bi + Bi_w(1 + Bih_N)}. \tag{2.50}$$

The limit $1/Bi = 0$ leads to the same conclusion with the ST case. However, the limit $Bi = 0$ is now different, as (2.49) shows that the dimensionless temperature on the free surface is $1/Bi_w$, and therefore depends on the heat transfer characteristics of the solid-gas interface. If in addition $Bi_w = 0$, corresponding to a wall perfectly insulated from the gas, the interfacial temperature diverges to infinity as the heat supplied by the wall to the liquid has nowhere else to go.

It should be emphasized that the surface temperature Θ_s depends on the film thickness only through the parameter Bi_w. If the latter vanishes, $\Theta_s = 1/Bi$ is independent of h_N and remains constant (on the other hand for ST, the quantity Θ_s always depends on h_N, provided that $Bi \neq 0$). In other words, the thermocapillary instability is suppressed for a wall perfectly insulated from the gas. In this case, the temperature gradient across the film layer is independent of h_N ($b_s = 1$), which implies that any elevation (depression) of the film thickness will be accompanied by an increase (decrease) of the wall temperature $\Theta|_{y=0} = (1 + Bih_N)/Bi$ so that the film surface temperature remains constant. Therefore, enabling heat losses at the wall through the Robin/mixed boundary condition (2.8) is the only way to enable the Marangoni instability when a uniform heat flux q_w is applied at the wall. For a nonuniformly heated wall, which is beyond the scope of this monograph, the thermocapillary instability leads to steady state deformations of the liquid-gas interface [138, 239, 254]. This thermocapillary effect is still present for the ST case with $Bi = 0$ and for the HF case with $Bi_w = 0$ [239].

Using the definition (2.47), which remains unaltered for the HF case but now $\Delta T = q_w l_\nu/\lambda$, the temperature scale for HF based on the viscous-gravity scaling, the modified film Marangoni number for HF has the form:

$$M^\star = \frac{MaBi}{h_N(Bi + Bi_w(1 + Bih_N))} = \frac{BM}{B + B_w(1 + B)}. \tag{2.51}$$

This film Marangoni number $M^\star$ based on the Nusselt flat film temperature difference between the wall and the free surface will be useful in Chap. 3, where we examine the linear stability of the Nusselt flat film solution.

Chapter 3
Primary Instability

We consider the linear stability of the base state (2.15a)–(2.15f), i.e., its stability with respect to infinitesimal perturbations, or equivalently its *primary instability*. For moderate Reynolds and Péclet numbers, the base state occurs soon after the inlet. The linear destabilization of this state is the first step of the evolution that eventually leads to the disordered spatio-temporal dynamic that typically characterizes falling film flows. It is imperative, therefore, that we analyze the primary instability and carefully examine the physical mechanisms responsible for its onset.

The linearity of the governing equations for the primary instability allows us to decompose the perturbations into *normal modes*, which greatly simplifies the subsequent analysis. This decomposition leads to the *Orr–Sommerfeld eigenvalue problem*, which plays a central role in the analysis of the primary instability.

3.1 Linearized Equations for the Disturbances

In terms of the Nusselt scaling in (2.16a)–(2.16f) the base state in (2.15a)–(2.15f) reads:

$$U(y) = y\left(1 - \frac{y}{2}\right), \tag{3.1a}$$

$$V(y) = W(y) = 0, \tag{3.1b}$$

$$P(y) = Ct(1 - y), \tag{3.1c}$$

$$\text{ST:} \quad \Theta(y) = 1 - \frac{By}{1 + B}, \tag{3.1d}$$

or

$$\text{HF:} \quad \Theta(y) = \frac{1 + B(1 - y)}{B + B_{\mathrm{w}}(1 + B)}. \tag{3.1e}$$

Substituting

$$\mathbf{v} = (U + \tilde{u}, \tilde{v}, \tilde{w}), \qquad T = \Theta + \tilde{T}, \qquad p = P + \tilde{p}, \qquad h = 1 + \tilde{h},$$

S. Kalliadasis et al., *Falling Liquid Films*, Applied Mathematical Sciences 176, DOI 10.1007/978-1-84882-367-9_3, © Springer-Verlag London Limited 2012

into the equations of motion and energy (2.17)–(2.21) and wall and free-surface boundary conditions (2.22)–(2.28), and linearizing for $\tilde{u}, \tilde{v}, \tilde{w}, \tilde{T}, \tilde{p}, \tilde{h} \ll 1$ yields the perturbation equations:

$$\partial_x \tilde{u} + \partial_y \tilde{v} + \partial_z \tilde{w} = 0, \tag{3.2a}$$

$$3Re(\partial_t \tilde{u} + U\partial_x \tilde{u} + DU\tilde{v}) + \partial_x \tilde{p} - \nabla^2 \tilde{u} = 0, \tag{3.2b}$$

$$3Re(\partial_t \tilde{v} + U\partial_x \tilde{v}) + \partial_y \tilde{p} - \nabla^2 \tilde{v} = 0, \tag{3.2c}$$

$$3Re(\partial_t \tilde{w} + U\partial_x \tilde{w}) + \partial_z \tilde{p} - \nabla^2 \tilde{w} = 0, \tag{3.2d}$$

$$3Pe(\partial_t \tilde{T} + U\partial_x \tilde{T} + D\Theta \tilde{v}) - \nabla^2 \tilde{T} = 0, \tag{3.2e}$$

subject to the boundary conditions:

- at the wall $y = 0$:

$$\tilde{u} = \tilde{v} = \tilde{w} = 0, \tag{3.3}$$

$$\text{ST:} \quad \tilde{T} = 0 \tag{3.4a}$$

  or

$$\text{HF:} \quad \partial_y \tilde{T} = B_{\mathrm{w}} \tilde{T} \tag{3.4b}$$

- at the undeformed film surface $y = 1$:

$$\tilde{v} = \partial_t \tilde{h} + U\partial_x \tilde{u}, \tag{3.5}$$

$$\tilde{p} = Ct\,\tilde{h} - (We - M\Theta)\nabla_{xz}^2 \tilde{h} + 2\partial_y \tilde{v}, \tag{3.6}$$

$$\tilde{h} = M(D\Theta\,\partial_x \tilde{h} + \partial_x \tilde{T}) + \partial_y \tilde{u} + \partial_x \tilde{v}, \tag{3.7}$$

$$0 = M(D\Theta\,\partial_z \tilde{h} + \partial_z \tilde{T}) + \partial_z \tilde{v} + \partial_y \tilde{w}, \tag{3.8}$$

$$\partial_y \tilde{T} = -B(D\Theta \tilde{h} + \tilde{T}), \tag{3.9}$$

where $D \equiv d/dy$, $\nabla_{xz}^2 = \partial_{xx} + \partial_{zz}$ is the two-dimensional Laplacian operator in the (x, z)-plane and $D\Theta$ is the slope of the base-state linear temperature distribution:

$$\text{ST:} \quad D\Theta = -\frac{B}{1+B} \tag{3.10a}$$

or

$$\text{HF:} \quad D\Theta = -\frac{B}{B + B_{\mathrm{w}}(1+B)}. \tag{3.10b}$$

The boundary conditions (3.5)–(3.9) have been obtained with the help of Taylor expansions at the undeformed free surface $h = 1$ since the interfacial boundary conditions are evaluated at $y = h = 1 + \tilde{h}$, i.e., variable X is expanded as $X|_h = X(1) + \tilde{x}|_1 + DX(1)\tilde{h}$. The resulting linearized interfacial boundary conditions have been simplified using $D^2 U = -1$, $D^2\Theta = 0$, $DP = -Ct$, $DU|_1 = 0$, $P|_1 = 0$ and the continuity equation (3.2a).

Note that for $1/B_{\mathrm{w}} = 0$, corresponding to very good heat transfer characteristics of the solid–gas interface, the temperature perturbation along the wall vanishes and

HF in (3.4b) reduces to ST in (3.4a). On the other hand, for $B_\mathrm{w} = 0$, corresponding to very poor heat transfer characteristics of the solid–gas interface or simply a wall insulated from the gas, the heat flux perturbation along the wall vanishes or, equivalently, we have the specified heat flux condition $\partial_y \tilde{T} = 0$. As was pointed out in Sect. 2.5 in the limit $B_\mathrm{w} = 0$ the thermocapillary instability is suppressed, and hence having $B_\mathrm{w} \neq 0$ in the HF thermal boundary condition enables the thermocapillary instability.

The system of equations (3.2a)–(3.2e) can be rearranged so that only the perturbations of the normal velocity $\tilde{v}$, the temperature $\tilde{T}$ and the film thickness $\tilde{h}$ appear. Let us first take the divergence of the linearized Navier–Stokes equations in vector form, i.e., $[\partial_x(3.2\mathrm{b}) + \partial_y(3.2\mathrm{c}) + \partial_z(3.2\mathrm{d})]$. With the use of the continuity equation (3.2a), the result is

$$\nabla^2 \tilde{p} = -6Re\,DU\,\partial_x\tilde{v}. \tag{3.11}$$

By applying now the two-dimensional Laplacian operator on the y-component of the linearized Navier–Stokes equation (3.2c) using (3.11) to eliminate the pressure and noting that $D^2 U = -1$, we obtain:

$$\nabla^2\big(3Re\partial_t\tilde{v} - \nabla^2\tilde{v}\big) + 3Re\big(1 + U\nabla^2\big)\partial_x\tilde{v} = 0. \tag{3.12}$$

We next consider the two-dimensional Laplacian operator of the linearized normal stress boundary condition (3.6). From (3.11), $\nabla^2 \tilde{p}|_1 = \nabla_{xz}^2 \tilde{p}|_1 + \partial_{yy}\tilde{p}|_1 = 0$. Differentiating then (3.2c) with respect to y,

$$\partial_{yy}\tilde{p} = -3Re(\partial_{yt}\tilde{v} + DU\,\partial_x\tilde{v} + U\partial_{yx}\tilde{v}) + \partial_y\nabla^2\tilde{v},$$

and evaluating the result at the undeformed free surface $y = 1$ gives

$$\nabla_{xz}^2\tilde{p}|_1 = 3Re(\partial_{yt}\tilde{v} + U\partial_{yx}\tilde{v}) + \partial_y\nabla^2\tilde{v}.$$

From ∇_{xz}^2 in (3.6), where $\nabla_{xz}^2(\tilde{p}|_1) = \nabla_{xz}^2\tilde{p}|_1$, we finally get:

$$Ct\nabla_{xz}^2\tilde{h} - (We - M\Theta)\nabla_{xz}^2\nabla_{xz}^2\tilde{h} + 3\nabla_{xz}^2\partial_y\tilde{v} + \partial_{yyy}\tilde{v} - 3Re(\partial_{yt}\tilde{v} + U\partial_{xy}\tilde{v}) = 0. \tag{3.13}$$

Taking now the divergence of the tangential stress boundary condition in vector form, i.e., $[\partial_x(3.7) + \partial_z(3.8)]$, and with the use of the continuity equation (3.2a), we obtain:

$$\partial_x\tilde{h} - M\big(D\Theta\nabla_{xz}^2\tilde{h} + \nabla_{xz}^2\tilde{T}\big) - \big(\nabla_{xz}^2 - \partial_{yy}\big)\tilde{v} = 0. \tag{3.14}$$

3.2 The Orr–Sommerfeld Eigenvalue Problem

We then seek the solution in the form of *normal modes*:

$$\begin{pmatrix} \tilde{v} \\ \tilde{T} \\ \tilde{h} \end{pmatrix} = \begin{pmatrix} \phi(y) \\ \tau(y) \\ \eta \end{pmatrix} \exp\{i(\mathbf{k}\cdot\mathbf{x} - \omega t)\}, \tag{3.15}$$

where $\mathbf{x} = (x, z)$, $\mathbf{k} = (k_x, k_z)$ is the *wavenumber vector* and ω is the *complex angular frequency* that contains the *complex wave velocity* $c = \omega/k$.

For *temporal stability analysis* we impose real wavenumber components $k_{x,z}$ (the disturbance travels in the direction $\tan^{-1}(k_z/k_x)$) and solve for the complex eigenvalue $\omega = \omega(\mathbf{k})$, a relation which is referred to as the *dispersion relation*. The *temporal growth rate* is ω_i (subscripts r and i are used to denote real and imaginary parts, respectively): if $\omega_i > 0$ the disturbance grows in time and the base state is unstable. $c_r = \omega_r/k$ is the *phase velocity* and $k = \sqrt{k_x^2 + k_z^2}$ is the modulus of the wavenumber vector. The *real angular frequency* is simply ω_r while the phase velocity along the x, z axes is $\omega_r/k_{x,z}$, respectively. There is a simple relation between the real angular frequency and the *ordinary frequency*: for a wave with velocity c_r and wavelength l, $f = c_r/l = c_r/(2\pi/k) = kc_r/(2\pi) = \omega_r/(2\pi)$ or $\omega_r = 2\pi f$. Hence, the real angular frequency is a simple multiple of the ordinary frequency.

For *spatial stability analysis*, we consider ω as real and seek complex $k_{x,z}$[1] so that $\mathbf{k}$ is now the eigenvalue. The simplest case is when one of the two components $k_{x,z}$ is real, e.g., in Chap. 7 we shall examine the spatial stability of the isothermal Nusselt flat film flow in two dimensions, i.e., $\mathbf{k} \cdot \mathbf{x} \equiv k_x x$. The *spatial growth rate* now is $-k_{xi}$: if $k_{xi} < 0$, the disturbance grows in space. The phase velocity is ω/k_{xr}. When both ω and k are complex, we have the concept of *generalized spatial/temporal stability analysis* to be defined in Chap. 7. In this chapter we perform a temporal stability analysis only.

Introducing (3.15) into the governing equations (3.12), (3.2e) and the boundary conditions (3.3)–(3.5), (3.9), (3.13) and (3.14) yields the system,

$$\left(D^2 - k^2\right)^2\phi + 3Rei\left[(\omega - k_x U)\left(D^2 - k^2\right) - k_x\right]\phi = 0, \tag{3.16a}$$

$$\left(D^2 - k^2\right)\tau - 3Pe\left[D\Theta\phi - i(\omega - k_x U)\tau\right] = 0, \tag{3.16b}$$

$$\phi(0) = 0, \tag{3.16c}$$

$$\text{ST:} \quad \tau(0) = 0 \tag{3.16d}$$

or

$$\text{HF:} \quad D\tau(0) = B_w\tau(0), \tag{3.16e}$$

$$\phi(1) + i\frac{1}{2}\eta(2\omega - k_x) = 0, \tag{3.16f}$$

$$\eta k^2\left[Ct + \left(We - M\Theta(1)\right)k^2\right] = \left[\left(D^2 - 3k^2\right) + i\frac{3}{2}Re(2\omega - k_x)\right]D\phi(1), \tag{3.16g}$$

---

[1]Expression (3.15) now requires the evaluation of $\mathbf{k} \cdot \mathbf{x}$ where $\mathbf{k}$ is a vector with complex components. The dot product $\mathbf{a} \cdot \mathbf{b}$ for two vectors $\mathbf{a}$ and $\mathbf{b}$ with real components can be easily generalized to vectors with complex components (see e.g., [108]). Assume $\mathbf{a} = (a_1, a_2, \ldots, a_n)$ and $\mathbf{b} = (b_1, b_2, \ldots, b_n)$. Then $\mathbf{a} \cdot \mathbf{b} = \sum_{i=1}^{n} a_i \bar{b}_j$ where the overbar denotes complex conjugation. Hence, in our case we simply have $\mathbf{k} \cdot \mathbf{x} = (k_{xr} + ik_{xi})x + (k_{zr} + ik_{zi})z$.

$$\left(D^2 + k^2\right)\phi(1) + M\left[\eta D\Theta + \tau(1)\right]k^2 + ik_x\eta = 0, \qquad (3.16h)$$

$$D\tau(1) + B\left[\eta D\Theta + \tau(1)\right] = 0, \qquad (3.16i)$$

where $U(1) = 1/2$ has been utilized.

The system of equations (3.16a)–(3.16i) is the *Orr–Sommerfeld eigenvalue problem* for a film falling down a heated wall subject to either the HF or ST conditions.[2] In Sects. 3.4 and 3.5 we shall discuss the solution of (3.16a)–(3.16i) for two limiting cases: *transverse perturbations* with $k_x = 0$ and $k_z \neq 0$ and *streamwise perturbations* with $k_x \neq 0$ and $k_z = 0$ (perturbations are quantified based on the direction in which the wavenumber is nonzero). But first it is instructive to define two types of instabilities typical of the heated falling film problem.

3.3 Oscillatory Versus Stationary (or Monotonic) Instabilities

For simplicity let us consider two-dimensional disturbances only with $k_x = k$ and $k_z = 0$. Let us assume that the dependence of the system on a control parameter, say Σ, is such that for $\Sigma = \Sigma_c$, $\omega_i(k = k_0) = 0$, for $\Sigma < \Sigma_c$, $\omega_i < 0$ $\forall k$ in a finite region of the k-axis around k_0, and for $\Sigma > \Sigma_c$, $\omega_i > 0$ $\forall k$ in a finite region of the k-axis, i.e., there is a band of unstable modes centered around the wavenumber k_0. k_0 then is the *critical wavenumber* corresponding to the *critical value* of the control parameter, Σ_c. In Sect. 3.4.3 we shall give an alternative definition of the critical value of a control parameter based on the "neutral curve."

We distinguish between two cases:

1: If $\omega_r(k_0) = 0$ for $\Sigma = \Sigma_c$ we have a *stationary instability* (also called "monotonic"). Both the S- and P-modes introduced in the Introduction belong to this category.
2: If $\omega_r(k_0) \neq 0$ for $\Sigma = \Sigma_c$ we have an *oscillatory instability* (also called "overstability"). This is the case described by (C.11) in Appendix C.5.

The case of long-wave instabilities, i.e., instabilities with $k_0 = 0$ and with both ω_r and ω_i vanishing at this wavenumber, requires special attention. This is precisely the case of the H-mode introduced in the Introduction and is described by (C.13a), (C.13b) in Appendix C.5. As we emphasize in the discussion following (C.13a), (C.13b), the limit $k \to 0$ is effectively degenerate as the disturbance reduces to a simple uniform shift of the base state (corresponding to the so-called *Goldstone mode*; see Appendix C.5). Finite-size effects remove the degeneracy and

---

[2]Some authors reserve the term "Orr–Sommerfeld eigenvalue problem" for the linear stability analysis of a parallel flow with respect to two-dimensional disturbances and use instead the term "generalized Orr–Sommerfeld eigenvalue problem" for three-dimensional disturbances—meaning the "generalization" of the Orr–Sommerfeld eigenvalue problem for two-dimensional disturbances to three-dimensional ones, e.g., [182]. Others use the term "Orr–Sommerfeld eigenvalue problem" for both two-dimensional and three-dimensional disturbances, e.g., [44].

forbid the mathematical artifact of infinite long wavelengths (in practice the smallest wavenumber is $k \sim 1/L$ with L the channel's length), so that a true "Hopf bifurcation" with $\omega_r \neq 0$ occurs.[3] In the linear regime the disturbance grows with a growth rate $\omega_i(k_{max})$ and at the same time it is periodic in space with wavenumber k_{max} and oscillates in time with frequency $-\lambda_i(k_{max})$. The combination of periodicity in space and oscillatory behavior in time leads to a "traveling wave" and the H-mode is clearly an oscillatory instability. Noteworthy is that the definition given in [62] that a long-wave instability is stationary if $\omega_r(k_0 = 0) = 0$ is confusing and would lead to the conclusion that the H-mode is stationary.

3.4 Transverse Perturbations: $k_x = 0$, $k_z = k$

3.4.1 Eigenvalue Problem

When considering transverse perturbations, i.e., $k_x = 0$ and $k_z = k$, system (3.16a)–(3.16i) becomes:

$$\left(D^2 - k^2\right)^2 \phi + 3Rei\omega\left(D^2 - k^2\right)\phi = 0, \tag{3.17a}$$

$$\left(D^2 - k^2\right)\tau - 3Pe[D\Theta\phi - i\omega\tau] = 0, \tag{3.17b}$$

$$\phi(0) = 0, \tag{3.17c}$$

$$\text{ST:} \quad \tau(0) = 0 \tag{3.17d}$$

or

$$\text{HF:} \quad D\tau(0) = B_w\tau(0), \tag{3.17e}$$

$$\eta = i\frac{\phi(1)}{\omega}, \tag{3.17f}$$

$$\eta k^2\left[Ct + \left(We - M\Theta(1)\right)k^2\right] = \left[\left(D^2 - 3k^2\right) + 3Rei\omega\right]D\phi(1), \tag{3.17g}$$

$$\left(D^2 + k^2\right)\phi(1) + M\left[\eta D\Theta + \tau(1)\right]k^2 = 0, \tag{3.17h}$$

$$D\tau(1) + B\left[\eta D\Theta + \tau(1)\right] = 0. \tag{3.17i}$$

One notices the symmetry $k \to -k$ of these equations, a consequence of the reflection symmetry $z \to -z$ of the full system. By comparing (3.16a)–(3.16i) and (3.17a)–(3.17i), one also notices that the base flow velocity is absent from (3.17a)–(3.17i), as the effect of the advection of the perturbations by the base flow

---

[3] The S-mode is also a long-wave variety but has $\omega_r = 0 \; \forall k$; on the other hand the P-mode is a short-wave variety but again with $\omega_r = 0 \; \forall k$.

$\propto k_x U$ has been suppressed. Hence, there should be an analogy between the heated falling film problem examined here and that of an horizontal film without a base flow heated from below with a heater that maintains the wall temperature at a constant value (ST) in which case the instabilities are of purely thermocapillary origin and stationary with $\omega_r = 0$ $\forall k$ [58, 107]. After all, apart from the role of mean flow in setting up the system and fixing the flat film thickness through the flow rate, mean flow does not influence the eigenvalue problem in (3.17a)–(3.17i). Of course, as noted in Sect. 2.2, our scaling does not allow the $\beta = 0$ limit. However, by identifying in Re and Pe, h_N with the dimensionless flat film thickness of the horizontal case and appropriately rescaling (3.17a)–(3.17i), i.e., by taking $g\bar{h}_N^2/3\nu$ for the velocity scale instead of $\bar{u}_N = g \sin \beta \bar{h}_N^2/3\nu$, replacing Re with $h_N^3/3$, setting $Ct = 1$ and replacing $g \sin \beta$ with g in the definitions of the dimensionless groups given in Chap. 2, the formulation of the linear stability eigenvalue problem in (3.17a)–(3.17i) is identical to that for the horizontal film (see, e.g., [58]).[4]

The linear instability of the ST case has been analyzed in detail in [107, 261]. In the next section we shall reproduce some of the previous results but we shall also analyze the HF case.

3.4.2 Neutral Stability Condition

For temporal stability analysis, a mode with $\omega_i = 0$ is said to be *neutrally stable*. On the other hand, for spatial stability analysis a neutrally stable mode is one with $k_i = 0$. The *neutral condition* is the condition that ensures that the dominant/least stable mode is neutrally stable. A *neutral curve* is a plot involving the parameters of the neutral condition, typically the pertinent dimensionless groups versus wavenumber (and hence by definition on a neutral curve the neutral condition is satisfied) and might have different branches.

Much like the ST case, for the HF case we also have stationary instabilities with $\omega_r = 0$ $\forall k$. Then, to obtain the neutral stability condition of stationary instabilities we must set $\omega_i = 0$ or $\omega = 0$ in (3.17a)–(3.17i). In this limit the system admits an analytical solution, which for HF reads:

HF:

$$M^\star = \left[\frac{16k[(B + B_w)k \cosh k + (B B_w + k^2) \sinh k](2k - \sinh 2k)}{3Pe[(k + 4k^3(B_w - 1)) \cosh k - k \cosh 3k - 4B_w \sinh^3 k + 4k^2(2 + k^2) \sinh k] - \dfrac{32k^5(B_w \cosh k + k \sinh k)}{Ct + k^2(We - M^\star/B)}} \right], \quad (3.18)$$

where $M^\star = -M D\Theta$ (see (2.51) and (3.10b)).

---

[4]For the horizontal case the film thickness is fixed by the amount of fluid; for the falling film the thickness is fixed through the flow rate.

By taking the limit $B_\mathrm{w} \to \infty$ in (3.18), i.e., for very good heat transfer characteristics of the solid–gas interface, we obtain the neutral condition for ST:

$$\text{ST:}\quad M^\star = \frac{4k(k\cosh k + B\sinh k)(2k - \sinh 2k)}{3Pe(k^3\cosh k - \sinh^3 k) - \frac{8k^5\cosh k}{Ct + k^2(We - M^\star/B)}}, \tag{3.19}$$

where again $M^\star = -MD\Theta$ (see (2.48) and (3.10a)).

Equation (3.19) can be related to the linear stability of a horizontal film ($\beta = 0$) heated uniformly from below [58, 255],[5] a situation that as we have mentioned earlier is formally forbidden by the nondimensionalization adopted in Sect. 2.2, which utilizes the average velocity, $\bar{u}_\mathrm{N} = g\sin\beta\bar{h}_\mathrm{N}^2/3\nu$. Nevertheless, as noted earlier, through appropriate rescaling the formulation of the linear stability eigenvalue problem in (3.17a)–(3.17i) is identical to that for the horizontal film and hence the neutral condition in (3.19) is identical to that for a horizontal film. This correspondence demonstrates the decoupling of the transverse Marangoni instability mechanism from the gravity-driven base flow. Recall that there is no mean flow in the transverse direction as we emphasized earlier. Other than fixing the film thickness through the flow rate, the flow does not play any role when $k_x = 0$.

Further, note that in the limits $Ct \to \infty$ and/or $We \to \infty$, (3.19) reduces to the neutral condition found by Pearson [206] for a horizontal layer with nondeformable interface heated uniformly from below (see Introduction). In these limits, (3.18) with $B_\mathrm{w} = 0$ reduces also to the neutral condition found by Pearson for a horizontal layer with a nondeformable interface.

Neutral stability curves for HF, i.e., solutions to (3.18) in the $(M^\star, k)$-plane are displayed in Fig. 3.1. The neutral stability curve for ST, i.e., solution to (3.19), is also plotted as a particular case of HF in the limit $B_\mathrm{w} \to \infty$. Note that (3.18) has two equal positive roots, one relevant for $k > 0$ and the other one for $k < 0$. In fact, (3.18) and its limiting case (3.19) have the symmetry $k \to -k$, as expected. The curves are the locus of $\omega = 0$, they separate stable regions from unstable ones and correspond to the onset of stationary instabilities. In the unstable region above the curves, $\omega_\mathrm{r} = 0$. The parameter values chosen correspond to a situation in which the film thickness and the gas–liquid system are held fixed. The set of control parameters then reduces to the heat transfer coefficient of the wall–gas interface and the heat flux supplied by the heater, i.e., the modified film Marangoni number $M^\star$ and the wall film Biot number B_w, or equivalently Ma and Bi_w (see Chap. 2).

---

[5]Though these works neglect the variation of surface tension with temperature in the normal stress boundary condition, it has been retained here. In the normal stress boundary condition (3.17g), the contribution of surface tension variation with temperature, $k^2 M\Theta(1)$, or equivalently the term $k^2 M^\star/B$ in the right hand sides of (3.18) and (3.19), can only be neglected compared to Ct and $k^2 We$ when k is small and We is large, respectively—note that in experimental observations We is large for most liquids, the so called "strong surface tension effect" (this is a crucial point for the remaining of the monograph and will be discussed in detail in Chaps. 4 and 5). In the general case, however, i.e., for finite wavenumbers, $k = \mathcal{O}(1)$, $M\Theta(1)$, or equivalently, $M^\star/B$ should be retained.

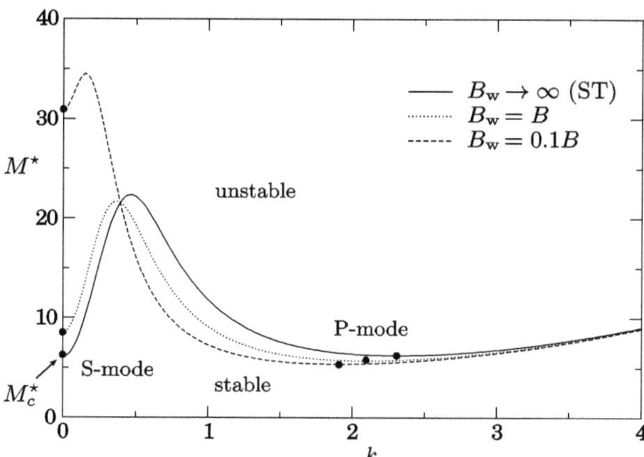

Fig. 3.1 Neutral stability curves from (3.18) (HF) in the $(M^\star, k)$-plane for different values of the wall Biot number, B_{w}. Parameter values: $Ct = 3.73$, $We = 120$, $Re = 1$, $Pe = Pr = 7$ and $B = 1.44$. The instability thresholds for the different modes are indicated by bullets: S-mode at $k = 0$ and P-mode at $k = \mathcal{O}(1)$. $M_{\mathrm{c}}^\star$ denotes the critical modified film Marangoni number for the S-mode

The curves have two minima corresponding to the onset of the long-wave S-mode at $k = 0$ and to the onset of the short-wave P-mode with $k = \mathcal{O}(1)$ [107]. The S-mode leads to long-scale free-surface deformations while the P-mode leads to "steady convection cells" with phase velocity $c_{\mathrm{r}} = \omega_{\mathrm{r}} = 0$ (sometimes referred to as "rolls") for which the deformation of the interface does not play an important role. As noted in the Introduction, this monograph is devoted to thin films and their associated long-wave instabilities. The short-wave P-mode then will not be considered in what follows and we shall focus on the long-wave H- and S-modes instead.

3.4.3 Critical Condition and Long-Wave Expansion

An alternative definition of the critical value of a control parameter to the one given in Sect. 3.3, used quite frequently in hydrodynamic stability, utilizes the concept of a neutral curve: a minimum/maximum value of a control parameter on a branch of a neutral curve such that instability occurs above/below the neutral curve and in a neighborhood around this value is a *critical* value of this parameter. For example, if a neutral curve is such that instability occurs above the curve and stability below and the curve exhibits more than one local minima, the values of the control parameter at the minima are its critical values. The value of the control parameter at the global minimum is the critical value of this parameter for all wavenumbers.

This definition based on the neutral curve allows for a systematic identification of the critical values of a control parameter in the (control parameter-wavenumber) parameter space as can be illustrated from Fig. 3.1. For the S-mode which is a long-wave variety, the critical value of $M^\star$ is obtained from the minimum of the neutral curve in the region $k \to 0$. For the ST case, the critical value of $M^\star$ for the S-mode coincides with the critical value of this parameter for all k. The critical value of the P-mode is obtained from the local minimum of the neutral curve at an $\mathcal{O}(1)$ value of the wavenumber.

It is straightforward to show that the above definition for a critical value of a control parameter is consistent with that given in Sect. 3.3. Let us consider for instance a situation with $\omega_i = Ak^2(\Sigma - \Sigma_c) - Bk^4$ with $A, B > 0$, as is the case e.g., with the H-mode. The neutral curve is then obtained by setting $\omega_i = 0$, which gives $k = 0$, a first branch of the neutral curve, and $\Sigma - \Sigma_c = (B/A)k_c^2$, a second branch of the neutral curve. The band of unstable modes for $\Sigma > \Sigma_c$ is $0 \le k \le k_c$ with k_c the *cut-off wavenumber*. Plotting $\Sigma - \Sigma_c = (B/A)k_c^2$ in the (k_c, Σ)-plane gives a parabola whose minimum is located at $k_c = 0$ and $\Sigma = \Sigma_c$ (hence, $k_c \equiv k_0 = 0$ is the critical wavenumber).

The critical modified film Marangoni number for the threshold of the S-mode can be easily obtained from (3.18) and (3.19) by taking the long-wave limit, i.e., as $k \to 0$:

$$\text{ST:}\quad M_c^\star = \frac{2(1 + B)Ct}{3} \tag{3.20a}$$

or

$$\text{HF:}\quad M_c^\star = \frac{2(B + B_w(1 + B))Ct}{3B_w}. \tag{3.20b}$$

Considering now a vertical geometry, $Ct = 0$, (3.20a)–(3.20b) indicates that the film is always unstable with respect to the long-wave S-mode, i.e., for all h_N (or equivalently all Re). Notice that the critical modified film Marangoni number (3.20b) goes to infinity for HF in the limit of a vanishing heat loss at the wall ($B_w \to 0$), which signals the loss of the long-wave thermocapillary S-mode for an insulated wall and an imposed heat flux (see also Sect. 2.5).

Close to the instability onset, i.e., just above the critical Marangoni number given by (3.20a)–(3.20b), a *long-wave expansion*[6] of the neutral stability conditions (3.18) and (3.19), i.e., a regular perturbation expansion of this condition for $k \ll 1$, yields the band of unstable wavenumbers $0 \le k \le k_c$ for which the linear growth rate of the infinitesimal disturbances is positive, with k_c the cut-off wavenumber. Further, by considering strong surface tension effects, i.e., large We, and assuming for simplicity that all remaining parameters are of $\mathcal{O}(1)$, simple expressions for the cut-off wavenumbers k_c are obtained:

$$\text{ST:}\quad k_c = \left(\frac{3(M^\star - M_c^\star)}{2We(1 + B)}\right)^{1/2} \tag{3.21a}$$

or

$$\text{HF:}\quad k_c = \left(\frac{3B_w(M^\star - M_c^\star)}{2We(B + B_w(1 + B))}\right)^{1/2}, \tag{3.21b}$$

---

[6]The long-wave expansion as a methodology for the reduction of the governing equations and associated wall and free-surface boundary conditions into simpler systems of equations will be outlined in detail in Chaps. 4 and 5.

corresponding to analytical representations of the neutral stability curves in Fig. 3.1 in the region of small wavenumbers.

Hence, the stabilizing effect of surface tension, which damps perturbations of relatively short wavelength, limits the range of unstable wavenumbers $0 \leq k \leq k_c$ close to the origin even away from the vicinity of the instability onset, i.e., for $M^\star - M_c^\star = \mathcal{O}(1)$, provided of course that We is sufficiently large.

3.5 Streamwise Perturbations: $k_x = k$, $k_z = 0$

3.5.1 Eigenvalue Problem

Yih extended "Squire's theorem" [83, 121] to isothermal free-surface flows [303] and he demonstrated that the stability of the primary flow with respect to two-dimensional perturbations determines also its stability with respect to three-dimensional ones. He also found that the most unstable perturbations, i.e., the ones with the largest maximum growth rate, consist of streamwise waves ($k_z = 0$). Although Squire's theorem does not apply when the Marangoni forces are present ($M \neq 0$) [139, 257, 258], and hence the fully three-dimensional Orr–Sommerfeld problem (3.16a)–(3.16i) cannot be transformed to an equivalent two-dimensional one, numerical integration of the three-dimensional Orr–Sommerfeld problem shows that the most unstable perturbations in the region of small wavenumbers correspond to streamwise waves [107] (in the region $k = \sqrt{k_x^2 + k_z^2} = \mathcal{O}(1)$, the most unstable disturbances consist of transverse rolls and are due to the P-mode, but once again in this monograph we restrict our attention to the H- and S-modes only).

We therefore focus on streamwise perturbations. As we are dealing with a two-dimensional flow it is convenient to make use of the streamfunction ψ, which satisfies $u = \partial_y \psi$ and $v = -\partial_x \psi$. Denoting with φ, the amplitude of the normal mode representation of the streamfunction perturbation—about the base state streamfunction, $\Psi(y) = \int U \, dy$—the amplitude of the normal mode for the perturbation of the cross-stream component of the velocity can be rewritten as

$$\phi(y) = -ik_x\varphi(y).$$

Setting $k_x = k$ and writing $\omega = kc$, with c the complex wave speed, the system (3.16a)–(3.16i) acquires the form:

$$\left(D^2 - k^2\right)^2 \varphi + 3Reik\left[(c - U)\left(D^2 - k^2\right) - 1\right]\varphi = 0, \tag{3.22a}$$

$$\left(D^2 - k^2\right)\tau + 3RePrik\left[D\Theta\varphi + (c - U)\tau\right] = 0, \tag{3.22b}$$

$$\varphi(0) = D\varphi(0) = 0, \tag{3.22c}$$

$$\text{ST:}\quad \tau(0) = 0 \tag{3.22d}$$

or

$$\text{HF:} \quad D\tau(0) = B_\mathrm{w}\tau(0), \tag{3.22e}$$

$$\eta = \frac{\varphi(1)}{c - 1/2}, \tag{3.22f}$$

$$\left[\left(D^2 - 3k^2\right) + 3Reik(c - 1/2)\right]D\varphi(1)$$
$$- i\eta k\left[Ct + \left(We - M\Theta(1)\right)k^2\right] = 0, \tag{3.22g}$$

$$\left(D^2 + k^2\right)\varphi(1) + ikM\left[\eta D\Theta(1) + \tau(1)\right] - \eta = 0, \tag{3.22h}$$

$$D\tau(1) + B\left[\eta D\Theta(1) + \tau(1)\right] = 0. \tag{3.22i}$$

Although an analytical solution of the system (3.22a)–(3.22i) can be found for $k \to 0$ as we shall see later, the full solution can only be obtained numerically. Appendix F.1 outlines a numerical procedure to solve (3.22a)–(3.22i) for the simpler isothermal case ($M = 0$) based on the continuation software AUTO-07P [79]. This procedure can easily be extended to the nonisothermal case, as done hereinafter and in Chap. 9, and for that matter to other problems where a numerical solution of an Orr–Sommerfeld eigenvalue problem in two dimensions is required.

Worth mentioning is a method presented by Anshus and Goren [10] to obtain approximate analytical solutions to (3.22a)–(3.22i) for the isothermal case ($M = 0$). As the instability occurs at the interface and energy is transferred from the base state to the perturbations at the interface (as shown in [147] and described in Sect. 3.6), Anshus and Goren approximated the base-state velocity distribution by its value at the free surface. Replacing U in (3.22a) with $U(1) = 1/2$ leads therefore to an ordinary differential equation with constant coefficients whose solution can easily be obtained analytically. This approximation is in excellent agreement with the full solution of (3.22a)–(3.22i) when the pure hydrodynamic instability ($M = 0$) is considered [10]. Though never attempted, it is likely that this method will also work when the hydrodynamic instability is coupled with the thermocapillary one ($M \neq 0$). Indeed, the transfer of energy from the base state to the perturbations also occurs at the interface when the thermocapillary S-mode is considered, since the mechanism of this mode is the generation of surface tension gradients induced by the deformation of the free surface and the linear distribution of temperature across the film.

3.5.2 Neutral Stability Condition

Neutral stability curves ($\omega_\mathrm{i} = 0$) obtained numerically from the full system (3.22a)–(3.22i) by using AUTO-07P are shown in Fig. 3.2 for the ST case. Parameter values are chosen to correspond to a situation in which β and the physical properties of the liquid–gas system are fixed (fixing the viscous-gravity scaling), i.e., we examine the influence of the flow rate and the temperature difference between the wall and

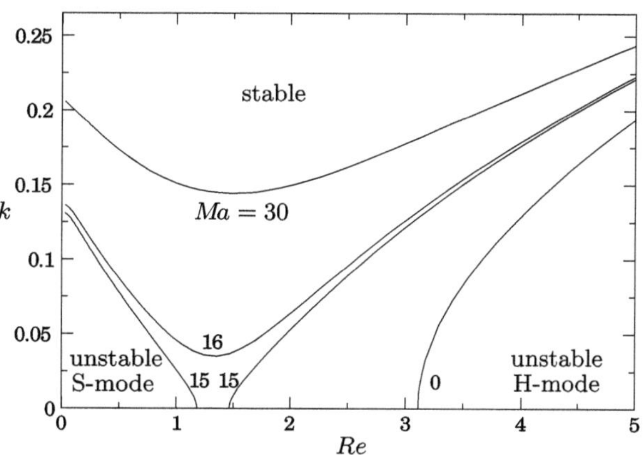

Fig. 3.2 Neutral stability curves for the specified temperature condition (ST) in the (Re, k)-plane for different values of the Marangoni number Ma. Parameter values: $\beta = 15°$ $(Ct = 3.73)$, $\Gamma = 250$, $Pr = 7$ and $Bi = 1$. Note that $k = 0$ is also a branch of the neutral curve

the ambient gas phase. Following our discussion at the end of Appendix D.1 then, the set of control parameters consists of the Reynolds number Re and Marangoni number Ma. We report the results as a function of Re, different Ma and fixed β, Pr, Γ and Bi.

The Nusselt flat film solution is found to be stable for large-wavenumber perturbations and unstable with respect to small-wavenumber perturbations. For $M = 0$, only the H-mode is present and because of the inclination, the hydrostatic pressure stabilizes small Reynolds number flows, more specifically for $Re < \frac{5}{6}Ct$ (see Sect. 3.5.4). For small positive M, the thermocapillary S-mode sets in for small Reynolds number flows, opening up a stability window defined by two curves emanating from the $k = 0$ axis, which eventually disappears for $Ma \geq 15.5$. Beyond this limit, the two stability curves merge and the two unstable regions unite, forming a single unstable domain, showing that the S- and H-modes reinforce each other. Notice that following the definition of a critical value of a parameter in Sect. 3.4.3, there is one critical value for Re when $Ma = 0$, two for $0 < Ma < 15.5$ and none for $Ma \geq 15.5$.

3.5.3 Long-Wave Expansion

The threshold at which the long-wave instability is triggered can be obtained with a long-wave analysis of the system (3.22a)–(3.22i) [304]. At zeroth-order in k (equivalently, for $k = 0$), the system (3.22a)–(3.22i) reduces to:

$$D^4 \varphi_0 = 0, \tag{3.23a}$$

$$D^2 \tau_0 = 0, \tag{3.23b}$$

$$\varphi_0(0) = D\varphi_0(0) = 0, \tag{3.23c}$$

$$\text{ST:} \quad \tau_0(0) = 0 \tag{3.23d}$$

or

$$\text{HF:}\quad D\tau_0(0) = B_{\mathrm{w}}\tau_0(0), \tag{3.23e}$$

$$D^3\varphi_0(1) = 0, \tag{3.23f}$$

$$D^2\varphi_0(1) - \frac{\varphi_0(1)}{c_0 - 1/2} = 0, \tag{3.23g}$$

$$D\tau_0(1) + B\left[\frac{\varphi_0(1)}{c_0 - 1/2}D\Theta(1) + \tau(1)\right] = 0. \tag{3.23h}$$

Integrating (3.23a) gives: $\varphi_0 = A_0 y^3 + B_0 y^2 + C_0 y + D_0$. Application of the no-slip condition at the wall (3.23c) and the normal stress condition at the free surface (3.23f) yields $A_0 = C_0 = D_0 = 0$. The tangential stress condition (3.23g) at the free surface then determines the eigenvalue c_0 at zeroth-order:

$$c_0 = 1.$$

The constant B_0 remains undetermined since the system is linear and homogeneous. For convenience, we normalize φ_0 by setting B_0 equal to unity:

$$\varphi_0 = y^2. \tag{3.24}$$

Thus, at zeroth-order in k the perturbed streamwise velocity is linear, $D\varphi_0 = 2y$, and its origin is the viscous shear stress. Since c_0 has no imaginary part, the perturbation is neither amplified nor damped, only advected with a constant phase speed. The axis $k = 0$ is a branch of the neutral stability curve and corresponds to a uniform change of the base state by a change in the flow rate, corresponding to what we have referred to as Goldstone mode. However, recall that finite-size effects do not allow the $k = 0$ limit and in fact in practice the smallest wavenumber in the system scales as $k \sim 1/L$, with L the channel's length.

Integrating (3.23b) gives:

$$\tau_0 = E_0 y + F_0. \tag{3.25}$$

The heat transfer conditions at the wall (3.23d) or (3.23e) and at the free surface (3.23h) then lead to

$$\text{ST:}\quad E_0 = 2D\Theta^2, \quad F_0 = 0, \tag{3.26}$$

or

$$\text{HF:}\quad E_0 = 2B_{\mathrm{w}}D\Theta^2, \quad F_0 = 2D\Theta^2, \tag{3.27}$$

where $c_0 = 1$ has been utilized. Notice that $D\Theta$ is here a constant given by (3.10a) and (3.10b) as the base flow temperature distribution (3.1d) and (3.1e) is linear.

Let us now expand φ, τ and c in asymptotic series in k,

$$c = c_0 + ikc_1 - k^2 c_2 - ik^3 c_3 + \mathcal{O}(k^4),$$

$$\varphi = \varphi_0 + ik\varphi_1 - k^2\varphi_2 - ik^3\varphi_3 + \mathcal{O}(k^4),$$

$$\tau = \tau_0 + ik\tau_1 - k^2\tau_2 - ik^3\tau_3 + \mathcal{O}(k^4),$$

where, since in (3.22a)–(3.22i) only odd powers of k have an imaginary coefficient, even terms of the above expansion are real while odd ones are purely imaginary. Substituting this expansion in (3.22a)–(3.22i) gives at first order:

$$\varphi_1 = Re\left(\frac{y^5}{20} - \frac{y^4}{4}\right) + \frac{Ct}{3}y^3 + B_1 y^2, \tag{3.28}$$

$$\tau_1 = Pe\left(-\frac{3}{40}E_0 y^5 - \frac{1}{8}(2D\Theta - 2E_0 + F_0)y^4\right.$$

$$\left. - \frac{1}{2}(E_0 - F_0)y^3 - \frac{3}{2}F_0 y^2\right) + E_1 y + F_1, \tag{3.29}$$

$$c_1 = \frac{2}{5}Re - \frac{1}{3}Ct - \frac{M}{4}(2D\Theta + E_0 + F_0). \tag{3.30}$$

The first-order term φ_1 of φ contains φ_0 within an arbitrary multiplicative constant B_1, corresponding to a redefinition of the original constant B_0 for φ_0 and thus replicating at first order the solution at zeroth order ($ikB_1 y^2 \propto \varphi_0$). By setting B_1 equal to zero so that φ_0 is suppressed at φ_1, the latter appears as a true correction to φ_0. Next, E_1 and F_1, the constants of integration of the second-order linearized energy balance (3.22b), are obtained using the heat transfer conditions at the wall (3.22d), (3.22e), (3.22h) and (3.22i),

$$\text{ST:}\quad E_1 = G, \qquad F_1 = 0,$$

or

$$\text{HF:}\quad E_1 = B_w G, \qquad F_1 = G,$$

where

$$G = 2D\Theta^2(Ct - Re) + \frac{2D\Theta^3 M}{1+B} - \frac{PeD\Theta}{40B}\left[10(4+B)D\Theta + (35+13B)E_0\right.$$

$$\left. + 5(16+9B)F_0\right],$$

thus completing the solution.

Noteworthy is that the above expansion yields the only root of the dispersion relation that can become unstable, and in fact throughout this chapter we focus on this root. There is also a countable infinite number of eigenvalues whose leading-order terms in the absence of the Marangoni effect assume the simple form $\sim -n^2\pi^2/(Rek)$. These are the *shear modes* associated with the semi-parabolic Nusselt velocity profile [42, 94, 207] and generated by the shear induced by the no-slip boundary conditions at the solid boundary. They are stable in the region of small-to-moderate Reynolds numbers considered here, but they can be destabilized for large Reynolds numbers leading to the "Tollmien–Schlichting" instability and transition to usual turbulence for a falling film.

3.5.4 Critical Condition

Since the first-order term of the velocity, ikc_1, is purely imaginary, k^2c_1 is the first contribution to the growth rate $\omega_i = kc_i$ of the instability. Hence, if $c_1 < 0$, the system will be linearly stable, whereas for $c_1 > 0$ the system will be linearly unstable. Therefore, the onset of the instability occurs at $c_1 = 0$, which yields the critical conditions:

$$\text{ST:} \quad Ct = \frac{6}{5}Re + \frac{3M^\star}{2(1+B)}, \tag{3.31a}$$

or

$$\text{HF:} \quad Ct = \frac{6}{5}Re + \frac{3B_w M^\star}{2(B + B_w(1+B))}. \tag{3.31b}$$

The terms on the right hand sides of (3.31a)–(3.31b) arise from inertia and thermocapillary forces, which are destabilizing, while the term on the left hand sides is due to the hydrostatic pressure which is stabilizing. For $M^\star = 0$ the above condition reduces to the well-known critical condition, $Re_c = (5/6)Ct$, for the onset of the H-mode in an isothermal falling film obtained by Benjamin [19] and Yih [304]. On the other hand, for $M^\star \neq 0$, the critical Reynolds number for, e.g., the ST case is $Re_c = (5/6)Ct - (5/4)(M^\star/(1+B))$ and thermocapillarity reduces the critical Reynolds number, a consequence of the destabilizing influence of the Marangoni effect.

As we shall see in Sect. 3.6, inertia leads to a phase shift between the interface location and the vorticity field that originates from the H-mode [256]. For transverse disturbances (e.g., rivulets aligned with the flow; see Fig. 1.11) there is no mechanism to allow energy transfer from the mean flow to the perturbation, so there is no term representing the mean shear. Indeed, if $Re = 0$ in (3.31a)–(3.31b), one recovers (3.20a)–(3.20b), which define the critical conditions for the thermocapillary instability for long-wave perturbations (S-mode). Hence, the critical $M^\star$ in the presence of flow ($Re \neq 0$) is smaller to that in the absence of flow ($Re = 0$). In other words, in the presence of flow the system can be unstable in a region that is otherwise stable with no flow. This result allows us to conclude that in the $k \to 0$ limit, streamwise perturbations giving rise to interfacial waves have higher growth rate than transverse ones leading to rivulets and hence can be considered more unstable (see also [107]). Another way to put it, long-wave transverse and streamwise modes are equally affected by thermocapillarity, but the latter gets an additional destabilizing boost from inertia; as was shown in Sect. 3.5.2, the H- and S-modes reinforce each other, at least in the linear regime. For finite amplitude waves, i.e., in the nonlinear regime, this point will be discussed in Chap. 9.

Let us now consider again the window of stability for the ST case in Fig. 3.2. It reveals the existence of two critical values of the Reynolds number, or equivalently the Nusselt flat film thickness h_N. This can be understood if we rewrite (3.31a) in terms of the viscous-gravity scaling, $M = Ma/h_N$, $B = Bih_N$ and $Re = h_N^3/3$:

$$Ct = \frac{2}{5}h_N^3 + \frac{3MaBi}{2h_N(1 + Bih_N)^2},$$

(3.32)

which can be recast into a polynomial for h_N, admitting, for certain values of Ma, two positive roots between which the film is stable. For the example of Fig. 3.2, this happens for $Ma < 15.5$. Hence, the parametrization of the Nusselt groups in terms of h_N and the viscous-gravity parameters has allowed us to unfold the two modes of instability.

3.5.5 Higher Order in the Long-Wave Expansion of the Dispersion Relation

We can proceed further with the wavenumber expansion of the system (3.22a)–(3.22i) in the same manner as for the zeroth and first orders. The complex eigenvalue for ST has the form:

$$\text{ST:} \quad c = 1 + ik\left[\frac{2}{5}Re - \frac{Ct}{3} + \frac{M^\star}{2(1 + B)}\right]$$

$$- k^2\left[1 - \frac{10}{21}CtRe + \frac{4}{7}Re^2 + \frac{M^\star}{80(1 + B)}\left(57Re + \frac{15 - 7B}{1 + B}Pe\right)\right]$$

$$- ik^3\left\{-\frac{3}{5}Ct + \frac{471}{224}Re - \frac{17363}{17325}CtRe^2 + \frac{75872}{75075}Re^3 + \frac{2}{15}Ct^2Re\right.$$

$$+ \frac{We}{3} - \frac{M^\star}{3B} + \frac{M^\star}{1 + B}\left[\frac{M^\star}{16(1 + B)}\left(5Re + \frac{3 - B}{1 + B}Pe\right) - \frac{49}{120}CtRe\right.$$

$$+ \frac{2707}{1792}Re^2 + \frac{(6 + 5B)}{6(1 + B)} - \frac{(-5435 + B(2090 + 749B))Pe^2}{44800(1 + B)^2}$$

$$\left.\left.+ \frac{(9605 - 3653B)PeRe}{22400(1 + B)} - \frac{CtPe(33 - 7B)}{240(1 + B)}\right]\right\}.$$

(3.33)

This expansion can then be used to obtain the temporal growth rate, $\omega_i = c_i k$ or $\omega_i = \mathcal{A}k^2 - \mathcal{B}k^4$, where $\mathcal{A} = (2/5)(Re - Re_c)$ is the coefficient of ik with $Re_c = (5/6)Ct - (5/4)(M^\star/(1+B))$ and $\mathcal{B}$ is the coefficient of ik^3. The relation for the growth rate expresses the balance between $\mathcal{A}k^2$ and $\mathcal{B}k^4$ in the linear regime and links the order of magnitude of k with those of $Re - Re_c$ and $\mathcal{B}$; it is precisely this balance that also determines the cut-off and maximum growing wavenumbers. From $\mathcal{A}k^2 \sim \mathcal{B}k^4$, $Re - Re_c \sim \sqrt{\mathcal{B}}k^2$ and the condition $k \ll 1$ for the expansion in (3.33) to be valid is satisfied when the distance from criticality is sufficiently small and/or $\mathcal{B}$ is sufficiently large.

As it was pointed out earlier, in experimental observations, $We \gg 1$ for most liquids (a strict relative order between We and k at this stage is not required). With this condition and assuming all other parameters in the coefficient of ik^3 in (3.33) to be of $\mathcal{O}(1)$, we obtain

$$\omega_i = k^2 \left(\frac{2}{5}Re - \frac{Ct}{3} + \frac{M^\star}{2(1+B)} \right) - k^4 \frac{We}{3}. \tag{3.34}$$

Let us now consider the neutral stability condition. We have already seen that the onset of instability occurs when the coefficient of k^2 vanishes, which yields the critical condition (3.31a)–(3.31b). Beyond onset, the neutral wavenumber, i.e., wavenumber with zero growth rate, occurs at $k = 0$, which is always a branch of the neutral curve as shown also in Fig. 3.2, and at the cut-off wavenumber, obtained by setting $\omega_i = 0$ in (3.34) or

$$\text{ST:} \quad \left(\frac{6}{5}Re - Ct + \frac{3M^\star}{2(1+B)} \right) - k_c^2 We = 0, \tag{3.35}$$

which again in terms of the viscous-gravity scaling and depending on the value of Ma can unfold into two roots. Equation (3.35) provides an analytical expression of the neutral stability curves in Fig. 3.2 for small wavenumbers. It also shows that k_c is small due to large We. In particular, with $Re - Re_c = \mathcal{O}(1)$, $k_c = \mathcal{O}(We^{-1/2})$. From (3.34) then ω_i is small. Hence, at onset the evolution is slow in both time and space.

Therefore, with the order of magnitude assignment $We \gg 1$, surface tension appears at the lowest possible order in the long-wave expansion, which in turn leads to simple expressions for the growth rate and cut-off wavenumbers; this is due to the coefficient of ik^3 in (3.33) being rather simple when $We \gg 1$. Similarly, the order of magnitude assignment $We = \mathcal{O}(\varepsilon^{-2})$ with ε the "gradient expansion parameter" simplifies substantially the formulation of the nonlinear problem, as we shall see in Chaps. 4 and 5. (With this assignment, the above discussion indicates that for $Re - Re_c = \mathcal{O}(1)$, $k \sim \varepsilon$. The question of connecting k to ε in general will be addressed in Sect. 5.1.4.) Moreover, large We is crucial for the validity of the boundary layer approximation.

If $We = \mathcal{O}(1)$, the coefficient $\mathcal{B}$ of k^4 is affected by inertia and hydrostatic head in addition to surface tension, two forces that limit the growth rate of short waves; on the other hand if We is small, the coefficient $\mathcal{B}$ is affected by inertia and hydrostatic head only. If, however, $We \gg 1$, surface tension is the only physical effect that limits the growth rate of short waves and in the linear regime it should balance the destabilizing inertia and thermocapillarity effects acting at long waves. It is precisely this balance that determines the temporal growth rate in (3.34) and hence the cut-off and maximum growing wavenumbers. Of course the role of strong surface tension is not limited to the linear regime; in the nonlinear regime and for sufficiently large We it is the only force that prevents the waves produced by inertia from forming shocks and breaking; this point is discussed in detail in Sects. 4.4, 4.6.

Not surprisingly, the criticality condition in (3.31a)–(3.31b) can also be obtained from (3.35) by simply setting $k_c = 0$. The *maximum growing wavenumber*, i.e., the wavenumber with the maximum growth rate, is obtained from $\partial \omega_i / \partial k \equiv \partial(k c_i)/\partial k = 0$, which gives $k = 0$ (such that at criticality or below, the maximum growth rate, which vanishes, has a wavenumber which vanishes also) and

$$\text{ST:} \quad \left(\frac{6}{5} Re - Ct + \frac{3M^\star}{2(1+B)} \right) - 2k_m^2 We = 0, \qquad (3.36)$$

or

$$k_m = \frac{k_c}{\sqrt{2}}.$$

Finally, the maximum linear growth rate reads:

$$\omega_{im} \equiv k c_i |_{k_m} = \frac{1}{12 We} \left(\frac{6}{5} Re - Ct + \frac{3M^\star}{2(1+B)} \right)^2. \qquad (3.37)$$

Similar results can be obtained for the HF case.

3.6 Mechanism of the Hydrodynamic Instability

The mechanism of the long-wave instability triggered by the S-mode can be fully understood with heuristic arguments, as done in the Introduction. However, the hydrodynamic instability of an isothermal falling film (H-mode) is more difficult to describe with the help of such arguments only. Indeed, the thermocapillary instability mechanism is governed by surface tension, independently of the presence of a flow, whereas the hydrodynamic instability mechanism is triggered by the flow, and in particular by the requirement that the perturbed flow satisfies the kinematic boundary condition and the stress balance at the free surface "simultaneously." Contrary to the S-mode, for which the growth of the perturbation is decoupled from its advection by the flow induced by the kinematic boundary condition, a crucial ingredient of the H-mode mechanism is precisely the advection by the flow of the perturbation. We thus complete below the arguments presented in the Introduction (see Fig. 1.3) in order to ascertain precisely how and why inertia promotes the instability. This is achieved by formulating the balances of energy and vorticity at the interface. For this purpose we utilize the linearized equations for the disturbances and the long-wave expansion.

We begin by setting $M = 0$ so that the fluid flow and thermal problems are decoupled. As pointed out in Sect. 3.5.1, Squire's theorem [83] then applies and we can limit ourselves to streamwise perturbations only. The system (3.2a)–(3.2e), (3.3) and (3.5)–(3.7) for the perturbation thus simplifies into

$$\partial_x \tilde{u} + \partial_y \tilde{v} = 0, \tag{3.38a}$$

$$3Re(\partial_t \tilde{u} + U \partial_x \tilde{u} + \tilde{v} DU) = -\partial_x \tilde{p} + \partial_{xx} \tilde{u} + \partial_{yy} \tilde{u}, \tag{3.38b}$$

$$3Re(\partial_t \tilde{v} + U \partial_x \tilde{v}) = -\partial_y \tilde{p} + \partial_{xx} \tilde{v} + \partial_{yy} \tilde{v}, \tag{3.38c}$$

with the no-slip and no-penetration conditions at the plate $y = 0$,

$$\tilde{u} = \tilde{v} = 0, \tag{3.38d}$$

and the kinematic and normal and tangential stress balances at the free surface $y = 1$,

$$\partial_t \tilde{h} + U(1)\partial_x \tilde{h} = \tilde{v}, \tag{3.38e}$$

$$\tilde{p} + (We\partial_{xx} - Ct)\tilde{h} - 2\partial_y \tilde{v} = 0, \tag{3.38f}$$

$$\tilde{h} D^2 U(1) + \partial_y \tilde{u} + \partial_x \tilde{v} = 0, \tag{3.38g}$$

where we have not substituted $D^2 U(1) = 1$, in order to keep track of the base-state shear stress in what follows.

3.6.1 Energy Balance of the Perturbation

Multiplying (3.38b) by $\tilde{u}$, (3.38c) by $\tilde{v}$ and summing up the resulting equations, one gets the "kinetic energy balance" of the perturbed state:

$$\frac{1}{2}(\partial_t + U\partial_x)(\tilde{u}^2 + \tilde{v}^2) = -DU\tilde{u}\tilde{v} - \frac{1}{3Re}(\tilde{u}\partial_x + \tilde{v}\partial_y)\tilde{p}$$

$$+ \frac{1}{3Re}\left[\tilde{u}(\partial_{xx} + \partial_{yy})\tilde{u} + \tilde{v}(\partial_{xx} + \partial_{yy})\tilde{v}\right]. \tag{3.39}$$

A perturbation localized in space can be decomposed into a sum of periodic functions through the Fourier transform. Thanks to the Parseval theorem, we decipher that the kinetic energy contained in the perturbation is the sum of the kinetic energies of all elements that make the Fourier basis. We can therefore restrict our energy balance to the kinetic energy of the normal mode of wavenumber k. Through integration across the film thickness over a wavelength $\lambda = 2\pi/k$, we get

$$\int_0^1 U(y) \int_0^\lambda \partial_x(\tilde{u}^2 + \tilde{v}^2)\, dx\, dy = \int_0^1 U\left[(\tilde{u}^2 + \tilde{v}^2)\right]_0^\lambda dy = 0. \tag{3.40}$$

Similarly, integrating by parts and with the help of the continuity equation (3.38a) one obtains:

$$-\int_0^\lambda \int_0^1 (\tilde{u}\partial_x + \tilde{v}\partial_y)\tilde{p}\,dy\,dx = -\int_0^1 [\tilde{u}\tilde{p}]_0^\lambda\,dy - \int_0^\lambda [\tilde{v}\tilde{p}]_0^1\,dx = -\int_0^\lambda \tilde{v}|_1\tilde{p}|_1\,dx. \tag{3.41}$$

Then, integrating twice by parts,

$$\int_0^\lambda \int_0^1 \left[\tilde{u}(\partial_{xx} + \partial_{yy})\tilde{u} + \tilde{v}(\partial_{xx} + \partial_{yy})\tilde{v}\right]dy\,dx$$

$$= \int_0^1 [\tilde{u}\partial_x\tilde{u} + \tilde{v}\partial_x\tilde{v}]_0^\lambda\,dy + \int_0^\lambda [\tilde{u}\partial_y\tilde{u} + \tilde{v}\partial_y\tilde{v}]_0^1\,dx$$

$$- \int_0^\lambda \int_0^1 \left\{(\partial_x\tilde{u})^2 + (\partial_y\tilde{u})^2 + (\partial_x\tilde{v})^2 + (\partial_y\tilde{v})^2\right\}dy\,dx, \tag{3.42}$$

from which, using (3.38f), $\tilde{p}|_1 = Ct\tilde{h} - We\partial_{xx}\tilde{h} + 2\partial_y\tilde{v}|_1$ and $\int_0^\lambda \tilde{v}|_1\partial_y\tilde{v}|_1\,dx = -\int_0^\lambda \tilde{v}|_1\partial_x\tilde{u}|_1\,dx = -[\tilde{v}|_1\tilde{u}|_1]_0^\lambda + \int_0^\lambda \tilde{u}|_1\partial_x\tilde{v}|_1\,dx$. We finally obtain:

$$\frac{1}{2\lambda}\frac{d}{dt}\int_0^\lambda \int_0^1 (\tilde{u}^2 + \tilde{v}^2)\,dy\,dx$$

$$= -\frac{1}{\lambda}\int_0^\lambda \int_0^1 \tilde{u}\tilde{v}DU\,dy\,dx + \frac{1}{\lambda}\int_0^\lambda \left[\frac{1}{3Re}(We\partial_{xx}\tilde{h} - Ct\tilde{h})\tilde{v}|_1\right.$$

$$\left. + \frac{1}{3Re}\tilde{u}|_1(\partial_y\tilde{u}|_1 - \partial_x\tilde{v}|_1)\right]dx$$

$$- \frac{1}{3Re\lambda}\int_0^\lambda \int_0^1 \left[2(\partial_x\tilde{u})^2 + (\partial_y\tilde{u})^2 + (\partial_x\tilde{v})^2\right]dy\,dx. \tag{3.43}$$

In this equation the "transport of vorticity perturbation" $(\partial_x\tilde{v} - \partial_y\tilde{u})|_1$ by the velocity perturbation field appears explicitly.

To make explicit the work of the viscous forces, denoted as DIS, where

$$\text{DIS} = -\frac{1}{3Re\lambda}\int_0^\lambda \int_0^1 \left[2(\partial_x\tilde{u})^2 + (\partial_y\tilde{u} + \partial_x\tilde{v})^2 + 2(\partial_y\tilde{v})^2\right]dy\,dx, \tag{3.44}$$

we take the square of the continuity equation (3.38a) and integrate it over the considered volume of fluid. The result is

$$0 = \int_0^\lambda \int_0^1 (\partial_x\tilde{u} + \partial_y\tilde{v})^2\,dy\,dx$$

$$= \int_0^\lambda \int_0^1 \left[(\partial_x\tilde{u})^2 + (\partial_y\tilde{v})^2\right]dy\,dx + 2\int_0^\lambda \int_0^1 (\partial_x\tilde{u}\partial_y\tilde{v})\,dy\,dx$$

$$= \int_0^\lambda \int_0^1 \left[(\partial_x \tilde{u})^2 + (\partial_y \tilde{v})^2\right] dy\,dx + 2\int_0^1 [\tilde{u}\partial_y \tilde{v}]_0^\lambda \, dy$$

$$- 2\int_0^\lambda [\tilde{u}\partial_x \tilde{v}]_0^1 \, dx + 2\int_0^\lambda \int_0^1 \partial_y \tilde{u} \partial_x \tilde{v}\,dy\,dx$$

$$= \int_0^\lambda \int_0^1 \left[(\partial_x \tilde{u})^2 + (\partial_y \tilde{v})^2 + 2\partial_y \tilde{u}\partial_x \tilde{v}\right] dy\,dx - 2\int_0^\lambda \tilde{u}|_1 \partial_x \tilde{v}|_1 \, dx.$$

This allows us to reformulate (3.43) in a more compact way as

$$\text{KIN} + \text{STE} + \text{HYD} = \text{SHE} + \text{REY} + \text{DIS}, \tag{3.45}$$

where KIN is the rate of change of the kinetic energy contained in the perturbation of wavenumber k,

$$\text{KIN} = \frac{1}{2\lambda}\frac{d}{dt}\int_0^\lambda \int_0^1 \left(\tilde{u}^2 + \tilde{v}^2\right) dy\,dx; \tag{3.46}$$

STE denotes the surface tension-driven rate of energy change,

$$\text{STE} = -\frac{We}{3Re\lambda}\int_0^\lambda \left[\tilde{v}|_1 (\partial_{xx}\tilde{h})\right] dx; \tag{3.47}$$

and HYD is the rate of change of the hydrostatic potential energy,

$$\text{HYD} = \frac{Ct}{3Re\lambda}\int_0^\lambda \tilde{h}(\tilde{v}|_1)\,dx. \tag{3.48}$$

SHE is the work of the shear stress at the interface,

$$\text{SHE} = \frac{1}{3Re\lambda}\int_0^\lambda \tilde{u}|_1 \left(\partial_y \tilde{u}|_1 + \partial_x \tilde{v}|_1\right) dx, \tag{3.49}$$

and REY is the work of the "Reynolds tensor",

$$\text{REY} = -\frac{1}{\lambda}\int_0^\lambda \int_0^1 \tilde{u}\tilde{v}DU\,dy\,dx. \tag{3.50}$$

Using (3.38e), STE can be recast into

$$\text{STE} = -\frac{We}{3Re\lambda}\int_0^\lambda \left(\partial_t \tilde{h} + U(1)\partial_x \tilde{h}\right)\partial_{xx}\tilde{h}\,dx. \tag{3.51}$$

Similarly, using the tangential stress balance at the free surface, one obtains

$$\text{SHE} = -\frac{1}{3Re\lambda}D^2 U(1)\int_0^\lambda \tilde{h}\tilde{u}|_1 \, dx, \tag{3.52}$$

where the work of the shear rate due to the displacement of the interface $D^2U(1)\tilde{h}$
now appears explicitly.

The Nusselt flat film flow is unstable if KIN is positive, and it is stable otherwise. STE and HYD are positive when the energy brought from the base flow to the perturbation is stored through surface tension and gravity. These three terms on the left hand side of (3.45) correspond to the distribution of the total energy of the perturbation. The terms appearing in the right hand side correspond to the work of the forces exerted on the fluid and thus to the extraction of energy from the base state to the perturbation. It is clear that the energy required for the perturbation to grow must be produced either through the work of the Reynolds tensor REY or by the work of the shear at the interface. Numerical integrations of (3.45) show that the perturbation extracts energy from the base state mostly through SHE with the Reynolds tensor playing a negligible role in the process [147].

Similar results can be obtained analytically close to the instability threshold using the long-wave expansion. Assuming for simplicity that $Re - Re_c = \mathcal{O}(1)$ and from Sect. 3.5.5, $Wek^2 = \mathcal{O}(1)$, one can easily obtain the expressions at $\mathcal{O}(k^2)$ of the different terms appearing in the balance (3.45):

$$\text{KIN} \approx \frac{2}{9}k^2\left(\frac{6}{5}Re - Ct - k^2We\right)E^2, \tag{3.53a}$$

$$\text{STE} \approx \frac{2}{9}k^4\frac{We}{Re}\left(\frac{6}{5}Re - Ct - k^2We\right)E^2, \tag{3.53b}$$

$$\text{HYD} \approx \frac{2}{9}k^2\frac{Ct}{Re}\left(\frac{6}{5}Re - Ct - k^2We\right)E^2, \tag{3.53c}$$

$$\text{SHE} \approx \left\{\frac{2}{3Re} + k^2\left[\frac{41}{180}Ct + \frac{20}{9Re} - \frac{2Ct^2}{3Re} + \frac{4321}{6720}Re\right.\right.$$
$$\left.\left. + k^2\left(\frac{41}{180}We - \frac{4}{3}\frac{CtWe}{Re}\right) - \frac{2}{3}k^4\frac{We^2}{Re}\right]\right\}E^2, \tag{3.53d}$$

$$\text{REY} \approx \frac{k^2}{180}\left(Ct - \frac{93}{112}Re + k^2We\right)E^2, \tag{3.53e}$$

$$\text{DIS} \approx \left[-\frac{2}{3Re} + k^2\left(-\frac{17}{90}Ct - \frac{20}{9Re} + \frac{4Ct^2}{9Re} - \frac{1249}{3360}Re\right.\right.$$
$$\left.\left. + k^2\left(-\frac{17}{90}We + \frac{8}{9}\frac{CtWe}{Re}\right) + \frac{4}{9}k^4\frac{W^2}{Re}\right)\right]E^2, \tag{3.53f}$$

where $E \equiv \exp(kc_i t)$.

Consequently, the balance (3.45) becomes, to leading order,

$$\text{SHE} = -\text{DIS} = \frac{2}{3Re}, \qquad \text{KIN} = \text{STE} = \text{HYD} = \text{REY} = 0. \tag{3.54}$$

The energy pumped from the base state is completely dissipated by viscosity and, since the kinetic energy of the perturbation does not change (KIN $= 0$), the axis $k = 0$ is part of the neutral stability curve (obtained earlier in this chapter) as implied by the dependence of the base state on the film thickness: a uniform elevation of the free surface must lead from one Nusselt solution to another one, and therefore is neither damped nor amplified (this is the Goldstone mode).

The above results show that the work of the Reynolds tensor (REY) is smaller than the work of the shear at the free surface (SHE). Recall the expression for the growth rate $\omega_i = kc_i$ in (3.34). Consequently, (3.53a)–(3.53f) shows that the kinetic energy of the perturbation (KIN), the gravity potential energy (HYD) and the surface energy (STE), to leading order all grow proportionally to the growth rate kc_i, in agreement with the numerical integrations of (3.45) [147]. This confirms that the energy is mainly transferred to the perturbation through the work of interfacial forces.

3.6.2 Vorticity Balance at the Perturbed Interface

Both the H- and S-modes are interfacial instability modes. As shown above the H-mode is an interfacial instability mode resulting from the unbalance of the perturbed shear stress at the interface. Let us therefore consider the stress balance at the interface in more detail. Below we follow the arguments by Hinch [114] and Kelly et al. [147].

We denote by $\tilde{\omega} = \partial_x \tilde{v} - \partial_y \tilde{u}$ the vorticity perturbation. By differentiating (3.38b) with respect to the cross-stream coordinate y and (3.38c) with respect to the stream-wise coordinate x, and subtracting the resulting equations, one obtains the balance of the vorticity perturbation:

$$(\partial_t + U \partial_x)\tilde{\omega} - \tilde{v} D^2 U(y) = \frac{1}{3Re}(\partial_{xx} + \partial_{yy})\tilde{\omega}. \tag{3.55}$$

Seeking the solution in the form of a normal mode, $\tilde{\omega} = \Omega(y)\exp(ik(x - ct))$, and expanding the amplitude Ω in the asymptotic series, $\Omega = \Omega_0 + ik\Omega_1 + \cdots$, the solution $\phi_0 = y^2$ at leading order of the long-wave expansion (see (3.24)) gives $\Omega_0 = D^2\phi_0 = -2$. The phase of the vorticity perturbation is therefore opposite to the displacement of the free surface η_0, which from (3.22f) reads $\eta_0 = 2$. Therefore, the displacement of the free surface is neither amplified nor damped at leading order.

At $\mathcal{O}(k)$ the shear acts opposite to the vorticity,

$$\Omega = -D^2\phi + \mathcal{O}(k^2), \tag{3.56}$$

so that we get:

$$D^2\Omega_1 = 3Re\big[(U - c_0)\Omega_0 + D^2 U \phi_0\big]. \tag{3.57}$$

The first term on the right hand side of the above expression corresponds to the advection of the vorticity perturbation by the motion of the fluid with respect to the

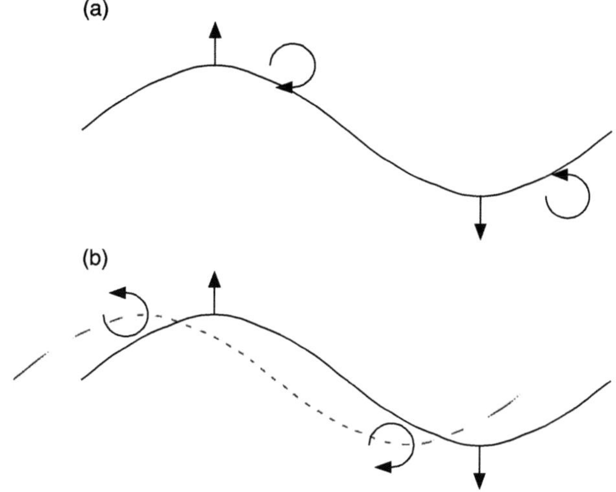

Fig. 3.3 (**a**) Displacement of the vorticity maximum at the interface by inertia when $Re \geq Ct$; (**b**) vorticity correction due to the base flow and the interface displacement when $Re \leq \frac{5}{3}Ct$

wave displacement. The second term originates from the advection by the vorticity perturbation of the base state, $-v_1 D^2 U(y)$. Since $c_0 > U$, $\forall y \in [0, 1]$ we have $(U - c_0)\Omega_0 = 2 - 2y + y^2 > 0$ which is partially compensated by $D^2 U \phi_0 = -y^2 < 0$ so that $D^2 \Omega_1 = 6Re(1 - y) > 0$. Integration of this equation requires two boundary conditions. The normal stress balance at the interface (3.22g) gives one condition:

$$D\Omega_1(1) = -\frac{Ct}{c_0 - 1/2}\varphi_0(1) + 3Re\left(c_0 - \frac{1}{2}\right)D\varphi_0(1). \qquad (3.58)$$

A second boundary condition is obtained through (3.56) and the choice $B_1 = 0$ in the expression of φ_1 in (3.28), as we did below (3.30). Since $\Omega_1(0) = 0$ and $D^2\Omega_1 > 0$, the correction of the vorticity at the interface $\Omega_1(1)$ increases with $D\Omega_1(1)$. On the right hand side of (3.58) there is a negative term originating from the hydrostatic pressure $(-Ct\tilde{h})$ and a positive term brought by the pressure perturbation promoted by the vorticity perturbation, which is also due to inertia. The sum of these terms gives $D\Omega_1(1) = 3Re - 2Ct$ so that $\Omega_1(1) = 2(Re - Ct)$ after a double integration of (3.57). The first-order correction $\Omega_1(1)$ is thus positive whenever $Re \geq Ct$. As the wavenumber k is small it introduces a phase shift, say v, through $\Omega_0 \exp(iv) \equiv \Omega(1) \approx \Omega_0(1) + ik\Omega_1(1) = -2[1 - ik(Re - Ct)]$, or $v = \mathcal{O}(k) < 0$. Consequently, the maximum in absolute value of the vorticity perturbation at the interface is somewhere ahead of the maximum of the displacement of the interface at lowest order, $\Re(\varphi_0(1)/(c_0 - 1/2) \exp[ik(x - ct)]) = 2\cos[k(x - ct)]$. Subsequently, the induced correction of the velocity field amplifies the interface deformation, as shown in Fig. 3.3a.

The balance of the tangential stress at the interface (3.22h) can be rewritten as

$$\Omega_1(1) + \frac{1}{c_0 - 1/2}\varphi_1(1) = \frac{1}{(c_0 - 1/2)^2}c_1\varphi_0(1), \qquad (3.59)$$

where, besides $\Omega_1(1)$, a second term corresponding to the displacement of the interface enforced by the correction to the stream function, $\varphi_1(1)/(c_0 - 1/2) = \frac{2}{3}Ct - \frac{2}{5}Re$, appears. When $Re \leq \frac{5}{3}Ct$, the interface is displaced upstream, suggesting an upstream transport of vorticity, $-D^2U\varphi_1(1)/(c_0 - 1/2) = \varphi_1(1)/(c_0 - 1/2) \geq 0$. This is destabilizing since $\varphi_0(1)/(c_0 - 1/2) = \eta_0 = 2$, and thus $c_1 > 0$ (see Fig. 3.3b).

The above discussion is limited to the effect of the displacement of the maximum of vorticity on the perturbed interface at lowest order, since the effect on the correction at $\mathcal{O}(k)$ of the interface position is a higher order one.

3.6.3 Summary of the Key Factors for the Hydrodynamic Instability

To summarize, there are three key elements responsible for the instability:

(i) The interface can be deformed, which enables the perturbation to pump energy from the base state through the tangential constraint at the interface.
(ii) To leading order in k, the perturbed stream function φ_0 is fully determined by the balance of viscosity and gravity acceleration and thus becomes slaved to the kinematics of the interface.
(iii) The displacement of the interface induced by φ_0 corresponds to a wave speed c_0 larger than the velocity of the base flow. This allows the displacement of the maximum of vorticity in a destabilizing fashion [256].

Point (iii) is probably the most significant one.

Chapter 4
Boundary Layer Approximation

The linear stability analysis performed in Chap. 3 shows that for large Weber numbers the onset of the evolution of a falling film is dominated by long-wave modes, which in turn suggests slow variations of the interface in time and space. This then motivates the introduction of a small parameter, ε, measuring the slow variations in time and space. This parameter forms the basis of a *gradient expansion* of the governing equations and wall and free-surface boundary conditions to let us obtain the different levels of modeling approaches and approximations utilized in the description of falling liquid films. The highest level of approximation is based on the elimination of the pressure. This is achieved by obtaining the pressure by integrating across the film the y component of the momentum equation with the inertial effects neglected. The resulting pressure distribution is then substituted into the x and z momentum equations while retaining the inertia terms in these equations. The resulting equations are referred to as the *boundary layer equations* since the assumptions leading to these equations are essentially the same with those in the derivation of the Prandtl equations of the boundary layer theory in aerodynamics. The boundary layer equations are obtained with only the gradient expansion and without overly restrictive stipulations on the orders of the different dimensionless groups with respect to ε. These equations serve as the first step toward subsequent approximations presented in this monograph.

4.1 Three-Dimensional Boundary Layer Equations

In Prandtl's boundary layer theory, the strong viscous diffusion in the cross-stream y direction balances inertia as well as the pressure gradient in the streamwise x direction, and it is this balance that gives rise to the Blasius profile [243]. The pressure distribution is typically obtained by integrating the y component of the momentum equation where convective/inertia effects are neglected. We shall demonstrate that the same approach in viscous film flows yields a set of equations very similar to the Prandtl equations in boundary layer theory. However, the flow is not, strictly

S. Kalliadasis et al., *Falling Liquid Films*, Applied Mathematical Sciences 176,
DOI 10.1007/978-1-84882-367-9_4, © Springer-Verlag London Limited 2012

speaking, a boundary layer flow, as there are no inner and outer regions which determine the thickness of the layer. In fact, unlike boundary layer theory, where the pressure gradient is imposed by the (outer) inviscid flow and is related to the velocity via Bernoulli's equation, in the case of a falling film the pressure gradient is self-induced and caused by the capillary forces at the interface and the hydrostatic head in the direction perpendicular to the wall.

The linear stability analysis performed in Chap. 3 has shown that, with the exception of the P-mode, which is beyond the scope of the monograph, the dominant modes, i.e., the H- and S-modes, are long-wave varieties. We then anticipate that, as is frequently the case with long-wave instabilities, in the nonlinear regime the free-surface waves will also be long, i.e., they are waves with a wavelength l long compared to the Nusselt flat film thickness $\bar{h}_N$ (see Fig. 2.1). This is especially so as surface tension is generally large, i.e., *We* is large and the flow rate is not too large. Hence the interface remains "smooth" at the scale of the film thickness $h(x, t)$. This then justifies a long wave assumption corresponding to a slowly varying interface in time and space, or equivalently, to slow time and space modulations of the Nusselt flat film solution (2.44).

To express the smallness of the interfacial slope and its slow variation in time, we introduce a small parameter, $\varepsilon \sim \partial_{x,z,t} \ll 1$. In the linear regime it must be related to the wavenumber of the infinitesimal perturbations, $k = 2\pi(\bar{h}_N/l)$, with the ratio $\bar{h}_N/l$ frequently referred to as the *film parameter*. This is also a small parameter for strong surface tension/large Weber numbers (see Sect. 3.5.5) and in fact in Chap. 3 we have already performed long-wave expansions for $k \ll 1$ to obtain, e.g., the instability threshold from the Orr–Sommerfeld eigenvalue problem.

One may be tempted to assume the relative order, $k \sim \varepsilon$, as is done quite frequently in the literature: after all, $\partial_{x,z} \sim \bar{h}_N/l \sim k$ and hence if $\partial_{x,z} \sim \varepsilon$, $k \sim \varepsilon$. And yet, as we shall demonstrate in Sect. 5.1.4, the above order of magnitude assignment between k and ε is true provided that $Re - Re_c = \mathcal{O}(1)$. We shall then refrain from assigning a relative order of magnitude between k and ε at this stage: ε will simply be treated as a perturbation parameter independent of k (a relative order between k and ε is not really required), allowing us to derive from the governing equations and wall and free-surface boundary conditions the different levels of modeling approaches and approximations used in the description of falling films.

As far as the nonlinear regime is concerned, the value of the film parameter cannot be assigned a priori but rather a posteriori as it is related to the characteristic scale of the distortions/nonlinear waves of the free surface, which have to be constructed from the particular model used to describe the evolution of the interface. Nevertheless, some estimates can be given a priori: for instance, we expect the characteristic wavelength of the waves in the nonlinear regime to be of the same order with that at onset. Hence, the film parameter in the nonlinear regime should be of the same order with that at onset.

Having defined ε, one then typically performs a long-wave expansion of the governing equations and associated boundary conditions, as is frequently done in the literature (e.g., [137, 201]). Here we adopt the term *gradient expansion* instead to denote the particular way we do the long-wave expansion:

(i) ε is introduced through the transformation,

$$(\partial_t, \partial_x, \partial_z) \to \varepsilon(\partial_t, \partial_x, \partial_z) \quad \text{and} \quad (\partial_{xx}, \partial_{zz}) \to \varepsilon^2(\partial_{xx}, \partial_{zz}), \tag{4.1}$$

i.e., it acts as an *ordering parameter* instead of defining it, e.g., from $\bar{h}_N/l$, with l an a priori unknown long scale in the streamwise and spanwise directions (which in turn would require nondimensionalization of the x, z directions with l).

(ii) the long-wave expansion with respect to ε is then carried out as usual; i.e., it is an asymptotic/perturbation expansion of all pertinent variables in powers of ε.

Let us assume that with the exception of the Weber number, all parameters Re, Pe, M, B, B_w are of $\mathcal{O}(1)$. With $Re = \mathcal{O}(1)$, the condition $Pe = \mathcal{O}(1)$ is equivalent to $Pr = \mathcal{O}(1)$. The Weber number is taken much larger, typically $We = \mathcal{O}(\varepsilon^{-2})$. The above orders of magnitude assignments are made for the sake of simplicity and in order to illustrate the main points of the derivation of the boundary layer equations. (As we shall also see in Sect. 5.1.4 it is precisely the order of magnitude assignment $We = \mathcal{O}(\varepsilon^{-2})$ that does not allow us to connect k and ε a priori.) As a matter of fact, strict orders of magnitude assignments for the different groups are not required and the above assignments used in the derivation of the boundary layer equations in this section can be relaxed. This point is discussed in Appendix D.2.

Let us first consider the continuity equation (2.17) which with (i) above becomes:

$$\varepsilon \partial_x u + \partial_y v + \varepsilon \partial_z w = 0.$$

The continuity then imposes that the y component of the velocity v is of $\mathcal{O}(\varepsilon)$. The transformation $v \to \varepsilon v$ is thus applied everywhere in the equations. The continuity equation finally reads

$$\partial_x u + \partial_y v + \partial_z w = 0, \tag{4.2a}$$

so that all terms in the equation are of the same order (as they should be to ensure mass conservation) and the continuation equation remains unaltered.

The momentum and energy equations (2.18)–(2.21) with (4.1) and $v \to \varepsilon v$ read

$$3\varepsilon Re(\partial_t u + u\partial_x u + v\partial_y u + w\partial_z u)$$
$$= -\varepsilon\partial_x p + \varepsilon^2\partial_{xx}u + \partial_{yy}u + \varepsilon^2\partial_{zz}u + 1, \tag{4.2b}$$

$$3\varepsilon^2 Re(\partial_t v + u\partial_x v + v\partial_y v + w\partial_z v) = -\partial_y p + \varepsilon\partial_{yy}v - Ct, \tag{4.2c}$$

$$3\varepsilon Re(\partial_t w + u\partial_x w + v\partial_y w + w\partial_z w)$$
$$= -\varepsilon\partial_z p + \varepsilon^2\partial_{xx}w + \partial_{yy}w + \varepsilon^2\partial_{zz}w, \tag{4.2d}$$

$$3\varepsilon Pe(\partial_t T + u\partial_x T + v\partial_y T + w\partial_z T) = \varepsilon^2\partial_{xx}T + \partial_{yy}T + \varepsilon^2\partial_{zz}T, \tag{4.2e}$$

where terms of $\mathcal{O}(\varepsilon^3)$ and higher have been neglected in the y component of the momentum equation (4.2c).

The dimensionless boundary conditions on the wall $y = 0$ are still given by (2.22), (2.23):

$$u = v = w = 0 \tag{4.2f}$$

$$\text{ST:} \quad T = 1 \tag{4.2g}$$

or

$$\text{HF:} \quad \partial_y T = -1 + B_{\mathrm{w}} T \tag{4.2h}$$

and at the free surface $y = h(x, z, t)$, the kinematic condition is still given by (2.24):

$$v = \partial_t h + u \partial_x h + w \partial_z h. \tag{4.2i}$$

The other boundary conditions at the free surface (2.25)–(2.28) acquire the form,

$$p = 2\varepsilon(\partial_y v - \partial_x h \partial_y u - \partial_z h \partial_y w) - \varepsilon^2 (We - MT)(\partial_{xx} h + \partial_{zz} h), \tag{4.2j}$$

$$\partial_y u = -\varepsilon M \partial_x \theta + \varepsilon^2 \big[\partial_z h (\partial_z u + \partial_x w) + 2\partial_x h (2\partial_x u + \partial_z w) - \partial_x v \big], \tag{4.2k}$$

$$\partial_y w = -\varepsilon M \partial_z \theta + \varepsilon^2 \big[\partial_x h (\partial_z u + \partial_x w) + 2\partial_z h (2\partial_z w + \partial_x u) - \partial_z v \big], \tag{4.2l}$$

$$\partial_y T = -BT - \varepsilon^2 \left(\frac{B}{2} T \big[(\partial_x h)^2 + (\partial_z h)^2 \big] - \partial_z h \partial_z T - \partial_x h \partial_x T \right), \tag{4.2m}$$

where terms of $\mathcal{O}(\varepsilon^3)$ and higher have been neglected and where the surface temperature $\theta = T|_{y=h}$ has been introduced using the relation

$$\big[(\partial_i + \partial_i h \partial_y) T \big] \big|_h \equiv \partial_i \theta \quad \text{with } i = x, z$$

(see Appendix C.4).

We then neglect the second-order inertia terms in the y component of the momentum equation (4.2c), we integrate the resulting equation across the film to obtain the pressure distribution in the film and we substitute this distribution into (4.2b) and (4.2d), thus eliminating the pressure as in Prandtl's boundary layer theory [243]: as the contribution of the pressure in (4.2b) and (4.2d) appears through the terms $\varepsilon \partial_{x,z} p$, neglecting terms of $\mathcal{O}(\varepsilon^2)$ and higher in (4.2c) will be necessary in order to keep the boundary layer equations *consistent* at $\mathcal{O}(\varepsilon^2)$; i.e., all neglected terms will be of higher order. The elimination of the pressure constitutes the main element of the boundary layer approximation.

It should be emphasized that this elimination would have been impossible at $\mathcal{O}(\varepsilon^3)$, because of the presence of the second-order inertia terms in (4.2c): going up to $\mathcal{O}(\varepsilon^3)$ is inconsistent with the spirit of the boundary layer theory as the y component of the Navier–Stokes equations is still present and we might as well use the full Navier–Stokes equations without resorting to any approximations.

Equations (4.2k), (4.2l) show that $\partial_y u$, $\partial_y w$ are of $\mathcal{O}(\varepsilon)$ at the free surface, so that (4.2j) becomes

$$p|_h = 2\varepsilon \partial_y v|_h - \varepsilon^2 We(\partial_{xx} h + \partial_{zz} h) + \mathcal{O}(\varepsilon^2), \tag{4.3}$$

where the $\mathcal{O}(\varepsilon^2)$ terms corresponding to surface tension effects, $\propto \partial_{xx}h + \partial_{zz}h$ representing the mean free-surface curvature in the long wave approximation, have been retained due to the assumption $\varepsilon^2 We = \mathcal{O}(1)$ while the variation of surface tension with temperature (MT) is a higher-order effect and thus it has been neglected. The order of magnitude assignment $\varepsilon^2 We = \mathcal{O}(1)$ brings the surface tension effects into the pressure distribution at $\mathcal{O}(1)$, i.e., the lowest possible order. Surface tension then contributes terms of $\mathcal{O}(\varepsilon)$ in the streamwise and spanwise momentum equations.

Integrating now (4.2c) across the film and utilizing the boundary conditions (4.3) and (4.2f), one obtains

$$p = Ct(h - y) - \varepsilon^2 We(\partial_{xx}h + \partial_{zz}h) + \varepsilon(\partial_y v + \partial_y v|_h) + \mathcal{O}(\varepsilon^2). \tag{4.4}$$

The first term on the right hand side is the hydrostatic pressure that vanishes for vertical walls $(Ct = 0)$ while the third term accounts for higher-order viscous effects. Then, substituting p into (4.2b) and (4.2d) yields

$$3\varepsilon Re\big[\partial_t u + \partial_x\big(u^2\big) + \partial_y(uv) + \partial_z(uw)\big]$$

$$= 1 + \partial_{yy}u - \varepsilon Ct\partial_x h + \varepsilon^2\big[2\partial_{xx}u + \partial_{zz}u + \partial_{xz}w - \partial_x(\partial_y v|_h)\big]$$

$$+ \varepsilon^3 We(\partial_{xxx}h + \partial_{xzz}h), \tag{4.5a}$$

$$3\varepsilon Re\big[\partial_t w + \partial_x(uw) + \partial_y(vw) + \partial_z\big(w^2\big)\big]$$

$$= \partial_{yy}w - \varepsilon Ct\partial_z h + \varepsilon^2\big[2\partial_{zz}w + \partial_{xx}w + \partial_{xz}u - \partial_z(\partial_y v|_h)\big]$$

$$+ \varepsilon^3 We(\partial_{xxz}h + \partial_{zzz}h), \tag{4.5b}$$

where use has been made of the identity, $u\partial_x\varphi + v\partial_y\varphi + w\partial_z\varphi = \partial_x(u\varphi) + \partial_y(v\varphi) + \partial_z(w\varphi) - \varphi(\partial_x u + \partial_y v + \partial_z w)$ and the continuity equation (4.2a).

Equations (4.5a)–(4.5b) together with the continuity equation (4.2a) and the energy equation (4.2e), the wall and free-surface boundary conditions (4.2f)–(4.2i), (4.2k)–(4.2m) are the *second-order boundary layer equations*, i.e., boundary layer equations where terms of $\mathcal{O}(\varepsilon^3)$ and higher have been neglected. The *first-order boundary layer equations* are simply obtained from the second-order ones by neglecting terms of $\mathcal{O}(\varepsilon^2)$ and higher.

Figure 4.1 shows a typical solitary wave for a vertical falling film obtained from direct numerical simulation (DNS) of full Navier–Stokes and wall and free-surface boundary condition [232] (the various DNS studies in the literature are typically based on numerical techniques such as "finite elements" or "volume of fluids method"). Solitary-wave solutions of the different low-dimensional models utilized in the description of falling films will be constructed and analyzed in detail in subsequent chapters. In anticipation of this construction we discuss here and elsewhere in the chapter some of the solitary-wave characteristics necessary for the understanding of the assumptions leading to the boundary layer equations and scaling arguments from these equations for solitary waves. Figure 4.1 also summarizes the typical terminology for solitary waves used hereinafter in this monograph.

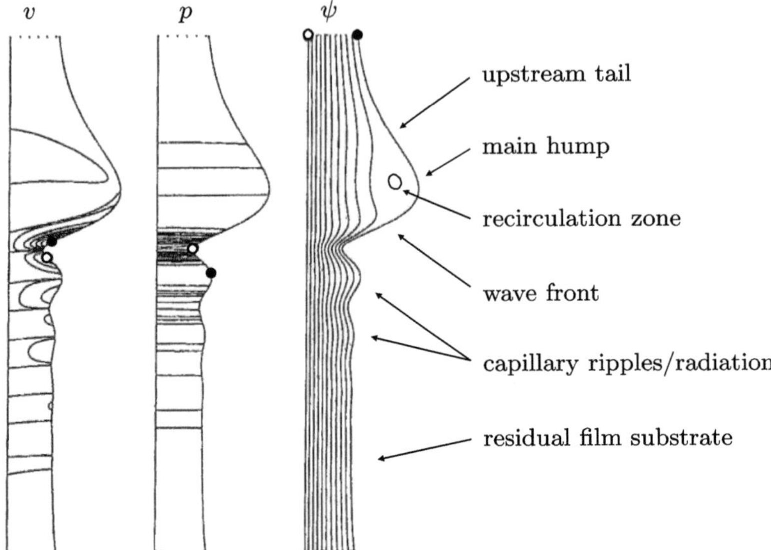

Fig. 4.1 Typical solitary wave for a falling film: contours of the y-component of velocity v, pressure p, and stream function ψ, of a solitary wave obtained from full Navier–Stokes. The parameter values are $Re = 6.1$, $We = 76.4$ and $Ct = M = 0$, corresponding to an experiment by Kapitza [141]. Global minima and maxima for each field are indicated by circles (o) and bullets (•), respectively. The solitary wave consists of a hump with a teardrop shape followed by small-amplitude capillary oscillations/radiation at its front. Reprinted with permission from Salamon, Armstrong and Brown, *Phys. Fluids*, 6(6):2202–2220, American Institute of Physics, 1994

The pressure variation in the direction perpendicular to the wall is clearly negligible in agreement with the boundary layer approximation; only small variations are observed at the trough right in front of the solitary hump and at the front-running capillary ripples, also called "radiation." The same conclusion was also drawn from the DNS study in [99, 176]. This reflects the fact that for the falling film problem, with the exception of gravity for an inclined plate, there is no mechanism that can modify the pressure distribution across the film, much like for boundary layers in aerodynamics.

Note that since the pressure distribution at the interface (4.3) does not depend on temperature (a direct consequence of neglecting the variation of surface tension with temperature in the normal stress balance), the second-order equations (4.5a)–(4.5b), and hence the corresponding first-order ones, are applicable irrespective of the presence of heating or not.

4.2 Two-Dimensional Boundary Layer Equations

For a two-dimensional flow ($w = 0$ and $\partial_z = 0$), the second-order boundary layer equations, i.e., the momentum equations (4.5a)–(4.5b) together with the continuity equation (4.2a), the wall and free-surface boundary conditions (4.2f)–(4.2i), (4.2k)–

(4.2m) reduce to

$$3\varepsilon Re\big(\partial_t u + \partial_x\big(u^2\big) + \partial_y(uv)\big) - \partial_{yy}u - 2\varepsilon^2\partial_{xx}u$$

$$= 1 - \varepsilon Ct\partial_x h + \varepsilon^3 We\partial_{xxx}h + \varepsilon^2\partial_x[\partial_x u|_h], \tag{4.6a}$$

$$3\varepsilon Pe(\partial_t T + u\partial_x T + v\partial_y T) = \big(\varepsilon^2\partial_{xx} + \partial_{yy}\big)T, \tag{4.6b}$$

$$\partial_x u + \partial_y v = 0, \tag{4.6c}$$

$$u|_0 = v|_0 = 0, \tag{4.6d}$$

$$\text{ST:}\quad T|_0 = 1 \tag{4.6e}$$

or

$$\text{HF:}\quad \partial_y T|_0 = -1 + B_{\mathrm{w}}T|_0, \tag{4.6f}$$

$$\partial_t h + u|_h\partial_x h = v|_h, \tag{4.6g}$$

$$\partial_y u|_h = 4\varepsilon^2\partial_x h\partial_x u|_h - \varepsilon^2\partial_x v|_h - \varepsilon M\partial_x[T|_h], \tag{4.6h}$$

$$\partial_y T|_h = -B\left(1 + \frac{1}{2}\varepsilon^2(\partial_x h)^2\right)T|_h + \varepsilon^2\partial_x h\partial_x T|_h. \tag{4.6i}$$

It is instructive now to compare the Orr–Sommerfeld eigenvalue problem in Sect. 3.5 with the corresponding eigenvalue problem obtained from a linear stability analysis of (4.6a)–(4.6i). The result is

$$D^3\varphi - ik\big(Ct + k^2 We\big)\frac{\varphi(1)}{c - 1/2} - k^2\big(2D\varphi + D\varphi(1)\big)$$

$$- 3Reik\big((U - c)D\varphi - \varphi DU\big) = 0, \tag{4.7}$$

together with (3.22b) and the boundary conditions (3.22c)–(3.22f), (3.22h), (3.22i). Following the same methodology as in Sect. 3.5, the expansion of the linear system of equations around $k = 0$ also leads to (3.33) for the ST case, except for the term $ik^3 M^\star/3B$, which is absent, and the coefficient $14513/6720 \approx 2.16$ instead of $471/224 \approx 2.10$. The small difference in the numerical value of the coefficient is a direct consequence of the approximated pressure gradient in the streamwise momentum balance.

4.3 On the Significance of the Second-Order Contributions

The second-order terms in the boundary layer approximation (4.6a)–(4.6i) originate from streamwise viscous diffusion, second-order contributions to the tangential stress at the free surface, streamwise thermal diffusion and second-order contributions to the heat losses at the free surface. We note the following:

(i) These terms are necessary to achieve good agreement with Orr–Sommerfeld in the region of moderate Reynolds numbers.

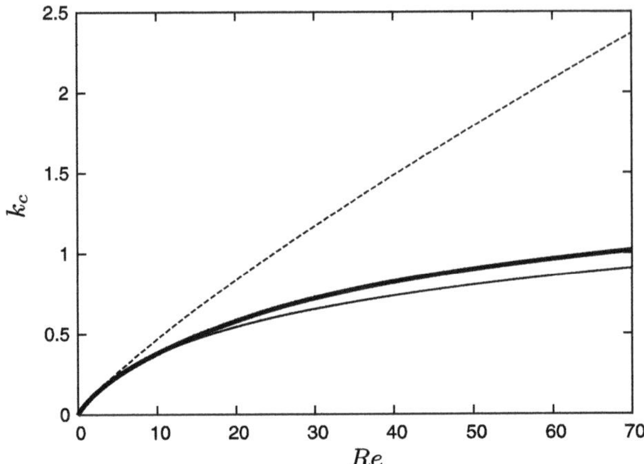

Fig. 4.2 Dimensionless cut-off wavenumber k_c as a function of the Reynolds number for an experiment performed by Kapitza [141] for a vertical wall and with alcohol as the working fluid: $\nu = 2 \times 10^{-6}\,\mathrm{m^2\,s^{-1}}$, and kinematic surface tension, $\sigma/\rho = 29 \times 10^{-6}\,\mathrm{m^3\,s^{-2}}$, so that the Kapitza number is $\Gamma = 528.8$. *Thick and thin solid lines* correspond to the solutions of the Orr–Sommerfeld problem and to the linear stability of the "simplified second-order model," whereas the *dashed line* corresponds to the "first-order model"—both models will be introduced in Chap. 6

(ii) They play a significant role on the wave profiles especially as far as the small-amplitude capillary ripples at the front of the solitary humps are concerned, which in turn are crucial for stationary wave selection in the spatio-temporal evolution of the film.

The first point is illustrated for the isothermal case in Fig. 4.2, where it is evident that the "simplified second-order model"—to be derived in Chap. 6 from the second-order boundary layer equations—follows closely the Orr–Sommerfeld result for sufficiently large Reynolds numbers. On the other hand, the "first-order model"—to be derived in Chap. 6 from the first-order boundary layer equations—performs well up to $Re \sim 5$, but then it starts to deviate rapidly from the exact result. The second point is illustrated for the isothermal case in Fig. 4.3 with the first-order and simplified second-order models. The figure demonstrates clearly that the inclusion of the second-order viscous terms has little effect on the main solitary hump, but it influences significantly the capillary ripples preceding the main hump. More precisely, for the relatively small value $Re = 6.07$ in Fig. 4.3 it is the amplitude of the capillary ripples that is mainly affected by the inclusion of the second-order terms, while for the moderate value $Re = 12$, both amplitude and frequency of the capillary ripples are affected. These terms are then crucial for a correct description of the capillary ripples, which in turn affect the wave interaction in the nonlinear regime and hence the nonlinear dynamics of the film. In fact this is a very interesting feature of the falling film problem: the capillary ripples are a linear effect but they have a significant influence on the nonlinear behavior of the film.

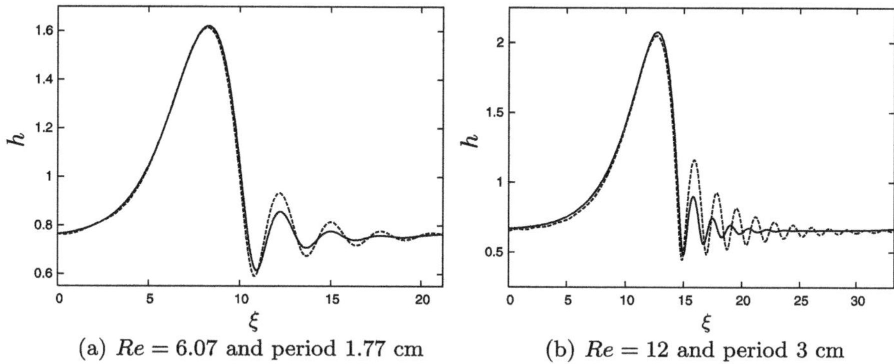

(a) $Re = 6.07$ and period 1.77 cm   (b) $Re = 12$ and period 3 cm

Fig. 4.3 Periodic wave profiles obtained with the "simplified second-order model" (*solid line*) and the "first-order model" (*dashed line*) for $We = 76.4$ in their moving frame ξ (moving with the speed of the pulse so that in this frame the pulse is stationary; we shall give the precise definition in Chap. 5). Contrast panel (**a**) with the corresponding DNS solution reproduced in Fig. 4.1

4.4 Strong Surface Tension Limit

The effect of surface tension has been retained in the momentum equations (4.5a)–(4.5b), though it appears of $\mathcal{O}(\varepsilon^3)$. This is due to the stipulation $\varepsilon^2 We = \mathcal{O}(1)$, which is in fact the simplest possible assumption on the order of magnitude of We. Based on discussions in Chap. 3 and earlier in the present chapter, we can now summarize the reasons for assuming a large We as follows:

(i) known experimental observations for most liquids correspond to large We;

(ii) convenience/simplicity, since in the linear regime surface tension effects appear at the lowest possible order in the long-wave expansion of the dispersion relation obtained from Orr–Sommerfeld; in the nonlinear regime and in the framework of the boundary layer equations, they appear at the lowest possible order in the pressure distribution, thus contributing at $\mathcal{O}(\varepsilon)$ in the boundary layer equations. The same is true with the *long-wave theory* to be developed in Chap. 5. In all cases the higher-order terms are rather involved;

(iii) it ensures the validity of the boundary layer approximation.

To further ascertain the effect of surface tension, let us first scale away the parameter ε from the system (4.6a)–(4.6i). This is possible as the transformations (4.1) and $v \to \varepsilon v$ can be easily reversed, reflecting the fact that ε acts as an ordering parameter: space and time gradients have a certain order with respect to ε defined by their order of differentiation, e.g., $\partial_{xx} u$ in (4.6a)–(4.6i), with ε not appearing explicitly, is of $\mathcal{O}(\varepsilon^2)$.

Let us consider now the steep front of a large-amplitude solitary pulse. Assume for simplicity a vertical plane and a two-dimensional flow. The streamwise gravity force and viscous drag promote the breaking of the wave at the front. As far as the viscous drag in particular is concerned, it is most active at the front region close to the wall where the front meets the flat film through the formation of a dimple, thus slowing down the liquid there and hence contributing to the steepening of the front. (Both gravity and viscous drag, represented by the terms 1 and $\partial_{yy}u$ in (4.6a), respectively, and acting in the positive and negative x directions, respectively, always balance, and in fact it is this balance, that gives rise to the Nusselt profile.) The only force that opposes this steepening is the pressure gradient induced by surface tension. As far as the role of the force of inertia is concerned we shall examine it in detail in the next section.

An estimate of the maximum slope $\partial_x h$ of a large-amplitude solitary wave (precisely at its front) can thus be obtained by balancing the mechanisms competing in the arrest of the wave-breaking, i.e., streamwise pressure gradient induced by surface tension, $We\partial_{xxx}h$, with the streamwise gravity acceleration and viscous drag, which are always of $\mathcal{O}(1)$ in (4.6a),

$$We\partial_{xxx}h \sim \frac{We}{\kappa^3} \sim 1 \quad \Longrightarrow \quad \kappa = We^{1/3}, \tag{4.8}$$

where κ is the aspect ratio between the streamwise characteristic length scale corresponding to this balance, say l_S in dimensional variables, which in turn corresponds to the characteristic length of the steep front of the waves, and the Nusselt film thickness: in terms of dimensional variables, $\sigma\,\partial_{xxx}h \sim \rho g \sin\beta$, and simple algebra shows that $l_S/\bar{h}_N \sim We^{1/3}$. With the order of magnitude assignment $We = \mathcal{O}(\varepsilon^{-2})$, we have $\bar{h}_N/l_S \sim \varepsilon^{2/3}$ so that l_S is much larger than the Nusselt film thickness $\bar{h}_N$ and the long-wave assumption is not violated. Equivalently, the long-wave assumption is sustained at the steep front of a large-amplitude solitary wave if $\partial_x h \ll 1$ there. In fact,

$$\max(\partial_x h) \sim \frac{\bar{h}_N}{\kappa\bar{h}_N} = \frac{1}{\kappa} \sim \varepsilon^{2/3}. \tag{4.9}$$

This estimate also shows that $\partial_x h$ at the front of a large-amplitude solitary wave is much larger than its formal order, $\mathcal{O}(\varepsilon)$; however, $\partial_x h$ never approaches unity at the front (unlike, e.g., the case of *hydraulic jumps*—to be discussed in Sects. 7.1.3 and 7.2.2.3). This also implies that we do not have a singular perturbation problem: the term $We\partial_{xxx}h$ with the highest spatial derivative is important throughout a solitary wave and not just in certain regions as in singular perturbation problems. We also note that the arguments presented here are based on a separation of scales between the front and back of the waves. As we shall demonstrate in Sect. 4.7, this is true only for $3Re > We^{1/3}$.

Hence, large We, which is defined as the *strong surface tension limit*, provides a clear physical explanation for the validity of the long-wave approximation at the steep front of a large-amplitude solitary pulse. The long-wave approximation there is satisfied precisely due to strong surface tension: large We becomes the cornerstone of the long-wave assumption for the boundary layer approximation. Relaxing the order of magnitude of We, but such that We is still large, also ensures the long-wave assumption. For example, if $We = \mathcal{O}(\varepsilon^{-1})$, then $\kappa = \mathcal{O}(\varepsilon^{-1/3})$. But by decreasing We further, i.e., $We = \mathcal{O}(1)$, the boundary layer equations are not applicable. A manifestation of this is that $We = \mathcal{O}(1)$ also causes $\kappa = \mathcal{O}(1)$ and the long-wave assumption in the framework of the boundary layer approximation is violated. We then need to proceed with full Navier–Stokes and Fourier equations, or with the long-wave theory for $We = \mathcal{O}(1)$. However, for $We = \mathcal{O}(1)$ the theory leads to a rather involved free-surface evolution equation as we discuss in Sect. 5.1.2.

Since $We = \Gamma / h_N^2$, the order of magnitude of We is actually related to the orders of magnitude of Γ and h_N. For $h_N = \mathcal{O}(1)$, large We corresponds to large Γ, representative of liquids with high surface tension and small kinematic viscosity such as water ($\Gamma \sim 3000$ at 25°C). On the other hand, there are certain liquids such as mineral oils [154, 155] and silicone oils [86, 231] whose surface tension is smaller than that of water, and kinematic viscosity is much larger than that of water. These liquids have $\Gamma \sim 1$. To sustain the boundary layer approximation now, we need to have $h_N \ll 1$, i.e., very small flow rates, resulting in a large We. Otherwise, if $h_N = \mathcal{O}(1)$, We and hence κ are of $\mathcal{O}(1)$, and the boundary layer equations are not applicable.

4.5 Dissipation

In addition to the influence of second-order viscous effects, Fig. 4.3 also reveals the effect of "dissipation."

Dissipation here refers to the damping of small scale wavy structures due to surface tension (in the strong surface tension limit), i.e. "dissipation of energy" at short scales. The presence of radiation is a signature of dissipation in the system. This type of dissipation should not be confused with viscous effects which (i) enable the steady Nusselt flat film solution and (ii) have a dispersive effect on the waves (see Sect. 7.1.1). Of course viscous dispersion does affect the amplitude and frequency of the capillary ripples/radiation at the front of solitary pulses. But a dissipative solitary pulse is always characterized by ripples at the front even in the absence of viscous dispersion, e.g., the solitary pulse of the KS equation (the equation and other weakly nonlinear prototypes are discussed in Appendix C.5). Viscous effects are connected with another type of dissipation, "viscous dissipation," which was instrumental in Kapitza's arguments as discussed in the Introduction.

Figure 4.3 reveals that by increasing Re both amplitude and frequency of oscillations in front of the primary solitary hump increase for both first-order and simplified second-order models. Hence, energy dissipation increases (large dissipation means either the amplitude or number of visible oscillations in front of the hump or both increase). This is because increasing inertia leads to more energy input to the system and this causes more dissipation of energy due to surface tension, as inertia must balance surface tension—surface tension remains fixed but the surface tension terms must change to accommodate the increased energy input; the precise way by which inertia balances surface tension in different parts of a solitary pulse is discussed in Sect. 4.7. Another way to put it, due to the opposing force of surface tension, some of the energy that goes into increasing the primary solitary hump by increasing Re, as Fig. 4.3 shows, also goes into increasing the amplitude and frequency of oscillations at its front.

On the other hand, increasing surface tension but keeping Re fixed also increases energy dissipation. But now the dissipated energy is distributed differently. Unlike the first case where Re increases but surface tension is constant, so that there is more energy overall and dissipation increases both amplitude and frequency of oscillations, now the frequency of oscillations increases but their amplitude decreases. Moreover, the amplitude of the primary hump decreases. This is to be expected as surface tension wants to flatten the film.

4.6 Shkadov Scaling

The significance of balancing gravity and viscous drag with surface tension for large-amplitude solitary waves was first identified by Shkadov [249], who introduced the scale ratio κ through (4.8). He proceeded to a compression of the streamwise coordinate by taking its scale as κ times the scale for y. He then introduced the following transformation for x, t, and v in the system (4.6a)–(4.6i), where ε has been scaled away:

$$x \rightarrow \kappa x, \quad t \rightarrow \kappa t \quad \text{and} \quad v \rightarrow v/\kappa. \tag{4.10}$$

The transformation (4.10) is referred to as the *Shkadov scaling*, to obtain the transformed system

$$\delta(\partial_t u + u\partial_x u + v\partial_y u) = 1 + \partial_{yy}u - \zeta\partial_x h + \partial_{xxx}h$$
$$+ \eta\big(2\partial_{xx}u + \partial_x[\partial_x u|_h]\big), \tag{4.11a}$$

$$Pr\delta(\partial_t T + u\partial_x T + v\partial_y T) = \eta(\partial_{xx}T + \partial_{yy}T), \tag{4.11b}$$

$$\partial_y u|_h = -\mathcal{M}\partial_x[T|_h] + \eta(4\partial_x h\partial_x u|_h - \partial_x v|_h), \tag{4.11c}$$

$$\partial_y T|_h = -B\left(1 + \frac{\eta}{2}(\partial_x h)^2\right)T|_h + \eta\partial_x h\partial_x T|_h, \tag{4.11d}$$

together with (4.6c)–(4.6f), and where we have made use of the following parameters:

$$\delta = \frac{3Re}{\kappa}, \qquad \zeta = \frac{Ct}{\kappa}, \qquad \eta = \frac{1}{\kappa^2}, \qquad \mathcal{M} = \frac{M}{\kappa}. \tag{4.12}$$

δ is the *reduced Reynolds number*, ζ is the *reduced inclination number* corresponding to the effect of the gravity component normal to the plate[1] and $\mathcal{M}$ is the *reduced film Marangoni number*. The parameter η appears along with every second-order streamwise viscous and thermal term in the momentum and energy equations and in the tangential stress balance and heat loss conditions at the interface. Hence by setting $\eta = 0$ we reduce (4.11a)–(4.11d) to the corresponding first-order boundary layer equations for a two-dimensional flow. In Chap. 7 we shall demonstrate that

---

[1]Actually, the reduced Reynolds number introduced by Shkadov is $\delta_{Sh} = \delta/45$. This numerical factor originates from a slightly different choice of variables. The present choice is preferred since it does not alter the numerical coefficients of the original boundary layer equations.

this parameter has a dispersive effect on the speeds of the linear waves, and thus it is called the *viscous dispersion number*.

It should be noted η is a small parameter. Even so, its effect is important and indeed, as we emphasized in Sect. 4.3, the second-order terms in the boundary layer equations, which are also small and $\propto \eta$ when these equations are rescaled with the Shkadov scaling, are important for a good agreement with Orr–Sommerfeld, and they play a significant role as far as the radiation in front of a solitary hump is concerned. Nevertheless, at times we shall be taking η to be of $\mathcal{O}(1)$ for illustration purposes; besides, the question of the behavior of different quantities of interest for large η is a valid one within the context of the boundary layer equations as "model equations".

The Shkadov scaling is inherent to the falling film problem in the region of moderate Reynolds numbers due to the separation of scales in the cross-stream and streamwise directions in this region that is in fact due to the strong effect of surface tension. Separation of scales also occurs in the region of small Reynolds numbers, but is different to that in the region of large ones. In fact we shall demonstrate shortly that solitary waves have quite different characteristic scales in the two regimes. As we shall see, the Shkadov scaling makes explicit the balance between all forces necessary to sustain large-amplitude nonlinear waves and their relative significance, such as inertia, gravity, viscosity and surface tension.

The Shkadov scaling has more advantages. The coefficient of streamwise surface tension $\partial_{xxx}h$ is exactly unity while in the region of moderate Reynolds numbers, $\delta \sim 1$, e.g., for the moderate value $Re \sim 10$ and the large values $We \sim 1000$ and $M \sim 10$, $\delta \approx 3$ and $\mathcal{M} \approx 1$, which is rather convenient from a numerical point of view. Further, by neglecting viscous dispersion, the Shkadov scaling brings all different effects into five parameters, δ, ζ, $\mathcal{M}$, Pr and B, thus reducing the number of governing parameters by one (from six to five).

4.7 Use of the Shkadov Scaling to Analyze the Balance of Different Forces on a Solitary Pulse

It is instructive at this stage to scrutinize the balance of the different forces on a pulse by using the Shkadov scaling. We distinguish between two different cases large amplitude and small-amplitude waves.

4.7.1 (i) Large-Amplitude Waves

We first examine the case of large Re,[2] or for simplicity, $\delta = \mathcal{O}(1)$. In this region we have large-amplitude waves.

---

[2]Recall that this monograph focuses on the regime of moderate Reynolds numbers (in the region $\sim$10–50). But occasionally, we shall be taking the Reynolds number to be large precisely for the same reasons we shall be taking η to be of $\mathcal{O}(1)$. We note, however, that strictly speaking for large Reynolds numbers the film flow behaves more like a "river flow".

The x component of the momentum equation (4.11a) in terms of the Shkadov scaling stipulates that δ must be at most of $\mathcal{O}(1)$ so that inertia never dominates the other terms in the equation. For simplicity, in all cases we assume $We = \mathcal{O}(\varepsilon^{-2})$. With $\delta = \mathcal{O}(1)$ then we have, $Re = \mathcal{O}(\varepsilon^{-2/3})$.

With $\delta = \mathcal{O}(1)$, (4.11a) stipulates that not only inertia is as important as surface tension, gravity and viscous drag at the front, but all forces balance throughout a solitary pulse and the estimate for the slope $\sim \varepsilon^{2/3}$ is valid throughout and not just at the front as suggested by (4.9). The same conclusion can be reached from (4.11a), where ε has been scaled away or (4.6a) where ε is present. Balancing inertia to gravity, viscous drag and surface tension in (4.11a) yields, $Re/\kappa \sim 1$ or $Re \sim \kappa$, so that $\delta \sim 1$. Consider now (4.6a). For clarity we replace the variables x, t there with $X = \varepsilon x$ and $T = \varepsilon t$, equivalent to the transformations $\partial_x = \varepsilon \partial_X$ and $\partial_t = \varepsilon \partial_T$ of the gradient expansion: X, T are slow scales while x, t are long ones, i.e., $x, t \sim \varepsilon^{-1}$ when $X, T \sim 1$. We also have $V = \varepsilon v$.

Equation (4.6a) for a two-dimensional flow on a vertical plane then takes the form,

$$3\varepsilon Re\big[\partial_T u + \partial_X (u^2) + \partial_y (uV)\big] = 1 + \partial_{yy} u + \varepsilon^3 We \partial_{XXX} h + \mathcal{O}(\varepsilon^2), \qquad (4.13)$$

where the $\mathcal{O}(\varepsilon^2)$ viscous dispersion terms have been neglected. Balancing gravity and viscous drag with surface tension,

$$1(\sim \partial_{yy} u) \sim \varepsilon^3 We \partial_{XXX} h \quad \Rightarrow \quad X \sim \varepsilon We^{1/3} \sim \varepsilon^{1/3},$$

while balancing inertia with gravity, viscous drag and surface tension,

$$\frac{\varepsilon Re}{X} \sim 1 \quad \Rightarrow \quad \varepsilon Re \sim \varepsilon^{1/3} \quad \Rightarrow \quad Re \sim \varepsilon^{-2/3},$$

which in turn also yields, $\delta \sim Re/\kappa \sim \varepsilon^{-2/3}/\varepsilon^{-2/3} \sim 1$. For $\delta \sim 1$, the length scale $X \sim \varepsilon^{1/3}$ is valid throughout the solitary wave and not just at its front. In terms of the original variable x, this corresponds to a length scale $x \sim \varepsilon^{-2/3}$, which is much shorter than the long-wave scale $x \sim \varepsilon^{-1}$, and a slope $\varepsilon^{2/3}$ throughout a pulse and not just at its front as suggested by (4.9); as we shall demonstrate shortly, for $\delta \sim 1$ there is no separation of scales between front and back of the wave.

Absorbing ε in (4.13) is equivalent to the transformations $X = \varepsilon \bar{X}$, $T = \varepsilon \bar{T}$ and $V = \varepsilon^{-1} \bar{V}$. These are then followed by the transformations $\bar{X} = \kappa \bar{\bar{X}}$, $\bar{T} = \kappa \bar{\bar{T}}$ and $\bar{V} = (1/\kappa) \bar{\bar{V}}$ in (4.13) due to the Shkadov scaling, giving

$$3Re\left[\frac{1}{\kappa} \partial_{\bar{\bar{T}}} + \frac{1}{\kappa} \partial_{\bar{\bar{X}}} (u^2) + \frac{1}{\kappa} \partial_y (u\bar{\bar{V}})\right] = 1 + \partial_{yy} u + \frac{We}{\kappa^3} \partial_{\bar{\bar{X}}\bar{\bar{X}}\bar{\bar{X}}} h + \mathcal{O}(\varepsilon^2), \qquad (4.14)$$

which is effectively (4.11a), rewritten here in terms of the variables $\bar{\bar{X}}$, $\bar{\bar{T}}$ and $\bar{\bar{V}}$. Once again, the estimate $\partial_x h \sim \varepsilon^{2/3}$ is easily confirmed:

$$\partial_x h = \varepsilon \partial_X h = \varepsilon \partial_{\bar{X}} h \frac{1}{\varepsilon} \equiv h_{\bar{\bar{X}}} \frac{1}{\kappa} \sim \varepsilon^{2/3}.$$

Let us now reexamine the balance between the different forces without substituting $Re \sim \varepsilon^{-2/3}$ and $We \sim \varepsilon^{-2}$, in order to maintain the explicit Re-, We-dependence of the different physical effects and hence uncover their dependence on δ.

- *Front*: At the front of the wave we balance gravity and viscous drag with surface tension:

$$1 \sim \frac{\varepsilon^3 We}{X_f^3} \quad \Rightarrow \quad X_f^3 \sim \varepsilon^3 We \quad \Rightarrow \quad \bar{X}_f \sim \kappa \quad \Rightarrow \quad \bar{\bar{X}}_f \sim 1,$$

where the subscript f is used to denote "front." The order of magnitude of the inertia term in this regime is

$$\frac{\varepsilon Re}{X_f} \sim \frac{Re}{\bar{X}_f} \sim \frac{Re}{\kappa \bar{\bar{X}}_f} \sim \frac{Re}{\kappa} \sim \delta.$$

Hence, for $\delta \sim 1$ inertia balances surface tension and gravity and viscous drag at the front, as we have seen earlier. This balance persists when δ increases from 1 provided that it remains of $\mathcal{O}(1)$. But as δ starts deviating from an $\mathcal{O}(1)$ value, inertia dominates over all other forces, not only at the front but throughout the wave as (4.11a) clearly indicates. The question then is which forces balance inertia in this regime.

As δ becomes large, a progressive deviation from the balance between viscous drag and surface tension at the front (used to obtain the Shkadov compression factor κ in Sect. 4.4) occurs. After all, $\kappa = We^{1/3}$ and involves gravity, viscous forces and surface tension. It must therefore be redefined for large δ to reflect a different balance, this time between inertia, viscous drag and surface tension: large δ implies a thick film, a "river flow." For such flows, surface tension is weak and one must balance inertia, viscous drag and gravity. It is precisely this balance that gives rise to the Saint–Venant equations to be discussed in Chap. 7. To see this balance in (4.13), we take the transformations, $X = \varepsilon \bar{X}$, $T = \varepsilon \bar{T}$ and $V = \varepsilon^{-1} \bar{V}$, followed by $\bar{X} = \kappa' \bar{\bar{X}}$, $\bar{T} = \kappa' \bar{\bar{T}}$ and $\bar{V} = (1/\kappa') \bar{\bar{V}}$ so that (4.14) becomes

$$3Re \left[\frac{1}{\kappa'} \partial_{\bar{\bar{T}}} + \frac{1}{\kappa'} \partial_{\bar{\bar{X}}} (u^2) + \frac{1}{\kappa'} \partial_y (u\bar{\bar{V}}) \right] = 1 + \partial_{yy} u + \frac{We}{\kappa'^3} \partial_{\bar{\bar{X}}\bar{\bar{X}}\bar{\bar{X}}} h + \mathcal{O}(\varepsilon^2), \quad (4.15)$$

where κ' is the new compression factor. Balancing inertia with gravity and viscous drag in (4.15) then gives

$$\kappa' = Re,$$

while the order of magnitude of the surface tension term on the right hand side of (4.15) is We/κ'^3. With $\delta > 1$ then we have $Re > We^{1/3}$ or $We/\kappa'^3 < 1$ and the role of surface tension diminishes as δ becomes large.

- *Back*: At the back of the wave we balance inertia with gravity and viscous drag:

$$\frac{\varepsilon Re}{X_b} \sim 1 \quad \Rightarrow \quad X_b \sim \varepsilon Re \quad \Rightarrow \quad \bar{X}_b \sim Re \quad \Rightarrow \quad \bar{\bar{X}}_b \sim \delta,$$

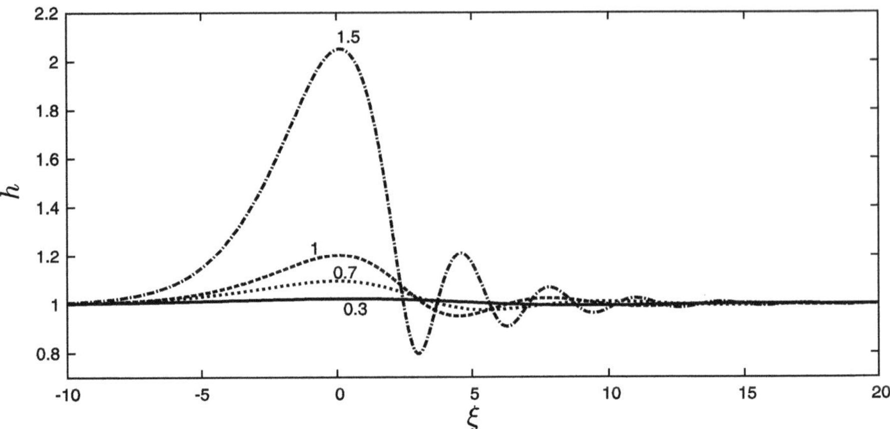

Fig. 4.4 Single-hump solitary wave solutions of the first-order model to be developed in Chap. 6 in their moving frame ξ for increasing δ: $\delta = 0.2, 0.3$ (both as *solid line*), $\delta = 0.7$ (*dotted line*), $\delta = 1$ (*dashed line*) and $\delta = 1.5$ (*dotted–dashed line*). The profiles for $\delta \lesssim 0.3$ are nearly indistinguishable. They are also hardly visible, as they correspond to very small deviations from the flat film thickness 1 (they can be seen clearly in Fig. 4.6 where the film thickness and moving coordinate are rescaled appropriately) but are topologically similar to the profile for $\delta = 1$, a consequence of the absence of scales separation between the back and front of the waves for $\delta \lesssim 1$. Separation of scales occurs only for $\delta > 1$

where the subscript b is used to denote "back." Hence, the upstream tail of the wave scales as δ so that as δ increases from 1 the wave is characterized by a long scale at the back ($\sim \delta$) followed by a short one at the front (~ 1), i.e., the wave has a long tail followed by a steep front (the tail is long relatively to the front). The order of magnitude of surface tension in this regime is:

$$\frac{\varepsilon^3 We}{X_b^3} \sim \frac{We}{\bar{X}_b^3} \sim \frac{We}{\kappa^3 \bar{\bar{X}}_b^3} \sim \frac{1}{\bar{\bar{X}}_b^3} \sim \frac{1}{\delta^3}.$$

For $\delta \sim 1$, surface tension balances all forces at the back (as a matter of fact all forces balance throughout a solitary pulse in this case, as we have seen earlier), but as δ increases from 1 the surface tension force decreases at the back.

Having determined the characteristic length scales of the front and back of the wave, the corresponding slope $\partial_x h$ in these regions in terms of δ can be easily estimated:

$$\partial_x h = \varepsilon \partial_X h = \varepsilon \partial_{\bar{X}} h \frac{1}{\varepsilon} = \partial_{\bar{\bar{X}}} h \frac{1}{\kappa} \sim \frac{1}{\kappa \bar{\bar{X}}} \sim \begin{cases} \dfrac{1}{\kappa \bar{\bar{X}}_f} \sim \dfrac{1}{\kappa} \sim \varepsilon^{2/3}; & \text{front} \\[2ex] \dfrac{1}{\kappa \bar{\bar{X}}_b} \sim \dfrac{1}{\kappa \delta} \sim \dfrac{\varepsilon^{2/3}}{\delta}; & \text{back.} \end{cases}$$

The above observations indicate an asymmetry between the front and back for large-amplitude solitary waves for $\delta > 1$ as confirmed in Fig. 4.4. The figure indicates a clear separation of scales between front and back for $\delta > 1$. The maximum slope does occur at the front consistent with (4.9). For $\delta = 1$ there is no separation of scales, the wave is almost symmetric and steep throughout.

- *Radiation oscillations*: The oscillations at the front correspond to a balance between inertia and surface tension:

$$\frac{\varepsilon Re}{X_o} \sim \frac{\varepsilon^3 We}{X_o^3} \quad \Rightarrow \quad X_o^2 \sim \frac{\varepsilon^2 We}{Re} \quad \Rightarrow \quad \bar{X}_o \sim \sqrt{\frac{We}{Re}} \quad \Rightarrow \quad \bar{\bar{X}}_o \sim \delta^{-1/2}.$$

Hence, as δ increases from 1, the wavelength of the oscillations at the front decreases as demonstrated in Fig. 4.4. The order of magnitude of the inertia and surface tension, terms in this regime is

$$\frac{\varepsilon Re}{X_o} \sim \frac{Re}{\bar{\bar{X}}_o} \sim \frac{Re}{\kappa \bar{\bar{X}}_o} \sim \delta^{3/2}.$$

Therefore, for $\delta > 1$ inertia and surface tension dominate over gravity and viscous drag in the oscillatory region in front of a large-amplitude solitary pulse.

4.7.2 (ii) Small-Amplitude Waves

We now examine the case of small Re, for simplicity $Re = \mathcal{O}(1)$, i.e., small δ. In this region the waves have small amplitude.

Small inertia leads to small-amplitude solitary pulses, i.e., a small deflection from the Nusselt flat film flow. As a matter of fact, in this regime the character of the flow should not be very different from the Nusselt flat film solution, where the dominant effects are gravity and viscous drag. Therefore, gravity and viscous drag balance to leading order while inertia and surface tension are higher-order effects. However, a permanent solitary wave requires the balance of inertia and surface tension:

$$\frac{\varepsilon Re}{X} \sim \frac{\varepsilon^3 We}{X^3} \quad \Rightarrow \quad \bar{\bar{X}} \sim \delta^{-1/2},$$

while the order of magnitude of inertia and surface tension is

$$\frac{\varepsilon Re}{X} \sim \delta^{3/2} \ll 1,$$

thus confirming that indeed inertia and surface tension are higher-order effects. Both front and back of a solitary wave now are of the same order without any separation of scales between the two, and the hump appears to be almost symmetric, as shown in Fig. 4.4. In terms of the X variable and with $Re \sim 1$, $We \sim \varepsilon^{-2}$, the above balance between inertia and surface tension yields $X \sim 1$.

These scaling arguments reveal that the Shkadov scaling is strictly speaking relevant for large-amplitude solitary waves, i.e., for $\delta = \mathcal{O}(1)$. It is precisely for this reason that the momentum equation (4.11a) might lead one to the conclusion that for small-amplitude solitary waves, i.e., for δ small, surface tension balances gravity and viscous drag. But clearly it is a higher-order effect as is also evident from the momentum equation (4.6a) prior to the introduction of the Shkadov scaling. Finally, we note that the computations in Fig. 4.4 indicate that for small δ all profiles are nearly indistinguishable, which must be due to the convergence of the first-order boundary layer approximation at small δ to the KS equation defined in Appendix C.5; this equation is free of parameters and hence it must have a single solitary-wave solution.

4.7.3 Behavior of the Eigenvalues of the Flat Film Solution of a Linearized Averaged Model

The above orders of magnitude estimates of the relative importance of the various physical effects in different regions of a solitary pulse can also be confirmed by examining the behavior of the eigenvalues of a linearized averaged model around the flat film fixed point. These terms are explained in detail in Chap. 7. Consider, e.g., the first-order model used in the computation of Fig. 4.4. For the purposes of this section it is sufficient to know that linearization of this model about its "fixed point," the flat film solution, gives a three-dimensional *dynamical system*, i.e. a system of three first-order ordinary differential equations, whose spectrum is described by a cubic characteristic equation. This equation has three roots, one real and positive, and a complex conjugate pair with negative real part. In the three-dimensional phase space then associated with the dynamical system, a solitary wave represents an orbit that departs from the fixed point along a one-dimensional "unstable manifold" spanned by the eigenvector associated with the real eigenvalue and returns back to the fixed point in an oscillatory fashion on the two-dimensional "stable manifold" spanned by the eigenvectors associated with the complex eigenvalues. Figure 4.5 depicts the real eigenvalue λ_1 corresponding to the unstable manifold and the imaginary part of the complex conjugate eigenvalues $\lambda_{2,3}$ associated with the stable manifold for the single-hump solitary-wave solutions of the first-order model as a function of δ.

The quantity $1/\lambda_1$ provides a measure of the characteristic scale of the upstream tail of a solitary wave. In fact, we can define $l_{\text{tail}} = 1/\lambda_1$. On the other hand $1/\Im(\lambda_{2,3})$ is a measure of the characteristic scale of the radiation oscillations in front of the solitary hump. The figure indicates that $\Im(\lambda_{2,3})$ varies as $\delta^{1/2}$ for both $\delta \lesssim 1$ and $\delta > 1$ so that the characteristic scale of the radiation oscillations varies as $\delta^{-1/2}$. This scale increases as δ decreases from 1 and decreases as δ increases from 1.

The figure also indicates that for $\delta \lesssim 1$, λ_1 varies as $\delta^{1/2}$ so that l_{tail} varies as $\delta^{-1/2}$, which is the same with the characteristic scale of the radiation oscillations at the front, a direct consequence of the absence of separation of scales between the front and back of the wave for $\delta \lesssim 1$ as concluded earlier. On the other hand, for $\delta > 1$, λ_1 varies like $1/\delta$ so that l_{tail} varies like δ and increases as δ increases.

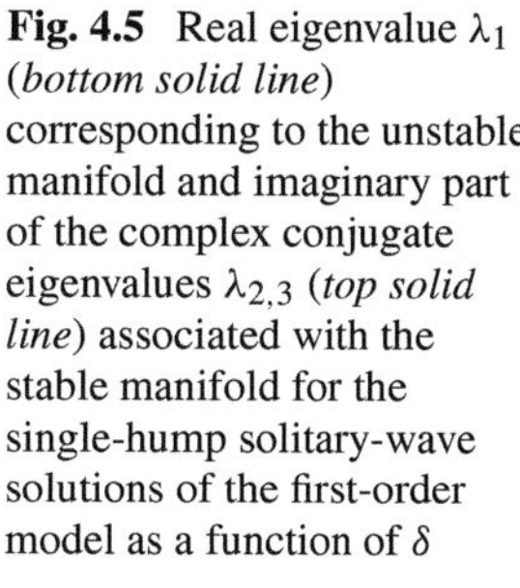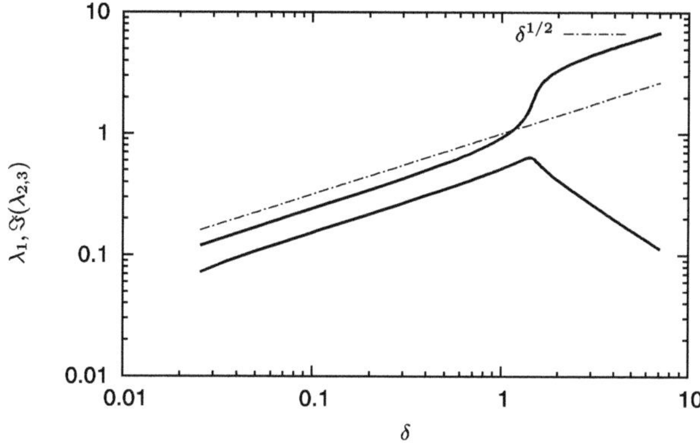

Fig. 4.5 Real eigenvalue λ_1 (*bottom solid line*) corresponding to the unstable manifold and imaginary part of the complex conjugate eigenvalues $\lambda_{2,3}$ (*top solid line*) associated with the stable manifold for the single-hump solitary-wave solutions of the first-order model as a function of δ

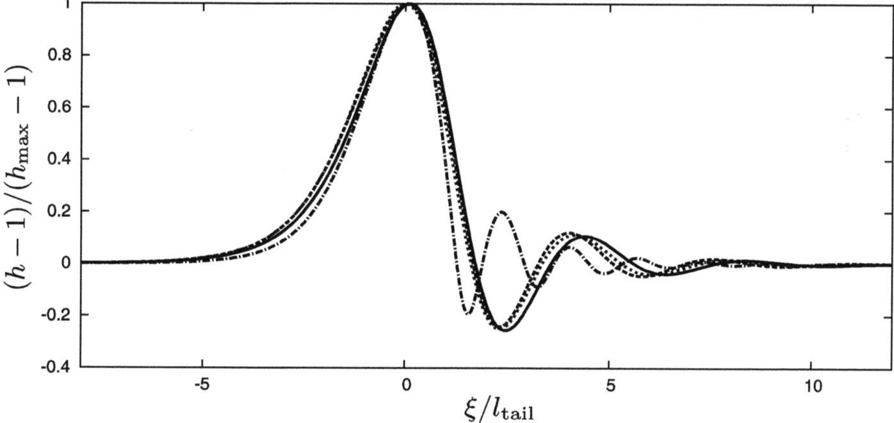

Fig. 4.6 Single-hump solitary-wave deviation height (difference from flat film) normalized with deviation maximum amplitude in its moving frame ξ normalized with the length tail, $l_{\text{tail}} = 1/\lambda_1$, as δ increases, for the profiles in Fig. 4.4

Also, since the sum of the three eigenvalues vanishes (again details will be given in Chap. 7), $\Re(\lambda_{2,3}) = -\lambda_1/2$ so that the characteristic length of the envelope of the radiation oscillations at the front varies like the scale l_{tail} at the back and increases as δ increases, whereas the characteristic length of the radiation oscillations decreases (due to the $\delta^{-1/2}$-dependence). As a consequence, the number of radiation oscillations increases as δ increases.

Finally, Fig. 4.6 compares the deviation from 1 of the solitary-wave height of the profiles in Fig. 4.4 rescaled with the deviation from 1 of the maximum amplitude as a function of the moving coordinate ξ rescaled with the length of the upstream tail, $l_{\text{tail}} = 1/\lambda_1$. The figure contrasts directly the back and front of the different profiles as δ increases. The profiles are nearly indistinguishable for $\delta \lesssim 0.3$ due to the convergence of small δ to the KS equation as pointed out earlier. For $\delta > 1$ an asymmetry between the front and back of the waves appears but the envelope of the radiation oscillations at the front is still almost symmetric to the back of the wave.

(This is because the real part of the eigenvalues of the stable manifold is equal to half the real eigenvalue of the unstable manifold, so that the characteristic length of the envelope of the radiation oscillations at the front varies like the scale l_{tail} at the back, as discussed above). However, a large number of oscillations (corresponding to a balance of inertia and surface tension) is now visible.

4.8 Cross-stream Inertia

4.8.1 On the Order of Magnitude of Cross-stream Inertia

Since the order of magnitude of the slow streamwise variable X changes from $\mathcal{O}(1)$ throughout a solitary pulse in case (ii)/small-amplitude waves for small δ to $\mathcal{O}(\varepsilon^{1/3})$ throughout a solitary pulse in case (i)/large-amplitude waves for $\delta = \mathcal{O}(1)$ so that the order of magnitude of streamwise inertia also changes, it is essential that we check the order of magnitude of cross-stream inertia that has been neglected in our boundary layer approximation.

Let us rewrite the cross-stream momentum equation (4.2c) for a two-dimensional flow on a vertical wall in terms of the variables X, T:

$$3\varepsilon^2 Re(\partial_T v + u \partial_X v + v \partial_y v) = -\partial_y p + \varepsilon \partial_{yy} v. \tag{4.16}$$

An important step in the derivation of the second-order boundary layer equations in Sect. 4.1 was to neglect cross-stream inertia in (4.16), which is of $\mathcal{O}(\varepsilon^2 Rev/X)$ (due to the Shkadov scaling, $v = (1/\kappa)V$, $v \partial_y v \ll \partial_T v \sim u \partial_X v$) compared to the ε-viscous term of $\mathcal{O}(\varepsilon v)$ or:

$$\frac{\varepsilon^2 Rev}{X} \ll \varepsilon v \quad \Rightarrow \quad \frac{\varepsilon Re}{X} \ll 1. \tag{4.17}$$

This condition is satisfied for case (ii) with $Re = \mathcal{O}(1)$:

$$\frac{\varepsilon^2 Rev}{X} \sim \varepsilon^2 Rev \sim \varepsilon^2 v \ll \varepsilon v.$$

However, for case (i) with $Re = \mathcal{O}(\varepsilon^{-2/3})$, (4.17) is not satisfied, since now we require $\varepsilon Re/X \sim 1$. On the other hand, for the first-order boundary layer equations where both cross-stream inertia and ε-viscous term of the y-momentum equation are neglected, (4.17) is not required.

The contribution of the left hand side of (4.16), denoted as CSI (cross-stream-inertia) in the following, to the pressure distribution obtained from this equation is $3\varepsilon^2 Re \int_y^h \text{CSI}\, dy$, which when substituted into the streamwise momentum equation becomes $3\varepsilon^3 Re\partial_x \int_y^h \text{CSI}\, dy$. (It is straightforward to confirm that this last term, of $\mathcal{O}(\varepsilon^3 Rev/X^2)$, is negligible compared to the streamwise inertia, surface tension and hydrostatic head terms of the streamwise momentum equation, with the exception of course of the viscous diffusive terms as we have just shown.) The only way then to rigorously justify the neglect of cross-stream inertia for large Re is for y very close to h, i.e., close to the interface. This then would necessarily restrict the region of validity of the second-order boundary layer equations close to the interface for large Re (after all the Shkadov scaling and all scaling arguments given earlier are centered on the interface of a solitary pulse), but these would then raise the question of validity of the second-order models obtained in Chap. 6 from averaging the second-order boundary layer equations across the film, i.e., from $y = 0$ to $y = h$. Nevertheless, as we shall see in the next section, in the linear regime energy arguments can be used to justify neglecting the contribution of $\int_y^h \mathsf{CSI}\, dy$. In the nonlinear regime the justification is made a posteriori, via comparison of the pressure distribution across the film obtained from the boundary layer equations with that obtained from DNS.

4.8.2 On the Region of Validity of the Boundary Layer Approximation

As emphasized earlier, the key assumption leading to the boundary layer equations is the neglect of the cross-stream inertia effects. It is useful at this point to outline the conditions on Re for which this is the case. As before, with the exception of $We = \mathcal{O}(\varepsilon^{-2})$, all other parameters are taken of $\mathcal{O}(1)$.

Appendix D.2 concludes that for the second-order boundary layer equations we must have $\varepsilon Re \ll 1$, which automatically ensures that $Re \ll We$ for the cross-stream inertia to be negligible compared to surface tension, and $Re \gg \varepsilon^2$. For the first-order boundary layer equations we must have εRe at most of $\mathcal{O}(1)$ (which automatically satisfies the condition $Re \ll We$) and $Re \gg \varepsilon$. As noted earlier, from the x component of the momentum equation in terms of the Shkadov scaling in (4.11a), δ must be at most of $\mathcal{O}(1)$ so that inertia never dominates the other terms in the equation. With $\delta \sim Re/We^{1/3} \sim \varepsilon^{2/3}Re$ at most of $\mathcal{O}(1)$, Re must be at most of $\mathcal{O}(\varepsilon^{-2/3})$, which then automatically satisfies the condition $\varepsilon Re \ll 1$ (which in turn automatically ensures $Re \ll We$) for the second-order boundary layer equations. Re at most of $\mathcal{O}(\varepsilon^{-2/3})$ also satisfies automatically the condition εRe at most of $\mathcal{O}(1)$ (which again automatically ensures $Re \ll We$) for the first-order boundary layer equations.

Interestingly, the upper bound on Re is satisfied automatically by the condition δ at most of $\mathcal{O}(1)$ even when the order of magnitude assignment $We = \mathcal{O}(\varepsilon^{-2})$ is relaxed. Appendix D.2 concludes that for the second-order boundary layer equations we must have $Re \ll \min\{We, \varepsilon^{-1}\}$, $Re \gg \varepsilon^2$ and $We \gg 1$, We at most of $\mathcal{O}(\varepsilon^{-2})$. The condition δ at most of $\mathcal{O}(1)$ translates to Re at most of $\mathcal{O}(We^{1/3})$. Consider two cases: (i) $We \ll \varepsilon^{-1}$ so that $\min\{We, \varepsilon^{-1}\} = We$ and $Re \ll We$. With Re at most of $\mathcal{O}(We^{1/3})$ we must also have $Re \ll We$ since $We^{1/3} \ll We$ due to $We \gg 1$; (ii) $We \gg \varepsilon^{-1}$ so that $\min\{We, \varepsilon^{-1}\} = \varepsilon^{-1}$ and $Re \ll \varepsilon^{-1}$. With Re at most of $\mathcal{O}(We^{1/3})$ we also have $We^{1/3} \ll \varepsilon^{-1}$ since We is at most of $\mathcal{O}(\varepsilon^{-2})$ and $\varepsilon^{-2/3} \ll \varepsilon^{-1}$.

In all cases, therefore, δ at most of $\mathcal{O}(1)$ ensures that the upper bound on Re required for the validity of the boundary layer equations is satisfied. This upper bound in turn ensures that cross-stream inertia is negligible compared to both surface tension (for both the first- and second-order boundary layer equations) and the

ε-viscous term in the y momentum equation (for the second-order boundary layer equations). Hence, the size of δ is crucial for the validity of the boundary layer equations. Let us leave aside for the time being the additional complication of the cross-stream inertia terms being of the same order of magnitude with the ε-viscous term in the y momentum equation for the second-order boundary layer approximation in the case of large-amplitude solitary pulses obtained in the region of large Re (case (i) in Sect. 4.7).

The condition δ at most of $\mathcal{O}(1)$ can be violated for sufficiently large Re. In fact, for a given liquid and inclination angle, i.e., for a given Kapitza number, the Weber number decreases when the Reynolds number increases, as can be seen from their definitions (2.39) and (2.37), respectively, so that δ written below in terms of the viscous-gravity scaling,

$$\delta = \frac{3Re}{We^{1/3}} = \frac{(3Re)^{11/9}}{\Gamma^{1/3}}, \tag{4.18a}$$

increases with Re. Of interest is also the size of the product $\eta\delta = 3Re/We$ since $Re \ll We$ ensures that surface tension dominates over cross-stream inertia (Appendix D.2). Clearly, the condition δ at most of $\mathcal{O}(1)$ automatically ensures $\delta\eta$ small since η is small. This is consistent with the above observation that δ at most of $\mathcal{O}(1)$ ensures that the upper bound on Re required for the validity of the boundary layer equations is satisfied. For large δ, the condition $\eta\delta \ll 1$ might not be satisfied automatically. Let us also express the product $\eta\delta$ in terms of the viscous-gravity scaling,

$$\eta\delta = \frac{3Re}{We} = \frac{(3Re)^{5/3}}{\Gamma}, \tag{4.18b}$$

which also increases with Re but less rapidly compared to δ.

In situations now where Re is sufficiently large so that the ratio $Re/We \sim \eta\delta$ is no longer small, e.g., $\eta\delta \sim 1$, cross-stream inertia effects are no longer negligible compared to surface tension. This automatically makes δ large, e.g., $\eta\delta \sim 1$ leads to $Re \sim We$ and $\delta \sim We^{2/3}$, which in turn demands that not only cross-stream inertia is not negligible compared to surface tension, but also compared to the ε-viscous term in the y momentum equation.

This raises the question of validity of the boundary layer assumption when the condition δ at most of $\mathcal{O}(1)$ is violated and hence of the models derived from this assumption when this condition is violated, such as the Kapitza–Shkadov equations and the weighted residuals models.

Figure 4.7 compares the neutral stability curves and the temporal growth rates obtained from the Orr–Sommerfeld eigenvalue problem to the corresponding linear stability problem obtained from the linearized second-order boundary layer equations ((4.7) together with (3.22b) and the boundary conditions (3.22c)–(3.22f), (3.22h), (3.22i)). f is the "film-forcing frequency" at the inlet and f_c is the "cut-off frequency" beyond which the film remains flat. (Above the cut-off frequency the film remains flat—details are given in Sect. 7.1.1. Film-forcing corresponds to "open flow conditions"—details are given in Sect. 5.3.1.) Parameter values are chosen to correspond to a glycerin–water liquid mixture ($\nu = 6.27 \times 10^{-6}$ m^2 s^{-1} and

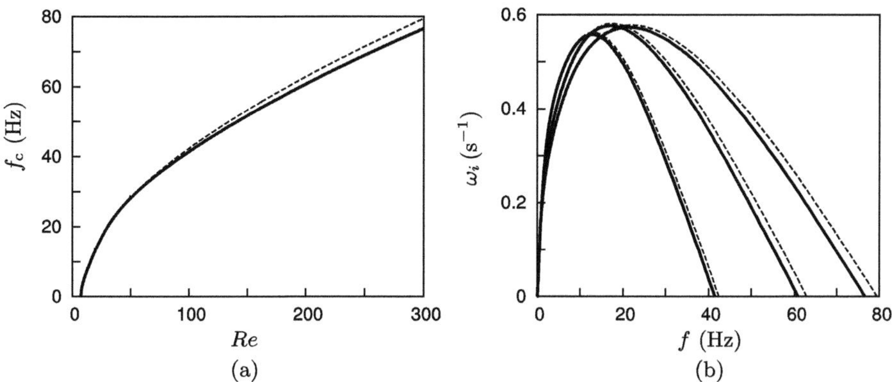

Fig. 4.7 Comparison of Orr–Sommerfeld analysis (*solid lines*) and eigenvalue problem obtained from the second-order boundary layer equations (*dashed lines*); (**a**) cut-off frequency as a function of the Reynolds number; (**b**) temporal growth rate as a function of the forcing frequency at $Re = 100, 200, 300$, for $\Gamma = 526$ and $Ct = 8.92$

$\sigma/\rho = 62.6 \times 10^{-6}$ m^3 s^{-2}) and the slope $\beta = 6.4°$ of the experiments reported in [168] for $\Gamma = 526$. The Orr–Sommerfeld eigenvalue problem was solved by continuation starting with the zero-wavenumber solution using the software AUTO-07P (see Appendix F.1).

For $\Gamma = 526$, the product $\eta\delta \sim 1$ corresponds to $Re \sim 14$ and therefore cross-stream inertia terms cannot be a priori neglected when compared to surface tension effects. At the same time, with $\Gamma = 526$ and $Re = 14$, from (4.18a) we have $\delta \sim 12$ and hence cross-stream inertia cannot only be a priori neglected compared to surface tension but also compared to the viscous term of the y momentum equation. Yet, results from the linearized boundary layer equations agree well with the results of Orr–Sommerfeld analysis up to rather high Reynolds numbers ($\sim$300). This indicates that, at least in the linear regime, the cross-stream inertia effects do not contribute to the instability. A plausible explanation for this can be obtained from energy arguments (see Sect. 3.6): The dominant contribution to the kinetic energy of the perturbation arises from the work done by the shear stress perturbation at the free surface. The instability mechanism resulting from inertia is directly associated with the vorticity perturbation relative to the surface displacement.

Consider for instance, the pressure distribution obtained by integrating the y component of the momentum equation (4.2c) across the film. In the isothermal case, $p = 3Re \int_y^h (Dv/Dt)\,dy + [\partial_y v + \partial_y v|_h] + Ct(h - y) - We\partial_{xx}h$, where ε has been scaled away. The contribution of the advection of the cross-stream velocity v by the flow is important only in the bulk far from the interface, i.e., $\int_y^h Dv/Dt\,dy$ contributes only for $0 < y \ll h$ and not close to the interface. Since the transfer of energy from the mean flow to the perturbation is weak in the bulk region, this may explain the observed agreement of the curves in Fig. 4.7.

As also emphasized in Sect. 4.1, DNS studies in the nonlinear regime have shown small deviations of the pressure distribution across the film [99, 176, 232]. In fact, deviations were noticeable only for strong free-surface deformations, e.g., in the case of solitary waves of large amplitude, and they occur mainly at the front of the primary solitary hump (see Fig. 4.1). This suggests that the pressure in the film is well approximated by $p \approx Ct(h - y) - We\partial_{xx}h$ so that the cross-stream inertia terms have little effect on the waves, even though, e.g., $\eta\delta = 2.1$ for the given parameter values (see (4.18b)). As noted in Sect. 4.1, with the exception of the hydrostatic head in the direction perpendicular to the wall, there is no mechanism that can modify the pressure distribution across the film, much like with boundary layers in aerodynamics. Hence, while the contribution of the cross-stream inertia terms $3Re \int_y^h (Dv/Dt)\,dy$ in the second-order boundary layer equations appears to be of the same order as the viscous term of the y momentum equation (for a large-amplitude solitary wave in the region of large Re (case (i) in Sect. 4.7) due to the lengthscale $X \sim \varepsilon^{1/3}$ of the wave in this regime), the cross-stream inertia effects have a small influence in both linear and nonlinear regimes.

Therefore, we may conclude that even though the conditions δ at most of $\mathcal{O}(1)$ and $\eta\delta \ll 1$ can be violated, and even though cross-stream inertia in the second-order boundary layer equations appears to be as important as the viscous term in the y momentum equation for large Re, the boundary layer equations preserve the structure of the Navier–Stokes and Fourier equations (with the exception of course of the elimination of the pressure) and are accurate outside their strict range of validity. In fact, they can be applied to a much wider range of parameter values than what could be expected a priori.

4.9 Reduction of the Boundary Layer Equations

Although the pressure has been eliminated, the second-order boundary layer equations have the same dimensionality as the full problem and hence do not constitute a drastic simplification. The complexity of the second-order boundary layer equations also makes them difficult to analyze. For instance, the search of nonlinear solutions of these equations requires elaborate numerical techniques to track the film interface. Several authors have developed such numerical schemes to simulate the time evolution of the local film thickness (see, e.g., [44]) but these full-scale computations are difficult to implement and time consuming. It is hence appropriate to obtain models of reduced dimensionality that also retain the essential dynamic characteristics of the full equations.

Such models are based on the method of weighted residuals and will be developed in Chap. 6. In these models, the dependence of the cross-stream y-coordinate is eliminated, a consequence of the strong "in-depth coherence" imposed by viscous diffusion in the y direction for both momentum and energy equations. The result is systems of equations for the evolution in space (x, z) and time t of the main physical quantities such as the film thickness h, the local flow rates q and p in the streamwise and spanwise directions, respectively, and the interfacial temperature θ. These equations reveal the existence of two distinct regimes for the velocity of solitary waves as a function of δ, discussed in the following section.

Fig. 4.8 Velocity c, amplitude $h_{\max}$ and length of the upstream tail, l_{tail}, of single-hump solitary waves versus the reduced Reynolds number δ, for an isothermal film flowing down a vertical wall ($\zeta = 0$), obtained with the simplified second-order model (to be derived in Chap. 6). For $\delta < 1$, $l_{\text{tail}} \propto \delta^{-1/2}$ is the characteristic scale of both back and front of the waves

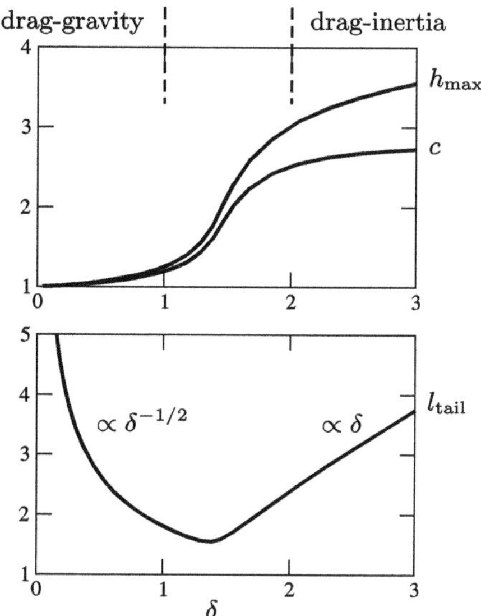

4.9.1 Drag-Gravity and Drag-Inertia Regimes

Figure 4.8 offers a plot of the speed c and the maximum amplitude $h_{\max}$ of single-hump solitary waves in their moving frame as a function of the reduced Reynolds number δ for an isothermal film falling down a vertical wall. The figure shows a steep increase of both speed and amplitude of the waves at $\delta \simeq 1$,[3] indicating the presence of two different regimes, which following Ooshida [196] are named as:

- The *drag-gravity regime*, where the component of gravity parallel to the flow, is mainly balanced by the viscous drag so that the character of the flow is close to that of the Nusselt flat film solution, with both inertia and surface tension playing effectively only a "perturbative" role throughout the wave. Hence, the cross-stream velocity distribution is very similar to the semiparabolic profile of the Nusselt flat film solution. In this regime both the back and front of the waves have the same length scales, $X \sim 1$, without any separation between the two (see Sect. 4.7.2).

  To obtain the dependence of the characteristic scale of the wave in terms of δ we simply balance in (4.11a) the two small effects, namely inertia $\propto \delta \partial_x$ and

---

[3]Most interestingly, the value $\delta \simeq 1$ is close to Kapitza's prediction of the threshold $\delta_c = 2.093$ for the occurrence of waves on a vertical wall (linear instability threshold) [140]. Even though his analysis was in error as pointed out in the Introduction—the threshold actually being $\delta_c = 0$ for a vertical wall as shown in Chap. 3—it was "supported" by his own experimental observations. In fact, below the sharp transition in speed (and amplitude) of the waves between the drag-gravity and drag-inertia regimes, the amplitude of the waves is so small that they are difficult to detect experimentally, and usually in practice the waves become visible only above the value predicted by Kapitza (see Fig. 4.8).

surface tension $\propto \partial_{xxx}$, which gives the scale $\propto \delta^{-1/2}$ as shown in Sect. 4.7 for both front and back. Hence the upstream tail at the back has the scale $l_{\text{tail}} \propto \delta^{-1/2}$, consistent with the computation in Fig. 4.8. This estimate is in line with the cutoff wavenumber $k_c \propto \sqrt{Re/We}$ for a vertical wall under isothermal conditions; see (3.35): with $x \to x/\kappa$, i.e., by inverting (4.10) in order to return to the original x variable utilized in Chap. 3, and with (4.8) and (4.12), $\partial_x \propto \delta^{1/2}$ becomes $\partial_x \propto (1/\kappa)\delta^{1/2} = \sqrt{Re/We} \propto k_c$. Since $k_c \sim \varepsilon$, the characteristic length scale of both front and back is $\sim \varepsilon^{-1}$, in agreement with the scaling arguments in Sect. 4.7. The estimate $\partial_x \propto k_c$ above also implies that the drag-gravity regime occurs close to the linear instability threshold, i.e., close to the onset of the waves resulting from the primary instability of the Nusselt flat film solution, consistent with our observation that in this regime inertia and surface tension are corrections to the Nusselt flow.

- The *drag-inertia regime*. Inertia and surface tension are no longer corrections to the Nusselt flow. This is case (i) analyzed in Sect. 4.7. As demonstrated there, for $\delta > 1$ there is a separation of scales between the front and the back of the wave with a steep front where gravity, viscous drag and surface tension balance and a long tail where gravity, viscous drag and inertia balance. From (4.11a) this balance is simply $\delta \partial_x \propto 1$, yielding $l_{\text{tail}} \propto \delta$ as first shown in Sect. 4.7 and consistent with the computation in Fig. 4.8. In such a regime, the basic assumption of considering inertia as a perturbation is clearly not true and the flow is radically different from the Nusselt flat film flow.

As noted above, the transition between the two regimes in Fig. 4.8 occurs at $\delta \simeq 1$, a universal result valid for all liquids. This is actually another advantage of the Shkadov scaling in addition to those discussed in Sect. 4.6, if not the main one: the Shkadov parameter δ is the relevant parameter for making the distinction between the drag-gravity and drag-inertia regimes. The key of course to allow the capture of the transition between the two regimes accurately is the development of models valid in both regimes, precisely the subject of Chap. 6.

4.9.2 Hierarchy of Models

The different levels of modeling approaches/simplifications of the Navier–Stokes and Fourier equations, depending on the flow regime being considered, are summarized in Fig. 4.9. We now discuss the different levels of approximations.

- *Low Reynolds number flow*: $Re = \mathcal{O}(1)$, $\delta \ll 1$.

  Indeed, from $\delta \sim Re/\kappa \sim Re/We^{1/3} \sim \varepsilon^{2/3} Re$ with $We = \mathcal{O}(\varepsilon^{-2})$, $\delta \ll 1$. This is the drag-gravity regime. A gradient expansion of the governing equations leads to a single evolution equation for the film thickness h. This is the long-wave theory that will be developed in Chap. 5. The theory is based on slaving of the dynamics of the flow to its kinematics, i.e., all variables are slaved to the film thickness h.

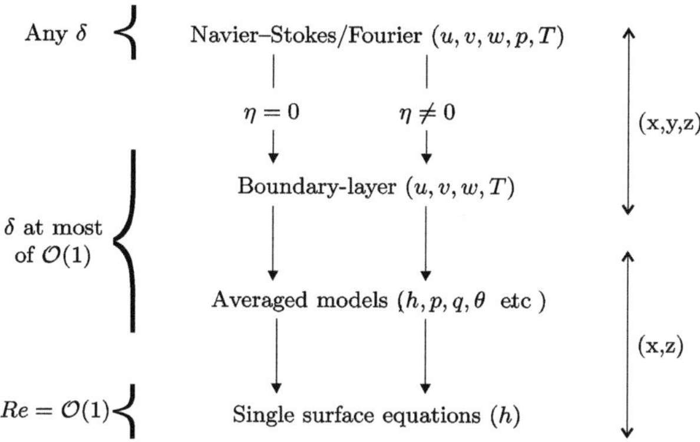

Fig. 4.9 Hierarchy of approximations from the Navier–Stokes and Fourier equations. For single surface equations, $Re = \mathcal{O}(1)$ represents a typical order of magnitude assignment in thin-film flow studies. The boundary layer equations are actually valid beyond the strict order of magnitude assignment, δ at most of $\mathcal{O}(1)$. But at the same time δ should not be large since in this region the film flow behaves more like a "river flow"

It was first introduced by Benney [21] for an isothermal film and later on extended by other authors to include a variety of other physical effects such as heating, surface active agents, evaporation, chemical reactions and topographical forcing as well as combinations of these (see, e.g., [137, 201]). The success of this approach is mainly due its simplicity. It is also relatively straightforward to associate each term of the evolution equation to a specific physical mechanism, such as gravity, inertia, surface tension, hydrostatic pressure and Marangoni effect, and hence to easily ascertain the influence of each physical effect on the dynamics of the film. We note that the assumption $Re = \mathcal{O}(1)$ represents a typical order of magnitude assignment in thin-film flow studies.

- *Moderate Reynolds number flow*: $\varepsilon Re \ll 1$ and $Re \gg \varepsilon^2$, δ at most of $\mathcal{O}(1)$.

  Recall that for the second-order boundary layer equations we have $\varepsilon Re \ll 1$ and $Re \gg \varepsilon^2$ (while for the first-order ones, εRe at most of $\mathcal{O}(1)$ and $Re \gg \varepsilon$). At the same time, the x component of the momentum equation of the boundary layer approximation in terms of the Shkadov scaling imposes that δ must be at most of $\mathcal{O}(1)$, or with $\delta \sim Re/\kappa \sim Re/We^{1/3} \sim \varepsilon^{2/3}Re$, where $We = \mathcal{O}(\varepsilon^{-2})$, Re is at most of $\mathcal{O}(\varepsilon^{-2/3})$ (which automatically satisfies the condition $\varepsilon Re \ll 1$ for the second-order boundary layer equations).

  This is the drag-inertia regime, which is accurately described by the boundary layer equations. In this regime it is still possible to reduce by one the number of dimensions of the boundary layer equations while keeping a model valid for δ at most of $\mathcal{O}(1)$. This is achieved by integrating the boundary layer equations across the layer following the Kármán–Pohlhausen averaging method in boundary layer theory in aerodynamics. As mentioned in the Introduction, in the context of the falling film problem, this method was introduced by Kapitza and Shkadov [140, 248]. The Kapitza–Shkadov's approach was further improved by Ruyer-Quil and Manneville [226–228] who combined the gradient expansion with a high order, weighted residuals approach. The main advantage of this methodology is that the resulting models agree with the exact linear behavior obtained from Orr–Sommerfeld and they also describe properly the nonlinear dynamics in a wide range of parameter values. This methodology will be discussed in detail in Chap. 6 and used in subsequent chapters.

In the case of high Reynolds number flow, $\varepsilon Re \gg 1$, $\delta \gg 1$, inertia effects become predominant and the flow is turbulent. The film now is so thick that surface tension should play a secondary role. The boundary layer equations fail to describe the wave dynamics in this regime and full-scale Navier–Stokes and Fourier equations must be considered. This regime is beyond the scope of this monograph.

Figure 4.9 also suggests that each level of simplification can be considered at first or second order:

- The $\mathcal{O}(\varepsilon)$-*models*, in which the viscous dispersion effects are not included or equivalently $\eta = 0$.
- The $\mathcal{O}(\varepsilon^2)$-*models*, in which the viscous dispersion effects are included or equivalently $\eta \neq 0$.

Finally, it should be emphasized that in the different levels of approximations we have implicitly assumed that εPe is at most of $\mathcal{O}(1)$, the maximum order on the right hand side of the energy equation (4.2e) (or in terms of the Shkadov scaling, δPr is at most of $\mathcal{O}(1)$). When εPe is large, the coupling between the velocity and the temperature fields may be strong enough to violate the approximation of small temperature gradients, $\partial_x T$, $\partial_z T \sim \varepsilon$; for instance, in this case thermal boundary layers might develop at the front stagnation point of a solitary pulse [279]. This limitation will be discussed in Chap. 9 when we address the dynamics of nonisothermal films.

4.10 Scalings: Three Sets of Parameters

Three scalings and corresponding dimensionless parameters have been introduced. They are formally equivalent although each of them has some advantages and drawbacks which have been discussed in previous chapters and are summarized below. For simplicity we focus on the ST case:

- *Viscous-gravity scaling*: $\{h_N, Ct, \Gamma, Ma, Bi, Pr\}$.

  This scaling expresses the balance between viscosity and gravity and is based on the length scale l_ν and the time scale t_ν. The Reynolds number appears implicitly through h_N, $Re = h_N^3/3$ (see (2.37)). Hence, for a fixed gas–liquid–solid system and β, the equations are effectively free of parameters, which is rather convenient, and the only free parameter is the Reynolds number (through the inlet condition), which is quite useful for the physical interpretation of the results and comparisons with experiments, where the film thickness is typically modified by changing the flow rate. In addition to the flow rate, any other physical parameters with which one might control an experiment, i.e., the temperature difference between the wall and the ambient gas phase, appears in just a single dimensionless group, also rather convenient from the point of view of physical interpretation of the results. The viscous-gravity scaling, however, has the drawback that the Nusselt flat film thickness h_N appears in the boundary condition $h \to h_N$ far from a local surface deformation, such as a solitary hump, which also corresponds to the inlet boundary condition. Hence, for numerical purposes there is a need for another scaling in which the film thickness is scaled out of the boundary conditions and the Nusselt flat film solution is fixed, thus allowing useful comparisons to be made.

- *Nusselt scaling*: $\{Re, Ct, We, M, B, Pr\}$.

  This scaling eliminates the drawback of the viscous-gravity scaling just discussed. It is based on h_N and it coincides with the viscous-gravity scaling for a film of thickness $\bar{h}_N$ equal to l_ν, i.e., for $h_N = 1$. The Nusselt scaling has the advantage that it explicitly scales out h_N from the equations of motion and energy and wall and free-surface boundary conditions. However, its drawback is that all parameters depend on the flow rate. Nevertheless, the parametrization of the Nusselt groups (2.35)–(2.41) is in terms of h_N and the parameters obtained from the viscous-gravity scaling, so that the Nusselt scaling distinguishes clearly between the flow and the properties of the gas–liquid–solid system and β.

- *Shkadov scaling*: $\{\delta, \zeta, \eta, \mathcal{M}, B, Pr\}$.

  As far as the study of nonlinear waves is concerned, this scaling locates the transition between the drag-gravity and drag-inertia regimes at $\delta \simeq 1$. In addition, the Shkadov scaling makes apparent the balance of all forces—i.e., inertia, gravity, viscosity and surface tension—necessary to sustain strongly nonlinear waves. The reason why this balance is not apparent in the Navier–Stokes equations is because the surface tension effect necessary for the balance is in the normal stress condition and is coupled to the Navier–Stokes equations through the pressure and velocity fields. This is precisely the reason why the Shkadov scaling is only defined in the context of the boundary layer approximation. At the same time, for the generic problem of an isothermal film ($\mathcal{M} = 0$) falling down a vertical wall ($\zeta = 0$) and by neglecting the viscous dispersion effects ($\eta = 0$), the Shkadov scaling brings into the single parameter δ all the effects crucial for the existence of small- or large-amplitude solitary waves. Further, it gathers all second-order viscous and thermal effects in the boundary layer equations under the parameter η. Hence, the truncation of these equations at first order reduces the number of independent parameters by one. Finally, a technical advantage of this scaling is that its parameters retain values close to unity in the region of moderate Reynolds numbers and for large Weber numbers, which is rather useful from the point of view of convergence of numerical schemes.

The Nusselt scaling is used for full Navier–Stokes and Fourier equations while the Shkadov scaling is appropriate for the boundary layer equations and hence for the averaged models to be developed from these equations in Chap. 6. The long-wave theory developed in Chap. 5 is first given in terms of the Nusselt scaling, as is often the case in the literature, and subsequently in terms of the Shkadov scaling.

Appendix D.3 summarizes the relations between the different sets of parameters for the ST case.

Chapter 5
Methodologies for Low-Reynolds Number Flows

We outline the methodologies to model the dynamics of a falling film for low flow rates, i.e., low Reynolds numbers, corresponding to the drag-gravity regime. The classical modeling approach for such flows is the long-wave theory based on a gradient expansion of the governing equations and wall and free-surface boundary conditions with respect to a small parameter ε measuring the slow variations of the free surface in time and space. The theory typically leads to a single evolution equation for the film thickness, frequently referred to as the BE (Benney equation). Weakly nonlinear expansions of this equation lead to different prototypes such as the KS and the Kawahara equations. These are followed by an extensive study of the validity of the BE in the entire parameter space. It is shown that in certain regions of the parameter space unbounded solutions occur. A Padé-like regularization method is subsequently developed to cure this deviant behavior.

5.1 Long-Wave Theory

In the drag-gravity regime, neither inertia nor surface tension changes the Nusselt flow structure significantly. Therefore, as long as the flow rate is low and in the general framework of the strong surface tension limit, the slope of the interface remains "smooth" at the scale of the film thickness $h(x, t)$. This is the basis of the gradient expansion defined in Sect. 4.1, which consists of the introduction of a small parameter ε through the transformation (4.1), followed by an asymptotic series expansion of all pertinent variables in powers of ε.

We then utilize the gradient expansion and seek the solution for the different variables in the form

$$
\begin{aligned}
u &= u^{(0)} + \varepsilon u^{(1)} + \varepsilon^2 u^{(2)} + \mathcal{O}(\varepsilon^3), \\
v &= v^{(0)} + \varepsilon v^{(1)} + \varepsilon^2 v^{(2)} + \mathcal{O}(\varepsilon^3), \\
w &= w^{(0)} + \varepsilon w^{(1)} + \varepsilon^2 w^{(2)} + \mathcal{O}(\varepsilon^3), \\
p &= p^{(0)} + \varepsilon p^{(1)} + \varepsilon^2 p^{(2)} + \mathcal{O}(\varepsilon^3), \\
T &= T^{(0)} + \varepsilon T^{(1)} + \varepsilon^2 T^{(2)} + \mathcal{O}(\varepsilon^3),
\end{aligned}
\tag{5.1}
$$

S. Kalliadasis et al., *Falling Liquid Films*, Applied Mathematical Sciences 176, DOI 10.1007/978-1-84882-367-9_5, © Springer-Verlag London Limited 2012

where the zeroth-order solution should correspond to the base state solution as shown below. Noteworthy is that time does not appear explicitly in (5.1). Instead, the different fields in (5.1) depend on time through the dependence of the free surface h on time, i.e., they are *adiabatically slaved* to h. This in turn is indicative of the adiabatic elimination of the short-time dynamics necessary to establish the gravity-viscosity balance. This balance occurs at the short viscous time scale, $t_{\mathrm{vis}} = \bar{h}_{\mathrm{N}}^2/\nu$, so that the base flow is fully developed before it undergoes any instability (see also Sect. 2.3). The first perturbations due to inertia are introduced at first-order in ε and occur in the long inertia time scale, $t_{\mathrm{ine}} \sim (\bar{h}_{\mathrm{N}}/\varepsilon)/\bar{u}_{\mathrm{N}}$.

Before we explore the long-wave theory, it is useful to recall the following points from Sect. 4.1:

- The film parameter $\bar{h}_{\mathrm{N}}/l$ and ε are of the same order only for $Re - Re_{\mathrm{c}} = \mathcal{O}(1)$; this point will be addressed in detail in Sect. 5.1.4. Much like with Chap. 4 we can simply proceed with the perturbation expansion without assigning a relative order between k and ε.
- The length scale l, or equivalently the wavenumber $k = 2\pi\bar{h}_{\mathrm{N}}/l$ of the perturbations, is unknown a-priori and can only be obtained after the interface is actually constructed by solving the problem.

To proceed we assume that with the exception of the Weber number all parameters, Re, Pe, M, B, B_{w}, Ct, are of $\mathcal{O}(1)$. The Weber number is taken to be of $\mathcal{O}(\varepsilon^{-2})$, corresponding to the strong surface tension limit. We then substitute (5.1) into the system of equations (4.2). At leading order in ε, the system reduces to

$$\partial_{yy}u^{(0)} = -1, \tag{5.2a}$$

$$\partial_y p^{(0)} = -Ct, \tag{5.2b}$$

$$\partial_{yy}w^{(0)} = 0, \tag{5.2c}$$

$$\partial_{yy}T^{(0)} = 0, \tag{5.2d}$$

$$\partial_y v^{(0)} + \partial_x u^{(0)} + \partial_z w^{(0)} = 0, \tag{5.2e}$$

at $y = 0$:

$$u^{(0)} = v^{(0)} = w^{(0)} = 0, \tag{5.2f}$$

$$\mathrm{ST:} \quad T^{(0)} = 1 \tag{5.2g}$$

or

$$\mathrm{HF:} \quad \partial_y T^{(0)} - B_{\mathrm{w}}T^{(0)} = -1, \tag{5.2h}$$

at $y = h$:

$$v^{(0)} - u^{(0)}\partial_x h - w^{(0)}\partial_z h = \partial_t h, \tag{5.2i}$$

$$p^{(0)} = -\varepsilon^2 We(\partial_{xx}h + \partial_{zz}h), \tag{5.2j}$$

$$\partial_y u^{(0)} = 0, \tag{5.2k}$$

$$\partial_y w^{(0)} = 0, \tag{5.2l}$$

$$\partial_y T^{(0)} + BT^{(0)} = 0, \tag{5.2m}$$

whose solution reads

$$u^{(0)} = h^2 \bar{y}\left(1 - \frac{\bar{y}}{2}\right), \tag{5.3a}$$

$$v^{(0)} = -h^2 \frac{\bar{y}^2}{2} \partial_x h, \tag{5.3b}$$

$$w^{(0)} = 0, \tag{5.3c}$$

$$p^{(0)} = Cth(1 - \bar{y}) - \varepsilon^2 We(\partial_{xx}h + \partial_{zz}h), \tag{5.3d}$$

$$\text{ST:} \quad T^{(0)} = \frac{1 + Bh(1 - \bar{y})}{1 + Bh}, \tag{5.3e}$$

or

$$\text{HF:} \quad T^{(0)} = \frac{1 + Bh(1 - \bar{y})}{B + B_{\mathrm{w}}(1 + Bh)}, \tag{5.3f}$$

the Nusselt flat film flow and where the reduced coordinate $\bar{y} = y/h$ has been introduced. The zeroth-order approximations for the film surface temperature (at $\bar{y} = 1$), which we will need later, are:

$$\text{ST:} \quad \theta^{(0)} = \frac{1}{1 + Bh}, \tag{5.4a}$$

or

$$\text{HF:} \quad \theta^{(0)} = \frac{1}{B + B_{\mathrm{w}}(1 + Bh)}. \tag{5.4b}$$

We now turn to the kinematic boundary condition (4.2i) or its integral version[1]

$$\partial_t h + \partial_x q + \partial_z p = 0, \tag{5.5}$$

where $q = \int_0^1 u\, d\bar{y}$ and $p = \int_0^1 w\, d\bar{y}$ are the streamwise and spanwise components of the flow rate. It is important to emphasize the role of this condition in the calculation procedure as compared to the other equations. It acts effectively as some

---

[1] This equation is also obtained by integrating the continuity equation (4.2a) across the film, using (4.2f) and (4.2i) as boundary conditions.

type of "solvability condition" at each level of the expansion. As a matter of fact, the solution for q and p (or equivalently u and w) at each level of the expansion is a functional of h and its successive space and time derivatives, which are still not related to each other. Once we obtain the solution for q and p (or equivalently u and w) at a given order, we insert it in (5.5), which then becomes a constraint relating h and its successive space and time derivatives or equivalently an *evolution equation* for h. This equation is formally one order higher than the solution found.

At leading order, (5.2i) furnishes the following nontrivial relation,

$$\partial_t h + h^2 \partial_x h = 0, \tag{5.6}$$

which describes the downwards propagation of waves driven by gravity at the film surface to leading order. (Because the heat transfer and the mechanical equilibrium of the flat film are decoupled to leading order, (5.6) does not involve the Marangoni effect. This effect appears at first order in the tangential stress conditions (4.2k), (4.2l).) Since there is no spanwise contribution at this order, the flow is purely two-dimensional. The equation can be contrasted to the "Burgers equation" [299]:

$$\partial_t h + \alpha h \partial_x h = \upsilon \partial_{xx} h. \tag{5.7}$$

Equation (5.6) has a nonlinear propagation term as in the Burgers equation, with the difference that $\alpha = \alpha(h)$ is not constant, but not a diffusive term as in the Burgers equation. In fact, (5.6) can be viewed as a nonlinear wave equation with the coefficient h^2 of $\partial_x h$ playing the role of a nonlinear wave velocity.

However, the Burgers equation in the limit $\upsilon \to 0$ is known to produce shocks and hence steep gradients incompatible with the long-wave assumption in our problem. This can be easily seen from the nonlinear wave velocity $\sim h^2$. Consider a propagating wave solution of (5.6). Points with different heights travel with different velocities, e.g., the crest of the wave travels faster than both the front and the back of the wave. Eventually a shock will develop. The effect of "viscosity" υ in the Burgers equation is then to smooth out the discontinuity leading to a finite shock width. Similarly, in our case we need a mechanism to smooth out the discontinuity appearing at leading order of the expansion. Continuing the expansion is thus necessary for obtaining the required saturating terms. In the strong surface tension limit, the dominant higher-order effects, which in the nonlinear regime should prevent the waves from forming shocks and breaking (and in such a way so as to ensure the validity of the long-wave assumption) are expected to arise from surface tension. Although, the inclusion of surface tension cannot a priori guarantee that the free surface will be well behaved in the nonlinear regime, we have no other choice really but to proceed to the next order of the expansion.

At first-order we obtain the following system:

$$\partial_{yy} u^{(1)} = 3Re\big(\partial_t u^{(0)} + u^{(0)}\partial_x u^{(0)} + v^{(0)}\partial_y u^{(0)} + w^{(0)}\partial_z u^{(0)}\big) + \partial_x p^{(0)}, \tag{5.8a}$$

$$\partial_y p^{(1)} = \partial_{yy} v^{(0)}, \tag{5.8b}$$

$$\partial_{yy} w^{(1)} = 3Re\big(\partial_t w^{(0)} + u^{(0)}\partial_x w^{(0)} + v^{(0)}\partial_y w^{(0)} + w^{(0)}\partial_z w^{(0)}\big) + \partial_z p^{(0)}, \tag{5.8c}$$

$$\partial_{yy}T^{(1)} = 3Pe\big(\partial_t T^{(0)} + u^{(0)}\partial_x T^{(0)} + v^{(0)}\partial_y T^{(0)} + w^{(0)}\partial_z T^{(0)}\big), \qquad (5.8\text{d})$$

$$\partial_y v^{(1)} + \partial_x u^{(1)} + \partial_z w^{(1)} = 0, \qquad (5.8\text{e})$$

at $\bar{y} = 0$:

$$u^{(1)} = v^{(1)} = w^{(1)} = 0, \qquad (5.8\text{f})$$

$$\text{ST:}\quad T^{(1)} = 0 \qquad (5.8\text{g})$$

or

$$\text{HF:}\quad \partial_y T^{(1)} - B_{\mathrm{w}} T^{(1)} = 0, \qquad (5.8\text{h})$$

at $\bar{y} = 1$:

$$v^{(1)} - u^{(1)}\partial_x h - w^{(1)}\partial_z h = 0, \qquad (5.8\text{i})$$

$$p^{(1)} = 2\big(\partial_y v^{(0)} - \partial_x h\partial_y u^{(0)} - \partial_z h\partial_y w^{(0)}\big), \qquad (5.8\text{j})$$

$$\partial_y u^{(1)} = -M\big(\partial_x T^{(0)} + \partial_x h\partial_y T^{(0)}\big), \qquad (5.8\text{k})$$

$$\partial_y w^{(1)} = -M\big(\partial_z T^{(0)} + \partial_z h\partial_y T^{(0)}\big), \qquad (5.8\text{l})$$

$$\partial_y T^{(1)} + BT^{(1)} = 0. \qquad (5.8\text{m})$$

The left hand sides of (5.8a)–(5.8e) and (5.2a)–(5.2m) are identical, and it turns out that this happens to all orders in ε. As such, instead of using the kinematic condition (5.2i), we could have used the standard *Fredhölm alternative* as suggested in [290]. In short, the Fredhölm alternative states that in order to solve the differential equation $\mathcal{L}F = G$ with $\mathcal{L}$ a linear differential operator with a null space, and hence to invert the singular operator $\mathcal{L}$, the right hand side of the equation G must have no components in the null space of the adjoint operator of $\mathcal{L}$ [96, 123]. This notion can be readily generalized to matrix-differential operators as is the case here.

The solution of the system at first order reads:

$$u^{(1)} = h^2 \bar{y}\left(1 - \frac{\bar{y}}{2}\right)\big[\varepsilon^2 We(\partial_{xxx}h + \partial_{xzz}h) - Ct\partial_x h\big]$$

$$\hspace{2cm} - Mh\bar{y}\partial_x\theta^{(0)} + Reh^5\partial_x h\left(\frac{\bar{y}^4}{8} - \frac{\bar{y}^3}{2} + \bar{y}\right), \qquad (5.9\text{a})$$

$$w^{(1)} = h^2 \bar{y}\left(1 - \frac{\bar{y}}{2}\right)\big[\varepsilon^2 We(\partial_{xxz}h + \partial_{zzz}h) - Ct\partial_z h\big] - Mh\bar{y}\partial_z\theta^{(0)},$$

$$\hspace{12cm} (5.9\text{b})$$

$$p^{(1)} = -h\partial_x h(\bar{y} + 1), \qquad (5.9\text{c})$$

$$\text{ST:} \quad T^{(1)} = \frac{3Pe\,Bh^4 \partial_x h}{(1+Bh)^2} \left[-\frac{Bh}{40} \bar{y}^5 + \frac{1+3Bh}{24} \bar{y}^4 - \frac{Bh}{6} \bar{y}^3 \right.$$

$$\left. - \frac{10 - Bh(5+4Bh)}{60(1+Bh)} \bar{y} \right], \tag{5.9d}$$

or

$$\text{HF:} \quad T^{(1)} = \frac{3Pe\,B\,B_{\mathrm{w}} h^4 \partial_x h}{[B + B_{\mathrm{w}}(1+Bh)]^2} \left[-\frac{Bh}{40} \bar{y}^5 + \frac{1+3Bh}{24} \bar{y}^4 \right.$$

$$- \frac{B}{6} \left(h - \frac{1}{B_{\mathrm{w}}} \right) \bar{y}^3 - \frac{B}{2B_{\mathrm{w}}} \bar{y}^2 + \left(\bar{y} + \frac{1}{B_{\mathrm{w}} h} \right)$$

$$\left. \times \frac{30B - 10B_{\mathrm{w}} + Bh[5(4B + B_{\mathrm{w}}) + 4B\,B_{\mathrm{w}} h]}{60[B + B_{\mathrm{w}}(1+Bh)]} \right], \tag{5.9e}$$

where the film surface temperature $\theta^{(0)} = T^{(0)}|_{\bar{y}=1}$ has been introduced (see (5.4a)–(5.4b)). The first-order corrections of the surface temperature read:

$$\text{ST:} \quad \theta^{(1)} = \frac{Pe\,Bh^4 \partial_x h(7Bh - 15)}{40(1+Bh)^3} \tag{5.10a}$$

or

$$\text{HF:} \quad \theta^{(1)} = \frac{Pe\,Bh^3 \partial_x h[60B - 20B_{\mathrm{w}} + B_{\mathrm{w}} h(35B - 15B_{\mathrm{w}} + 7B\,B_{\mathrm{w}} h)]}{40[B + B_{\mathrm{w}}(1+Bh)]^3}, \tag{5.10b}$$

and the first-order contribution to the cross-stream velocity is found from the continuity equation, i.e., $v^{(1)} = -\int (\partial_x u^{(1)} + \partial_z w^{(1)})\,dy$, which is a fifth-order polynomial in $\bar{y}$. To obtain the above solutions, the time-derivatives $\partial_t h$ and $\partial_{xt} h$ have been replaced using the zeroth-order expression (5.6), i.e., $\partial_t h = [\partial_t h]^{(0)} + \varepsilon [\partial_t h]^{(1)} \equiv -h^2 \partial_x h + \mathcal{O}(\varepsilon)$, where the superscript indicates the order in the ε-expansion. This rule also applies for $\partial_t \theta^{(0)}$ since $\theta^{(0)}$ is slaved to the film thickness through (5.4a)–(5.4b). To obtain (5.9a)–(5.9e) and (5.10a)–(5.10b) the second-order terms introduced by $[\partial_t h]^{(1)}$ are dropped out and have to be included at next order (see Sect. 5.1.2).

5.1.1 The Evolution Equation for the Film Thickness

Substituting now $u^{(1)}$, $v^{(1)}$, $w^{(1)}$ into the kinematic condition (5.8i) yields the first-order evolution equation of the film thickness that in its full three-dimensional form reads:

$$\partial_t h + h^2 \partial_x h + \varepsilon Re \frac{2}{5} \partial_x \left(h^6 \partial_x h \right)$$

$$+ \varepsilon \nabla_{xz} \cdot \left[-Ct \frac{h^3}{3} \nabla_{xz} h - M \frac{h^2}{2} \nabla_{xz} \theta^{(0)} + \varepsilon^2 We \frac{h^3}{3} \nabla_{xz} \nabla_{xz}^2 h \right] = 0. \quad (5.11)$$

The second term in this equation is the convective term due to mean flow, the third term arises from inertia, the fourth term is due to the hydrostatic head in the direction perpendicular to the wall, the fifth term is due to the Marangoni effect and the sixth term contains the streamwise and spanwise curvature gradients associated with surface tension. We note that instead of starting from the full Navier–Stokes and Fourier equations and associated wall and free-surface boundary conditions, (5.11) can be also obtained by a gradient expansion of the boundary layer equations, or even from the averaged models that will be developed in Chap. 6 (which by construction they yield the long-wave evolution equation). Note also that ε can be scaled out of (5.11) much like we did with the boundary layer equations.

Equation (5.11) may be written in a "conservative form"[2] as

$$\partial_t h + \nabla_{xz} \cdot \mathbf{q} = 0,$$

where $\mathbf{q} = (q, p)$ is the local flow rate vector. Let us now consider a two-dimensional flow. The spanwise component of the flow rate then vanishes, i.e., $p = 0$. The streamwise component of the flow rate written first for the Burgers equation (5.7) reads, $q = (1/2)\alpha h^2 - \varepsilon \upsilon \partial_x h$. Comparison with the evolution equation (5.11) yields h-dependent coefficients in the form $\alpha(h) = (2/3)h$ and $\upsilon(h) = -(2/5)Reh^6 + (1/3)Cth^3 - (1/2)MBh^2/(1+Bh)^2$, where $\theta^{(0)}$ from (5.4a) has been used. Since $\upsilon(h)$ accounts for diffusive effects, its positive terms will be stabilizing while its negative ones will be destabilizing. In addition to the terms pertinent to the Burgers equation, the flow rate q contains a higher order "diffusive" term $\varepsilon \beta \partial_{xxx} h$ with $\beta = (1/3)\varepsilon^2 We h^3$. This term will still be stabilizing if positive. Based on this high-order Burgers equation prototype, we can easily ascertain the role of the different terms in (5.11): The third term that originates from inertia is responsible for the hydrodynamic instability (H-mode), the fourth one represents the stabilizing effects of the hydrostatic pressure, the fifth one is responsible for the thermocapillary instability (S-mode) and the last one accounts for the stabilizing effect of surface tension.

Notice that the Marangoni effect and associated S-mode are captured through the dependence of the flat film temperature at the free surface $\theta^{(0)}$ on the film thickness

---

[2]The form refers to an arrangement of an equation or system of equations in the form

$$\partial_t \mathbf{H} + \nabla \cdot \mathbf{Q}(\mathbf{H}) = \mathbf{R}(\mathbf{H}),$$

where $\mathbf{R}(\mathbf{H})$ is a "source" term and $\mathbf{Q}(\mathbf{H})$ is the "flux" associated with the quantity $\mathbf{H}$. The right hand side of the above form should not involve any first-order spatial derivatives; these should be contained in the flux term of the left hand side.

in (5.4a)–(5.4b). As emphasized in Sect. 2.5, in the case of HF and for an insulated wall ($B_w \to 0$), the dependence of the flat film temperature at the free surface $\theta^{(0)}$ on the film thickness h is lost (see (5.4b)) and the long-wave S-mode vanishes (which does not mean that the short-wave P-mode vanishes, but this mode is not captured in the gradient expansion framework).

The two-dimensional version ($\partial_z = 0$) of (5.11) in isothermal conditions ($M = 0$) has the form

$$\partial_t h + h^2 \partial_x h + \varepsilon \partial_x \left\{ \frac{2}{5} Re h^6 \partial_x h - Ct \frac{h^3}{3} \partial_x h + \varepsilon^2 We \frac{h^3}{3} \partial_{xxx} h \right\} = 0, \qquad (5.12)$$

an equation often called the BE (see Introduction). Though Benney [21] did develop the systematic long-wave expansion procedure that we used here to obtain (5.12), he omitted surface tension, which was included by Gjevik [102]. Even so, the term "BE" is quite generic in thin film studies and is often used to designate any long-wave evolution equation, either in two dimensions or three dimensions, used to describe the dynamics of thin films in different settings, as, e.g., the case with (5.11), which is the BE with the Marangoni effect (see, e.g., [128]).

The approach we have just outlined is known as *long-wave theory*. It is based on an asymptotic reduction of the governing equations and boundary conditions that convert the highly nonlinear boundary value problem that describes the evolution of the film to a sequence of solvable perturbation problems. The final result is a single nonlinear partial differential equation for the film thickness. Notice that unlike the boundary layer equations that are also derived with the gradient expansion but without overly restrictive stipulations on the orders of magnitude of the different dimensionless groups, the long-wave theory requires certain orders of magnitude assignments for the dimensionless groups.

For the sake of clarity and simplicity, we shall restrict the analysis for the remainder of this chapter to the ST case only.

5.1.2 Higher-Order Terms in the Gradient Expansion

For a two-dimensional flow ($w = 0$, $\partial_z = 0$), it can be shown by iteration that at each order n of the expansion the velocity field contribution $u^{(n)}$ can be written in the form of a polynomial in y, h and its derivatives $\partial_x^m h$, where $m = m(n)$ an integer. It can also be shown that for $n \geq 2$, the term of highest degree in y appearing in $u^{(n)}$ has a power $4n$ and originates from the inertia interactions between the Nusselt flow profile $u^{(0)}$ and its correction at order $n - 1$, $u^{(n-1)}$, via $\varepsilon^n 3Re(u^{(0)} \partial_x u^{(n-1)} + v^{(n)} \partial_y u^{(0)}) = \varepsilon^n 3Re(u^{(0)} \partial_x u^{(n-1)} - \partial_y u^{(0)} \int_0^y \partial_x u^{(n-1)} \, dy)$. Further, it can be seen that, if c_n is the coefficient of the term y^{4n} in $u^{(n)}$, then

$$c_{n+1} = -\frac{3(4n - 1)}{2(4n + 1)(4n + 3)(4n + 4)} c_n \quad \text{for } n \geq 1,$$

so that $c_2 = -9/4480$, $c_3 = 1/506880$, etc., demonstrating that the contributions of these highest-degree terms become quickly negligible in the evolution equation for h at order n, $\partial_t h + \partial_x(q^{(0)} + \cdots + \varepsilon^n q^{(n)}) = 0$.

As shown in Sect. 4.1, the inertia terms in the y-component of the Navier–Stokes equation should be neglected in order to obtain the underlying boundary layer equations. It is therefore not possible in the context of the boundary layer approximation to proceed up to third order at which the elimination of the pressure from the governing equation is no longer permitted. It is also important to be able to contrast the long-wave theory with the boundary layer equations. For these reasons, we shall proceed with the asymptotic expansion only up to second order. The second-order system is identical to (5.8a)–(5.8e) incremented by one in the expansion order and such that the left hand side will involve the unknown second-order contributions and the right hand side the known first-order contributions (5.9a)–(5.9e). Because the algebra at this order is cumbersome, only the results for a two-dimensional flow are presented.

The second-order BE reads

$$\partial_t h + \partial_x\left(q^{(0)} + \varepsilon q^{(1)} + \varepsilon^2 q^{(2)}\right) = 0, \tag{5.13}$$

where the different terms of the flow rate $q(x,t)$ are given by:

$$q^{(0)} = \frac{1}{3}h^3, \tag{5.14a}$$

$$q^{(1)} = \left(\frac{2}{5}Reh^6 - \frac{1}{3}Cth^3 + \frac{MBh^2}{2(1+Bh)^2}\right)\partial_x h + \frac{1}{3}\varepsilon^2 Weh^3 \partial_{xxx}h, \tag{5.14b}$$

$$\begin{aligned}
q^{(2)} = \Bigg\{ & \frac{7}{3}h^3 - \frac{8}{5}CtReh^6 + \frac{127}{35}Re^2h^9 + \frac{BMh^5}{(1+Bh)^4}\left[\frac{3}{4}Pe + \frac{33}{20}Re\right. \\
& + Bh\left(\frac{1}{840} - \frac{1}{4}Pe + \frac{131}{70}Re\right) + B^2h^2\left(\frac{1}{840} - \frac{7}{40}Pe + \frac{31}{140}Re\right)\Bigg]\Bigg\}(\partial_x h)^2 \\
& + \Bigg\{h^4 - \frac{10}{21}CtReh^7 + \frac{4}{7}Re^2h^{10} + \frac{BMh^6}{(1+Bh)^3}\left[-\frac{1}{1680} + \frac{3}{16}Pe + \frac{5}{7}Re\right. \\
& + Bh\left(-\frac{1}{1680} - \frac{7}{80}Pe + \frac{5}{7}Re\right)\Bigg]\Bigg\}\partial_{xx}h + Re\varepsilon^2 We\left(\frac{10}{21}h^7\partial_{xxxx}h\right. \\
& + \frac{12}{5}h^6(\partial_{xx}h)^2 + 4h^6\partial_x h\partial_{xxx}h + \frac{24}{5}h^5(\partial_x h)^2\partial_{xx}h\Bigg).
\end{aligned} \tag{5.14c}$$

Noteworthy is that the surface tension term that involves the third derivative of h, $\varepsilon^2 Weh^3\partial_{xxx}h$, though it appears of $\mathcal{O}(\varepsilon^3)$, is included in the first-order expression for the flow rate (5.14b), a consequence of the stipulation $We = \mathcal{O}(\varepsilon^{-2})$. As with the linear stability analysis and derivation of the boundary layer equations, one of the reasons for this order of magnitude assignment for We is convenience and simplicity.

Surface tension appears at the leading-order pressure distribution (5.3d), which in turn brings the surface tension effects in $q^{(1)}$ given by (5.14b), i.e., at $\mathcal{O}(\varepsilon)$, and one notices that $q^{(1)}$ is in fact a much simpler expression than $q^{(2)}$ in (5.14c).

Of course there is also the contribution of the second-order surface tension effects in the last row of $q^{(2)}$ in (5.14c) (four terms in total) and in fact the first of these terms contains the fourth derivative $\partial_{xxxx}h$, which in turn gives a fifth derivative $\partial_{xxxxx}h$ in (5.13), i.e., it increases the order of the differential equation (5.12) already of fourth order by one, which in practice then makes the numerical analysis of the second-order BE cumbersome. However, it is possible to simplify its numerical analysis by assuming $We = \mathcal{O}(\varepsilon^{-1})$: interestingly, the term $Weh^3\partial_{xxx}h$ in $q^{(1)}$ then moves to $q^{(2)}$ and the four surface tension terms in $q^{(2)}$ move up to $q^{(3)}$, so that $\varepsilon Weh^3\partial_{xxx}h$ is the only surface tension term that remains in (5.13).

Reducing the order of We further moves the term $Weh^3\partial_{xxx}h$ to even higher orders of the flow rate, i.e., with $We = \mathcal{O}(1)$, $Weh^3\partial_{xxx}h$ moves up to $q^{(3)}$, which is rather lengthy and is not given here. However, now the term $\partial_{xxx}h$ is multiplied by other terms as well, in addition to Weh^3. These terms are due to hydrostatic head and inertia so that for $We = \mathcal{O}(1)$ surface tension is not the only force that prevents the waves from breaking. Recall also from the long-wave expansion of the Orr–Sommerfeld dispersion relation in Sect. 3.5.5 that for $We = \mathcal{O}(1)$ in addition to surface tension, hydrostatic head and inertia also limit the growth rate of short waves.

5.1.3 Primary Instability for the First-Order BE

The linear stability analysis of the Nusselt flat film flow by using the first order BE (5.12) is done in the same way as in Sect. 3.1: A perturbation of the base-state film thickness is imposed in the form of the normal mode,

$$h = 1 + \varsigma \exp\{i(kx - \omega t)\}, \tag{5.15}$$

where ς and k are both real and represent the amplitude and the wavenumber of the perturbation, respectively, while ω is the complex angular frequency. Inserting the normal mode representation (5.15) into (5.11) and linearizing for $\varsigma \ll 1$ yields the linear phase speed and growth rate:

$$c \equiv \frac{\omega_{\mathrm{r}}}{k} = 1, \tag{5.16}$$

$$\omega_{\mathrm{i}} = \varepsilon\left[\left(\frac{2}{5}Re - \frac{1}{3}Ct\right)k^2 + \frac{1}{2}\frac{BM}{(1+B)^2} - \frac{1}{3}\varepsilon^2 Wek^4\right]. \tag{5.17}$$

The perturbation will grow for $\omega_{\mathrm{i}} > 0$, i.e., for perturbation wavenumbers smaller than the cut-off wavenumber:

$$k_{\mathrm{c}} = \frac{1}{(\varepsilon^2 We)^{1/2}}\left(\frac{6}{5}Re - Ct + \frac{3}{2}\frac{BM}{(1+B)^2}\right)^{1/2}, \tag{5.18a}$$

from which the criticality condition is easily obtained to be

$$\frac{6}{5}Re - Ct + \frac{3}{2}\frac{B\,M}{(1+B)^2} = 0.$$ (5.18b)

We have hence resolved the primary instability of the Nusselt flat film solution using the long-wave theory. This instability corresponds to a *Hopf bifurcation* from the Nusselt flat film solution. The emerging branch of oscillatory solutions will be called *supercritical (subcritical)* if it bifurcates toward the region where $k < k_c$ ($k > k_c$), respectively (the reader should consult one of the numerous texts on bifurcation theory, e.g., [111], but only basic elements of this theory are required for this monograph).

Equation (5.17) can be written in general form as, $\omega_i(k, \mu) = \mu k^2 - k^4 + \mathcal{O}(k^6)$ by suitably rescaling time and k; μ denotes the "control parameter." It is then not difficult to see that $\omega_i(0, \mu) = 0 \; \forall \mu$ while $\omega_i(k, \mu)$ remains small at small k. Such a long-wave mode is a Goldstone mode that has been already discussed in previous chapters and is linked with a particular conservation law (see Appendix C.5). It corresponds to a shift of the height of the interface (from one Nusselt solution to another one), which is neutrally stable as long as the shift is uniform. This explains that the axis $k = 0$ in the (μ, k)-plane (with $\omega_i = 0$) is always part of the neutral stability curve.

5.1.4 On the Relative Order Between the Wavenumber of Interfacial Disturbances and the Gradient Expansion Parameter

The transformation $\partial_t \to \varepsilon \partial_t \; \partial_x \to \varepsilon \partial_x$ in the gradient expansion is a crucial step in the development of the long-wave theory earlier in this chapter. This transformation is equivalent to the introduction of slow time and space variables, $\mathsf{T} = \varepsilon t$ and $\mathsf{X} = \varepsilon x$ or $\partial_t = \varepsilon \partial_\mathsf{T}, \partial_x = \varepsilon \partial_\mathsf{X}$, so that x, t are long scales, i.e., $x, t \sim \varepsilon^{-1}$ for $\mathsf{X}, \mathsf{T} \sim 1$. Notice that for purposes of clarity the symbols x, t are here reserved for the space and time variables prior to the introduction of the gradient expansion used to obtain the BE.

Let us now examine the linear stability of the second-order BE (5.13). For simplicity consider the isothermal case, $M = 0$. With $h = 1 + \varsigma \exp\{i(\mathsf{K}\mathsf{X} - \Omega \mathsf{T})\}$ where Ω is the complex angular frequency; again for purposes of clarity we use symbols different to ω and k which are here reserved for the complex angular frequency and wavenumber, respectively, prior to the introduction of the gradient expansion. The linear phase speed and growth rate are found to be

$$\mathsf{C} \equiv \frac{\Omega_r}{\mathsf{K}} = 1 + \varepsilon^2\left(1 - \frac{10}{21}ReCt + \frac{4}{7}Re^2\right),$$ (5.19)

$$\Omega_i = \varepsilon\left[\frac{2}{5}(Re - Re_c)\mathsf{K}^2 - \frac{1}{3}\varepsilon^2 We\mathsf{K}^4\right],$$ (5.20)

where $Re_c = (5/6)Ct$, as defined from the Orr–Sommerfeld analysis, see (3.31). Hence, the growth rate is identical to that in (5.17) obtained from the first-order BE, but the phase velocity has an $\mathcal{O}(\varepsilon^2)$ correction that makes it wavenumber-dependent, i.e., the $\mathcal{O}(\varepsilon^2)$ terms in the long-wave theory introduce dispersion at onset.

The cut-off wavenumber is

$$\mathsf{K}_c = \sqrt{\frac{6(Re - Re_c)}{5\varepsilon^2 We}}, \tag{5.21}$$

and clearly its order of magnitude is affected by the order of magnitude of $Re - Re_c$, i.e., by the distance from criticality. We distinguish between two specific cases here:

(i) $Re - Re_c = \mathcal{O}(1)$ which leads to $\mathsf{K}_c = \mathcal{O}(1)$. From $\mathsf{X} = \varepsilon x$, $1/\mathsf{K}_c = \varepsilon/k_c$ or $k_c = \varepsilon \mathsf{K}_c \sim \varepsilon$, consistent with the relative order $k \sim \varepsilon$ as one might have expected from the outset (see Sect. 4.1).

(ii) $Re - Re_c = \mathcal{O}(\varepsilon^2)$, a convenient order of magnitude assignment due to the square root dependence of K_c on $Re - Re_c$. Now $\mathsf{K}_c = \mathcal{O}(\varepsilon)$, which suggests the introduction of a second slow scale, $\chi = \varepsilon \mathsf{X}$ or $\partial_\mathsf{X} = \varepsilon \partial_\chi$. Equivalently, X is a long scale, i.e., $\mathsf{X} \sim \varepsilon^{-1}$ for $\chi \sim 1$.

Let us first examine if we can assume $Re - Re_c = \mathcal{O}(\varepsilon^2)$ in the linear stability of the second-order BE, i.e., if the neglected terms of $\mathcal{O}(\varepsilon^3)$ in (5.13) are indeed negligible in the linear regime described by the second-order BE when $Re - Re_c = \mathcal{O}(\varepsilon^2)$. With $h = 1 + \hat{h}$, the linearized third-order BE becomes [185]

$$\partial_T \hat{h} + \partial_\mathsf{X} \hat{h} + \frac{2}{5}\varepsilon(Re - Re_c)\partial_{\mathsf{XX}}\hat{h} + \frac{1}{3}\varepsilon^3 We \partial_{\mathsf{XXXX}}\hat{h}$$

$$+ \varepsilon^2\left(1 - \frac{10}{21}CtRe + \frac{4}{7}Re^2\right)\partial_{\mathsf{XXX}}\hat{h} + \varepsilon^3 C\partial_{\mathsf{XXXX}}\hat{h} = 0, \tag{5.22}$$

where C is a relatively lengthy function of Re and Ct. With $\mathsf{X} \sim \varepsilon^{-1}$, the destabilizing inertia and stabilizing surface tension terms in (5.22) are of the same order, $\varepsilon^5 \hat{h}$; this is to be expected as for large We the two terms represent the physical effects that determine the emerging pattern at onset—after all, the long scale ε^{-1} was suggested by $\mathsf{K}_c = \mathcal{O}(\varepsilon)$, which expresses the balance between inertia and surface tension. The order of magnitude of the term responsible for dispersion in (5.22) is $\varepsilon^5 \hat{h}$, and it balances the instability and stability terms—it is precisely this balance that will allow us to obtain the Kawahara equation from the second-order BE in Sect. 5.2.1. The order of magnitude of the last term in (5.22) is $\varepsilon^7 \hat{h}$, and it can be safely neglected. Hence, we can safely assume $Re - Re_c = \mathcal{O}(\varepsilon^2)$ in the linear stability of the second-order BE.

However, this is no longer the case with the first-order BE where the instability and stability terms are of $\mathcal{O}(\varepsilon^5 \hat{h})$ but the neglected terms responsible for dispersion in the linear regime are also of $\mathcal{O}(\varepsilon^5 \hat{h})$. This is to be expected since $Re - Re_c = \mathcal{O}(\varepsilon^2)$, i.e., very close to criticality, the wave amplitudes are small. We then need to go to a higher order BE to resolve small amplitudes.

Let us now consider the introduction of the second slow scale. From $\mathsf{X} = \varepsilon x$, $1/\mathsf{K}_c = \varepsilon/k_c$ or $k_c = \varepsilon \mathsf{K}_c \sim \varepsilon^2$. Thus, we can no longer stipulate $k = 2\pi(\bar{h}_\mathrm{N}/l) \sim \varepsilon$ from the outset. The relative order $k_c \sim \varepsilon^2$ is a direct consequence of $Re - Re_\mathrm{c} = \mathcal{O}(\varepsilon^2)$ and large We (as opposed to $Re - Re_\mathrm{c} = \mathcal{O}(1)$ and large We in case (i)). The waves now are much longer.

Setting $We = \mathcal{O}(\varepsilon^{-2})$, effectively using We to define ε and at the same time stating $k \sim \varepsilon$ from the outset, is equivalent to solving the linear problem and determining the dispersion relation before actually deriving the long-wave model (it is the dispersion relation that determines the wavelength selected by the system and hence the relative order between k and ε). Also from $\mathsf{K} \sim \varepsilon$ or $k_c \sim \varepsilon^2 \ll k \sim \varepsilon$, so that disturbances with $k \sim \varepsilon$ are stable. But it is incorrect to stipulate a scale $k \sim \varepsilon$ that is stable! After all, the characteristic scale must be that corresponding to an observable pattern, which can only be obtained from an instability.

This apparent paradox is resolved with the second scale mentioned above: $\partial_\mathsf{X} = \varepsilon \partial_\chi$ or $\partial_x = \varepsilon \partial_\mathsf{X} = \varepsilon^2 \partial_\chi$ and the two variables x and χ are connected through, $\chi = \varepsilon^2 x$. Therefore, if $\mathcal{K}$ is the wavenumber in the slow scale χ, $1/\mathcal{K}_c = \varepsilon^2/k_c$. But since $k_c \sim \varepsilon^2$, the cut-off wavenumber in the final slow scale is $\mathcal{K}_c \sim 1$, as in case (i).

In other words, ε^2 and not ε is now the appropriate small parameter. As a matter of fact, introducing from the outset the slow scales $\tau = \varepsilon^2 t$ and $\chi = \varepsilon^2 x$ for the gradient expansion, would give (5.13) but with $q^{(1)}$ multiplied by ε^2, We in $q^{(1)}$ multiplied by ε^4 and $q^{(2)}$ multiplied by ε^4. A linear stability now with $h = 1 + \varsigma \exp\{i(\mathcal{K}\chi - \xi\tau)\}$ would give for the growth rate,

$$\xi_i = \varepsilon^2 \left[\frac{2}{5}(Re - Re_\mathrm{c})\mathcal{K}^2 - \frac{1}{3}\varepsilon^4 We \mathcal{K}^4 \right],$$

which with $Re - Re_\mathrm{c} = \mathcal{O}(\varepsilon^2)$ yields $\mathcal{K} \sim 1$, as expected.

5.1.5 Comparison with Orr–Sommerfeld

We now compare the expression for the cut-off wavenumbers obtained from the BE with that obtained in Sect. 3.5 from a long-wave expansion of the Orr–Sommerfeld eigenvalue problem (see (3.35) with $M^\star = 0$). For both cases (i) and (ii) $k_c = \varepsilon \mathsf{K}_c$, so that from (5.21),

$$k_c = \sqrt{\frac{6(Re - Re_\mathrm{c})}{5We}}, \tag{5.23}$$

which is identical to that obtained from a long-wave expansion of the Orr–Sommerfeld problem.

For $M^\star \neq 0$ the expression for the cut-off wavenumber in (5.18a) obtained from the long-wave theory after appropriately transforming it is also identical to that obtained in Sect. 3.5 (see (3.35)) from a long-wave expansion of the Orr–Sommerfeld eigenvalue problem; hence it gives precisely the same critical condition with (3.31a) (see (5.18b)).

It is now evident why prechoosing a priori a precise relative order between k and ε when $We = \mathcal{O}(\varepsilon^{-2})$ (effectively used to define ε as pointed out earlier) can be deceptive. The exact dispersion relation of the problem, i.e., the one obtained from Orr–Sommerfeld, imposes certain orders between Re, We and k_c and in fact restricts the relative orders of these parameters: assuming $\partial_x \to \varepsilon \partial_x$ followed by a perturbation in ε (the two steps we have called "gradient expansion") leads to the BE whose dispersion relation suggests that for $Re - Re_c \sim \varepsilon^2$ we need an additional multiple-scale transformation, $\partial_x \to \varepsilon \partial_x (\to \varepsilon^2 \partial_x)$. Instead, considering from the outset the Orr–Sommerfeld analysis, more specifically the expression for the cut-off wavenumber from (3.35), leads to the conclusion that for the long-wave theory the most general small parameter should be $k_c (\sim \bar{h}_{\mathrm{N}}/l) \sim \sqrt{(Re - Re_c)/We} \sim \varepsilon\sqrt{Re - Re_c}$, since $We = \mathcal{O}(\varepsilon^{-2})$. This in turn suggests the introduction of a single slow scale only, $\partial_x = k_c \partial_X$, followed by a gradient expansion in k_c.

That we would have the same k_c obtained from the BE and a long-wave expansion of Orr–Sommerfeld could have been anticipated: (a) the long-wave theory is obtained from a regular perturbation expansion of the full Navier–Stokes and Fourier equations or the boundary layer equations (whose linear stability properties agree with those of the full Navier–Stokes and Fourier equations as emphasized in Sect. 4.8.2); (b) the instability onset occurs at wavenumber $k = 0$, as opposed to a finite k. With the same expression for k_c we also have the same criticality condition, i.e., the same value of Re_c which is automatically the case in (5.23) as Re_c is by definition the one found from Orr–Sommerfeld ($= (5/6)Ct$). The reverse is not necessarily true: for Re_c we need to have the same coefficient for k^2 in the growth rate, while for k_c we also need the same coefficient for k^4.

Actually the BE not only captures exactly the critical conditions but also the neutral conditions and interfacial quantities in the vicinity of criticality (but clearly, due to the underlying long-wave assumption it fails to describe accurately the dynamics within the film, e.g., development of convection cells due to the Marangoni effect). More specifically, the neutral stability curve obtained from the long-wave theory agrees close to criticality with that obtained from Orr–Sommerfeld, i.e., for $Re - Re_c$ up to an $\mathcal{O}(1)$ value, but deviates from the Orr–Sommerfeld one as Re increases further, i.e., for $Re - Re_c > 1$. By including higher-order terms in the BE, this deviation is progressively delayed to higher Re. However, the same is not true for interfacial quantities such as h, i.e., the validity of the BE as far as interfacial quantities is concerned cannot be extended far from criticality by including higher-order terms in the BE: indeed, we shall demonstrate later in this chapter that the long-wave theory effectively fails to describe the nonlinear dynamics of the film at an $\mathcal{O}(1)$ value of the Reynolds number.

We then declare that BE is *exact close to criticality* as far as critical, neutral conditions and interfacial quantities are concerned.

We close this section by noting that the long-wave theory fails to predict the shear modes obtained from the Orr–Sommerfeld eigenvalue problem. Their destabilization at very large Reynolds numbers underlines the intrinsic dynamics of these modes, which the long-wave theory fails to capture. There is a countable infinite number of such modes due to the y-dependence of the Orr–Sommerfeld eigenvalue problem. The long-wave theory eliminates this dependence by slaving all variables to h, yielding finally a single evolution equation and hence only a single mode (which in the presence of the Marangoni effect can be unfolded into two through the parametrization of the Nusselt scaling in terms of the viscous-gravity parameters).

5.2 Weakly Nonlinear Models

Weakly nonlinear prototypes represent a significant development in nonlinear dynamics and pattern formation. But in the context of the falling film problem, weakly nonlinear approaches impose certain a priori assumptions about the wave dynamics, the principal one being small-amplitude waves. As such, the resulting equations are only appropriate in a limited regime of the parameter space consistent with the imposed assumptions. However, if their physical relevance might be questionable, their elegance and simplicity are not. The latter make them amenable to mathematical and numerical scrutiny allowing us to decipher rapidly some of the falling film characteristics in the region of small-amplitude waves.

5.2.1 Models in Two Dimensions

The derivation of weakly nonlinear prototypes from the BE combines a weakly nonlinear expansion and multiple scale-type arguments similar to those utilized in Appendix C.5, for, e.g., the derivation of the *BKdV equation* in (C.16) from the generic amplitude equation in (C.15a)–(C.15e), and in Sect. 5.1.4, where we discussed the relative order of the gradient parameter ε and the wavelength of the perturbations.

Deriving weakly nonlinear models is, in general, considered to be a relatively simple process, but it involves some subtle points and must be done with care. In fact, the literature contains an abundance of such derivations in the context of the falling film and other thin film problems, but often one gets the impression of some uncertainty regarding the assumptions and precise conditions as well as regions in the parameter space where the different models are applicable (e.g., [44])

The first basic step in the derivation of weakly nonlinear models from the BE is the assumption of small-amplitude interfacial disturbances. We stipulate that the system is close to criticality, i.e., $Re - Re_c$ is small, or $Re - Re_c = \mathcal{O}(1)$. The second step is a weakly nonlinear expansion combined with a multiple-scale analysis. This procedure has the advantage that it retains only a single nonlinearity, which is the simplest possible nonlinearity for the system, i.e., the quadratic dominant nonlinearity $h\partial_x h$, which arises effectively from a nonlinear correction to the phase speed, a nonlinear kinematic effect that captures how larger waves moves faster than smaller ones. Thus, the weakly nonlinear expansion suppresses the strong nonlinearity introduced by inertia effects that is responsible for the unorthodox finite-time blow up behavior of the traveling wave solutions of the BE as detailed in Sect. 5.4.

5.2.1.1 Starting from the First-Order BE

For simplicity we assume an isothermal film. The derivation given here can then be easily extended to the nonisothermal case. We first obtain the simplest of the models, the KS equation. Consider the first-order BE (5.12) for a vertical plane, $Ct = 0$. Substitute $h \sim 1 + \varepsilon \tilde{h}$, $\xi = x - t$, $\tau = \varepsilon t$, utilize the chain rule $(\partial_t)_x = (\partial_\xi)_\tau (\partial_t \xi)_x + (\partial_\tau)_\xi (\partial_t \tau)_x \equiv \varepsilon (\partial_\tau)_\xi - (\partial_\xi)_\tau$, and neglect terms of $\mathcal{O}(\varepsilon^3)$ and higher to obtain

$$\varepsilon^2 \partial_\tau \tilde{h} - \varepsilon \partial_\xi \tilde{h} + \varepsilon \partial_\xi \tilde{h} + 2\varepsilon^2 \tilde{h} \partial_\xi \tilde{h} + \frac{2}{5} \varepsilon^2 Re \partial_{\xi\xi} \tilde{h} + \frac{1}{3} \varepsilon^4 We \partial_{\xi\xi\xi\xi} \tilde{h} = 0. \qquad (5.24)$$

Hence, the moving coordinate transformation $\xi = x - t$ allows us to remove the linear term $\varepsilon \partial_\xi \tilde{h}$ resulting from the weakly nonlinear expansion of the mean flow term $(1/3)\partial_x(h^3)$ in (5.12). Note that all terms in (5.24), time-dependent term, nonlinearity, inertia and surface tension, are of $\mathcal{O}(\varepsilon^2)$. Note also that with $h \sim 1 + \varepsilon \tilde{h}$ the neglected terms of $\mathcal{O}(\varepsilon^2)$ in the first-order BE (5.12) become of $\mathcal{O}(\varepsilon^3)$ and can be safely neglected compared to the retained terms of $\mathcal{O}(\varepsilon^2)$ in (5.24). Dividing now with ε^2 and proceeding to the following change of variables,

$$\xi = \sqrt{\frac{5\varepsilon^2 We}{6Re}} X; \qquad \tau = \frac{25\varepsilon^2 We}{12Re^2} T; \qquad \tilde{h} = \frac{1}{5}\sqrt{\frac{6Re^3}{5\varepsilon^2 We}} H,$$

we obtain the KS equation defined in Appendix C.5,

$$\partial_T H + H \partial_X H + \partial_{XX} H + \partial_{XXXX} H = 0, \qquad (5.25)$$

describing the weakly nonlinear evolution of the vertical falling film for $Re = \mathcal{O}(1)$ and strong surface tension, $We = \mathcal{O}(\varepsilon^{-2})$—recall that these are the orders of magnitude assignments under which the BE was derived.

It is worth noting that quite frequently in the literature, the KS equation in the context of the falling film problem is given with a coefficient of 4 for the nonlinearity. This is a direct consequence of the coefficient of 4 in (5.24), instead of a 2 there, which in turn is due to the different scaling for the velocity and hence time employed in several thin film studies, which use the interfacial velocity, $U_0 = g \bar{h}_N^2 \sin \beta / (2\nu)$, instead of that used here, i.e., $\bar{h}_N^2/(t_\nu l_\nu)$ (see (2.16b)) or $2U_0$. Hence in these studies the time scale is $\bar{h}_N/U_0$ instead of the one used here, $\bar{h}_N/2U_0$ (see (2.16a)), which then leads to a coefficient $2/3$ in front of $\partial_x(h^3)$ instead of $1/3$ here (see (5.12)). But in fact, any coefficient in front of the nonlinearity can be easily rescaled to unity, e.g., if the coefficient is 4 this is achieved with the transformation $H \to (1/4)H$.

Finally, it is noted that the slow time scale $\tau = \varepsilon t$ is consistent with the dispersion relation obtained from the full BE (5.20). Clearly, the conclusion in

Sect. 5.1.4 that for $Re - Re_c = \mathcal{O}(1)$, $\mathsf{K}_c = \mathcal{O}(1)$, applies for a vertical film as well ($Re_c = 0$ and $Re = \mathcal{O}(1)$). More precisely, $\mathsf{K}_c = \sqrt{6Re/(5\varepsilon^2 We)}$. In addition, $\mathsf{K}_{max} = \sqrt{3Re/(5\varepsilon^2 We)}$, also an $\mathcal{O}(1)$ quantity. Hence, the order of magnitude of the maximum growth rate is $(\Omega_i)_{max} \sim \varepsilon$ and the disturbances grow as $\hat{h} \sim \exp\{(\Omega_i)_{max}t\} \sim \exp\{\varepsilon t\}$, consistent with the introduction of the slow time scale εt.

5.2.1.2 Starting from the Second-Order BE

Let us now obtain the weakly nonlinear model for the inclined plane case. Once again for simplicity we assume an isothermal film. We utilize the second-order BE (5.13) rewritten here for clarity:

$$\partial_t h + \partial_x\left(q^{(0)} + \varepsilon q^{(1)} + \varepsilon^2 q^{(2)}\right) = 0, \tag{5.26}$$

where

$$q^{(0)} = \frac{1}{3}h^3, \tag{5.27a}$$

$$q^{(1)} = \left[\left(\frac{2}{5}Reh^6 - \frac{1}{3}Cth^3\right)\partial_x h + \frac{1}{3}\varepsilon^2 Weh^3\partial_{xxx}h\right], \tag{5.27b}$$

$$q^{(2)} = \left[\left(\frac{7}{3}h^3 - \frac{8}{5}CtReh^6 + \frac{127}{35}Re^2h^9\right)(\partial_x h)^2\right.$$

$$+ \left(h^4 - \frac{10}{21}CtReh^7 + \frac{4}{7}Re^2h^{10}\right)\partial_{xx}h$$

$$+ Re\varepsilon^2 We\left(\frac{10}{21}h^7\partial_{xxxx}h + \frac{12}{5}h^6(\partial_{xx}h)^2 + 4h^6\partial_x h\partial_{xxx}h\right.$$

$$\left.\left. + \frac{24}{5}h^5(\partial_x h)^2\partial_{xx}h\right)\right]. \tag{5.27c}$$

Now substitute into (5.26) $h \sim 1 + s\tilde{h}$ with $s \ll 1$ and approximate the different terms of this equation as follows:

$$\partial_x\left(h^3\right) = \partial_x\left(1 + 3s\tilde{h} + 3s^2\tilde{h}^2\right) + \mathcal{O}(s^3/x), \tag{5.28a}$$

$$\varepsilon\partial_x\left[\left(\frac{2}{5}Reh^6 - \frac{1}{3}Ct\right)h^3\partial_x h\right] = \varepsilon\left(\frac{2}{5}Re - \frac{1}{3}Ct\right)s\partial_{xx}\tilde{h} + \mathcal{O}(\varepsilon s^2/x^2), \tag{5.28b}$$

$$\frac{1}{3}\varepsilon^3 We\partial_x\left(h^3\partial_{xxx}h\right) = \frac{1}{3}\varepsilon^3 Wes\partial_{xxxx}\tilde{h} + \mathcal{O}(\varepsilon s^2/x^4), \tag{5.28c}$$

$$\varepsilon^2\partial_x\left[\left(\frac{7}{3}h^3 - \frac{8}{5}CtReh^6 + \frac{127}{35}Re^2h^9\right)(\partial_x h)^2\right] = \mathcal{O}(\varepsilon^2 s^2/x^3), \tag{5.28d}$$

$$\varepsilon^2 \partial_x \left[\left(h^4 - \frac{10}{21} CtRe h^7 + \frac{4}{7} Re^2 h^{10} \right) \partial_{xx} h \right]$$

$$= \varepsilon^2 \left(1 - \frac{10}{21} ReCt + \frac{4}{7} Re^2 \right) s \partial_{xxx} \tilde{h} + \mathcal{O}\left(\varepsilon^2 s^2 / x^3 \right), \tag{5.28e}$$

$$Re\varepsilon^4 We \partial_x \left(\frac{10}{21} h^7 \partial_{xxxx} h + \frac{12}{5} h^6 (\partial_{xx} h)^2 + 4h^6 \partial_x h \partial_{xxx} h + \frac{24}{5} h^5 (\partial_x h)^2 \partial_{xx} h \right)$$

$$= \frac{10}{21} Re\varepsilon^4 We s \partial_{xxxxx} \tilde{h} + \mathcal{O}\left(\varepsilon^2 s^2 / x^5 \right), \tag{5.28f}$$

where the orders of magnitude of the neglected terms take into account a possible order of magnitude assignment of x with respect to ε later on and hence the presence of x there. Note that the only nonlinearity is $\tilde{h} \partial_x \tilde{h}$ originating from the mean flow term $h^3 \partial_x h$. Note also that in the weakly nonlinear regime the second-order surface tension effects in (5.27c) contribute a fifth-order dispersion term.

We first consider the more involved case (ii) in Sect. 5.1.4, i.e., $Re - Re_c = \mathcal{O}(\varepsilon^2)$, with $Re_c = (5/6)Ct$. We balance the instability with the stability terms in (5.28b), (5.28c):

$$\varepsilon(Re - Re_c) s \partial_{xx} \tilde{h} \sim \varepsilon^3 We s \partial_{xxxx} \tilde{h} \quad \Rightarrow \quad x \sim \varepsilon^{-1},$$

i.e., for the two terms to balance x must be a long scale. To determine the order of magnitude of s, we balance the nonlinearity in (5.28a) with the third-order dispersion term in (5.28e):

$$s^2 \tilde{h} \partial_x \tilde{h} \sim \varepsilon^2 s \partial_{xxx} \tilde{h} \quad \Rightarrow \quad s = \varepsilon^4.$$

The order of magnitude of the instability, stability, nonlinearity and third-order dispersion terms then is

$$\varepsilon(Re - Re_c) s \partial_{xx} \tilde{h} \sim \varepsilon^9; \qquad \varepsilon^3 We s \partial_{xxxx} \tilde{h} \sim \varepsilon^9;$$

$$s^2 \tilde{h} \partial_x \tilde{h} \sim \varepsilon^9; \qquad \varepsilon^2 s \partial_{xxx} \tilde{h} \sim \varepsilon^9;$$

and all these terms balance, while the order of magnitude of the higher-order surface tension terms in (5.28f) is

$$\varepsilon^2 s \partial_{xxxxx} \tilde{h} \sim \varepsilon^{11}$$

and can be safely neglected. We can now confirm that the neglected terms in (5.28a)–(5.28f) are indeed negligible: $\mathcal{O}(s^3/x) = \mathcal{O}(\varepsilon^{13})$, $\mathcal{O}(\varepsilon s^2/x^2) = \mathcal{O}(\varepsilon^{11})$, $\mathcal{O}(\varepsilon s^2/x^4) = \mathcal{O}(\varepsilon^{13})$, $\mathcal{O}(\varepsilon^2 s^2/x^3) = \mathcal{O}(\varepsilon^{13})$ and $\mathcal{O}(\varepsilon^2 s^2/x^5) = \mathcal{O}(\varepsilon^{15})$ and can be safely neglected compared to the $\mathcal{O}(\varepsilon^9)$ nonlinearity, instability, stability and third-order dispersion terms. Also, the neglected terms of $\mathcal{O}(\varepsilon^3)$ in (5.26) are indeed negligible: from (5.22), the order of these terms is $\varepsilon^3 s/x^4 \sim \varepsilon^{11}$.

The time scale on which the time derivative term $\partial_t h$ in (5.26) balances the non-linearity, instability, stability and third-order dispersion terms can now be obtained as follows:

$$\partial_t h \sim s \partial_t \tilde{h} \sim \varepsilon^9 \quad \Rightarrow \quad t \sim \varepsilon^{-5},$$

which is a long time scale. The existence of this scale can also be easily confirmed from the expression for the growth rate in (5.20). With $K_c \sim \varepsilon$, the order of magnitude of the maximum growth rate is $(\Omega_i)_{\max} \sim \varepsilon^5$ and the disturbances grow with a rate $\hat{h} \sim \exp\{(\lambda_r)_{\max}t\} \sim \exp\{\varepsilon^5 t\}$, which in turn suggests the introduction of the slow time scale $\varepsilon^5 t$.

With $\chi = \varepsilon(x - t)$ and $\tau = \varepsilon^5 t$ then we obtain the weakly nonlinear model,

$$\varepsilon^9 \partial_\tau \tilde{h} - \varepsilon^5 \partial_\chi \tilde{h} + \varepsilon^5 \partial_\chi \tilde{h} + 2\varepsilon^9 \tilde{h} \partial_\chi \tilde{h} + \frac{2}{5}(Re - Re_c)\varepsilon^7 \partial_{\chi\chi}\tilde{h}$$

$$+ \frac{1}{3}\varepsilon^{11} We \partial_{\chi\chi\chi\chi}\tilde{h} + \varepsilon^9\left(1 - \frac{10}{21}ReCt + \frac{4}{7}Re^2\right)\partial_{\chi\chi\chi}\tilde{h} = 0, \quad (5.29)$$

and once again the moving coordinate transformation $x - t$ in the definition of the slow length scale allows us to remove the linear term $\varepsilon^5 \partial_\chi \tilde{h}$ resulting from the weakly nonlinear expansion of the mean flow term $(1/3)\partial_x(h^3)$. Note that the dispersive term in (5.29) is due to: (a) second-order viscous diffusive effects, which contribute the term $h^4 \partial_{xx}h$ in (5.27c) whose weakly nonlinear expansion yields a coefficient of unity for $\partial_{\chi\chi\chi}\tilde{h}$ in (5.29); (b) second-order inertia effects, which contribute the term $[-(10/21)ReCth^7 + (4/7)Re^2 h^{10}]\partial_{xx}h$ in (5.27c), whose weakly nonlinear expansion yields a coefficient of $-(10/21)ReCt + (4/7)Re^2$ for $\partial_{\chi\chi\chi}\tilde{h}$ in (5.29). However, $-(10/21)ReCt + (4/7)Re^2 = (4/7)Re(Re - Re_c)$ so that the contribution of the second-order inertia effects to dispersion is of order $\varepsilon^9(Re - Re_c)\partial_{\chi\chi\chi}\tilde{h} \sim \varepsilon^{11}$ and must be neglected since in (5.28a)–(5.28f) the neglected terms are of $\mathcal{O}(\varepsilon^{11})$ and higher. Hence, (5.29) is simplified to

$$\partial_\tau \tilde{h} + 2\tilde{h}\partial_\chi \tilde{h} + \frac{2}{5}\frac{Re - Re_c}{\varepsilon^2}\partial_{\chi\chi}\tilde{h} + \frac{1}{3}\varepsilon^2 We \partial_{\chi\chi\chi\chi}\tilde{h} + \partial_{\chi\chi\chi}\tilde{h} = 0, \quad (5.30)$$

which with the change of variables

$$\chi = \sqrt{\frac{5\varepsilon^4 We}{6(Re - Re_c)}}X; \qquad \tau = \frac{25\varepsilon^6 We}{12(Re - Re_c)^2}T; \qquad \tilde{h} = \frac{1}{5}\sqrt{\frac{6(Re - Re_c)^3}{5\varepsilon^8 We}}H,$$

i.e., the same with that used to derive (5.25) but with Re replaced with $(Re - Re_c)/\varepsilon^2$, yields the Kawahara equation defined in Appendix C.5,

$$\partial_T H + H\partial_X H + \partial_{XX}H + \delta_K \partial_{XXX}H + \partial_{XXXX}H = 0, \quad (5.31)$$

where

$$\delta_K = \sqrt{\frac{15}{2}\frac{1}{(Re - Re_c)We}}.$$

With $Re - Re_c = \mathcal{O}(\varepsilon^2)$ and $We = \mathcal{O}(\varepsilon^{-2})$, all coefficients in the above change of variables as well as δ_K are of $\mathcal{O}(1)$.

It is worth noting that in terms of the generic prototype (C.15a)–(C.15e) for a system with a conservation law considered in Appendix C.5, the above derivation is equivalent to $\mathcal{F}(\alpha) = \delta_1 \alpha^2$, $\gamma_{2,4} = \mathcal{O}(\epsilon)$, $\gamma_3 = \mathcal{O}(\epsilon^2)$ and $\Sigma - \Sigma_c = \mathcal{O}(\epsilon^2)$—these orders of magnitude assignments emulate the presence of the gradient parameter ε in (5.26) and the orders $Re - Re_c = \mathcal{O}(\varepsilon^2)$, $We = \mathcal{O}(\varepsilon^{-2})$. To see this, let us for the purposes of clarity rewrite (C.15a)–(C.15e) with $\mathcal{F}(\alpha) = \delta_1 \alpha^2$:

$$\partial_t \alpha = \delta_1 \partial_x (\alpha^2) + \gamma_2 (\Sigma - \Sigma_c) \partial_{xx} \alpha + \gamma_4 \partial_{xxxx} \alpha + \gamma_3 \partial_{xxx} \alpha.$$

Balancing instability with stability, $\gamma_2 (\Sigma - \Sigma_c)/x^2 \sim \gamma_4/x^4$ or $x \sim \epsilon^{-1}$. Balancing the nonlinearity with instability and stability, $\alpha \partial_x \alpha \sim \alpha \gamma_2 (\Sigma - \Sigma_c)/x^2$ or $\alpha \sim \epsilon^4$. The order of instability, stability and nonlinearity terms then is $\alpha \partial_x \alpha \sim \epsilon^9$ and that of the dispersion term, $\gamma_3 \partial_{xxx} \alpha \sim \epsilon^9$. All terms on the right hand side of the above equation are of the same order.

We now turn to the simpler case (i) in Sect. 5.1.4, i.e., $Re - Re_c = \mathcal{O}(1)$. Based on the above analysis we anticipate that the pertinent weakly nonlinear model is the KS equation. As a matter of fact, the situation here is identical to that of a vertical falling film with $Re = \mathcal{O}(1)$. Balancing the instability and stability terms in (5.28b), (5.28c) now gives $x \sim \mathcal{O}(1)$. With this order of magnitude assignment for x, balancing the nonlinearity with the instability and stability terms in (5.28a)–(5.28c) gives $s = \varepsilon$. The nonlinearity, instability and stability terms then are of $\mathcal{O}(\varepsilon^2)$ and all these terms balance, while the dispersion term in (5.28e) is of $\mathcal{O}(\varepsilon^3)$ so that dispersion now is a higher-order effect and can be neglected. The higher-order surface tension terms in (5.28f) are also of $\mathcal{O}(\varepsilon^3)$ and can be neglected. We can then readily confirm that the neglected terms in (5.28a)–(5.28f) are indeed negligible. Also, the neglected terms of $\mathcal{O}(\varepsilon^3)$ in (5.26) are indeed negligible. The KS equation is then obtained with the same change of variables used to derive (5.25) but with Re replaced with $Re - Re_c$.

In terms of the generic prototype (C.15a)–(C.15e) for a system with a conservation law considered in Appendix C.5, the above derivation of the KS equation is equivalent to $\mathcal{F}(\alpha) = \delta_1 \alpha^2$, $\gamma_{2,4} = \mathcal{O}(\epsilon)$, $\gamma_3 = \mathcal{O}(\epsilon^2)$ and $\Sigma - \Sigma_c = \mathcal{O}(1)$. Indeed, balancing instability with stability, $\gamma_2/x^2 \sim \gamma_4/x^4$ or $x \sim 1$. Balancing the nonlinearity with instability and stability, $\alpha \partial_x \alpha \sim \alpha \gamma_2/x^2$ or $\alpha \sim \epsilon$. The order of instability, stability and nonlinearity then is $\alpha \partial_x \alpha \sim \epsilon^2$, while that of dispersion is $\gamma_3 \partial_{xxx} \alpha \sim \epsilon^3$.

It is possible also to obtain the Kawahara equation in (5.31) but with $\delta_K \gg 1$. Balancing nonlinearity with third-order dispersion in (5.28a)–(5.28f) gives $s \sim \varepsilon^2/x^2$. With this order of magnitude assignment for s, balancing instability with stability in (5.28a)–(5.28f) gives $x \sim (Re - Re_c)^{-1/2}$. The order of the nonlinearity then is $\varepsilon^4 (Re - Re_c)^{5/2}$, of the instability and stability terms $\varepsilon^3 (Re - Re_c)^3$ and of the second-order surface tension terms $\varepsilon^4 (Re - Re_c)^{7/2}$. For the nonlinearity to balance the instability and stability terms, $\varepsilon^4 (Re - Re_c)^{5/2} \sim \varepsilon^3 (Re - Re_c)^3$ or $Re - Re_c \sim \varepsilon^2$, which then leads to the Kawahara equation with $\delta_K = \mathcal{O}(1)$ as above. On the other hand, for the nonlinearity to dominate over the instability and stability terms, $\varepsilon^4 (Re - Re_c)^{5/2} \gg \varepsilon^3 (Re - Re_c)^3$ or $Re - Re_c \ll \varepsilon^2$. Once again we can also confirm that the neglected terms in (5.28a)–(5.28f) and the neglected terms of $\mathcal{O}(\varepsilon^3)$ in (5.26) are indeed negligible. Now the instability and stability terms are of higher order compared to the nonlinearity and third-order dispersion terms, i.e., the appropriate model is a perturbed BKdV equation, or a "driven-dissipative BKdV equation" (see Appendix C.5), which can be easily converted to the Kawahara equation in (5.31) but with $\delta_K \gg 1$. Notice, e.g., that the perturbed BKdV equation $\partial_T H + H \partial_X H + \alpha \partial_{xx} H + \beta \partial_{xxx} H + \gamma \partial_{xxxx} H = 0$ with $\alpha, \gamma \ll 1$ is equivalent to a Kawahara equation $\partial_T H + H \partial_X H + \partial_{xx} H + \delta_K \partial_{xxx} H + \partial_{xxxx} H = 0$ with $\delta_K = \beta/(\alpha\gamma)^{1/2} \gg 1$.

For a vertical film with $Re = \mathcal{O}(1)$ and $We = \mathcal{O}(\varepsilon^{-2})$ we have already seen that the pertinent weakly nonlinear model is the KS equation. It is possible to obtain the Kawahara equation for a vertical film, but the orders of magnitude assignments $Re = \mathcal{O}(1)$ and $We = \mathcal{O}(\varepsilon^{-2})$ will have to be relaxed. More specifically, let us assume $Re = \mathcal{O}(\varepsilon)$ and $We = \mathcal{O}(\varepsilon^{-1})$. Recall from Sect. 5.1.2 that decreasing the order of magnitude assignment of We simply moves the surface tension term $Weh^3 \partial_{xxx} h$ in (5.14b) to a higher-order term of the flow rate, e.g., with $We = \mathcal{O}(\varepsilon^{-1})$, $Weh^3 \partial_{xxx} h$ moves from $q^{(1)}$ to $q^{(2)}$. Similarly, changing the order of magnitude of Re also moves the inertia term $Reh^6 \partial_x h$ in (5.14b) to a higher-order term of the flow rate, e.g., with $Re = \mathcal{O}(\varepsilon)$, $Reh^6 \partial_x h$ moves from $q^{(1)}$ to $q^{(2)}$ while all terms in $q^{(2)}$ involving Re would move to higher orders. The same with the surface tension terms in $q^{(2)}$. For simplicity we set $Re = \varepsilon Re_0$ and $We = \varepsilon^{-1} We_0$ where $Re_0, We_0 = \mathcal{O}(1)$. The second-order BE equation (5.13) then becomes:

$$\partial_t h + \frac{1}{3}\partial_x\left(h^3\right) + \varepsilon^2 \partial_x\left(\frac{7}{3}h^3(\partial_x h)^2 + h^4 \partial_{xx} h + \frac{2}{5}Re_0 h^6 \partial_x h + \frac{1}{3}We_0 h^3 \partial_{xxx} h\right).$$

$$(5.32)$$

With scaling arguments similar to those we used before, it can be shown that the pertinent weakly nonlinear prototype is the Kahawara equation with $\delta_K = \mathcal{O}(1)$,

$$\partial_T H + H \partial_X H + \partial_{XX} H + \delta_K \partial_{XXX} H + \partial_{XXXX} H = 0,$$

where

$$\delta_K = \sqrt{\frac{15}{2}\frac{1}{Re_0 We_0}}$$

with $Re_0 We_0 \equiv ReWe$.

We can also obtain the Kawahara equation with $\delta_K \gg 1$ (perturbed BKdV equation) but assuming $Re = \mathcal{O}(\varepsilon^2)$ and $We = \mathcal{O}(1)$, which would move the inertia and surface tension terms $Reh^6 \partial_x h$ and $Weh^3 \partial_{xxx} h$, respectively, from $q^{(1)}$ in (5.14b) to $q^{(3)}$. This is rather complicated and hence we refrain from detailing the derivation here.

5.2.1.3 Summary of Weakly Nonlinear Prototypes in Two Dimensions for an Isothermal Falling Film

To summarize, depending on the order of magnitude of $Re - Re_c$, while maintaining $Re = \mathcal{O}(1)$ and $We = \mathcal{O}(\varepsilon^{-2})$, we have the following weakly nonlinear prototypes for an inclined film:

(i) $Re - Re_c \ll \varepsilon^2$: Kawahara equation with $\delta_K \gg 1$ (perturbed BKdV equation).
(ii) $Re - Re_c = \mathcal{O}(\varepsilon^2)$: Kawahara equation with $\delta_K = \mathcal{O}(1)$.
(iii) $Re - Re_c = \mathcal{O}(1)$: KS equation.

For a vertical film, depending on the order of magnitude of Re, We, we have the following prototypes:

(i) $Re = \mathcal{O}(\varepsilon^2)$, $We = \mathcal{O}(1)$: Kawahara equation with $\delta_K \gg 1$ (perturbed BKdV equation).
(ii) $Re = \mathcal{O}(\varepsilon)$, $We = \mathcal{O}(\varepsilon^{-1})$: Kawahara equation with $\delta_K = \mathcal{O}(1)$.
(iii) $Re = \mathcal{O}(1)$, $We = \mathcal{O}(\varepsilon^{-2})$: KS equation.

5.2.1.4 Some Properties of the KS and Kawahara Equations

Both KS and Kawahara equations have been extensively studied in the literature (e.g., [47, 51, 144, 145]). They have been reported for a wide variety of systems such as flame propagation, solitary vortices in plasma, magmons in magma segregation in Earth's mantle and localized rolls in nematic crystals and, in general, in nonlinear media with energy supply and dissipation. They provide paradigmatic models for the study of low-dimensional spatio-temporal chaos or *weak/dissipative turbulence* (as defined by Manneville [177, 189]). The Kawahara equation in particular can be utilized as a generic prototype for the study of the influence of dispersion in nonlinear systems with energy supply and dissipation with the coefficient δ_K characterizing the relative importance of dispersion. In fact, for small δ_K the large-time behavior of the Kawahara equation is similar to that of the KS equation, i.e., spatio-temporal chaos/turbulent-like dynamics. On the other hand, Kawahara [144] demonstrated that sufficiently large δ_K tends to arrest the spatio-temporal chaos in favor of spatially periodic cellular structures, each of which approaching the BKdV soliton as δ_K increases. After all, the Kawahara equation with large δ_K is equivalent to a perturbed BKdV equation as earlier discussed.

To understand the influence of the different terms on the dynamic behavior of the solution of the Kawahara equation (5.31), consider the linear stability of this equation. By substituting in (5.31), $H = 1 + \varsigma \exp\{i(kX - \omega T)\}$ with $\varsigma \ll 1$, we find that even X derivatives contribute to the growth rate, $\omega_i = k^2 - k^4$, and odd X derivatives contribute to the phase speed, $c \equiv \omega_r/k = 1 - \delta_K k^2$. More specifically, the term $H\partial_{XX}H$ is responsible for instability/energy supply and the term $\partial_{XXXX}H$ is stabilizing and corresponds to energy dissipation. The above expression for the growth rate indicates that small-amplitude sinusoidal disturbances are linearly unstable for long wavelengths and stable for short wavelengths. The term $\partial_X H$ determines the wave propagation of the infinitesimal waves, equal to 1, and the term $\partial_{XXX}H$ is responsible for the presence of $-\delta_K k^2$ in the expression for the phase velocity, thus making the phase speed wavenumber dependent, a signature of the presence of dispersion. The Kawahara equation then retains the fundamental elements of any nonlinear process that involves wave evolution in two dimensions and as such it is a very useful prototype for the study of nonlinear phenomena that involve wave evolution in two dimensions.

As far as solitary wave solutions of the Kawahara equation are concerned, their numerical construction reveals that for large δ_K they become large in amplitude (however, in terms of the variables in (5.30) the amplitude is still small) and fairly close to the symmetric sech^2-soliton shape for the BKdV equation [145]. But, for smaller values of δ_K the pulses become asymmetric and develop an oscillatory structure (radiation) at the front of the primary hump. This structure is enhanced by decreasing δ_K further. On the other hand, the width of the pulses is almost the same for all δ_K, as the coefficients of the second and forth order derivatives in (5.31) are unity, which fixes the wavenumber that gives the maximum linear growth rate.

Reducing the oscillatory structure in front of the primary solitary hump by increasing δ_K appears to contradict our discussion in Sect. 4.5: By decreasing surface tension, i.e., the coefficient, say γ, of the fourth derivative in the Kawahara equation (prior to its rescaling to introduce δ_K in front of the third derivative) should amplify the oscillations in front of the hump; but decreasing γ is equivalent to increasing δ_K. It turns out that in the Kawahara equation prior to its rescaling, when γ is reduced the ratio of the wave amplitude to the maximum amplitude of the oscillations at the front increases.

5.2.1.5 Other Prototypes

An additional nonlinearity is also possible in certain systems. This is the case, for example, when the Marangoni effect is considered on the surface of a shallow horizontal layer heated from below with its upper boundary a free surface open to the ambient air and in the presence of buoyancy (the so called "Bénard–Marangoni convection"). A weakly nonlinear analysis of the Boussinesq–Fourier equations then shows that the pertinent weakly nonlinear prototype is of the form [55, 100]:

$$\partial_t H + \alpha_1 H \partial_x H + \alpha_2 \partial_{xx} H + \alpha_3 \partial_{xxx} H + \alpha_4 \partial_{xxxx} H + \alpha_5 \partial_x (H \partial_x H) = 0. \quad (5.33)$$

The additional term $\partial_x (H \partial_x H)$ plays a stabilizing role (interestingly, the same nonlinearity appears in the weakly nonlinear equation describing the instability of a contact line driven by gravity [135]). This equation has been referred to in the literature as the *Korteweg–de Vries–Kuramoto–Sivashinsky–Velarde equation* (KdV-KS-V) and reduces to the KS equation when $\alpha_3 = \alpha_5 = 0$ and to the BKdV equation when $\alpha_2 = \alpha_4 = \alpha_5 = 0$ [55]. Hence, for small α_3 and α_5, (5.33) becomes a perturbed KS equation [150], while for small α_2, α_4 and α_5, (5.33) becomes a perturbed BKdV equation, or a "driven-dissipative BKdV equation" (see Appendix C.5).

In the falling film problem the nonlinearity of the type $\partial_x (H \partial_x H)$ in (5.33) originates from a term $\sim \varepsilon s^2 \tilde{h} \partial_x \tilde{h}$ in (5.28b), which is of $\mathcal{O}(\varepsilon s^2 / x^2)$ and was neglected. In the absence of the Marangoni effect, the precise form of this term is found from (5.27b) to be, $\varepsilon[(12/7)Re - Ct]s^2 \tilde{h} \partial_x \tilde{h}$. As we saw earlier, when $Re - Re_c = \mathcal{O}(\varepsilon^2)$, from balancing instability and stability we have $x \sim \varepsilon^{-1}$, while from balancing the mean flow nonlinearity with dispersion, $s = \varepsilon^4$. Hence, in the weakly nonlinear equation, the mean flow nonlinearity, instability, stability and dispersion are all of $\mathcal{O}(\varepsilon^9)$ but the nonlinearity $\sim \varepsilon s^2 \partial_x (\tilde{h} \partial_x \tilde{h})$ is of $\mathcal{O}(\varepsilon^{11})$ and must be neglected (still it is the dominant term from the neglected terms in (5.28a)–(5.28f); all other neglected terms are of $\mathcal{O}(\varepsilon^{13})$). In fact, even if the condition $Re - Re_c = \mathcal{O}(\varepsilon^2)$ is relaxed, i.e., $Re - Re_c \ll \varepsilon^2$ or $Re - Re_c = \mathcal{O}(1)$, the nonlinearity $\varepsilon s^2 \partial_x(\tilde{h}\partial_x\tilde{h})$ is always a higher-order term.

Finally, it is worth noting that even though strong nonlinearities are not present in the above weakly nonlinear models, they can still yield singularities (like with the BE), depending on the initial conditions. In [200] for instance, the driven-dissipative BKdV equation is shown to blow up for sufficiently smooth and small-amplitude initial conditions provided that α_5 is smaller than a critical value. When α_5 is larger than this critical value small-amplitude initial data evolves into finite-amplitude irregular patterns. In [150] it is shown that the driven-dissipative BKdV equation can blow up for localized, finite-amplitude initial conditions. On the other hand, the KS and Kawahara equations are always bounded for sufficiently smooth, small-amplitude initial conditions. Problems related to singularity formation in weakly nonlinear models raise the question of applicability and relevance of these models to the original physical problem they are supposed to describe.

5.2.2 Models in Three Dimensions

We anticipate that the conditions under which dispersion is important in two dimensions are the same with those in three dimensions. Moreover, the same orders of magnitude assumptions used to obtain the KS or Kawahara equation in two dimensions should also be true in three dimensions, leading to three-dimensional model equations which are extensions to three dimensions of the two-dimensional KS or Kawahara equation and which possibly contain additional terms due to the transverse variation; after all, we should simply be able to recover the appropriate two-dimensional model from a three-dimensional one by simply dropping the transverse dependence.

Let us consider the three-dimensional first-order BE in (5.11) rewritten here for clarity in its expanded form:

$$\partial_t h + h^2 \partial_x h + \frac{2}{5}\varepsilon Re \partial_x(h^6 \partial_x h) - \frac{1}{3}\varepsilon Ct\left[\partial_x(h^3 \partial_x h) + \partial_z(h^3 \partial_z h)\right]$$

$$+ \frac{1}{3}\varepsilon^3 We\left\{\partial_x\left[h^3(\partial_{xxx}h + \partial_{zzx}h)\right] + \partial_z\left[h^3(\partial_{xxz}h + \partial_{zzz}h)\right]\right\} = 0, \quad (5.34)$$

where once again for simplicity we focus on the isothermal case. As with the models developed in two dimensions, we perform a weakly nonlinear expansion with $h \sim 1 + s\tilde{h}$ where $s \ll 1$ to obtain the following weakly nonlinear equation:

$$s\partial_t\tilde{h} + s\partial_x\tilde{h} + 2s^2\tilde{h}\partial_x\tilde{h} + \frac{2}{5}\varepsilon(Re - Re_c)s\partial_{xx}\tilde{h} - \frac{1}{3}\varepsilon Ct s\partial_{zz}\tilde{h} + \frac{1}{3}\varepsilon^3 We s\nabla_{xz}^4\tilde{h} = 0.$$

$$(5.35)$$

Contrasting (5.35) with (5.24) shows that the nonlinearity and instability terms remain unaltered and the stability term has been extended to account for the variation of the three-dimensional mean curvature of the surface, $\partial_{xx}h + \partial_{zz}h$, to both streamwise and spanwise directions. We also note the presence in (5.35) of the stabilizing term $-(1/3)\varepsilon Ct s \partial_{zz}\tilde{h}$ due to the hydrostatic part of the pressure. Indeed, a linear stability analysis of (5.35) where $\tilde{h}$ is sought in the form of the normal mode $\tilde{h} = \exp\{\lambda t + i(kx + mz)\}$ with k, m the wavenumbers in the x, z directions, respectively, gives for the temporal growth rate,

$$\lambda_r = \frac{1}{3}\varepsilon\left[\frac{6}{5}(Re - Re_c)k^2 - Ct\,m^2 - \varepsilon^2 We\left(k^4 + 2k^2m^2 + m^4\right)\right]. \tag{5.36}$$

Let us now set $Re - Re_c \sim \alpha^2$ and balance the instability and stability terms in (5.35):

$$\varepsilon(Re - Re_c)s\partial_{xx}\tilde{h} \sim \varepsilon^3 We\,s\,\nabla_{xz}^4\tilde{h} \quad \Rightarrow \quad x, z \sim \alpha^{-1}.$$

Balance nonlinearity with instability and stability:

$$s^2\tilde{h}\partial_x\tilde{h} \sim \varepsilon(Re - Re_c)s\partial_{xx}\tilde{h} \quad \Rightarrow \quad s = \varepsilon\alpha^3.$$

The nonlinearity, instability and stability terms then are of $\mathcal{O}(\varepsilon^2\alpha^7)$. The hydrostatic term is of $\mathcal{O}(\varepsilon^2\alpha^5)$. All terms then balance if $\alpha \sim 1$. The time-dependent term in (5.35) balances all other terms on the long time scale, $t \sim \varepsilon^{-1}$, suggesting the introduction of the slow time scale, $\eta = \varepsilon t$, consistent with the expression for the growth rate in (5.36). Introducing also the moving coordinate transformation, $\xi = x - t$, converts (5.36) to

$$\partial_\eta\tilde{h} + 2\tilde{h}\partial_\xi\tilde{h} + \frac{2}{5}(Re - Re_c)\partial_{\xi\xi}\tilde{h} - \frac{1}{3}Ct\partial_{zz}\tilde{h} + \frac{1}{3}\varepsilon^2 We\nabla_{xz}^4\tilde{h} = 0, \tag{5.37}$$

which with the change of variables

$$\xi = \sqrt{\frac{5\varepsilon^2 We}{6(Re - Re_c)}}X; \qquad z = \sqrt{\frac{5\varepsilon^2 We}{6(Re - Re_c)}}Z; \qquad \eta = \frac{25\varepsilon^2 We}{12(Re - Re_c)}T,$$

$$\tilde{h} = \frac{1}{5}\sqrt{\frac{6(Re - Re_c)^3}{5\varepsilon^2 We}}H,$$

becomes

$$\partial_T H + H\partial_X H + \partial_{XX}H - \chi\partial_{ZZ}H + \nabla_{xz}^4 H = 0, \tag{5.38}$$

the *Nepomnyashchy equation* [188], where

$$\chi = \frac{Re_c}{Re - Re_c}$$

is a "parameter of inclination." For a vertical plane, $Ct = 0$, (5.38) becomes a three-dimensional KS equation, which in the absence of transverse variation is reduced to the two-dimensional KS equation in (5.25).

The linear stability analysis of the two-dimensional solitary wave solutions of (5.38)—which are two-dimensional KS solitary waves—in the transverse direction was examined in [75]. It was shown that there exists a critical value of χ corresponding to a critical inclination angle below which two-dimensional solitary pulses are stable, in agreement with the experiments by Liu et al. [170] for small inclination angles.

The natural question now is: Can we obtain a three-dimensional Kahawara-type equation? For this purpose we must include the third-order dispersion in the three-dimensional version of $q^{(2)}$ in (5.14c). The weakly nonlinear equation (5.35) then becomes

$$
s\partial_t \tilde{h} + s\partial_x \tilde{h} + 2s^2 \tilde{h}\partial_x \tilde{h} + \frac{2}{5}\varepsilon(Re - Re_c)s\partial_{xx}\tilde{h}
$$

$$
- \frac{1}{3}\varepsilon Ct\, s\partial_{zz}\tilde{h} + \frac{1}{3}\varepsilon^3 We\, s\, \nabla^4_{xz}\tilde{h} + \varepsilon^2 s\partial_x \nabla^2_{xz}\tilde{h}. \tag{5.39}
$$

The only way to balance dispersion with the other terms is to relax the requirement, $x \sim z$. Let us then balance the instability and fourth-order x derivative associated with the stabilizing surface tension term in (5.39):

$$
\varepsilon(Re - Re_c)s\partial_{xx}\tilde{h} \sim \varepsilon^3 We\, s\partial_{xxxx}\tilde{h} \quad \Rightarrow \quad x \sim \alpha^{-1}.
$$

We also balance the instability term with the hydrostatic head:

$$
\varepsilon(Re - Re_c)s\partial_{xx}\tilde{h} \sim \varepsilon Ct\, s\partial_{zz}\tilde{h} \quad \Rightarrow \quad z \sim \alpha^{-2}.
$$

Hence, for $\alpha \ll 1$, i.e., very close to criticality, the characteristic length scale in the spanwise direction is much longer than that in the streamwise one. In other words, very close to criticality the spanwise length of the developed three-dimensional structures is much longer than their streamwise length, i.e., three-dimensional structures very close to criticality have a very slow variation in the transverse direction, and the dynamics of the inclined film in this region are effectively determined by the streamwise direction. On the other hand, as the distance from criticality increases, more specifically for $Re - Re_c = \mathcal{O}(1)$, $z \sim x$ and the transverse length scale has grown to match the streamwise one, which in turn implies that the three-dimensional structures are now localized.

We also balance nonlinearity with instability,

$$
s^2 \tilde{h}\partial_x \tilde{h} \sim \varepsilon(Re - Re_c)s\partial_{xx}\tilde{h} \quad \Rightarrow \quad s \sim \varepsilon\alpha^3,
$$

and instability with the third-order x derivative associated with dispersion,

$$
\varepsilon(Re - Re_c)s\partial_{xx}\tilde{h} \sim \varepsilon^2 s\partial_{xxx}\tilde{h} \quad \Rightarrow \quad s = \varepsilon^4.
$$

Finally, to obtain the time scale on which the time-dependent term in (5.39) balances the nonlinearity, instability, fourth-order x derivative associated with surface tension and third-order x derivative associated with dispersion,

$$s\partial_t \tilde{h} \sim s^2 \tilde{h} \partial_x \tilde{h} \quad \Rightarrow \quad t \sim \varepsilon^{-5},$$

a long time scale suggesting the introduction of the slow scale $T = \varepsilon^5 t$ (consistent with the expression for the growth rate in (5.36)). By introducing now the slow variables $X = \varepsilon(x - t)$ and $Z = \varepsilon^2 z$, (5.39) becomes

$$\partial_T \tilde{h} + 2\tilde{h}\partial_X \tilde{h} + \frac{2}{5}\frac{Re - Re_c}{\varepsilon^2}\partial_{XX}\tilde{h} - \frac{1}{3}Ct\partial_{ZZ}\tilde{h} + \frac{1}{3}\varepsilon^2 We\big(\partial_{XXXX}\tilde{h}$$
$$+ 2\varepsilon^2 \partial_{XXZZ}\tilde{h} + \varepsilon^4 \partial_{ZZZZ}\tilde{h}\big) + \partial_{XXX}\tilde{h} + \varepsilon^2 \partial_{XZZ}\tilde{h} = 0,$$

so that to leading order in ε the only Z-dependence comes from the hydrostatic term $\partial_{ZZ}\tilde{h}$:

$$\partial_T \tilde{h} + 2\tilde{h}\partial_X \tilde{h} + \frac{2}{5}\frac{Re - Re_c}{\varepsilon^2}\partial_{XX}\tilde{h} - \frac{1}{3}Ct\partial_{ZZ}\tilde{h} + \frac{1}{3}\varepsilon^2 We\partial_{XXXX}\tilde{h} + \partial_{XXX}\tilde{h} = 0,$$

$$(5.40)$$

which with $\partial_Z = 0$ is identical to (5.30).

For a vertically falling film we have demonstrated in Sect. 5.2.1 that for $Re = \mathcal{O}(\varepsilon)$ and $We = \mathcal{O}(\varepsilon^{-1})$ the pertinent weakly nonlinear model is the two-dimensional Kawahara equation with a dispersion parameter $\delta_K = \mathcal{O}(1)$. In three dimensions and with the same orders of magnitude assignments for Re, We, we obtain the three-dimensional Kawahara equation. As before, we set $Re = \varepsilon Re_0$ and $We = \varepsilon^{-1} We_0$ where Re_0, $We_0 = \mathcal{O}(1)$. The three-dimensional weakly nonlinear equation then reads

$$s\partial_t \tilde{h} + s\partial_x \tilde{h} + 2s^2 \tilde{h} \partial_x \tilde{h} + \varepsilon^2\left(\frac{2}{5}Re_0 s\partial_{xx}\tilde{h} + \frac{1}{3}We_0 s\nabla_{xz}^4 \tilde{h} + s\partial_x \nabla_{xz}^2 \tilde{h}\right) = 0. \quad (5.41)$$

By balancing the instability with the stability terms in this equation,

$$Re_0 s\partial_{xx}\tilde{h} \sim We_0 s\nabla_{xz}^4 \tilde{h} \quad \Rightarrow \quad x, z \sim 1,$$

and for a vertical film the developed three-dimensional structures very close to criticality have the same length scales in both streamwise and spanwise directions. With $x \sim 1$, the instability and stability terms also balance the dispersion term $s\partial_x \nabla_{xz}^2 \tilde{h}$. By balancing the nonlinearity with these terms,

$$s^2 \tilde{h}\partial_x \tilde{h} \sim \varepsilon^2 s\partial_{xx}\tilde{h} \quad \Rightarrow \quad s = \varepsilon^2,$$

while to obtain the time scale on which the time derivative in (5.41) balances nonlinearity, dispersion, instability and stability,

$$s\partial_t \tilde{h} \sim s^2 \tilde{h}\partial_x \tilde{h} \quad \Rightarrow \quad t \sim \varepsilon^{-2},$$

which suggests the introduction of the slow time scale $\tau = \varepsilon^2 t$. Finally, the term $s\partial_x \tilde{h}$ can be removed with the moving coordinate transformation, $\xi = x - t$. Equation (5.41) then becomes

$$\partial_\tau \tilde{h} + 2\tilde{h}\partial_\xi \tilde{h} + \frac{2}{5}Re_0\partial_{\xi\xi}\tilde{h} + \frac{1}{3}We_0\nabla_{\xi z}^4 \tilde{h} + \partial_\xi \nabla_{\xi z}^2 \tilde{h} = 0, \qquad (5.42)$$

which with the change of variables

$$\xi = \sqrt{\frac{5We_0}{6Re_0}}X; \qquad z = \sqrt{\frac{5We_0}{6Re_0}}Z; \qquad \tau = \frac{25We_0}{12Re_0^2}T; \qquad \tilde{h} = \frac{1}{5}\sqrt{\frac{6Re_0^3}{5We_0}}H$$

becomes the three-dimensional Kawahara equation,

$$\partial_T H + H\partial_X H + \partial_{XX} H + \delta_K \partial_X \nabla_{XZ}^2 H + \nabla_{XZ}^4 H = 0, \qquad (5.43)$$

where

$$\delta_K = \sqrt{\frac{15}{2}\frac{1}{Re_0 We_0}},$$

an $\mathcal{O}(1)$ dispersion parameter, which is identical with that in the two-dimensional case. With $\partial_Z = 0$, the equation reduces to the two-dimensional one (5.31).

Finally, much like with the two-dimensional case, we can obtain the three-dimensional Kawahara equation with $\delta_K \gg 1$ (perturbed three-dimensional BKdV equation) but assuming $Re = \mathcal{O}(\varepsilon^2)$ and $We = \mathcal{O}(1)$ which would move the inertia and surface tension terms $Reh^6\partial_x h$ and $Weh^3\partial_{xxx}h$, respectively, from $q^{(1)}$ in (5.27b) to $q^{(3)}$, which is lengthy and so details of the derivation are not given here.

5.3 Traveling Waves

Although the precise methodology for the search of traveling wave solutions will be outlined in Chap. 7 in the framework of dynamical systems theory, we discuss here some of the results corresponding to the drag-gravity regime with the important distinction between "closed" and "open flow conditions."

5.3.1 Closed and Open Flow Conditions

Traveling waves are computed as stationary solutions in a reference frame moving at the speed of the wave, denoted c. To obtain the equations governing the traveling wave solutions we introduce the *moving coordinate transformation, $\xi = x - ct$,* in the time-dependent models whose traveling wave solutions we seek, and we set

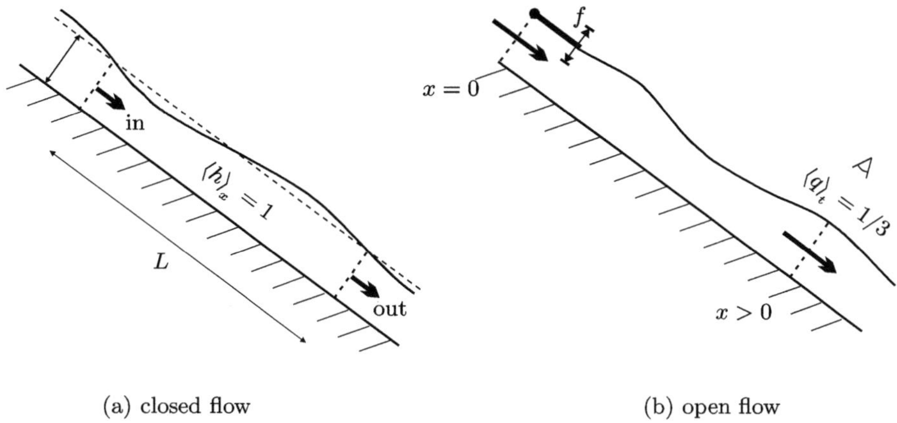

(a) closed flow   (b) open flow

Fig. 5.1 Two different flow conditions: (**a**) the mass is conserved in the domain of length L such that the amount of liquid leaving the domain (out) is exactly equal to the amount of liquid entering the domain (in); (**b**) considering a periodic forcing of frequency f at the inlet $x = 0$, the quantity of liquid flowing at any point $x > 0$ during a period $\tau = 2\pi/f$ is conserved

$\partial_t = -c\partial_\xi$ for the waves to be stationary in the moving frame. Of particular interest in this section are periodic traveling waves.

The most usual boundary conditions imposed are the periodic ones. However, these boundary conditions correspond to the situation of a *closed flow*, i.e., one for which the liquid flowing out of the domain is reinjected at the inlet as depicted in Fig. 5.1a. Therefore, these conditions do not correspond to an *open flow* in an actual experiment. Let us describe in detail these two flow conditions.

For a closed flow:

$$X|_{x=L} = X|_{x=0} \quad \forall t, \tag{5.44}$$

where X refers to any flow variable and L is the length of the closed domain. But because the flow is continuously driven by gravity in the streamwise direction, (5.44) cannot be achieved experimentally. As a matter of fact, in experiments the flow is open and often the film is forced at a given frequency, say f. We then presume that a *synchronization* between the flow at any location in space and the inlet forcing exists so that the developed waves maintain their periodicity in time (see Fig. 5.1b). Evidently,

$$X|_{t=\tau} = X|_{t=0} \quad \forall x, \tag{5.45}$$

where τ is the period of the oscillations, i.e., $2\pi/f$.

Let us now utilize the integral version of the kinematic boundary condition (5.5) for a two-dimensional flow:

$$\partial_t h + \partial_x q = 0. \tag{5.46}$$

In the case of a closed flow situation we define the spatial average of any quantity X from, $\langle X \rangle_x = L^{-1} \int_0^L X \, dx$. Combining then the spatial average of (5.46) with

(5.44) furnishes

$$\frac{d}{dt}\langle h\rangle_x = 0. \tag{5.47}$$

Therefore, the spatial average of the film thickness, i.e., the amount of liquid in the domain L, is constant at any time t and must equal its value at the initial time, i.e., unity. We next presume that the final step of the time-dependent computation is the formation of a regular wavetrain of traveling waves.[3] The computational domain L containing a certain number of waves, say n, can be written as, $L = n\lambda$, where λ is the wavelength of the traveling waves. In the moving frame $\xi = x - ct$,

$$\langle h\rangle_x = \frac{1}{L}\int_0^L h\,dx = \frac{1}{L}\int_{-ct}^{L-ct} h\,d\xi,$$

since $dx = d\xi$ by keeping t constant. A consequence of periodic boundary conditions is that we can define a periodic extension of h outside the domain $[0, L]$ and in intervals of length L. We then have $\int_{-ct}^{L-ct} h\,d\xi = \int_0^L h\,d\xi$ so that,

$$\frac{1}{L}\int_0^L h\,dx = \frac{1}{n\lambda}\int_0^{n\lambda} h\,d\xi = \frac{1}{\lambda}\int_0^\lambda h\,d\xi,$$

where we have made use of the periodicity of the traveling waves in their frame of reference. Consequently, $\langle h\rangle_x = \langle h\rangle_\xi \equiv \lambda^{-1}\int_0^\lambda h\,d\xi$, so that

$$\langle h\rangle_\xi = 1, \tag{5.48}$$

which will be denoted hereinafter as the *closed flow condition* for the computation of traveling wave solutions.

Turning to a time periodic modulation of the film surface, we define the time average of any quantity X over the period τ of the inlet forcing from, $\langle X\rangle_t = \tau^{-1}\int_0^\tau X\,dt$. By taking then the time average of (5.46) and utilizing the condition (5.45), we obtain:

$$\frac{d}{dx}\langle q\rangle_t = 0. \tag{5.49}$$

The time average flow rate is independent of the location x and therefore equal to its inlet value, $\langle q\rangle_t = \bar{q}_N/[(\bar{h}_N/(\lambda_\nu t_\nu))\bar{h}_N] = 1/3$, where the expression for $\bar{q}_N$ in (2.36) has been utilized. Again, applying this condition to traveling waves,

$$\langle q\rangle_t = \frac{1}{\tau}\int_0^\tau q\,dt = -\frac{1}{c\tau}\int_x^{x-c\tau} q\,d\xi = \frac{1}{c\tau}\int_{x-c\tau}^x q\,d\xi,$$

---

[3] This is mostly the case, however, Ramaswamy et al. [218] reported the formation of oscillatory modes made of irrationally related periodic oscillations in time at the end of some of their DNS computations with the periodic boundary condition (5.44) (see also Sect. 7.2.4).

since $dt = -c^{-1} d\xi$ by keeping x constant. With an argument similar to that used above for the spatial case, i.e., because of the periodicity in time, we can define a periodic extension of q with period τ so that $\int_{x-c\tau}^{x} q\, d\xi = \int_{0}^{c\tau} q\, d\xi$. But the synchronization of the flow with the forcing frequency at the inlet and the presence of a periodic wavetrain in space[4] suggest that if λ is the period of the traveling waves in the moving frame, $c\tau = \lambda$ so that $\langle q \rangle_t = \lambda^{-1} \int_0^\lambda q\, d\xi$, or

$$\langle q \rangle_\xi = \frac{1}{3}. \tag{5.50}$$

Equation (5.50) will be denoted hereinafter as the *open flow condition* for the computation of traveling wave solutions. Such solutions are computed using the continuation software AUTO-07P [79] and details are given in Appendix F.2. These computations also give λ (which in fact is a continuation parameter).

Let us now consider the conservation equation (5.46) in the moving frame of reference of a traveling wave, i.e., with $\xi = x - ct$, and integrate it once to obtain

$$q = ch + q_0, \tag{5.51}$$

where the integration constant q_0 represents the (negative) constant flow rate in the moving frame of reference, i.e., underneath the wave and in the opposite direction (q_0 is also a continuation parameter in AUTO-07P). Further, let us average (5.51):

$$\langle h \rangle_\xi = \frac{\langle q \rangle_\xi - q_0}{c}. \tag{5.52}$$

By imposing the open flow condition (5.50), (5.52) shows that the average film thickness is given by

$$\langle h \rangle_\xi = \frac{1/3 - q_0}{c}, \tag{5.53}$$

and thus it will be influenced by the wave features c and q_0. Both (5.48) and (5.53) are integral constraints used later on in this chapter and in Chap. 7 for the computation of traveling waves by continuation.

The requirements, $\langle h \rangle_x = 1$ and $\langle q \rangle_t = 1/3$, for open and closed flow conditions, respectively, can be related to time-dependent computations. In fact, in such computations the requirement $\langle h \rangle_x = 1$ is automatically imposed from the periodicity in space, i.e., condition (5.44). To compare time-dependent computations with periodic boundary conditions in a domain of length L to the traveling wave results obtained with AUTO-07P by using (5.48), we must adjust L to $n\lambda$, i.e., the computational domain must be equal to a number of spatial wavelengths of the traveling waves.

---

[4]The implicit assumption here is that a spatially periodic wavetrain results from a time periodic forcing at the inlet. However, this is not always the case, e.g., Fig. 7.39 shows that for $x \lesssim 2\mathrm{m}$ we have a wavetrain that is periodic in time but modulated in space so that the wavelength changes locally. As a consequence, we cannot relate the period in space with that in time.

Again, this procedure presumes that the final step of the time-dependent computation will be the formation of a wavetrain of traveling waves propagating with a constant shape and speed, which after all is not guaranteed. We can also do the reverse: obtain λ from the time-dependent computations and then impose λ in AUTO-07P: We start with an initial condition a wave profile obtained with AUTO-07P for a given λ, and then we alter λ by continuation until we match the one obtained from the time-dependent computations.

For the requirement $\langle q \rangle_t = 1/3$ there is no guarantee that it will be automatically satisfied in open flow-time-dependent computations (such as those described in Appendix F.3): It is a direct consequence of the inlet forcing and the synchronization condition of time periodicity (5.45) all along the plane. For this to be true the forcing has to be able to overcome the noise that is always present in computations. If the noise is sufficiently strong, the time periodicity along the plane might be lost. We also need a convectively unstable system, i.e., one which is a noise amplifier as first noted in the Introduction. If the system is absolutely unstable, it will oscillate with its own intrinsic frequency.

Provided that the above conditions are satisfied, connection of the traveling wave solutions obtained with AUTO-07P by imposing the open flow condition (5.53) to those resulting from time-dependent computations requires not only that the inlet flow rate in the time-dependent computations be set to 1/3 but also that the corresponding time period in the laboratory frame is λ/c and this must be set equal to $2\pi/f$ with f the inlet forcing frequency. The same requirement, $\lambda/c = 2\pi/f$, applies for comparison purposes of traveling wave solutions (obtained with AUTO-07P by imposing the open flow condition (5.53)) with experiments—usually experiments report the wavelength or forcing frequency and speed of the waves from which one can compute the wavelength (if the flow synchronizes to the inlet forcing).

As far as comparing time-dependent simulations using periodic boundary conditions with experiments with periodic inlet forcing is concerned, we need to have synchronization in space of the solution to the numerical simulations, i.e., a regular wave pattern with a given wavelength. The wavelength selected by the system then is $\lambda = L/n$ with L the numerical domain length and n an integer. To compare open flow-time-dependent computations (such as those described in Appendix F.3) we ensure that the wavelength L/n matches the experimental one. However, this condition is not sufficient. In fact, in experiments, once synchronization in space to a regular train of traveling waves has occurred, $\langle h \rangle_t$ is generally different from its inlet value. As a consequence, careful comparisons between numerical and experimental results require that the initial spatial mean thickness $\langle h \rangle_x$ of the numerical simulations be adjusted to the experimental value $\langle h \rangle_t$ corresponding to the observed traveling waves (a value that is generally not reported in the experimental studies, see Sect. 8.4 where this point is further discussed).

If there is no forcing at the inlet, an integral condition still applies for an open flow. For this purpose, we must change the earlier definition of the time average to $\langle X \rangle_t = T^{-1} \int_0^T X \, dt$ where T is some large time corresponding to the time of the experiment. By taking the time average then of the integral version of the kinematic boundary condition, (5.46),

$$\frac{1}{T}[h|_T - h|_0] + \frac{d}{dx}\langle q \rangle_t = 0,$$

which for large T, and provided that $\langle q \rangle_t = T^{-1} \int_0^T q\, dt$ can be defined, i.e., it is finite and independent of T, yields $\langle q \rangle_x = \text{const}$ and therefore is equal to the inlet value of the flow rate, $1/3$. But in this case we might not be able to connect experiments or time-dependent computations with the results for traveling waves obtained from AUTO-07P, as we might have a combination of small and large waves in experiments or time-dependent computations and a wavelength that varies substantially with space.

But clearly, between the two conditions, open and closed flow, the open flow one is more suitable for comparisons with experiments. However, several studies in the literature have utilized the closed flow condition. It is, therefore, not surprising that quite frequently discrepancies between theory and experiments are observed. This is, e.g., the case with the study by Salamon et al. [232], who imposed the closed flow condition (5.48) for the computation of traveling waves.

We note finally that in experiments it is usually observed that the formation of large-amplitude waves is accompanied by a significant decrease of the time-average film thickness over the time T of the experiments [3]. This can be understood as follows: Under the crest of large amplitude waves, inertia effects can significantly reduce the wall shear stress, say τ_w, from its value for an undeformed surface (the Nusselt flow solution gives $\tau_w \propto h$; see the work by Tihon et al. [273], especially Figs. 11 and 13 in that reference). A consequence of this decrease of the wall friction is that the fluid particles travel faster on average when the flow is wavy than for the corresponding Nusselt flow. Since the average flow rate is conserved, a reduction of the average thickness ensues.

5.3.2 Traveling Wave Solutions in the Drag-Gravity Regime

For an isothermal vertical film, we can distinguish between two main *families of waves* [48, 50]. The first one, referred to as the *family of slow waves*, is denoted by γ_1 and terminates at small wavenumbers as a slow solitary wave with a dominant depression, a *negative-hump solitary wave* ("negative polarity"). The second family, referred to as the *family of fast waves*, is denoted by γ_2 and terminates at small wavenumbers as a fast solitary wave with a dominant elevation, a *positive-hump solitary wave* ("positive polarity"). Appendix F.2 outlines in detail the computational methodology used to compute such wave families. It should be emphasized that unless, specifically stated, when we refer to waves, e.g., solitary waves in this monograph, we mean positive waves.

The existence of negative waves can be demonstrated with the KS equation (5.25), which was derived earlier from both the first- and second-order BE for small amplitude waves and $Re - Re_c = \mathcal{O}(1)$. Carrying out a moving coordinate transformation, $X = X - CT$, and setting $\partial_T = -C\partial_X$ for the waves to be stationary in the moving frame, we obtain their governing equation:

$$H^{\mathrm{IV}} + H'' + HH' - CH' = 0,$$

where the primes denote differentiation with respect to X. This equation is invariant under the transformation $H \to -H$, $C \to -C$ and $X \to -X$, so-called "reversible symmetry" [49]. Thus, for every (periodic or solitary) wave propagating in one direction, there exists a counter propagating dual one with an inverted profile. These hollow negative waves do not actually propagate backward. Recall that the KS equation (5.25) has been derived in a frame moving with the kinematic wave speed 1. Hence negative waves travel with a negative speed relative to the critical speed 1.

A reversible symmetry applies for the Kawahara equation (5.31) as well, obtained earlier from the second-order BE with $Re - Re_c = \mathcal{O}(\varepsilon^2)$. In the moving frame,

$$H^{\mathrm{IV}} + \delta_{\mathrm{K}} H''' + H'' + HH' - CH' = 0,$$

which is invariant under the transformation $H \to -H$, $C \to -C$, $\delta_{\mathrm{K}} \to -\delta_{\mathrm{K}}$ and $X \to -X$ [49]. Thus, for every (periodic or solitary) wave propagating in one direction for $\delta_{\mathrm{K}} > 0$, there exists a counter propagating dual one with an inverted profile for $\delta_{\mathrm{K}} < 0$. Negative waves for the KS equation have been constructed in [50] and for the Kawahara equation in [49].

Of course for both isothermal and heated films problems, $\delta_{\mathrm{K}} > 0$, which breaks the reversible symmetry of the KS equation. Nevertheless, we anticipate that at least for sufficiently small δ_{K}, negative waves are still present, which is in fact the case as we shall demonstrate in Sect. 7.2.3. In other words, it is not necessary to have $\delta_{\mathrm{K}} < 0$ to obtain negative waves from the Kawahara equation. For sufficiently large δ_{K} and as the Kawahara equation approaches the perturbed BKdV equation, negative waves with $\delta_{\mathrm{K}} > 0$ no longer exist, as demonstrated in Appendix C.6, where we discuss negative polarity occurring in the BKdV equation.

The bifurcation diagram in Fig. 5.2a shows as solid lines the first two wave families of the first-order BE in terms of the maximum wave thickness $h_{\max}$ and the phase speed c as a function of the wavenumber k. Typical traveling wave solutions corresponding to the γ_2 family are shown in Fig. 5.2b. The γ_2 family starts at cut-off wavenumber k_c corresponding to the Hopf bifurcation point (HB) from the Nusselt flat film solution. This bifurcation is supercritical, so that the wavelength of solutions is larger than the cut-off value $2\pi/k_c$. It yields stationary waves whose amplitude and phase speed increase as k decreases. In the limit $k \to 0$, the solutions correspond to *homoclinic orbits*—i.e., traveling waves with infinite wavelength— in the phase space (Chap. 7 provides a detailed description of homoclinicity) or solitary waves in real space (they satisfy the conditions $h \to 1$ as $x \pm \infty$). The γ_1 family bifurcates by *period doubling* (PD) from the family of $n = 2$ harmonic solutions (by "harmonic solution" here we mean the solution of the linearized equation for h, $h = 1 + A\sin(2n\pi x/l)$ where $l = 2\pi/k$ is the period of the waves; n is

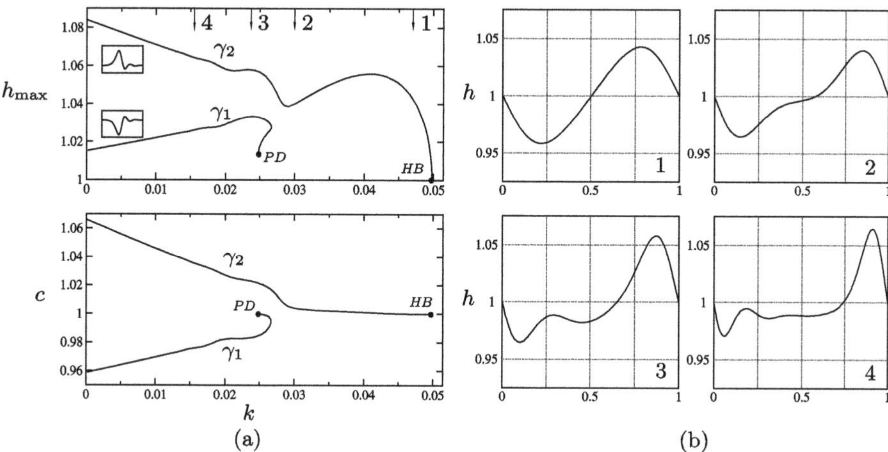

Fig. 5.2 (a) Bifurcation diagram and (b) some typical periodic stationary solutions (i.e., traveling waves in their moving reference frame of size $2\pi/k$) of the γ_2 family obtained with the first-order BE (5.12). The locations of the profiles 1, 2, 3 and 4 on the γ_2 branch are also indicated. Parameter values are $Ct = 0$ (vertical and isothermal wall), $Re = 2.0667$ and $We = 1000$ (i.e., $\Gamma = 3375$). The closed flow condition has been enforced. *HB*: Hopf bifurcation; *PD*: period doubling

the "harmonic parameter" f used in Appendix F.2).[5] The corresponding waves are negative-hump waves whose phase speed slows down as k decreases.

Figure 5.2 also indicates that the two families of slow and fast waves bifurcate from a family of stationary waves that travel at exactly three times the average velocity of the flow through an *imperfect pitchfork bifurcation*. Recall that for small-amplitude waves on a vertical plane and $Re = \mathcal{O}(1)$, the pertinent weakly nonlinear prototype obtained from the first-order BE is the KS equation. This equation has the reversible symmetry as pointed out earlier and as a consequence the corresponding bifurcation for the two wave families is a *perfect bifurcation* (see also Sect. 7.2.3, Fig. 7.35). The first-order BE does not have the reversible symmetry $(h - 1 \to -(h - 1),\ x \to -x$ and $c \to -c)$ and hence the bifurcation in Fig. 5.2 is an imperfect one. The natural expectation here is that for Fig. 5.2 we do have $Re = \mathcal{O}(1)$ and so one might expect a perfect pitchfork instead. After all, we are not far from the region of validity of the KS equation. But actually, the imperfection is an effect of the neglected terms in the multiple-scale expansion that gives the KS equation, e.g., the inertia term $\varepsilon^2 \partial_\xi (\tilde{h} \partial_\xi \tilde{h})$, which after one integration gives a

---

[5]Indeed, as the wavenumber decreases, higher harmonics become linearly unstable at $k_n = k_c/n$ with $n = 2, 3, \ldots$. The resulting families $\gamma_{1,2}^{(n)}$ for $n > 1$ correspond therefore to trains of n identical negative- or n identical positive-hump traveling wave solutions. Their maximum heights $h_{\max}^{(n)}(k)$ are not displayed in Fig. 5.2 because they are homothetic in k, i.e., given that $h_{\max}^{(n)}(k_n) = h_{\max}(k_c)$ it follows that $h_{\max}^{(n)}(k/n) = h_{\max}(k)$. The individual solutions correspond simply to n identical solutions of the $n = 1$ family placed in a domain of size $2\pi n/k$.

term $\sim \tilde{h}\partial_\xi\tilde{h}$, breaking the symmetry $\tilde{h} \to -\tilde{h}$, $\xi \to -\xi$. We then expect that by reducing Re, i.e., by making the wave amplitude smaller, we will start approaching a perfect pitchfork. The bifurcation structure of slow and fast waves of the Kawahara equation will be discussed in Sect. 7.2.3.

5.4 Validity Domain of the BE

As shown in Sect. 5.1.3, the linear stability analysis of BE is in agreement with Orr–Sommerfeld. The BE also allows for periodic traveling wave solutions that approach solitary waves, i.e., for $k \to 0$ with $2\pi/k$ the period of the waves. But as we shall demonstrate in this section, the BE can lead to finite-time blow up in time-dependent simulations when the Reynolds number exceeds a limiting value. Accordingly, no solitary waves can be observed beyond this value.

Formally, the BE is a particular case of the generic evolution equation,

$$\partial_t h + \partial_x \left(h^3 + \Phi h^m \partial_x h + h^3 \partial_{xxx}h\right) = 0, \qquad (5.54)$$

where m is a positive integer and Φ a positive parameter. For $m = 6$, the structure of the first-order BE corresponding to an isothermal vertical film is recovered. Equation (5.54) with $m = 3$ applies to the problem of thin film flowing down a vertical fiber [95, 136], and numerical simulations in this case do show an accelerated growth of the amplitude of solitary waves that was associated with the drop formation process on the film [136], but not a true finite-time blow up. On the contrary, simulations of (5.54) with $m = 6$ show that it leads to finite-time blow up obtained first by Pumir et al. [216] and illustrated in Fig. 5.3a (the numerical scheme for the time-dependent evolution of the BE equation is based on the scheme described in Appendix F.3).

Figure 5.3b shows that the solitary waves' branch exhibits a turning point, say at $\Phi = \Phi^\star$, and branch multiplicity (with two branches, a lower branch and an upper one) for $m > 3$. It means that for $\Phi > \Phi^\star$, (5.54) does not have any stationary solitary wave solutions. Numerical evidence suggests that the deviant finite blow up behavior of the BE occurs in the region where solitary waves do not exist. Actually, Fig. 5.3a shows that for $m = 6$ blow up occurs at $\Phi = 0.36$, which is smaller than $\Phi^\star \simeq 0.4$ from Fig. 5.3a. Although in general blow up always occurs for $\Phi > \Phi^\star$, the precise value, somewhere in the vicinity of $\Phi^\star$, where this happens depends on the details of the particular computation being performed, in particular coalescence events, which in turn depend on other factors such as domain size and initial condition. For the particular case considered in Fig. 5.3a, due to coalescence events, the "local Reynolds number" based on the *substrate thickness* ("substrate" refers to the portions of the flat film separating the solitary waves, see also Fig. 4.1)[6] of the

---

[6]In general, the local Reynolds number can be defined by assuming that locally the flow is a Nusselt one, i.e., by replacing in the Reynolds number based on the Nusselt flat film thickness (2.35) $\bar{h}_N$

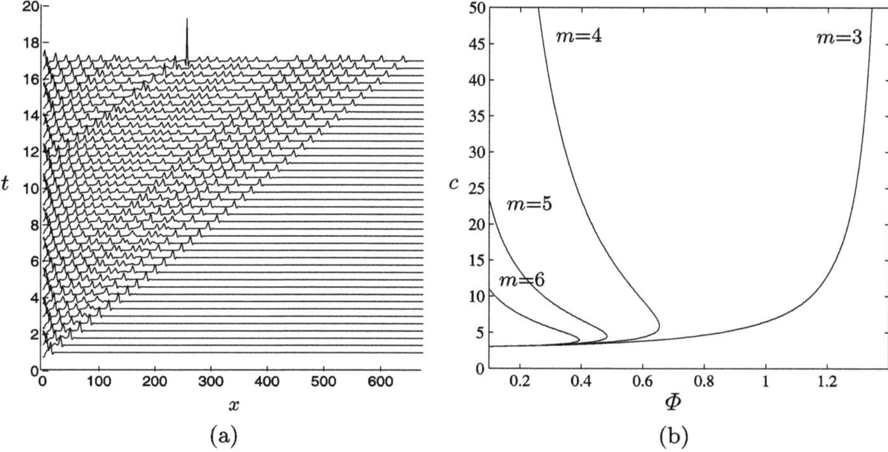

Fig. 5.3 (**a**) Space–time plot showing the evolution of the solution of the BE (5.54) with $m = 6$ for $\Phi = 0.36$ in the isothermal vertical case. The film thickness is plotted at regular time intervals. The flow is open and oriented *from left to right*. A periodic forcing with noise is imposed at $x = 0$. (**b**) Branches of single-hump solitary wave solutions to (5.54) for the phase speed c versus Φ for different values of m

dominant structure is actually larger than the value input in the computation; so, Φ is "locally larger" than 0.36. This subtle issue is discussed further in [241]. Reference [117] conjectured that (5.54) has a finite-time blow up behavior whenever $m > 3$. In [22] this criterion was refined, proving that nonlinearities with powers $m < 5$ can allow for bounded solutions under certain conditions. In any case, the above observations reveal that the strong nonlinearity $\partial_x(h^6\partial_x h)$ due to inertia is the cause of the peculiar singularities found with the BE. We now study in more detail the blow up behavior of the BE.

5.4.1 Blow up Versus Wavenumber

Because the positive-hump solutions have a larger maximum film thickness $h_{\max}$ than the negative-hump ones, they will be subject to blow up at lower Reynolds number. This can be understood from the nature of the strongly nonlinear term $\sim \partial_x(h^6\partial_x h)$ in (5.13) responsible for singularity formation. Similarly, at a given

---

with $\bar{h}$ or Reh^3. But the local flow rate is $q = \bar{q}/[(\bar{h}_N^2/(t_\nu l_\nu))\bar{h}_N] = \bar{u}\bar{h}/[(\bar{h}_N^2/(t_\nu l_\nu))\bar{h}_N] = h^3/3$. The local Reynolds number then is $3q Re$. Hence, the local Reynolds number based on the substrate thickness, say h_s, is $\sim Reh_s^3$. For a single soliton, the substrate thickness is almost the same with the inlet one, $h_s \sim h_N$. For many solitons, $h_s < h_N$. A physical explanation is given at the beginning of Sect. 7.2.3 (the reduction is not related to the arguments given at the end of Sect. 5.3.1 on the time-average film thickness). With coalescence, the number of waves goes down, which then leads to h_s increasing and hence the local Reynolds number increases.

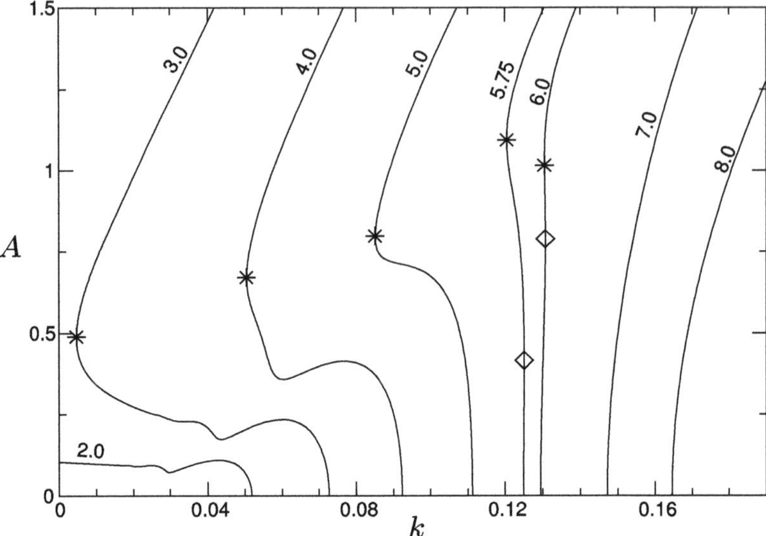

Fig. 5.4 Bifurcation diagram showing branches of single-hump traveling wave solutions for various Re computed with the first-order BE and $Ct = M = 0$, $\Gamma = 2950$ (i.e., isothermal vertical film of water at 15°C, see Appendix D.4) like in the experiments reported in [141]. The closed flow condition is enforced. $A = (h_{max} - h_{min})$ is the wave amplitude. *Asterisks and diamonds* indicate *saddle-node* bifurcations, the loci of which are followed through the parameter space in Fig. 5.5

k the single-hump solutions have larger amplitudes than multi-hump ones, hence they blow up first. In fact, Pumir et al. [216] demonstrated that during evolution toward blow up, single-hump solutions are always present. The singularity starts from a solitary wave that suddenly exhibits a catastrophic growth and blows up in finite time for some $Re > Re^*$, with Re^* the value of the limit points of the bifurcation diagrams for the speed c of the waves as a function of Re.[7] Hence, in the following we will only focus on single-hump traveling wave solutions—i.e., belonging to the γ_2-family of waves—in order to discuss the validity domain of the BE in terms of the existence of single-hump waves. We shall see that the occurrence of unbounded γ_2-wave solutions to the BE is closely related to the finite-time blow up observed in time-dependent simulations.

The bifurcation diagram in Fig. 5.4 depicts the families of γ_2 traveling wave solutions computed with the first-order BE (5.13) for several Re. As we discuss at the end of Appendix D.1, the Nusselt scaling distinguishes clearly between the flow and the properties of the gas–liquid–solid system and β. It then makes sense to report the results obtained from the full equations in (2.17)–(2.28), or equivalently

---

[7]As already pointed out, intriguingly, the KS and Kawahara equations obtained from a weakly nonlinear expansion of the first- and second-order BE, respectively, remain bounded for sufficiently smooth and small-amplitude initial conditions. As a matter of fact, they predict solitary wave solutions past the limit values Re^*. Disappointingly, the advantage of the more complex BE over the KS and Kawahara equations is rather limited.

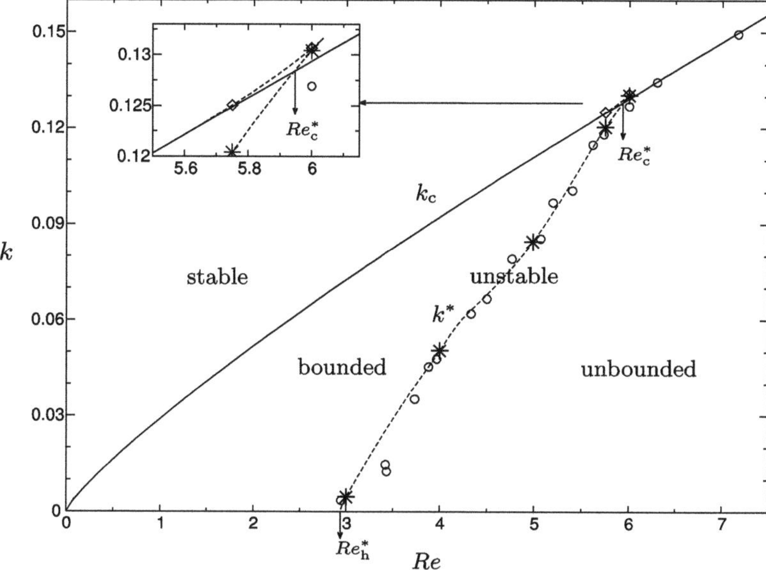

Fig. 5.5 Stability diagram of the first-order BE (parameter values are the same with Fig. 5.4). The Nusselt flat film solution is linearly unstable for wavenumbers smaller than the cut-off wavenumber k_c (*solid line*) and stable otherwise. The blow up boundary (*dashed line*) separates the bounded from the unbounded traveling wave solutions and coincides with the results of time-dependent numerical simulations (circles) [198]. Re_h^* and Re_c^* indicate the Reynolds numbers at which homoclinic orbits ($k = 0$) blow up and at which all the linearly unstable modes blow up, respectively. *Asterisks and diamonds* correspond to the saddle node bifurcation points as shown in Fig. 5.4. The *inset* is a zoom on the subcritical region

the BE obtained from a regular perturbation expansion of the full equations, in terms of Re (or equivalently h_N) and the viscous-gravity parameters (whose combination makes up the Nusselt parameters). In the particular example considered here, we fix $\beta = 0$ and the liquid, i.e., Γ. The only free parameter is then Re (or equivalently h_N).

The closed flow condition (5.48) is enforced. For $3 \lesssim Re \lesssim 5$, the wave families feature a *saddle-node bifurcation* at k^* indicated by an asterisk. This implies that for $k < k^*$ and $3 \lesssim Re \lesssim 5$, the BE has no stationary solution of the γ_2-type, while for $k > k^*$ two stationary *solution branches* coexist. References [216] and [198] have shown that only the lower branch of small amplitude corresponds to bounded solutions. The bifurcation at k_c is supercritical for $Re \lesssim 5$. However, the interval $[k^*, k_c]$ shrinks with increasing Re until it vanishes. For larger Re the Hopf bifurcation becomes subcritical. Noteworthy is the intrinsic structure of the families for $Re \approx 5.75 - 6$, where a second saddle node is present as indicated by the diamonds in Fig. 5.4.

Figure 5.5 depicts the locus of k^* for the γ_2 family as a function of Re. Reference [198] gives time-dependent simulations of traveling wave solutions for various wavenumbers k and constructs a boundary for finite-time blow up as marked out by the circles in Fig. 5.5. This boundary matches the dashed line in Fig. 5.5, indicating the connection between the saddle node bifurcation point of the BE for the γ_2 wave family and the finite-time blow up observed in computer simulations. Consequently, the traveling waves are bounded to the left of this boundary and unbounded to the right. We will refer to this boundary in the following as the *blow up boundary*.

We can now define two limit values of the blow up boundary as indicated in Fig. 5.5: Re_h^* for $k^* \to 0$ at which only homoclinic orbits become singular, i.e., precisely the Re value where the bifurcation diagram for the speed of the solitary waves as a function of Re turns and above which no solitary waves exist,[8] and Re_c^* for $k^* = k_c$ at which all the linearly unstable modes lead to singularities.

Notice that the inset of Fig. 5.5 reveals the existence of a subdomain of solutions near threshold for wavenumber above k_c, indicating the existence of a subcritical bifurcation. Consequently, the corresponding solutions are always unstable. However, this subcritical behavior appears to be unphysical as we shall see.

5.5 Parametric Study for Closed and Open Flows

We now study systematically the blow up features of the BE for a falling film. We impose both closed and open flow conditions. We start with a vertical isothermal film and we subsequently analyze the influence of inclination and the Marangoni effect for the ST case.

5.5.1 The BE with the Shkadov Scaling

Our analysis so far was based on the Nusselt scaling. However, keeping track of the domain boundaries in parameter space where stationary solutions exist is quite involved. As a matter of fact, six parameters can be varied, namely the inclination of the plate (Ct), the surface tension (We), the temperature difference between the wall and the ambient gas phase (M), the heat transfer coefficient at the interface by changing the ambient gas phase (B), the liquid (Pr) and finally the inlet flow rate (Re).

On the other hand, the Shkadov scaling has certain advantages over the Nusselt one as discussed in the previous chapter. Recall, however, that the Shkadov scaling is strictly speaking valid for large-amplitude waves, i.e., in the region of moderate Reynolds numbers, but for the sake of convenience and simplicity we use it for small-amplitude ones as well, i.e., for small δ. We then apply the Shkadov transformation, $x \to \kappa x$, $t \to \kappa t$, where $\kappa = (\varepsilon^2 We)^{1/3}$, to the first-order BE (5.11) for a

---

[8]For traveling waves the saddle node bifurcation corresponds precisely to the turning point of the solution branches. For time-dependent computations things are slightly different; even for slightly smaller values than Re^* we can have blow up (see also Sect. 5.4).

two-dimensional flow and for the ST case

$$\partial_t h + \partial_x \left(\frac{h^3}{3} + \frac{2}{15}\delta h^6 \partial_x h - \zeta \frac{h^3}{3}\partial_x h + \frac{h^3}{3}\partial_{xxx}h + \frac{h^2}{2}\frac{\mathcal{M}B\partial_x h}{(1+Bh)^2} \right) = 0, \quad (5.55)$$

where the Shkadov parameters are given again for clarity:

$$\delta = \frac{3Re}{\kappa}, \qquad \zeta = \frac{Ct}{\kappa} \quad \text{and} \quad \mathcal{M} = \frac{M}{\kappa}.$$

Notice that by rescaling time t with the factor $1/3$, (5.55) is formally equivalent to (5.54) with $\zeta = 0$ and $\mathcal{M} = 0$ when $\Phi = 2\delta/5$ and $m = 6$.

Equation (5.55) for isothermal conditions ($\mathcal{M} = 0$) and for a vertical plane ($\zeta = 0$) has also been written by Nakaya [186], who arrived at the same scaling with Shkadov (by aiming at decreasing the number of parameters in this equation, from two in the isothermal vertical case, Re and We, to only one, δ), but from the BE itself and not from physical arguments related to the separation of scales inherent to the falling film problem in the region of moderate Reynolds numbers. Further, with the Shkadov scaling the coefficient of surface tension in the BE is unity, while the remaining coefficients either have values close to unity or smaller, which is rather convenient from a numerical point of view.

With the Shkadov scaling, the wavenumber is also rescaled as $k \to k/\kappa$ and the cut-off wavenumber becomes

$$k_c = \left(\frac{2}{5}\delta - \zeta + \frac{3}{2}\frac{\mathcal{M}B}{(1+B)^2} \right)^{1/2}. \tag{5.56}$$

It is now useful to contrast (5.56) for the cut-off wavenumber obtained with the Shkadov scaling and (5.21) obtained with the Nusselt scaling, or equivalently, the critical Reynolds number Re_c readily obtained from (5.21) by setting $k_c = 0$ while δ_c is easily found from (5.56), also by setting $k_c = 0$. Although δ, ζ and $\mathcal{M}$ involve surface tension, i.e., κ, the transformation from δ_c to Re_c implies only a multiplication of all terms with κ, which then cancels out so that the critical Reynolds number Re_c does not depend on surface tension. This transformation from δ_c to Re_c, or for that matter from a dispersion relation in terms of the Shkadov scaling to one in terms of the Nusselt scaling, will be a necessary step, e.g., when we analyze the linear stability of the averaged models (which are written in terms of the Shkadov scaling, and hence their linear stability characteristics are expressed in terms of this scaling), each time a comparison is needed, either with full Navier–Stokes and Fourier equations (Orr–Sommerfeld) or with experiments.

5.5.2 Isothermal Vertical Films: Closed and Open Flows

Figure 5.6 shows the stability diagram obtained with the BE (5.55). The dashed and dot-dashed lines are the blow up boundaries computed using, respectively, the

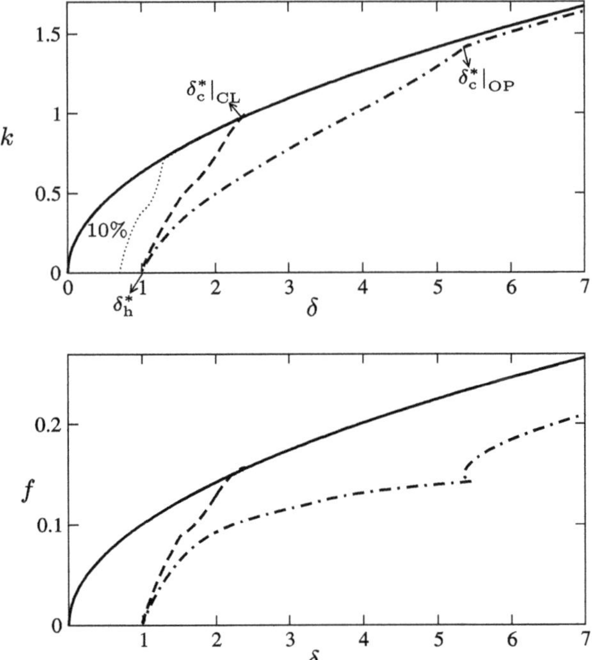

Fig. 5.6 Stability diagrams in the (k, δ), and (f, δ)-planes for an isothermal vertical film, i.e., $\zeta = \mathcal{M} = 0$: neutral stability curve (*solid line*) computed with (5.56), blow up boundaries indicated by the *dashed and dot-dashed lines* obtained with the closed and open flow conditions, respectively, and "accuracy" curve (*dotted line*), at which the maximum amplitude of the solutions with the BE for both flow conditions differs by 10% from the amplitude obtained by the first-order model to be introduced in Chap. 6 ($< 10\%$ *to the left* and $> 10\%$ *to the right*)

closed flow condition, $\langle h \rangle_\xi = 1$, and the open flow condition, $\langle h \rangle_\xi = (1/3 - q_0)/c$. The major difference with the open flow condition is that the Hopf bifurcation is supercritical for all δ, while for the closed flow condition the bifurcation is always subcritical for $\delta > \delta_c^*$, the value corresponding to Re_c^*. This then shows that close to criticality the BE should always give bounded solutions with the open flow condition. However, Fig. 5.6 also shows that for $\delta > \delta_c^*|_{OP}$ the corresponding region of k is very small. Interestingly, the stability diagram for the frequency $f = ck/2\pi$ versus δ (c is the phase speed of either the infinitesimal perturbations ($c = 1$) or the nonlinear traveling wave solutions) shows a wider band of bounded solutions with the open flow condition because of the phase speed c decreasing for $\delta > \delta_c^*|_{OP}$. In any case, comparison of the wave amplitudes computed with the BE and with the first-order model (which is valid in both the drag-gravity and the drag-inertia regime; Chap. 6), shows good agreement, i.e., less than 10% discrepancy, only in a small region of the stability diagram in Fig. 5.6.

The blow up features as marked in Fig. 5.6 read

$$\delta_h^* = 0.986, \qquad \delta_c^*|_{CL} = 2.358 \quad \text{and} \quad \delta_c^*|_{OP} = 5.401, \tag{5.57}$$

where the subscripts CL and OP indicate the corresponding flow condition. Note that since homoclinic orbits are solutions of infinite wavelength, they do not depend on the flow condition. δ_h^* being very close to unity demonstrates that indeed δ is the natural parameter to discriminate between drag-gravity and drag-inertia regimes, as first noted in Sect. 4.9.1 and to also show that BE blows up in the drag-inertia

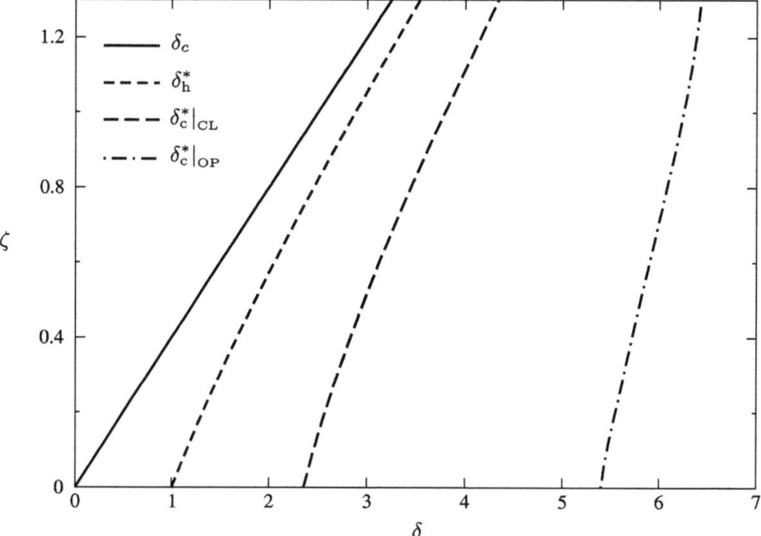

Fig. 5.7 Stability diagram in the (ζ, δ)-plane computed with (5.55) for an isothermal film, i.e., $\mathcal{M} = 0$, and showing the different blow up limits δ^*, along with the condition that defines the critical value δ_c

regime, i.e., for $\delta > 1$. Thus, $\delta = 1$ satisfactorily quantifies the transition between the two regimes, an estimate which was also made in Sect. 4.9.1 based on the behavior of the wave tail (see Fig. 4.8). As an example, the blow up features in the case of a water film at 20°C (physical properties are given in Appendix D.4) are $Re_h^* = 3.0$, $Re_c^*|_{CL} = 6.2$ and $Re_c^*|_{OP} = 12.2$. Computations done in terms of the Shkadov scaling δ for an isothermal vertical film are straightforwardly reverted back to the Nusselt scalings as follows: from $\delta = 3Re/\kappa = 3Re/We^{1/3}$ where $We = \Gamma/h_N^2$ and $h_N = (3Re)^{1/3}$, by fixing the liquid, hence Γ, a δ value can be easily converted to an h_N value, which in turn can be used to obtain the value of Re.

It appears that the range of Re for which solutions are bounded, at least for $k > k^*$, is larger with the open flow condition than with the closed one. However, even though they are bounded, these solutions overestimate by far the amplitude of the actual ones obtained from the first-order model to be introduced in Chap. 6 (compare with the "accuracy" limit drawn in Fig. 5.6).

Finally, from the definition of δ, we can also infer that the range of validity of the BE, i.e., the range of Re for which solitary waves (homoclinic orbits) are bounded, increases with the Kapitza number as $\Gamma^{3/11}$.

5.5.3 Influence of Inclination

Figure 5.7 depicts the stability diagram of BE in the (ζ, δ)-plane computed with (5.55) for an isothermal film, i.e., $\mathcal{M} = 0$. The solid line corresponds to the criticality condition $k_c = 0$, i.e., $\zeta = 2\delta_c/5$, above which the Nusselt flat film solution is stable. The short-dashed line indicates the boundary where homoclinic orbits blow up (δ_h^*), whereas at the long-dashed and the dotted-dashed lines all the linearly unstable modes blow up (δ_c^*) for closed (CL) and nearly all for open (OP) flow conditions, respectively. In between, the solutions are bounded only in the range of wavenumbers $[k^*, k_c]$ and unbounded in the range $[0, k^*]$.

As an example, the blow up features in the case of a water film at 20°C (again physical parameters are given in Appendix D.4) for a plate inclined at 10° from the horizontal are $Re_h^* = 5.6$, $Re_c^*|_{CL} = 8.0$ and $Re_c^*|_{OP} = 13.3$. These are computed by utilizing the vertical Kapitza number $\Gamma_\perp$, which is independent of the inclination angle, i.e., $\Gamma = \Gamma_\perp/(\sin\beta)^{1/3}$, to isolate the effect of inclination. Using (5.56), this leads to the relation

$$\zeta = \frac{\cos\beta}{(\sin\beta)^{10/11}}\left(\frac{\delta^2}{\Gamma_\perp^3}\right)^{1/11}, \tag{5.58}$$

from the definition of ζ. Increasing ζ reduces therefore the range of validity of the BE in the linearly unstable domain. Again, reverting back to the Nusselt scales is straightforward: from $\delta = 3Re/\kappa = 3Re/We^{1/3}$ where $We = \Gamma_\perp/[(\sin\beta)^{1/3}h_N^2]$ and $h_N = (3Re)^{1/3}$, by fixing the liquid, hence $\Gamma_\perp$ and β, a δ value can be easily converted to an h_N value, which in turn can be used to obtain the value of Re.

5.5.4 Influence of the Marangoni Effect in the Small Biot Number Limit: $B \ll 1$

For common liquids, the Biot number is usually in the range 10^{-2} to 10^{-3} (see Appendix D.4). Therefore, we can safely use the approximation

$$\frac{\mathcal{M}B}{(1+Bh)^2} \approx \mathcal{M}B \tag{5.59}$$

in (5.55) to interrogate the influence of the Marangoni effect with $\mathcal{M}B$ as a single parameter. In this limit, the generic equation (5.54) shows that the Marangoni term behaves as an "$m = 2$" term and does not lead by itself to singularity even though it can promote it.

Figure 5.8 represents the stability diagram of the BE in the $(\mathcal{M}B, \delta)$-plane for a vertical wall, i.e., $\zeta = 0$. Here, a range of unstable wavenumbers exists for all δ. The relation with physical properties is as follows:

$$\mathcal{M}B = Ma\,Bi\left(\frac{1}{\Gamma^4\delta}\right)^{1/11}, \tag{5.60}$$

obtained by eliminating Re from $\mathcal{M}B$ and δ using (5.56). The domain of existence for homoclinic orbits nearly vanishes for large $\mathcal{M}B$. However, from Appendix D.4, the product $Ma\,Bi$ remains small for common liquids. For instance, for a vertically

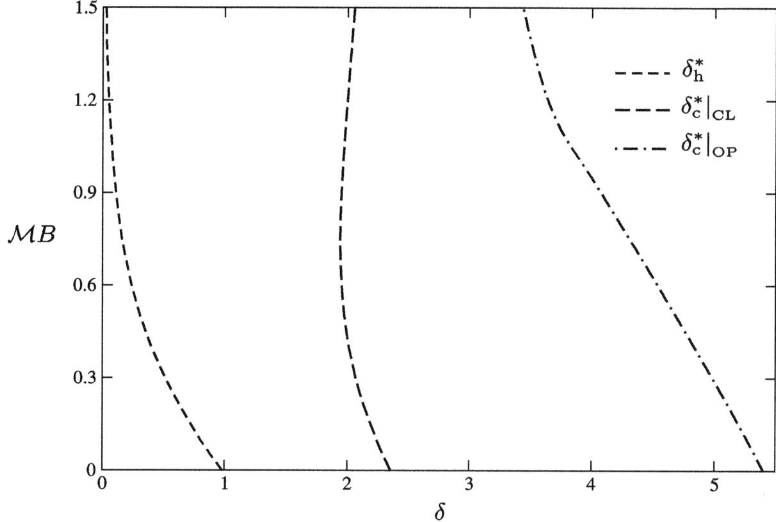

Fig. 5.8 Stability diagram in the $(\mathcal{MB}, \delta)$-plane for a vertically falling film, i.e., $\zeta = 0$. The legend is the same as for Fig. 5.7

falling water film at 20°C the value $Ma\,Bi = 2$ corresponds to $\Delta T = 28$ K when $\alpha = 100$ W m^{-2} K. In this case, the BE can be used with satisfactory accuracy up to $Re = 1.8$ ($\delta_h^{10\%} = 0.52$).

5.5.5 Subcritical Behavior of the BE

For the isothermal case, we have seen in Fig. 5.4 that the Hopf bifurcation computed with the BE is always subcritical for $Re > Re_c^*$, at least for the closed flow condition. In fact, the bifurcation becomes already subcritical for Reynolds numbers slightly smaller than Re_c^* as shown by the inset of Fig. 5.5: In the subdomain bounded from below by the solid line and by the dashed curves otherwise, the family of solutions has not only one but two saddle nodes or turning points. Let us here analyze how this branch behavior is influenced by the Marangoni effect and whether the subcritical bifurcation in this case may have a physical meaning.

Figure 5.9 displays the stability diagram for different values of $\mathcal{MB}$ and the blow up boundary when the closed flow condition is enforced. It shows that the subdomain where subcritical bifurcations are found extends toward smaller δ ($< \delta_c^*$) for increasing $\mathcal{MB}$. Notice that no difference was observed in the curves of Fig. 5.9 when we do not make the small Biot number limit approximation (5.59), but still using the common values of the Biot number given in Appendix D.4.

The question here is whether this subcritical behavior is really a meaningful consequence of the Marangoni effect. Actually, for a horizontal layer the Marangoni instability is subcritical and leads to a singularity associated with a touch down of

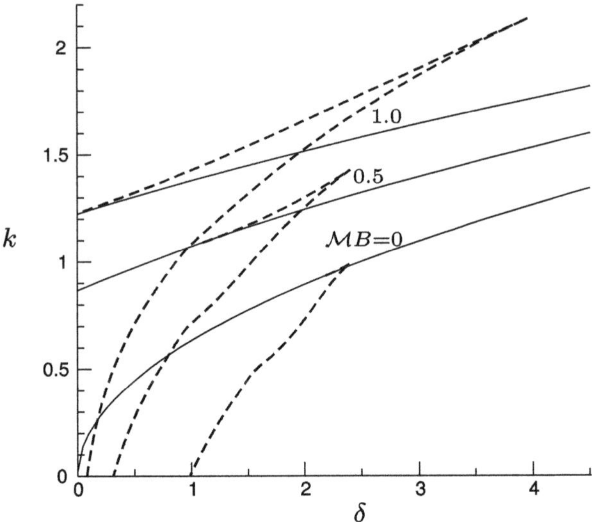

Fig. 5.9 Stability diagram in the (k, δ)-plane for a vertically falling film, i.e., $\zeta = 0$, and different values of $\mathcal{M}B$: cut-off wavenumbers k_c (*solid lines*) and blow up boundaries for closed flow condition (*dashed lines*)

the interface in finite time [289]. For the heated falling film problem, the studies in [139, 279] demonstrated that for sufficiently small Re the solitary wave characteristics (amplitude, phase speed) diverge to infinity. It was conjectured in these studies that this seemingly anomalous behavior is not associated with the absence of a solution, and hence it does not correspond to a true singularity formation, as forces of nonhydrodynamic origin, namely, van der Waals (long-range attractive) intermolecular forces that have not been included are predominant in the region of small Re (small film thicknesses), thus arresting the above divergence. In fact, it is quite likely that for the heated falling film problem, the inclusion of such forces will lead to a formation of a series of drops separated by "dry" patches. We can then infer here that the subcritical behavior of the BE including the Marangoni effect described above is unphysical, in contrast to the subcritical instabilities known for horizontal liquid layers. Let us give two more arguments in this direction: (i) as shown in [199], the second-order BE (5.13) does not exhibit such subcritical behavior; (ii) we find that for vertically falling films, more sophisticated models (presented in Chap. 6) never yield subcritical behavior.

5.5.6 Concluding Remarks on the Validity Domain of the BE

We have demonstrated that the blow up of solutions to the BE is related to the absence of homoclinic solutions. By tracking the transition between bounded and unbounded solutions, we defined the validity domain of the BE up to $\delta = 1$, a limiting value which is decreased by the Marangoni effect. Unfortunately, the addition of higher-order terms reduces the range of validity even more as the asymptotic series used to obtain the BE have usually poor convergence properties [232]. It is precisely the long-wave expansion that brings in increasingly high-order nonlinearities such as the inertia term $\partial_x(h^6 \partial_x h)$ in (5.12), which is the origin of the unphysical features of the BE. Hence, one attempt to overcome the limitations of the BE, but to remain in the framework of the long-wave assumption, is to "regularize" the inertia term, as shown in the next section.

5.6 Regularization with Padé Approximants

To remedy the singularity formation observed with the BE, Ooshida [196] developed a resummation technique inspired from the *Padé approximants* method (see, e.g., [18, 115]), appropriately extended to differential operators (the classical Padé technique is outlined in Appendix C.7). More specifically, for the case of an isothermal film he introduced a "regularization operator," $\mathcal{G} = \mathcal{I} + \varepsilon \mathcal{G}^{(1)} + \varepsilon^2 \mathcal{G}^{(2)}$, where $\mathcal{I}$ is the identity operator, $\mathcal{G}^{(1)} = G^{(1)}(h)\partial_x$ and $\mathcal{G}^{(2)} = G^{(2)}(h)\partial_{xx}$, so the expression of q as a function of h and its derivatives obtained from the long-wave expansion, formally written as $q \equiv \mathcal{Q}(h)$, is rewritten as $\mathcal{G}^{-1}\mathcal{F}$.

Let us initially implement the idea at first order since the singular behavior to be corrected is already present in (5.12). We need to choose $\mathcal{G}$ in the form, $\mathcal{I} + \varepsilon G^{(1)}(h)\partial_x$, with $G^{(1)}(h)$ to be determined so as to eliminate the dangerous terms in $\mathcal{F}$ when evaluating $\mathcal{G}\mathcal{Q}$:

$$\left(\mathcal{I} + \varepsilon G^{(1)} \partial_x\right)\left[\frac{1}{3}\left(h^3 + \frac{6}{35}\varepsilon Re\partial_x\left(h^7\right) - \frac{1}{4}\varepsilon Ct\partial_x\left(h^4\right) + \varepsilon^3 We h^3 \partial_{xxx} h\right)\right]$$

$$= \frac{h^3}{3} + \varepsilon\left[h^2 G^{(1)} \partial_x h + \frac{2}{35}Re\partial_x\left(h^7\right) - \frac{1}{12}Ct\partial_x\left(h^4\right) + \frac{1}{3}\varepsilon^2 We h^3 \partial_{xxx} h\right]$$

$$+ \mathcal{O}\left(\varepsilon^2\right), \tag{5.61}$$

where $G^{(1)}$ is adjusted to $(-2/5)Reh^4$ so that the terms that contain Re on the right hand side of (5.61) vanish. Evaluating the regularized identity $\partial_x(\mathcal{G}\mathcal{Q}) = \partial_x\mathcal{F}$ then gives

$$\partial_x \mathcal{Q} - \varepsilon\partial_x\left[\frac{2}{5}Reh^4\partial_x\mathcal{Q}\right] = \frac{1}{3}\partial_x\left[h^3 - \varepsilon Cth^3\partial_x h + \varepsilon^3 We h^3 \partial_{xxx} h\right].$$

Replacing $\partial_x \mathcal{Q}$ with $-\partial_t h$ (from (5.5) with $p = 0$) yields

$$\partial_t h + h^2 \partial_x h + \varepsilon\partial_x\left\{-\frac{2}{25}Re\partial_t\left(h^5\right) - \frac{Ct}{3}h^3\partial_x h + \frac{1}{3}\varepsilon^2 We h^3 \partial_{xxx} h\right\} = 0. \tag{5.62}$$

At second order, Ooshida chose rather to adjust the "coefficients" $\mathcal{G}^{(1)}$ and $\mathcal{G}^{(2)}$ in $\mathcal{G}$ so that $\mathcal{F}\,(=\mathcal{G}\mathcal{Q})$ could be reduced to $q^{(0)} + \varepsilon\mathcal{F}^{(1)}$, i.e., $\mathcal{F}^{(2)} \equiv 0$, which yields

$$\mathcal{G} = 1 - \frac{10}{7}\varepsilon Reh^4\partial_x - \varepsilon^2 h^2\partial_{xx}. \tag{5.63}$$

Computation of the regularized identity $\partial_x(\mathcal{G}\mathcal{Q}) \equiv \partial_x\mathcal{F}$ with the replacement of $\partial_x\mathcal{Q}$ with $-\partial_t h$ led Ooshida to the equation

$$\partial_t h + h^2\partial_x h + \varepsilon\partial_x\left\{-\frac{2}{7}Re\partial_t\left(h^5\right) - \frac{36}{245}Re\partial_x\left(h^7\right) - \frac{1}{3}Cth^3\partial_x h\right.$$

$$\left. + \frac{1}{3}\varepsilon^2 Weh^3\partial_{xxx}h\right\} - \varepsilon^2\partial_x\left(h^2\partial_{xt}h\right) = 0, \tag{5.64}$$

which can be rewritten using the Shkadov scaling:

$$\partial_t h + \frac{1}{3}\partial_x\left\{h^3 - \frac{2}{7}\delta\partial_t\left(h^5\right) - \frac{36}{245}\delta\partial_x\left(h^7\right)\right.$$

$$\left. - \frac{1}{4}\zeta\partial_x\left(h^4\right) + h^3\partial_{xxx}h - 3\eta h^2\partial_{xt}h\right\} = 0. \tag{5.65}$$

5.7 Generalization of the Single-Equation Model Obtained with Regularization

The main difference between the BE with $M = 0$ and the Ooshida equation (5.64) or (5.62) is that the inertia terms originating from $q^{(1)}$ appear as combinations of $\partial_t\left(h^5\right)$ and $\partial_x\left(h^7\right)$ with different weights. In fact, using the equivalence $\partial_t h = -h^2\partial_x h + \mathcal{O}(\varepsilon)$ one can write

$$h^6\partial_x h = \partial_x\left(\frac{1}{7}h^7\right) = -\partial_t\left(\frac{1}{5}h^5\right) + \mathcal{O}(\varepsilon). \tag{5.66}$$

Let us then investigate all combinations of $\partial_t\left(h^5\right)$ and $\partial_x\left(h^7\right)$ when second-order viscous effects are neglected, and thus consider

$$\partial_t h + h^2\partial_x h + \varepsilon\partial_x\left\{-\frac{2}{25}Re\Delta\partial_t\left(h^5\right) - \frac{2}{35}Re(\Delta - 1)\partial_x\left(h^7\right)\right.$$

$$\left. - \frac{1}{3}Cth^3\partial_x h + \frac{1}{3}\varepsilon^2 Weh^3\partial_{xxx}h\right\} = 0, \tag{5.67}$$

which can be rewritten in terms of the Shkadov scaling as follows:

$$\partial_t h + \frac{1}{3}\partial_x\left\{h^3 - \frac{2}{25}\delta\Delta\partial_t\left(h^5\right) - \frac{2}{35}\delta(\Delta - 1)\partial_x\left(h^7\right)\right.$$

$$\left. - \frac{1}{4}\zeta\partial_x\left(h^4\right) + h^3\partial_{xxx}h\right\} = 0, \tag{5.68}$$

and where Δ is a free parameter allowing us to recover different model equations, e.g., the BE is recovered when $\Delta = 0$. On the other hand, the combination

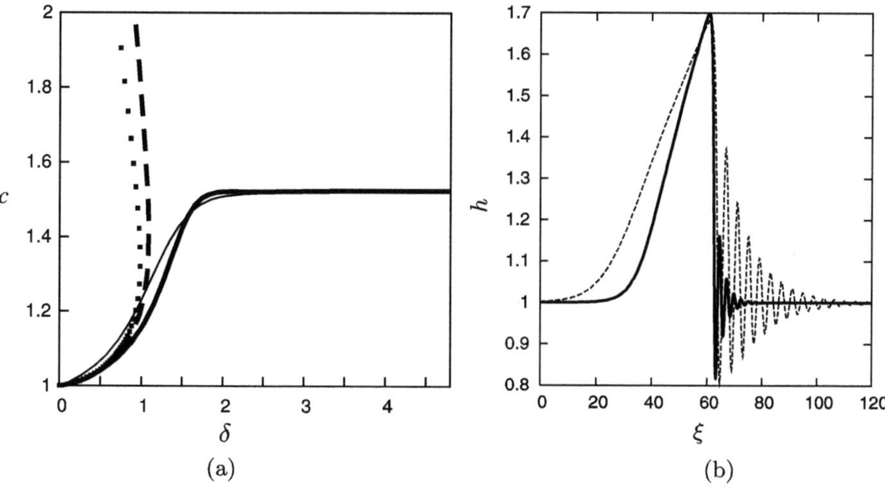

Fig. 5.10 (a) Speeds of single-hump homoclinic solutions of (5.68) as functions of the rescaled Reynolds number δ for a vertical wall ($\zeta = 0$): $\Delta = 0$ (*dotted line*), $\Delta = 1$ (*dashed line*), $\Delta = 25/7$ (*thick solid line*) and solutions to Ooshida's equation (5.65) with $\eta = 0.1$ (*thin solid line*). (**b**) Single-hump homoclinic solutions to (5.65) at $\delta = 5$ and $\zeta = 0$: $\eta = 0.1$ (*solid line*) and $\eta = 0$ (*dashed line*)

of the first-order terms $\partial_t(h^5)$ and $\partial_x(h^7)$ appearing in Ooshida's second-order equation (5.65) corresponds to $\Delta = 25/7 \simeq 3.6$ while Ooshida's first-order equation (5.62) is recovered when $\Delta = 1$.

Recall that the blow up behavior of the BE occurs at $\delta = 1$, which also corresponds to the transition between the drag-gravity and drag-inertia regimes. This then suggests that the reduced Reynolds number δ is the natural parameter for validation purposes/assessment of the validity of a model that aims to describe the falling film dynamics. Figure 5.10 displays the speeds of solitary waves as functions of the reduced Reynolds number δ for the different values of the parameter Δ. These solutions have been computed by continuation using AUTO-07P. The computational methodology for tracking solitary wave solutions is detailed in Appendix F.2 for the BE in terms of the Nusselt scaling. The associated codes given can be readily modified to tackle (5.68) so that the continuation is done with respect to δ with the other parameters fixed, i.e., ζ for the case considered here. For given liquid and Ct as would be the case in a real experiment, the continuation can be done through Re as both δ and ζ depend on Re. This way, plots of the form $c = c(Re)$ for a given Ct are automatically generated. Such plots are important for comparisons with experiments and DNS.

It can be seen that the solution branches for the homoclinic orbits of the BE (5.55) (dotted line) exhibit a turning point at $\delta \approx 1$, in accordance with (5.57), so that solitary wave solutions are not expected for larger values of δ. The curve corresponding to $\Delta = 1$, i.e., (5.62), remains close to the one obtained with $\Delta = 0$ and similarly exhibits a turning point. By contrast, when $\Delta = 25/7$ (thick solid line) the curve does not turn back as δ increases. Equation (5.68) with $\Delta = 25/7$ differs from Ooshida's equation (5.65) only through the absence of the viscous term $-\eta\partial_x(h^2\partial_{xt}h)$. However, this viscous contribution does not play a significant role in the regularization process. Comparisons of the speed of the solitary waves (thick and thin solid lines in Fig. 5.10a) in fact show little difference. Viscous dispersion does not affect significantly the maximum amplitude of the waves, however, it does have a significant effect on the overall shape of the wave envelope and the number of visible oscillations preceding the main hump as seen in Fig. 5.10b for $\eta = 0.1$, and as we first announced in Sect. 4.3. Notice here that although, strictly speaking, homoclinic solutions are solutions of infinite wavelength, in practice their wavelength/computational domain is taken sufficiently large until we achieve solutions that are domain-independent. The computation of homoclinic solutions is done using AUTO-07P with the HOMCONT option for tracing homoclinic orbits.

A consequence of the regularization of the free-surface equation, is the existence of homoclinic solutions for all Reynolds numbers, which in turn alleviates the singularity formation observed with BE. In fact, computation of homoclinic solutions to (5.68) shows that the regularization can be achieved without much algebra by simply modifying the inertia terms entering the equation with the help of the equivalence, $\partial_t h = -h^2\partial_x h + \mathcal{O}(\varepsilon)$. Unfortunately, the regularization procedure is not sufficient to obtain quantitative agreement with experiments and DNS in the region of moderate Reynolds numbers. As already noted by Ooshida [196], the amplitudes and speeds of solitary waves in this region are grossly underestimated by (5.65). This is a direct consequence of slaving the dynamics of the film to its kinematics. On the other hand, as we shall demonstrate in Chap. 7, in the drag-gravity regime the regularized equation performs well but so does the BE. Similarly, (5.68) fails to predict accurately the solitary waves in the region of moderate Reynolds numbers for all values of Δ we explored (besides the specific values previously quoted). This calls for a different approach that would allow accurate modeling in the widest possible range of Reynolds numbers. The quest for such an approach is described in the next chapter.

Chapter 6
Modeling Methodologies for Moderate Reynolds Number Flows

Nearly all low-dimensional models for isothermal films at moderate Reynolds numbers found in the literature rely on a fundamental closure assumption for the streamwise velocity field: a simple self-similar velocity profile with the variables (x, t) and y/h separated. This is the basis for the classical Kapitza–Shkadov model. The velocity profile in this model is a self-similar semi-parabolic velocity profile in which the variables are separated as above and which trivially satisfies the x component of the momentum equation at zero Reynolds number (in which case the interface is flat). In this chapter we discuss a systematic methodology to relax the self-similar assumption while maintaining separation of variables as in the long-wave theory: it is based on a combination of an expansion for the velocity field in terms of polynomial test functions, the gradient expansion and an elaborate averaging technique that utilizes the method of weighted residuals. The averaging can be justified by the *in-depth coherence* of the flow ensured by the action of viscosity. The result is two "optimal" models in the sense that the models are always the same independently of the particular averaging methodology employed (provided of course that a sufficiently large number of test functions is used in each case). The two models are: A two-equation system consistent at $\mathcal{O}(\varepsilon)$, referred to as the *first-order model*, and a four-equation system consistent at $\mathcal{O}(\varepsilon^2)$, referred to as the *full second-order model*. An ad-hoc compromise between the two in both complexity and accuracy is provided by the *simplified second-order model*, whereas a regularization procedure enables us to reduce the dimension of the four-equation system and to obtain a two-equation model consistent at $\mathcal{O}(\varepsilon^2)$ which we refer to as the *regularized model*. The weighted residuals formulation developed here is compared to the center-manifold analysis by Roberts. The momentum equation in Robert's model contains all the terms of the momentum equation of the first-order model but with different coefficients. However, it also contains additional terms including high-order nonlinearities which then necessarily restrict the applicability of the model in the drag-gravity regime. On the other hand, the average models we obtain from the weighted residuals formulation, are capable of describing the drag-inertia regime, even though the formulation presumes that inertia effects are weak corrections to the balance of viscous drag and gravity, which strictly speaking holds only in the drag-gravity regime. The reason

S. Kalliadasis et al., *Falling Liquid Films*, Applied Mathematical Sciences 176,
DOI 10.1007/978-1-84882-367-9_6, © Springer-Verlag London Limited 2012

that the average models are capable of describing the drag-inertia regime is that the nonlinearities in these models do not lead to the unphysical loss of the solitary-wave branch of solutions at $\delta > 1$ and hence they also cure the deficiencies of the long-wave theory/BE (Benney equation) in the drag-inertia regime. Traveling-wave solutions of the averaged models are compared favorably with DNS demonstrating that indeed low-dimensional modeling of films flows in the drag-inertia regime can be achieved in terms of a small number of coupled evolution equations.

6.1 Background and Motivation

In Chap. 5 we outlined the long wave theory, which leads to the BE. But the BE does have its shortcomings, the main one being an unphysical behavior in the drag-inertia regime, i.e., for $\delta > 1$. The regularization method proposed by Ooshida [196] enables us to obtain an evolution equation without any blow up in finite time, but it does not reproduce quantitative features of the drag-inertia regime. In this regime inertia plays a significant role, as opposed to the drag-gravity regime (which corresponds to a balance between viscous drag on the wall and gravity acceleration) where inertia plays a perturbative role. It is precisely because inertia terms were considered as first-order perturbations in the gradient expansion parameter ε of the drag-gravity balance, which in turn corresponds to the Nusselt flat film flow, that the BE can be derived.

However, even though strictly speaking in the drag-inertia regime, inertia terms cannot be simply considered as perturbations, experiments [4] and DNS [176] clearly prove that departures of the streamwise velocity distribution from the Nusselt flat film parabolic profile are still small, with the exception of the steep front of a pulse. This remarkable *in-depth coherence* of the flow is due to the action of viscosity and suggests that the elimination of the cross-stream y dependence of the equations is still possible in the drag-inertia regime. This is achieved by an averaging procedure across the film combined with a projection in terms of amplitude functions that depend only on the location on the plate x and time t. The outcome is a small number of coupled evolution equations for the amplitude functions. As we shall demonstrate in this chapter and in Chap. 7, these equations capture the dynamics of the film both in the drag-gravity and drag-inertia regimes. The fundamental reason for the inability of the BE to describe nonlinear waves far from criticality (even though departures from the streamwise velocity distribution from the Nusselt flat film parabolic profile are small) is the slaving of all flow variables to the film thickness h. Introducing more degrees of freedom through an averaging procedure as done in this chapter enables us to move from the description of the motion of a fluid particle on the interface to that of the motion of a column of fluid between the wall and the interface, a consequence of the in-depth coherence mentioned above.

In the majority of cases studied in the literature, interface equations for film flows are based on the mass conservation equation (5.5) rewritten here for clarity,

$$\partial_t h + \partial_x q = 0, \tag{6.1}$$

which is the integral version of the kinematic boundary condition, along with a closure on the flow rate, $q = \mathcal{Q}(h)$, which could be an explicit function of h such as $q = \frac{1}{3}h^3$ (which in turn leads to an evolution equation typical of *kinematic waves* [299]), a power series of ε involving h and its derivatives, as in the long wave theory/BE, or a combination of a series and a differential operator as in Ooshida's regularization technique. In all cases the closure $q = \mathcal{Q}(h)$ expresses the slaving of the velocity field to the evolution of the film thickness. However, it seems that the only way to correctly handle the film flow dynamics in the drag-inertia regime, i.e., in the region of moderate Reynolds numbers, is to relax this slaving, and to recognize that in this regime, q, and possibly other quantities become genuine degrees of freedom at moderate Reynolds numbers. This idea is at the core of probably all efforts to model the film flow dynamics at moderate Reynolds numbers. Most of the corresponding models are based on *in-depth averaging* of the original equations, i.e., averaging of the equations across the film, and certain assumptions on the functional form of the dependence of the velocity on the film thickness.

The starting point of such averaging approaches is the boundary layer approximation developed in Chap. 4. The velocity field is then projected onto a set of polynomial test functions followed by averaging of the resulting equations using the method of weighted residuals. Recall that in the boundary layer approximation inertia terms can be taken of the same order as the gravity and viscous drag, i.e., δ is at most of $\mathcal{O}(1)$. However, to make progress we utilize a gradient expansion for the velocity field, which can only be justified rigorously when inertia plays a weak/perturbative role; that is, in the drag-gravity regime (δ less than unity). Following then the long wave/BE approach, our highly nonlinear problem is converted into a sequence of solvable problems. Hence, the velocity field is obtained explicitly, at each step in terms of functions of h but with amplitudes that are independent of h e.g., to leading order, the amplitude of the velocity field contains the flow rate q (note that it is only asymptotically close to the instability threshold that the amplitudes are connected to h so that the BE can be recovered).

Therefore, contrary to the long wave theory/BE expansion where all variables are slaved to h, the averaging approach gives a higher level of flexibility to the velocity, thus allowing it to have its own evolution. This is the fundamental difference between the averaging procedure and classical long wave theory. Yet, the in-depth coherence of the flow is still ensured by gravity and viscosity with inertia playing a perturbative role. Surface tension is also assumed small. Therefore, strictly speaking the derivation is based on the drag-gravity regime and we violate our basic assumption of small inertia and surface tension when we start investigating the drag-inertia regime. But the averaged models perform well there for the following reasons:

(i) The velocity profile is close to the semiparabolic one obtained from the balance between viscous drag and gravity, a consequence of the in-depth coherence of the flow as mentioned above. A significant deviation, however, from the semiparabolic profile occurs close to the steep front of a solitary pulse. On the other hand, for the derivation of the simplest of the averaged models, the so-called Kapitza–Shkadov model, inertia and surface tension are not considered small. But the basic assumption of this model, a single test function—the semi-parabolic profile of the drag-gravity regime—is only verified in the inertia-less limit, $Re \to 0$. Hence, almost by definition, this model cannot capture inertia correctly.

(ii) The averaged models capture the back of a solitary pulse, capillary ripples at its front and the amplitude and speed of the pulses. Hence they resolve what is really important for the dynamics (for example, recall that the capillary ripples are crucial in the wave interaction problem). The steep front of a solitary pulse is not crucial for the dynamics, as it is, e.g., with the Euler equations in gas dynamics, which capture the shock dynamics without resolving the shock front. It is possible to describe correctly the front of a pulse but then we would need to relax a key technical assumption in our derivation process: the dependence of the deviation amplitudes to the amplitude of the leading-order test function. But then the averaging methodology would become equivalent to a full scale numerical projection, and we might as well use full Navier–Stokes instead of the low-dimensional approximations we aim to obtain.

In the models obtained from the averaging procedure we then scale away ε. Subsequently the models are recast by using the Shkadov scaling. Recall that $\delta \approx 1$ nicely demarcates the two regimes, drag-inertia and drag-gravity. In fact, the reduced Reynolds number δ is the natural parameter for validation purposes/ assessment of the validity of a model that aims to describe the falling film dynamics (the BE blows up in finite time precisely at $\delta = 1$). A primary test then for any new model consists of the construction of solitary waves as far as possible in the drag-inertia regime for $\delta > 1$.

The compression factor κ of the Shkadov scaling "freezes" the ordering parameter ε to $1/\kappa$ and brings at order unity the streamwise pressure gradient induced by surface tension $\propto \partial_{xxx}h$. While this is advantageous from a numerical point of view as noted in Sect. 4.6, it can also be a drawback. Recall from Sect. 4.7 where we analyzed the balance of different forces for a solitary pulse, that the pressure gradient induced by surface tension is always comparable to inertia. Surface tension of order unity thus implies inertia effects of order unity. For this reason, the choice of the compression factor κ is strictly speaking relevant only for large-amplitude waves, i.e., when $\delta \sim 1$. Yet, our modeling approach is based on the in-depth coherence of the flow enabled by the balance between gravity and viscosity and assumes small inertia effects, i.e., δ small, or equivalently small-amplitude waves. For this reason, our formulation will be presented using the Nusselt scaling and ε, thus making explicit the order of magnitude of the different terms on ε. The Shkadov scaling is not appropriate for the derivation of the different models as already noted in Sect. 4.7; starting from the boundary layer equations in terms of the Nusselt scaling and ε allows us to clearly identify the relative orders of magnitude of the different terms, which is essential for the derivation of the averaged models.

The Shkadov scaling is then introduced only at the final stage. Once again, the Shkadov scaling is strictly speaking valid for large-amplitude waves, i.e., when inertia and surface tension are not small. But technically, the introduction of the Shkadov scaling does not invalidate our assumption of small inertia and surface tension.

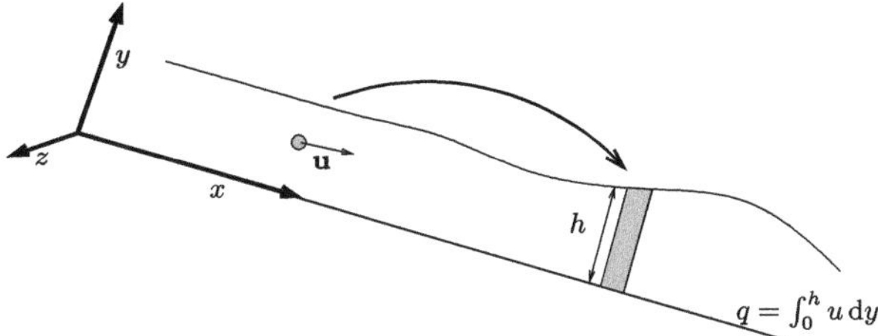

Fig. 6.1 Sketch of a falling liquid film: from local original quantities to in-depth averaged quantities

Rescaling the equations with the Shkadov scaling does not change their solution; it is the way we present the solution that changes.

For simplicity this chapter focuses on isothermal films. The methodologies developed here will be extended to nonisothermal flows in Chap. 9.

6.2 Averaged Two-Equation Models

In this section we present most of the available models using only two coupled evolution equations for the film thickness h and the local flow rate q. The starting point of the derivation of such models is typically the in-depth averaging of the basic equations across the fluid layer. This simple idea is an application of the Kármán–Polhausen technique used in boundary layer theory in aerodynamics (see, e.g., [243]) and was first applied to film flows by Kapitza [140].

6.2.1 Kapitza–Shkadov Model

The key to describing the dynamics of the film via additional degrees of freedom instead of using a single variable h is to turn our attention from the description of a fluid particle on the surface to that of a "slice" of fluid of infinitesimal thickness, for which equations that determine the evolution of "integral quantities" can be sought. The first of these quantities is an obvious one, the thickness of the film h associated with the mass of the fluid slice. Another one is the flow rate, $q = \int_0^h u \, dy$, associated to its momentum (see Fig. 6.1).

Numerical simulations by Chang et al. [50] have shown that the basic features of the different wave regimes on the falling film can be satisfactorily captured by the first-order boundary layer equations derived in Chap. 4. For purposes of clarity and convenience we recall here the main steps of the derivation of the boundary layer equations. When we consider slow space and time modulations, a small parameter ε is introduced formally along with each derivative in time or space, $\partial_{x,t} \propto \varepsilon$. The assumed slow space variation implies that the velocity component normal to the plate v is much smaller than the streamwise component u, as imposed by the continuity equation rewritten below in two dimensions:

$$\partial_x u + \partial_y v = 0. \tag{6.2}$$

Consequently, the inertia terms in the y component of the momentum equation are of higher order and can be neglected. The resulting equation is a linear one and is subject to the normal stress balance, which to leading order contains the effect of surface tension only. It is then integrated once to yield the pressure distribution across the film. After substitution of the latter into the x component of the momentum equation and truncation at $\mathcal{O}(\varepsilon)$, the streamwise momentum equation reads

$$3\varepsilon Re(\partial_t u + u\partial_x u + v\partial_y u) = 1 + \partial_{yy} u - \varepsilon Ct\partial_x h + \varepsilon^3 We\partial_{xxx} h. \tag{6.3}$$

Boundary conditions are the no slip and no penetration condition on the wall

$$u|_0 = v|_0 = 0, \tag{6.4}$$

the (exact) kinematic boundary condition at the free surface

$$\partial_t h + u|_h \partial_x h = v|_h, \tag{6.5}$$

which is equivalent to the mass balance (6.1), and the tangential stress balance, which at $\mathcal{O}(\varepsilon)$ reduces to

$$\partial_y u|_h = 0. \tag{6.6}$$

Equations (6.3)–(6.6) are the two-dimensional, first-order boundary layer equations. They are consistent at $\mathcal{O}(\varepsilon)$, i.e., the neglected terms are all of higher order.

Integration of the continuity equation leads to the mass balance (6.1) for the fluid slice. Turning to the momentum balance of the fluid slice, the streamwise boundary layer momentum equation is integrated across the depth from $y = 0$ to $y = h$ to furnish

$$\int_0^h \left[3\varepsilon Re(\partial_t u + u\partial_x u + v\partial_y u) - \partial_{yy} u\right] dy = h - \varepsilon Cth\partial_x h + \varepsilon^3 Weh\,\partial_{xxx} h. \tag{6.7}$$

Recasting the inertia terms as

$$\int_0^h (\partial_t u + u\partial_x u + v\partial_y u)\, dy = \int_0^h (\partial_t u + u\partial_x u - u\partial_y v)\, dy + [uv]_0^h,$$

and using the continuity equation (6.2), the boundary condition (6.4) and with the help of the kinematic boundary condition (6.5), the inertia terms can be rewritten as

$$\int_0^h (\partial_t u + 2u\partial_x u)\, dy + u|_h \partial_t h + u^2|_h \partial_x h.$$

Thus (6.7) reads

$$3\varepsilon Re\left[\partial_t \int_0^h u\, dy + \partial_x \int_0^h u^2\, dy\right] = h - \varepsilon Cth\partial_x h + \varepsilon^3 Weh\partial_{xxx} h - \partial_y u|_0, \tag{6.8}$$

or equivalently

$$3\varepsilon Re\left[\partial_t q + \frac{\partial}{\partial x}\left(\frac{\Upsilon q^2}{h}\right)\right] = h - \varepsilon\,Cth\partial_x + \varepsilon^3 Weh\partial_{xxx} - \tau_{\mathrm{w}}, \qquad (6.9)$$

where $\tau_{\mathrm{w}} = \partial_y u|_0$ is the shear at the wall and Υ is a *shape factor* defined as [214]

$$\Upsilon = \frac{h}{q^2}\int_0^h u^2\,dy, \qquad (6.10)$$

which relates the "first moment" of u, $\int_0^h u\,dy$, to its "second moment" $\int_0^h u^2\,dy$. The set of two equations (6.1) and (6.9) is closed provided that expressions of the shape factor Υ and the wall shear τ_{w} as respective functions of h and q are known.

Kapitza [140] and Shkadov [248] solved this problem by applying a closure hypothesis which is quite common in boundary layer theory in aerodynamics (e.g., in the treatment by Blasius of the semi-infinite stationary boundary layer, see [243], Chap. XII), although we actually do not have a true boundary layer. Kapitza and Shkadov assumed the velocity field to be *self-similar*. This means that two velocity distributions, as functions of the cross-stream coordinate y at two different locations on the plate x_1 and x_2, can be related through

$$\frac{u(x_1, y/Y_1)}{U_1} = \frac{u(x_2, y/Y_2)}{U_2}, \qquad (6.11)$$

where U_1, U_2, Y_1 and Y_2 are constants depending on the considered locations. The flow geometry imposes $Y_1 = h(x_1)$ and $Y_2 = h(x_2)$. Kapitza and Shkadov therefore assumed the velocity distribution to coincide locally with its Nusselt flat film solution:

$$u(x, y, t) = 3\frac{q(x, t)}{h(x, t)}\left[\frac{y}{h(x, t)} - \frac{1}{2}\left(\frac{y}{h(x, t)}\right)^2\right]. \qquad (6.12)$$

This ad hoc but convenient assumption is actually supported by experimental results [4] and by DNS [176] that show that departures of the velocity from a parabolic distribution across the layer are weak. Still, the long wave theory developed in Chap. 5 that leads to the BE shows that even at first order in ε, corrections to (6.12) do exist. In fact, if one is to obtain rigorously a parabolic distribution in the form (6.12) from the boundary layer equations, the inertia terms in the momentum balance (6.3) must be neglected. Consequently, the implicit assumption underlying (6.12) is that a semiparabolic velocity profile that satisfies trivially the x component of the momentum equation for zero Reynolds numbers persists, even in the region of moderate Reynolds numbers when the interface is no longer flat. This is possible through the action of viscosity and because the pressure distribution can be assumed to be constant across the fluid layer (again this is for a vertical plane; for a horizontal one the pressure is affected by the hydrostatic head). These can then ensure that the distribution of the velocity across the layer adapts instantaneously to the deformations of the free surface and to changes of the flow rate. However, the velocity profile is not totally slaved to the film thickness h unlike what the long wave theory posits: the quadratic function in the brackets of (6.12) is slaved to h but not its coefficient, which depends on q.

Inserting (6.12) into the definition (6.10) of the shape factor gives the constant value, $\Upsilon = 6/5$. Similarly, the wall shear is simply $\tau_{\mathrm{w}} = 3q/h^2$. We thus get a closed system of two evolution equations. These equations are written below using the Shkadov scaling through the formal transformation $\{3\varepsilon Re \to \delta,\ \varepsilon Ct \to \zeta,\ \varepsilon^3 We \to 1$ and $\varepsilon^2 \to \eta\}$:

$$\partial_t h = -\partial_x q, \tag{6.13a}$$

$$\delta\partial_t q = h - 3\frac{q}{h^2} - \delta\frac{12}{5}\frac{q}{h}\partial_x q + \left(\delta\frac{6}{5}\frac{q^2}{h^2} - \zeta h\right)\partial_x h + h\partial_{xxx}h, \tag{6.13b}$$

the Kapitza–Shkadov model first discussed in the Introduction.[1] We note, however, that the system (6.13a), (6.13b) is also frequently referred to as the *integral-boundary-larger* (IBL) model (see Introduction also), thus underlying the analogy between its derivation and the Blasius boundary layer theory. The inertia terms in this model appear in the averaged momentum balance (6.13b) through the parameter δ. The first two terms of the right hand side correspond to the streamwise gravity acceleration and the viscous drag, whereas the two last ones account for the stabilizing effects of gravity ($\propto \zeta$) and surface tension, respectively.

Solutions to (6.13a), (6.13b) agree qualitatively with both experimental data and DNS [50, 250, 253]. Yet, the model does have some limitations the main of which we can identify by simply performing a gradient expansion for the flow rate in the form, $q = q^{(0)} + \varepsilon q^{(1)} + \ldots$, then truncating it at first order and solving for $q^{(1)}$. This yields an expression for q that is identical with the one predicted from the long wave/BE expansion of Chap. 5, with the exception of a coefficient $1/9$ for the inertia-h^6 term of $q^{(1)}$ instead of the correct factor $2/15$.[2] This 20% error is analogous to the error introduced by the Kármán–Polhausen averaging approach in the case of a boundary layer along a semi-infinite plate, where the simplest polynomial velocity distribution verifying the boundary condition, a simple linear profile, leads to a 13% error for the prediction of the shear stress at the plate.

---

[1] In fact, using the self-similar assumption (6.12), Kapitza formulated a system of equations in terms of the thickness h and the averaged velocity $\bar{u} = \frac{1}{h}\int_0^h u\,dy$ that is nearly equivalent to (6.13a), (6.13b). However, as pointed out in the Introduction, Kapitza's original study [140] focused on stationary waves only. In addition, he omitted the term $v\partial_y u$, assuming it is much smaller than $u\partial_x u$. This erroneous derivation was actually included in the first edition of Levich's monograph *Physicochemical Hydrodynamics* [163]. Shkadov [248] corrected Kapitza's derivation and he was the first to apply the boundary layer approximation to two-dimensional nonstationary film flows. The extension of the boundary layer approximation to three-dimensional flows was done by Demekhin and Shkadov [71].

[2] Although the BE breaks down at $\delta = 1$, it is very useful as a benchmark equation: Any model attempting to cure the deficiencies of the BE for $\delta > 1$, should yield the BE with an appropriate gradient expansion, thus confirming the validity of the model close to criticality.

The first visible consequence of the inaccuracy of the h^6-term coefficient is an erroneous estimation of the linear instability threshold, i.e., $Re_c^{(KS)} = Ct$ instead of the exact $Re_c = \frac{5}{6}Ct$, as obtained from both Orr–Sommerfeld (Chap. 3) and the BE (Chap. 5). This underestimation of the critical Reynolds number of the primary instability limits the use of the Kapitza–Shkadov model to the vertical wall case $(Ct = 0)$, where it predicts the correct result that the flow is unstable at all Reynolds numbers; still its limitations will show up at the nonlinear stage.

6.2.2 Higher-Level Models Based on the Self-similar Closure

Following the Kapitza–Shkadov approach, several studies have been devoted to film flows by using the self-similar closure assumption (6.12), mostly aiming at a remedy to the limitations of the Kapitza–Shkadov model. For instance, (6.13b) was augmented with second-order viscous contributions in the bulk ($\propto \eta \partial_{xx} u$) and at the interface ($\propto \eta[2\partial_x h(\partial_x u|_h - \partial_y v|_h) - \partial_x v|_h]$) [214]. Applying the Kármán–Polhausen technique on the second-order boundary layer equations, which are slightly modified to also account for supplementary terms coming from surface tension and produced by the gradient expansion of the surface curvature, $\partial_{xx} h/[1 + (\partial_x h)^2]^{3/2}$, one obtains the following expression for the pressure at the interface:

$$p|_h = 2\varepsilon^2 \partial_y v|_h - \varepsilon^3 We\left[\left(1 - \varepsilon^2 \frac{3}{2}(\partial_x h)^2\right)\partial_{xxx} h - 3\varepsilon^2 \partial_x h(\partial_{xx} h)^2\right].$$

Written in terms of the Shkadov scaling, the resulting averaged momentum balance is

$$\delta \partial_t q = h - 3\frac{q}{h^2} - \frac{12}{5}\delta\frac{q}{h}\partial_x q + \left(\frac{6}{5}\delta\frac{q^2}{h^2} - \zeta h\right)\partial_x h$$

$$+ h\left[\left(1 - \frac{3}{2}\eta(\partial_x h)^2\right)\partial_{xxx} h - 3\eta \partial_x h(\partial_{xx} h)^2\right]$$

$$+ \eta\left[6\frac{q}{h^2}(\partial_x h)^2 - \frac{6}{h}\partial_x q \partial_x h - 6\frac{q}{h}\partial_{xx} h + 5\partial_{xx} q\right], \tag{6.14}$$

to be used in conjunction with the mass conservation equation (6.1). Though introducing $\mathcal{O}(\varepsilon^2)$ corrections, the system (6.1)–(6.14) suffers from the same drawback as the Kapitza–Shkadov model (6.13a), (6.13b) as it still leads to an erroneous critical Reynolds number, which in turn limits its use to vertical walls. Even worse, the extra surface tension terms, $\propto h(\partial_x h)^2 \partial_{xxx} h$ and $\propto h\partial_x h(\partial_{xx} h)^2$, are strongly nonlinear. The computation of solitary pulse solutions to (6.1)–(6.14) reveals the presence of a limiting Reynolds number above which solitary wave branches of solutions disappear. This is unacceptable on physical grounds, much like with the BE (Chap. 5).

Other approaches have been devoted to relaxing some of the underlying assumptions of the boundary layer approximation, especially the neglect of the cross-stream inertia terms, $Dv/Dt \equiv (\partial_t + u\partial_x + v\partial_y)v$, in the y component of the momentum balance [9, 161, 288]. The idea then is to retain cross-stream momentum in the y component of the momentum equation and to substitute there the self-similar closure assumption (6.12). However, this leads to rather involved expressions for the pressure as function of h and q, which we shall not write here. After averaging the streamwise momentum equation, a system of two equations is obtained consisting of the streamwise momentum balance and the mass balance (6.1). Unfortunately, due to the nonlinearities in such models, unphysical singularities appear in certain regions of the parameter space when one studies traveling wave solutions. Moreover, even though the models are more elaborate, the self-similar closure assumption is still maintained. Alas! The self-similar semiparabolic profile (6.12) holds partly because the pressure distribution can be assumed to be constant across the fluid layer (once again for a vertical layer). At the same time, as emphasized in Sect. 4.8.2, comparisons with DNS suggest that the pressure in the film is well-approximated by $p \approx Ct(h - y) - \varepsilon^2 We\partial_{xx}h$ so that the cross-stream inertia terms have little effect on the waves. Therefore, one may question the purpose of relaxing the approximation of constant pressure distribution across the film while at the same time making use of the closure assumption (6.12).

6.3 Center Manifold Analysis

One way to avoid the self-similar parabolic closure (6.12) has been proposed by Roberts [221, 223]. The idea comes from the "center manifold" analysis for finite-dimension systems of nonlinear ordinary differential equations (see, e.g., [38, 111]), and its extension to partial differential equations may be sketched as follows. Assume that the evolution of a set of physical variables $\mathbf{u}$ is governed by the differential equation

$$\partial_t \mathbf{u} = \mathcal{L}\mathbf{u} + \mathbf{N}(\mathbf{u}, \boldsymbol{\epsilon}), \tag{6.15}$$

where $\boldsymbol{\epsilon}$ is a vector of parameters, $\mathcal{L}$ is a linear matrix-differential operator that describes the flow dynamics close to the origin, $(\mathbf{u}, \boldsymbol{\epsilon}) = (\mathbf{0}, \mathbf{0})$, and $\mathbf{N}$ is a nonlinear functional of $\mathbf{u}$ and $\boldsymbol{\epsilon}$. Several physical systems are described by systems of equations of this form.

Assume now that the linear operator $\mathcal{L}$ has n eigenvalues with zero real part and all other eigenvalues have negative real parts. $\mathbf{u}$ can be projected onto the eigenfunctions of $\mathcal{L}$. If $\mathbf{a}$ is the vector of the amplitudes or "modes" of the eigenfunctions of $\mathcal{L}$ with zero real part in the projection for $\mathbf{u}$, the dynamics of the "flow" in a small neighborhood of the origin in the $(\mathbf{u}, \boldsymbol{\epsilon})$-space is governed by the n modes, i.e., by $\mathbf{a}$. This means that the n-dimensional vector $\mathbf{a}$ of the associated amplitudes satisfies in the small neighborhood of the origin

$$\partial_t \mathbf{a} = \mathbf{G}(\mathbf{a}, \boldsymbol{\epsilon}) \quad \text{such that} \quad \mathbf{u} = \mathbf{U}(\mathbf{a}, \boldsymbol{\epsilon}), \tag{6.16}$$

where the "hypersurface" $\mathcal{C}$ of equation $\mathbf{u} = \mathbf{U}(\mathbf{a})$ is the center manifold and $\partial_t \mathbf{a} = \mathbf{G}$ is the n-dimensional model of the dynamics. The existence of the center manifold is then assured by the convergence of the solution to $\mathcal{C}$: Suppose a solution $\mathbf{u}(t_0)$ of (6.15) lies at time t_0 in a small neighborhood at the origin in the $(\mathbf{u}, \boldsymbol{\epsilon})$-space. Then there exists a trajectory $\mathbf{U}(\mathbf{a})$ on the center manifold that verifies

$$\left\| \mathbf{u}(t_0 + t) - \mathbf{U}\big(\mathbf{a}(t_0 + t)\big) \right\| = \mathcal{O}\big(\exp(-\alpha t)\big) \quad \text{for } t > 0 \tag{6.17}$$

where $\| \cdot \|$ is an appropriately chosen norm and $-\alpha$ is some upper bound on the negative real part of the eigenvalues of $\mathcal{L}$. In Appendix C.8 we sketch the center manifold projection for a scalar equation.

Equation (6.17) underlines a first limitation of the approach. Indeed, the center manifold is of no use if a sufficiently small upper bound $-\alpha$, i.e., a sufficient separation between the eigenvalues of zero real part and those of negative real part, cannot be found in the spectrum of $\mathcal{L}$. The second and most important limitation of the applicability of the center manifold theory to film flows is that the theory holds for systems with a finite dimension, that is, for finite-dimension systems of ordinary differential equations [38]. When partial differential equations are considered, which can be viewed as systems of ordinary differential equations of "infinite dimension," such as Navier–Stokes, a rigorous theory is still under construction and very few results are available [98]. Thus, Roberts' approach must be viewed as a derivation technique of models whose range of validity in the parameter space must be checked a posteriori by comparing their solutions to those of the primitive equations.

In fact, the approach relies on the linear viscous modes of a uniform film in the zero-wavenumber limit and in the absence of gravity. Slow modulations of the free surface and gravity effects are thus considered as small perturbations around the motionless state. Let us then temporarily change our scaling and rewrite the primitive equations using space and times scales based on the film thickness $\bar{h}_{\mathrm{N}}$ and the viscous time scale $\bar{h}_{\mathrm{N}}^2/\nu$. The Navier–Stokes equation (2.4) thus reads

$$\frac{D\mathbf{v}}{Dt} = -\nabla p + \nabla^2 \mathbf{v} + Ga\mathbf{F}, \tag{6.18}$$

where $\mathbf{F} = (\sin\beta, -\cos\beta, 0)$ is the body force and $Ga = g\bar{h}_{\mathrm{N}}^3/\nu^2 = 3Re/\sin\beta$ is the "Galileo number". Therefore, the assumption of small gravity effects as compared to viscous damping implies $Ga \ll 1$ and is equivalent to the small inertia assumption $Re \ll 1$ (provided that the nearly horizontal case ($\beta \ll 1$) is avoided). Hence, the set of parameters $\boldsymbol{\epsilon}$ is made of the gradient expansion parameter ε and Ga (or, equivalently, the Reynolds number Re).

Linearizing now around the motionless film of uniform thickness, i.e., introducing perturbations of the form $h = 1 + \tilde{h}$, $u = \tilde{u}$, $v = \tilde{v}$, $p = \tilde{p}$ in the zero-wavenumber and zero-gravity limits, $\varepsilon = Ga = 0$, gives

$$\partial_y \tilde{v} = 0, \qquad \partial_t \tilde{u} - \partial_{yy}\tilde{u} = 0, \qquad \partial_t \tilde{v} + \partial_y \tilde{p} - \partial_{yy}\tilde{v} = 0, \tag{6.19}$$

completed by the no-slip and no-penetration conditions at the wall $y = 0$,

$$\tilde{u} = \tilde{v} = 0, \tag{6.20}$$

the kinematic boundary condition and the continuity of the stress at the free surface $y = 1$,

$$\partial_t \tilde{h} - \tilde{v} = 0, \qquad 2\partial_y \tilde{v} - \tilde{p} = 0, \qquad \partial_y \tilde{u} - (1 - \gamma)\tilde{u} = 0. \qquad (6.21)$$

Note that the linearized tangential stress condition at the free surface has been arbitrarily modified through the introduction of an artificial parameter γ such that the primitive linear system is recovered for $\gamma = 1$. Systems (6.19), (6.20) and (6.21) can be written in the form $\partial_t \mathbf{u} = \mathcal{L}_\gamma \mathbf{u}$, which defines the linear matrix-differential operator $\mathcal{L}_\gamma$ parameterized by γ and acting on the set of variables $\mathbf{u} = (\tilde{u}, \tilde{v}, \tilde{p}, \tilde{h})$. The eigenmodes are the Goldstone mode with $\tilde{h} = \text{const}$, $\tilde{u} = \tilde{v} = \tilde{p} = 0$ corresponding to the zero eigenvalue (these are the four components of the zero eigenfunction), and the family of decaying viscous modes, $\tilde{u} \propto \sin(ly) \exp(\lambda t)$, $\tilde{v} = \tilde{p} = 0$ and $\tilde{h} = \text{const}$, whose eigenvalue λ and cross-film wavenumber l satisfy

$$\lambda = -l^2, \qquad l \cot l = (1 - \gamma). \qquad (6.22)$$

Notice that both Goldstone and viscous modes have $\tilde{h} = \text{const}$ but the difference is that the Goldstone mode corresponds to $\lambda = 0$.

Let us now consider the eigenvalues of the unmodified operator $\mathcal{L}_1$, i.e., 0 and $-(2n + 1)^2 \pi^2 / 4$. $\mathcal{L}_1$ has only one zero eigenvalue and the corresponding center manifold $\mathcal{C}_1$ is of "codimension" (codim) 1, meaning that the dynamics on $\mathcal{C}_1$ is governed by a unique evolution equation for the amplitude associated with the Goldstone mode, which is naturally the film thickness h. The velocity field thus remains enslaved to h. The "spectral gap," say s, between the zero eigenvalue and the largest nonzero one is only $\pi^2 / 4$. Therefore, the advection time of the fluid particles, whose ratio to the viscous time scale is precisely Re^{-1}, becomes comparable to the relaxation time s^{-1} of the trajectories in the phase space of the codim 1 center manifold $\mathcal{C}_1$ when Re is of order unity, which necessarily limits the approach to low Reynolds numbers, as with the classical BE long wave expansion analyzed in Chap. 5. As a matter of fact, the construction of $\mathcal{C}_\infty$ performed by Roy et al. [224] leads exactly to the BE (5.12).

The modification of the boundary conditions (6.21) enables the first nonzero eigenvalue to shift to zero. For $\gamma = 0$, the set of zero eigenvalues is augmented by one and the corresponding center manifold $\mathcal{C}_0$ is of codim 2. The spectral gap s is then larger than 20. We then hope that the range of Reynolds numbers for which the approach gives reliable results may extend to the region of moderate or even large Reynolds numbers. The set of associated amplitudes is then made of the film thickness h and the averaged velocity $\bar{u} = (1/h) \int_0^h u \, dy$, which becomes a genuine degree of freedom. After construction of the center manifold $\mathcal{C}_0$ at $\gamma = 0$, the difficulty is to come back to the initial problem and thus to find the "extension" $\mathcal{C}_1'$ at $\gamma = 1$ of the center manifold. Notice that, if the exponential convergence property (6.17) holds for the codim 2 center manifold $\mathcal{C}_0$, there is no indication of such a result for the extension $\mathcal{C}_1'$.

In practice, $\mathcal{C}'_1$ may be constructed step-by-step through the expansion

$$\mathbf{u}(x, y, t) = \mathbf{U} = \sum_{m,n,p=0}^{\infty} \gamma^m \varepsilon^n Ga^p \mathbf{U}^{(m,n,p)}\big(h(x,t), \bar{u}(x,t)\big) \tag{6.23}$$

and the evolution equations

$$\partial_t h = G_h = \sum_{m,n,p=0}^{\infty} \gamma^m \varepsilon^n Ga^p G_h^{(m,n,p)}\big(h(x,t), \bar{u}(x,t)\big), \tag{6.24a}$$

$$\partial_t \bar{u} = G_{\bar{u}} = \sum_{m,n,p=0}^{\infty} \gamma^m \varepsilon^n Ga^p G_{\bar{u}}^{(m,n,p)}\big(h(x,t), \bar{u}(x,t)\big). \tag{6.24b}$$

The ansatz in (6.23) and (6.24a)–(6.24b) is then substituted into the dynamical system

$$\frac{\partial \mathbf{U}}{\partial h} G_h + \frac{\partial \mathbf{U}}{\partial \bar{u}} G_{\bar{u}} = \mathcal{L}_\gamma \mathbf{U} + \mathbf{N}(\mathbf{U}), \tag{6.25}$$

where G_h, $G_{\bar{u}}$ represent the components of the vector $\mathbf{G}$ in (6.16) (the center manifold is two-dimensional) and the left hand side is simply $\partial_t \mathbf{U}$ (so that (6.25) has the generic form in (6.15)). The asymptotic expansions of $\mathbf{U}$, G_h and $G_{\bar{u}}$ lead to a hierarchy of equations that can be solved order after order. The center manifold is then approached through the "approximation theorem": If the governing equations are satisfied to some order of accuracy, then the center manifold will have been found with the same degree of accuracy. Yet, solving (6.25) can be in practice a formidable task that requires an iterative strategy [222] that heavily relies on symbolic manipulation software (Roberts used the computer algebra package REDUCE that can be downloaded for free from http://www.reduce-algebra.com). The main difficulty of the approach is then to check that the γ series in (6.23) and (6.24a)–(6.24b) have convergence radii larger than one.

Finally, the two-equation system governing the evolution on $\mathcal{C}'_1$, or Roberts' model, is recast below in terms of the film thickness h and the flow rate $q = h\bar{u}$:

$$\delta \partial_t q \approx \frac{\pi^2}{12}\left(h - \zeta h \partial_x h + h \partial_{xxx} h - 3\frac{q}{h^2}\right) + \delta\left(-2.504\frac{q}{h}\partial_x q + 1.356\frac{q^2}{h^2}\partial_x h\right)$$

$$+ \eta\left(3.459\frac{q}{h^2}(\partial_x h)^2 - 3.353\frac{\partial_x h \partial_x q}{h} - 4.676\frac{q}{h}\partial_{xx} h + 4.093\partial_{xx} q\right)$$

$$+ \frac{1}{100}\left[\delta\big(1.727hq\partial_x h + 0.7983h^2\partial_x q\big) + \delta^2\left(-0.1961\frac{q^3}{h^2}(\partial_x h)^2\right.\right.$$

$$\left.\left. - 1.78\frac{q^2}{h}\partial_x h \partial_x q + 0.1226q(\partial_x q)^2 - 1.792\frac{q^3}{h}\partial_{xx} h + 0.7778q^2\partial_{xx} q\right)\right]$$

$$+ \frac{\zeta\delta}{100}\left(-1.357hq(\partial_x h)^2 - 1.012h^2\partial_x h \partial_x q - 1.713h^2 q \partial_{xx} h\right.$$

$$+ 0.4821 h^3 \partial_{xx} q) + \frac{\delta}{100} \left(-10.98 \frac{q}{h} (\partial_x h)^4 + 7.12 (\partial_x h)^3 \partial_x q \right.$$

$$+ 10.68 q (\partial_x h)^2 \partial_{xx} h - 4.451 h \partial_x h \partial_x q \partial_{xx} h - 1.113 h q (\partial_{xx} h)^2$$

$$- 2.225 h (\partial_x h)^2 \partial_{xx} q + 0.6404 h^2 \partial_{xx} h \partial_{xx} q + 0.244 h q \partial_x h \partial_{xxx} h$$

$$+ 1.225 h^2 \partial_x q \partial_{xxx} h + 0.4269 h^2 \partial_x h \partial_{xxx} q + 1.713 h^2 q \partial_{xxxx} h$$

$$\left. - 0.4821 h^3 \partial_{xxxx} q \right), \tag{6.26}$$

taken together with the mass balance (6.1) and written using Shkadov's scaling.

One can easily recognize in the first line of (6.26) all terms present in the Kapitza–Shkadov averaged momentum balance (6.13b), but with different coefficients. Second-order terms gathered under the parameter η in the second line correspond to the effects of viscous dispersion. Such terms were already present in (6.14). The additional terms correspond to ε^2 corrections due to inertia ($\propto \delta^2$), gravity ($\propto \zeta \delta$) and inertia arising from corrections to the velocity profile due to capillary effects (last bracketed terms, $\propto \delta$). The coefficient $-\pi^2/4$ of the viscous diffusion term q/h^2 corresponds to the first nonzero eigenvalue of the spectrum of the diffusion linear operator $\mathcal{L}_1$, as expected, since the whole procedure is basically a reduction of the slow time and space evolution of the film to the two first eigenmodes $(h, u) \propto (1, 0)$ and $(h, u) \propto (0, \sin(\pi y/2))$ of $\mathcal{L}_1$.

A linear stability analysis around the Nusselt flat film solution shows that the correct critical Reynolds number is in fact recovered using (6.1) and (6.26). However, (6.26) contains nonlinearities of order as high as seven. Computations of the solitary wave branches of solutions to this model then leads to turning points in the parameter space such that no solitary wave solutions can be found above a certain value of δ. As with the BE where the higher-order inertia term is responsible for its unphysical behavior, the loss of solutions with the Roberts model seems to be directly related to the additional high-order nonlinearities contained in (6.26) as compared to (6.13b).

There is no question that Roberts' center manifold approach is a powerful tool for the derivation of low-dimensional models when the effects of inertia [221], or a slowly varying wall [224], or both are considered [223]. Its main advantages are that it offers a systematic derivation procedure, one that can be applied fruitfully to several other problems, provided that the underlying assumptions for its applicability are met, and that no closure assumption on the velocity distribution is necessary. In fact, the dependence of the velocity field on the amplitude h and average velocity $\bar{u}$ is obtained as a result of the derivation process and is not assumed from the outset.

However, this technique suffers from its complexity (which may explain that the introduction of further degrees of freedom has not been considered by Roberts himself), and also from the necessity of a "numerical trick" through the introduction of an artificial parameter γ, which is finally set equal to unity: As already pointed out, one may wonder if the extension $\mathcal{C}_1'$ of the codim 2 center manifold $\mathcal{C}_0$ shares the convergence property (6.17), which is rigorous only for dynamical systems of finite dimension. In addition, the application of the center manifold approach depends on the degree of complexity of the linear operator $\mathcal{L}$. For the problem of a film on a planar wall one can easily construct the eigenfunctions and eigenvalues of the operator as done earlier. But this is not always the case in other systems. For example, for the problem of a film falling down a vertical fiber, the operator involves $1/r$ with r the radial distance from the fiber centerline [231]. As a result, the eigenfunctions contain logarithmic terms, which in turn make the applicability of the center manifold approach cumbersome. These logarithmic terms can be simplified by considering small aspect ratios $\bar{h}_N/R$ with R the fiber radius. This is precisely the approach followed by Roberts and Li [223]. However, quite frequently in experiments, $\bar{h}_N \sim R$ [86, 149].

In the remainder of this chapter a different strategy from Roberts' approach is proposed. It is based on a combination of a gradient expansion and an extension of the Kapitza–Shkadov averaging approach by an expansion of the velocity field in terms of polynomial test functions. After all, the gradient expansion shows that in the asymptotic limit $\varepsilon \to 0$, the cross-stream distribution of the velocity is polynomial, a property that enables one to obtain models that are both accurate and consistent at order ε or ε^2 with relatively simple algebra.

6.4 Relaxing the Self-similar Assumption

Shkadov was the first to propose the relaxation of the self-similar parabolic-profile assumption and expansion of the velocity field on a basis of functions of the reduced cross-stream variable $\bar{y} = y/h(x,t)$ [248]. In fact, at least close to the instability threshold, $Re - Re_c \ll 1$, the long wave theory indicates that the first-order correction to the velocity profile (5.9a) has the form

$$u^{(1)} = \frac{1}{3}(3Re)h^5\partial_x h f^{(1)}(y/h) + \left(-Cth^2\partial_x h + \varepsilon^2 Weh^2\partial_{xxx}h\right)f^{(0)}(y/h) \quad (6.27)$$

$(\varepsilon^2 We = \mathcal{O}(1))$ where $f^{(0)}(\bar{y}) = \bar{y} - (\bar{y}^2/2)$ corresponds to the parabolic profile and $f^{(1)}(\bar{y}) = \bar{y} - (\bar{y}^3/2) + (\bar{y}^4/8)$. The velocity distribution can then be rewritten in the form

$$u(x, y, t) = a_0(x, t)f^{(0)}(\bar{y}) + \varepsilon a_1(x, t)f^{(1)}(\bar{y}), \quad (6.28)$$

so that close to onset, the distribution of the velocity field across the layer is utterly determined by the two amplitudes a_0 and a_1 only and such that the variable $\bar{y}$ is separated from the variables x, t. We note that for simplicity, and in order to illustrate the main points of the derivation process of the models we will develop in this chapter, we shall treat Re as an $\mathcal{O}(1)$ parameter, which allows for an easy "bookkeeping" of the orders of magnitude of the different terms. A similar assumption was made in Sect. 4.1 in the derivation of the boundary layer equations.

To relax the self-similar assumption while keeping the idea of separation of variables, one can assume $u(x, t) = 2a(x, t)y + 3b(x, t)y^2$ and obtain three coupled equations for h, a and b [190]. However, inserting this ansatz into the tangential stress condition for the first-order boundary layer equations $\partial_y u|_h = 0$ decreases the number of degrees of freedom by linking a and b through the condition $2a + 6bh = 0$. Therefore, the velocity profile actually reads $u = 2a(y - y^2/(2h)) = 2ha(\bar{y} - \frac{1}{2}\bar{y}^2)$, leading to the erroneous prediction of the instability threshold, $Re_c^{(KS)} = Ct$. A similar attempt consists of the projection of u onto a set of five polynomials adding collocation conditions at the wall and free surface to (6.3) and (6.1) [306].

Another idea [226] is to correct the parabolic velocity profile with the polynomials that appear in the long wave theory, or, more precisely, by linearly independent combinations of these polynomials appropriately chosen for the ease of algebraic calculations. This way a full agreement with the long wave theory can be expected by construction. Let us then expand u as

$$u(x, y, t) = b_0(x, t)g^{(0)}(\bar{y}) + b_1(x, t)g^{(1)}(\bar{y}), \tag{6.29}$$

where $g^{(0)} \equiv f^{(0)}$ and $g^{(1)} \equiv (1/6)((1/4)\bar{y}^4 - \bar{y}^3 + \bar{y}^2)$. Simple algebra shows that $g^{(1)}$ is a linear combination of $f^{(0)}$ and $f^{(1)}$. Further, the unknown fields b_0 and b_1 are supposed to be slowly varying functions of x and t, and from the definition of q they satisfy the relationship

$$q = \int_0^h u(y)\,dy = \frac{1}{3}h\left(b_0 + \frac{1}{15}b_1\right). \tag{6.30}$$

At this stage we have three unknowns, h, b_0, and b_1, but from (6.30) it is evident that h, q and b_1 can be viewed as unknowns, with b_0 given by $b_0 = (3q/h) - \frac{1}{15}b_1$. Substituting this into $\tau_w \equiv \partial_y u|_0 = b_0/h$ gives $\tau_w = (3q/h^2) - (b_1/15h)$. Then, b_1 appears as a correction to the shear at the plate that would be created by a parabolic velocity profile corresponding to a film with thickness h and flow rate q. To see this, redefine b_1 as $b_1 = -15h\tau$ so that

$$\tau_w = \frac{3q}{h^2} + \tau. \tag{6.31}$$

What we have just done is pass from the original algebraic variables b_0 and b_1 to the more physically sound variables q and τ. Accordingly, (6.29) can be transformed to

$$u = \left(\frac{3q}{h} + h\tau\right)g^{(0)}(\bar{y}) - 15h\tau g^{(1)}(\bar{y}). \tag{6.32}$$

A useful condition can be obtained from the momentum balance (6.3) differentiated with respect to y,

$$3\varepsilon Re[\partial_{ty}u + u\partial_{xy}u + v\partial_{yy}u] - \partial_{yyy}u = 0,$$

and evaluated at $y = 0$ giving

$$3\varepsilon Re\partial_t(\tau_w) - \partial_{yyy}u|_0 = 0. \tag{6.33}$$

This equation shows that the shear perturbations at the wall are directly linked to the presence of corrections to the velocity profile departing from the parabolic shape for which they vanish identically. Using (6.32) in (6.33) we get

$$\tau = \varepsilon \frac{1}{15} h^2 \partial_t \tau_{\mathrm{w}}.$$

Hence, τ is a first-order correction, a result of the derivation and not an a priori assumption, so that the term $\varepsilon \partial_t \tau$ is in fact of second order and can be dropped from (6.33). Thus, τ_{w} reads

$$\tau_{\mathrm{w}} = \frac{3q}{h^2} + 3\varepsilon Re \frac{h^2}{5} \partial_t \left[\frac{q}{h^2} \right]. \tag{6.34}$$

Substituting (6.34) into the averaged momentum balance (6.9), using (6.1) to eliminate $\partial_t h$ and the zeroth-order estimate $\Upsilon = 6/5$, which is enough here, we obtain

$$3\varepsilon Re \partial_t q = \frac{5}{6} h - \frac{5}{2} \frac{q}{h^2} - 3\varepsilon Re \frac{7}{3} \frac{q}{h} \partial_x q + \varepsilon \left(3Re \frac{q^2}{h^2} - \frac{5}{6} Ct \right) \partial_x h + \frac{5}{6} \varepsilon^3 We h \partial_{xxx} h.$$

$$\tag{6.35}$$

This equation together with (6.1) is a model consistent at first order, and with the same structure as the Kapitza–Shkadov one (6.13a), (6.13b) but with slightly different coefficients. The differences arise from a better account of the perturbations of τ_{w} introduced via the third derivative term in (6.33) by the $\bar{y}^3$ term in $g^{(1)}$.

Let us now consider a gradient expansion of (6.1) and (6.35) of the form $q = q^{(0)} + \varepsilon q^{(1)} + \cdots$, relevant for small-amplitude waves. This gives $q^{(0)} = h^3/3$ (as expected) and $q^{(1)} = (2/15)(3Re)h^6 \partial_x h - (1/3)Ct h^3 \partial_x h + (1/3)h^3 \varepsilon^2 We \partial_{xxx} h$, which, when substituted into $\partial_t h + \partial_x [q^{(0)} + \varepsilon q^{(1)}] = 0$, gives us back the first-order BE (5.12). Hence, by construction the near-onset behavior is correctly predicted by this modified Shkadov model, which no longer underestimates the value of the instability threshold. Noteworthy is that the inability of the Kapitza–Shkadov model (6.13a), (6.13b) to accurately predict the instability threshold is not due to a wrong estimate of the averaged streamwise acceleration, i.e., the shape factor Υ, but to the neglect of the inertia corrections to the shear at the wall due to the delay of the velocity distribution to adjust to the free-surface deformation.

At zeroth-order in ε, q is slaved to h and τ_{w} does not fluctuate: Inserting a parabolic profile for u in (6.33) gives $\partial_t (\partial_y u |_0) = 0$. At $\mathcal{O}(\varepsilon)$, q and h are two slowly varying effective degrees of freedom linked to τ_{w} through (6.34). Proceeding to the next step, for a consistent modeling at $\mathcal{O}(\varepsilon^2)$, four additional fields must be introduced associated with the velocity corrections induced by the gradient expansion at second order. Supplementary conditions at the wall analogous to (6.33) can be derived to determine these additional fields. Further, assuming these four fields to be of $\mathcal{O}(\varepsilon^2)$ corrections to the velocity field, which is the case as long as the long-wave expansion holds, they can be expressed as functions of the variables h, q and τ. After substitution, we are left with three equations for three unknowns, h, q and τ, the latter playing the role of an additional independent effective degree of freedom [226].

6.5 Method of Weighted Residuals

Several conclusions can now be drawn from the previous sections:

(i) The origin of the discrepancy between the results from the Kapitza–Shkadov approach and the BE/long wave expansion rests on the treatment of the $\mathcal{O}(1)$ terms of the streamwise momentum balance and, in particular, the viscous diffusion term $\partial_{yy}u$ there.

(ii) A drastic reduction of the complexity of the system of equations to be solved can be achieved through an appropriate elimination of the amplitudes/fields of the velocity profile that are effectively slaved to the true degrees of freedom of the system.

(iii) To account for the wave dynamics deeply into the drag-inertia regime, one has to be careful with the order of the nonlinearities as high-order nonlinearities may trigger an unphysical behavior of the solutions similar to the unorthodox finite-time blow up behavior encountered with the BE.

Every model presented up to now in this chapter, with the exception of the Roberts model (6.1) and (6.26), is derived by presuming u to be a sum of polynomials in the cross-stream coordinate y, or equivalently the reduced one $\bar{y} = y/h$, and by performing an across-the-layer averaging of the momentum balance. These ideas are now developed further and made more explicit: It can be shown that by writing u as a polynomial in $\bar{y}$ and through a general averaging formulation, at each specific level of truncation with respect to the gradient expansion parameter ε "optimal" models can be obtained in the sense that the resulting models are the same independently of the particular averaging methodology that is adopted [227, 228]. Also, at each level of truncation the best choices for the scalar products and weight functions are sought out, i.e., the ones that lead to the final results with a minimum of algebra and which in turn help us to select the averaging technique. In that respect the *Galerkin* projection is the most efficient one.

This formulation presumes that inertia effects are weak corrections to the balance of viscous drag and gravity. Of course, strictly speaking this holds only in the drag-gravity regime. Nevertheless, our hope is that the obtained optimal models can be accurate outside their region of validity and are thus capable of describing the drag-inertia regime. This can only be tested by comparison of the linear wave regime with experiments and Orr–Sommerfeld and of the nonlinear wave characteristics with experiments and DNS of the full Navier–Stokes equations (as we do later in this chapter and in Chap. 7). At the same time optimal models should ensure the existence of solitary wave solutions for the widest possible range of parameters thus preventing the occurrence of unphysical blow ups.

We now focus on the description of the averaging methodology. In general, most physical problems can be formally written as $\mathcal{E}(\mathcal{U}) = 0$ for some set of field variables $\mathcal{U}$ in a space $\mathcal{S}$. In the particular case examined here $\mathcal{E}$ corresponds to the momentum balance and $\mathcal{U}$ to the streamwise velocity field. The methodology developed here can be extended to more involved and sophisticated problems, e.g., when spanwise dependence (see Chap. 8) and/or Marangoni effects (see Chap. 9) are included. In the weighted residuals method, the solution to $\mathcal{E}$ is sought in the form of a series expansion $\mathcal{U} = \sum_{j=0}^{j_{\max}} a_j f_j$, where the f_j, $j = 0, \ldots, j_{\max}$, are chosen test functions that form a complete basis for $\mathcal{S}$ and the amplitudes a_j have to be determined [92]. $\mathcal{S}$ is supposed to be equipped with an inner product, noted $\langle \cdot | \cdot \rangle$. Weight functions w_j, $j = 0, \ldots, j_{\max}$, are next chosen as the main ingredients of a projection rule to define the *residuals*: $\mathcal{R}_{j'} = \langle w_{j'} | \mathcal{E}(\sum a_j f_j) \rangle$, $j' = 0, \ldots, j_{\max}$. The vanishing residuals $\mathcal{R}_{j'} = 0$ thus yield a system for the amplitudes a_j.

This is a method of weighted residuals. It is used quite frequently in a wide variety of applications and problems. "Finite differences," "finite elements," and "spectral" methods are essentially byproducts of the above general idea. Choice of the w_j fixes the particular weighted residuals method being used. "Collocation," "subdomain" and Galerkin methods are the most commonly used weighted residual methods. The corresponding weight functions are: Dirac delta functions in the case of the collocation method, "hat functions" in the case of the subdomain method, and finally the test functions themselves in the case of the Galerkin method. Convergence to the solution is generally achieved quickly by increasing the number of residuals $j_{\max}$, which explains the widespread use of the weighted residual methods. In particular, the finite element method and the finite volume method are often used in DNS studies of wavy film flows [99, 116, 175, 176, 218, 232]. However, the aim here is not to apply the weighted residuals approach as a numerical methodology, but rather to employ analytically the basic ingredients of this approach as a means to reduce the complexity of the original set of equations. Accordingly, the weighted residuals technique will be combined with the long wave approximation.

Once again we emphasize that although the BE fails to capture the dynamics far from criticality, it is a rather useful as a template model such that any new model attempting to cure the deficiencies of the BE model far from criticality must agree with the BE model close to criticality. For this reason, the consistency with the long wave theory is imposed at each step of our approximation so that a gradient expansion of the reduced equations coincides with the corresponding gradient expansion of the original set of equations, i.e., with the corresponding long wave theory at the same level of approximation. We will see that the complexity of the equations to be solved increases rapidly with the order of truncation. Consequently, we shall restrict our attention to low-order approximations and the corresponding equations will contain terms at most of $\mathcal{O}(\varepsilon^2)$. Therefore, the proposed approach is somewhere in between a direct application of the weighted residuals method and the gradient expansion.

Let us now expand $\mathcal{E} = 0$ and the variables $\mathcal{U}$ as an asymptotic series in the gradient expansion parameter ε, $\mathcal{E}^{(0)} + \varepsilon \mathcal{E}^{(1)} + \varepsilon^2 \mathcal{E}^{(2)} + \cdots = 0$ and $\mathcal{U}^{(0)} + \varepsilon \mathcal{U}^{(1)} + \varepsilon^2 \mathcal{U}^{(2)} + \cdots = 0$. Assume that in the projection of the variables $\mathcal{U}$ onto the test functions f_j it is possible to assign an order with respect to ε for the amplitudes a_j so that $\mathcal{U}^{(0)} = \sum_0^{j_0} a_j^{(0)} f_j$, $\mathcal{U}^{(1)} = \sum_{j_0}^{j_1} a_j^{(1)} f_j$ and so on. One could therefore solve a sequence of problems $\mathcal{E}^{(0)} = 0$, $\mathcal{E}^{(1)} = 0, \ldots$, as is typically the case with asymptotic expansions, and thus sequentially determine the amplitudes $a_j^{(0)}, a_j^{(1)}, \ldots$. This would lead back again to a complete slaving of $\mathcal{U}$ to the film thickness h and thus to the classical long wave theory interface equations.

Instead, we may truncate $\mathcal{E} = 0$ at a given order, for example, $\mathcal{O}(\varepsilon)$, so that $\mathcal{E}^{(0)} + \varepsilon\mathcal{E}^{(1)} = 0$, i.e., we treat ε as a mere index/ordering parameter and not as a strict perturbation parameter; effectively performing only the first step of the gradient expansion as introduced in Sect. 4.1 but without the second step, i.e., the asymptotic/perturbation expansion in series of ε. The resulting residuals are next solved for the amplitudes $a_j^{(0)}$ and $a_j^{(1)}$. Since time and space derivatives of the amplitudes $a_j^{(1)}$ are of higher order than ε, it is then possible to show that these amplitudes are slaved to the evolution of h and $a_j^{(0)}$, leading to a drastic simplification of the system of equations to be solved. The basic idea is therefore to make some use of the gradient expansion in the averaging procedure without pushing it to its utmost consequences, thus allowing a certain level of flexibility for the variables $\mathcal{U}$.

6.6 First-Order Formulation

We are now ready to apply the weighted residuals method. For the first-order formulation the starting point is the first-order boundary layer equations (6.3)–(6.5). The streamwise velocity component is expanded as

$$u(x, y, t) = \sum_{j=0}^{j_{\max}} a_j(x,t) f_j\big[y/h(x,t)\big], \qquad (6.36)$$

while the cross-stream velocity component can be readily obtained through integration of the continuity equation (6.2), $v = -\int_0^y \partial_x u \, dy$. The slow space-time evolution of the film suggests a natural separation of variables with the (x, t) dependence included in the amplitudes a_j and the cross-stream dependence accounted for by test functions in terms of the reduced variable $\bar{y} = y/h$. As already mentioned, Shkadov did propose an expansion of the velocity field in terms of test functions of the cross-stream variable y which satisfy the boundary conditions (6.4) and (6.6) [248]. But he did not pursue it. The use of $\bar{y}$, instead of y, means that from the start the test functions are locally slaved to the film thickness modulations; however, the amplitudes of the test functions can be independent of h (e.g., to leading order, the amplitude of the velocity field contains the flow rate q) and hence the above expansion provides a certain degree of flexibility for the velocity, allowing it to have its own evolution.

Instead, Shkadov considered only a single test function

$$f_0 = \bar{y} - \frac{1}{2}\bar{y}^2, \qquad (6.37)$$

which corresponds to the self-similar parabolic velocity assumption (6.12). Following Shkadov, we demand that the test functions verify the boundary conditions (6.4), (6.6):

$$f_j'(1) = 0, \qquad f_j(0) = 0. \qquad (6.38)$$

We now recall the following important points: (i) the Nusselt flat film velocity profile is parabolic; (ii) in the BE/long wave theory, corrections to this profile are polynomials in $\bar{y}$; (iii) the set of polynomials of increasing order forms a complete basis for the space of sufficiently smooth functions in $[0, 1]$ which satisfy the conditions (6.4) and (6.6). Based on these points, it is then reasonable to take polynomials as test functions:

$$f_j = \bar{y}^{j+1} - \frac{j+1}{j+2}\bar{y}^{j+2}. \tag{6.39}$$

Item (i) is the most important one. When the film is uniform, the velocity distribution is semiparabolic and every amplitude a_j vanishes except for a_0. Therefore, the a_j are departures from the semiparabolic velocity profile induced by the deformations of the free surface. The following proof confirms that the amplitudes a_j are also slowly varying in time and space: by differentiating $j_{\max}$ times (6.3) we get

$$\partial_{y^{j_{\max}+2}}u = 3\varepsilon Re\,\partial_{y^{j_{\max}}}\left(\partial_t u + u\partial_x u + v\partial_y u\right). \tag{6.40}$$

Now $\partial_{y^{j_{\max}+2}}u = -(j_{\max}+1)^2 j_{\max}(j_{\max}-1)\ldots 2a_{j_{\max}}$ and the right hand side of (6.40) is of $\mathcal{O}(\varepsilon)$ so that $a_{j_{\max}}$ is also slowly varying. Differentiating next, $(j_{\max}-1)$ times (6.3) shows that $a_{j_{\max}-1}$ is also of $\mathcal{O}(\varepsilon)$ or higher, and so on until a_1. As a consequence, derivatives of the fields a_j, $j \geq 1$, can be neglected in the evaluation of the right hand side of (6.40). Therefore, since f_0 is a polynomial of degree two, $\partial_t u + u\partial_x u + v\partial_y u$ is a polynomial in $\bar{y}$ of degree four at most, and the right hand side of (6.40) vanishes for $j_{\max} \geq 5$. Thus, $a_j = 0$ for $j \geq 5$, which shows that the amplitudes a_j are of order higher than ε for $j \geq 5$.

In practice, after having defined the weights and written down the residuals, one obtains at $\mathcal{O}(\varepsilon)$,

$$\mathbf{MA} = \varepsilon\mathbf{B}, \tag{6.41}$$

where $\mathbf{A} = (a_{1 \leq j \leq j_{\max}})$ is a vector of dimension $j_{\max}$, $\mathbf{B}$ is a vector of dimension $j_{\max} + 1$ function of h, a_0, $\partial_t h$, $\partial_x h$, $\partial_t a_0$ and $\partial_x a_0$ and $\mathbf{M}$ is a $j_{\max} + 1 \times j_{\max}$ matrix. The inversion of the linear system (6.41) gives explicit expressions for the amplitudes a_j as functions of the film thickness h, a_0 and their derivatives, as well as the solvability condition that both h and a_0 must fulfill. This condition, together with the mass conservation equation (6.1) and the flow rate

$$q = \int_0^h u\,dy = h\sum_0^{j_{\max}} \frac{2}{(j+2)(j+3)}a_j, \tag{6.42}$$

gives a system of two equations for the two unknowns h and a_0 for the film flow evolution. We shall return to this point shortly.

Until now, we have avoided specifying particular weights. It is easy to show that the results obtained so far are independent of the choice of the weights provided that $j_{\max} \geq 4$. Indeed, requiring (6.3) to be satisfied everywhere—and not simply on average—and inserting into this equation the expansion (6.36), leads to the cancellation of one polynomial in the reduced normal coordinate $\bar{y}$, say $\mathcal{P}$. Truncation at $\mathcal{O}(\varepsilon)$ of the advective terms $\partial_t u + u \partial_x u + v \partial_y u$ involves only the parabolic profile corresponding to a_0, and the corresponding polynomial in $\bar{y}$ is of degree four only. Therefore, the monomial of highest degree appearing in $\mathcal{P}$ originates from the term $\partial_{yy} u$ so that $\mathcal{P}$ is of degree $j_{\max}$. The number of independent conditions on the unknowns a_j provided by the cancellation of $\mathcal{P}$ is thus $j_{\max} + 1$ and it is equal to the number of the residuals (6.41). In this case, any choice of the weight functions would lead to equivalent systems of equations (much like, e.g., invertible linear transformations can lead from one system to another) and then to the same model for the evolution of the flow.

Requiring the fulfillment of (6.3) by identifying all the coefficients of this polynomial sequentially in order of increasing degree yields

$$\frac{1}{h^2}(-a_0 + 2a_1) = -1 + \varepsilon Ct \partial_x h - \varepsilon^3 We \partial_{xxx} h, \tag{6.43a}$$

$$\frac{1}{h^2}(-4a_1 + 6a_2) = 3\varepsilon Re\left[\partial_t a_0 - \frac{a_0}{h}\partial_t h\right], \tag{6.43b}$$

$$\frac{1}{h^2}(-9a_2 + 12a_3) = 3\varepsilon Re\left[-\frac{1}{2}\partial_t a_0 + \frac{a_0}{h}\partial_t h + \frac{1}{2}a_0\partial_x a_0 - \frac{a_0^2}{2h}\partial_x h\right], \tag{6.43c}$$

$$\frac{1}{h^2}(-16a_3 + 20a_4) = 3\varepsilon Re\left[-\frac{1}{3}a_0\partial_x a_0 + \frac{2a_0^2}{3h}\partial_x h\right], \tag{6.43d}$$

$$\frac{1}{h^2}(-25a_4 + 30a_5) = 3\varepsilon Re\left[\frac{1}{12}a_0\partial_x a_0 - \frac{a_0^2}{6h}\partial_x h\right], \tag{6.43e}$$

$$\frac{1}{h^2}\left(-(j+1)^2 a_j + (j+1)(j+2)a_{j+1}\right) = \mathcal{O}(\varepsilon^2) \quad \text{for } 5 \le j \le j_{\max} - 1, \tag{6.43f}$$

and

$$-\frac{(j_{\max}+1)^2}{h^2}a_{j_{\max}} = \mathcal{O}(\varepsilon^2). \tag{6.43g}$$

Inversion of the linear system (6.43a)–(6.43g) gives

$$a_0 = h^2 + 3\varepsilon Re\left[-\frac{1}{3}h^2\partial_t a_0 + \frac{1}{6}ha_0\partial_t h - \frac{1}{10}h^2 a_0\partial_x a_0 + \frac{1}{30}ha_0^2\partial_x h\right]$$
$$- \varepsilon Cth^2\partial_x h + \varepsilon^3 Weh^2\partial_{xxx} h, \tag{6.44a}$$

$$a_1 = 3\varepsilon Re\left[-\frac{1}{6}h^2\partial_t a_0 + \frac{1}{12}ha_0\partial_t h - \frac{1}{20}h^2 a_0\partial_x a_0 + \frac{1}{60}ha_0^2\partial_x h\right], \tag{6.44b}$$

$$a_2 = 3\varepsilon Re\left[\frac{1}{18}h^2\partial_t a_0 - \frac{1}{9}ha_0\partial_t h - \frac{1}{30}h^2 a_0\partial_x a_0 + \frac{1}{90}ha_0^2\partial_x h\right], \tag{6.44c}$$

$$a_3 = 3\varepsilon Re\left[\frac{1}{60}h^2 a_0 \partial_x a_0 - \frac{1}{30}ha_0^2 \partial_x h\right], \tag{6.44d}$$

$$a_4 = 3\varepsilon Re\left[-\frac{1}{300}h^2 a_0 \partial_x a_0 + \frac{1}{150}ha_0^2 \partial_x h\right], \tag{6.44e}$$

and one recovers $a_i = \mathcal{O}(\varepsilon^2)$ for $i \geq 5$. Equation (6.44a) is the solvability condition for h, a_0 we mentioned earlier. After substitution of (6.44b)–(6.44e) into (6.42) we have

$$q = \frac{1}{3}ha_0 + 3\varepsilon Re\left[-\frac{1}{45}h^3 \partial_t a_0 + \frac{1}{360}h^2 a_0 \partial_t h - \frac{3}{280}h^3 a_0 \partial_x a_0 + \frac{1}{504}h^2 a_0^2 \partial_x h\right]. \tag{6.45}$$

The solvability condition (6.44a) and the mass conservation equation (6.1) with the flow rate q given by (6.45) form a closed system for the two unknowns h and a_0. It is then tempting to neglect second-order terms appearing in (6.1) and to write simply

$$\partial_t h = -\frac{1}{3}\partial_x(ha_0). \tag{6.46}$$

The set of equations (6.44a) and (6.46) remains consistent at $\mathcal{O}(\varepsilon)$. With the help of the change of variables $\tilde{q} = ha_0/3$, (6.44a) and (6.46) are transformed into

$$\partial_t h = -\partial_x \tilde{q}, \tag{6.47a}$$

$$3\varepsilon Re\partial_t \tilde{q} = h - 3\frac{\tilde{q}}{h^2} + 3\varepsilon Re\left[\frac{\tilde{q}}{h}\partial_t h + \frac{6}{5}\frac{\tilde{q}^2}{h^2}\partial_x h - \frac{7}{5}\frac{\tilde{q}}{h}\partial_x \tilde{q}\right]$$
$$- \varepsilon Cth\partial_x h + \varepsilon^3 Weh\partial_{xxx}h. \tag{6.47b}$$

Substitution of $-\partial_x \tilde{q}$ with $\partial_t h$ and introducing the Shkadov scaling leads back to the Kapitza–Shkadov model (6.13a), (6.13b). However, the mass conservation equation (6.1) shows, as we shall demonstrate in Chap. 7, that waves on falling film flows are basically *kinematic waves*, and hence (6.1) seems to have a special status. It is therefore preferable to keep (6.1) (which is exact), and to make approximations based on the flow rate q, which is an intrinsic variable with a solid physical meaning and does not depend on the choice of the test functions. Equation (6.45) then gives

$$a_0 = 3\frac{q}{h} + 3\varepsilon Re\left[\frac{1}{15}h^2 \partial_t a_0 - \frac{1}{120}ha_0 \partial_t h + \frac{9}{280}h^2 a_0 \partial_x a_0 - \frac{1}{168}ha_0^2 \partial_x h\right], \tag{6.48}$$

which combined with (6.44a) gives:

$$h^2 - 3\frac{q}{h} + 3\varepsilon Re\left[-\frac{6}{5}h\partial_t q + \frac{69}{40}q\partial_t h - \frac{333}{280}q\partial_x q + \frac{108}{70}\frac{q^2}{h}\partial_x h\right]$$
$$- \varepsilon Cth^2 \partial_x h + \varepsilon^3 Weh^2 \partial_{xxx}h = 0. \tag{6.49}$$

With a first-order expansion we can then replace a_0 with $3q/h$ in (6.49) and by using the identity $\partial_t h = -\partial_x q$ we obtain

$$3\varepsilon Re \partial_t q = \frac{5}{6}h - \frac{5}{2}\frac{q}{h^2} - 3\varepsilon Re \frac{17}{7}\frac{q}{h}\partial_x q + \varepsilon\left(3Re\frac{9}{7}\frac{q^2}{h^2} - \frac{5}{6}Cth\right)\partial_x h$$

$$+ \frac{5}{6}\varepsilon^3 Weh\partial_{xxx}h, \tag{6.50}$$

which can be rewritten in terms of the Shkadov scaling through the formal transformation $\{3\varepsilon Re \to \delta,\ \varepsilon Ct \to \zeta,\ \varepsilon^3 We \to 1 \text{ and } \varepsilon^2 \to \eta\}$,

$$\delta\partial_t q = \frac{5}{6}h - \frac{5}{2}\frac{q}{h^2} - \delta\frac{17}{7}\frac{q}{h}\partial_x q + \left(\delta\frac{9}{7}\frac{q^2}{h^2} - \frac{5}{6}\zeta h\right)\partial_x h + \frac{5}{6}h\partial_{xxx}h. \tag{6.51}$$

Again the number of parameters has been reduced from three to only two—δ and ζ. Equation (6.51), together with (6.1), constitutes a closed and self-consistent model, which will be referred to as the *first-order model*.

We now perform a gradient expansion of the form $q = q^{(0)} + \varepsilon q^{(1)}$ in the system (6.1) and (6.50). At zeroth order we obtain, $0 = \frac{5}{6}h - \frac{5}{2}q^{(0)}/h^2$, therefore $q^{(0)} = \frac{1}{3}h^3$, as expected. At first order we get

$$3\varepsilon Re \partial_t q^{(0)} = -\frac{5}{2}\frac{q^{(1)}}{h^2} - \frac{17}{7}3\varepsilon Re\frac{q^{(0)}}{h}\partial_x q^{(0)} + \varepsilon\left(3Re\frac{9}{7}\left(\frac{q^{(0)}}{h}\right)^2 - \frac{5}{6}Cth\right)\partial_x h$$

$$+ \frac{5}{6}\varepsilon^3 Weh\partial_{xxx}h.$$

Making use of the expression for $q^{(0)}$ and substituting $-\partial_x q^{(0)}$ with $\partial_t h$ we obtain, $q^{(1)} = (3Re\frac{2}{15}h^6 - \frac{1}{3}Cth^3)\partial_x h + \frac{1}{3}\varepsilon^2 Weh^3\partial_{xxx}h$, which in turn leads back to the first-order BE (5.55) when inserted into the integral version of the kinematic boundary condition, $\partial_t h + \partial_x(q^{(0)} + \varepsilon q^{(1)}) = 0$.

Therefore, the gradient expansion of the system (6.1) and (6.51) fully agrees with the classical long wave theory, as can be expected from the consistency at $\mathcal{O}(\varepsilon)$ of its derivation as all neglected terms are of order higher than ε, i.e., by construction. This agreement ensures that the system (6.1) and (6.51) does not suffer from the major limitation of the Kapitza–Shkadov model (6.13a), (6.13b), i.e., an erroneous prediction of the instability threshold. Most impressively, the model in (6.1) and (6.51) is *optimal* at first order in the sense that once an expansion for the velocity field in terms of polynomials has been performed and the amplitudes of this expansion have been assigned certain orders of magnitude with respect to ε (all amplitudes are of $\mathcal{O}(\varepsilon)$ and higher except the first one, which is of $\mathcal{O}(1)$), the model is independent of the particular projection methodology, i.e., it is always the same, independent of the averaging technique employed (any other approximation based on weighted residuals and polynomial test functions will thus converge to it when the number of the test functions is increased). This will be demonstrated in the next section. At the same time, the polynomial velocity field reconstructed from the coefficients a_j fulfills (6.3) exactly and not only on average, i.e., once the amplitudes have been ordered with respect to ε, then (6.3) is satisfied exactly.

6.7 Comparison of Weighted Residuals Methods

As stated earlier, weighted residuals methods differ from each other by different definitions of the weights $w_j(\bar{y})$. The residuals are obtained by integrating (6.3) over the layer depth

$$\int_0^h w_{j'}(y/h)\big[3\varepsilon Re\big(\partial_t u + u\partial_x u + v\partial_y u\big) - \partial_{yy}u\big]\,dy$$

$$= h\big(1 - \varepsilon Ct\partial_x h + \varepsilon^3 We\partial_{xxx}h\big)\int_0^1 w_{j'}(\bar{y})\,d\bar{y}. \tag{6.52}$$

As discussed in Sect. 6.6, neglecting inertia at second-order demands setting the derivatives of the amplitudes a_j, $1 \le j \le j_{\max}$, to zero in these equations. This leaves us with a system that can be solved for the amplitudes, and from which an equation for q is finally derived. Whatever the weighting strategies and the approximation levels, the equation expressing momentum conservation in all two-equation models for h and q obtained in this way will always have the same functional form as (6.51) but with different coefficients depending on the approximation level. Comparison between approximation levels can thus be made on the basis of the coefficients κ_i of the averaged momentum equation written below in terms of the Shkadov scaling

$$\delta\partial_t q = \kappa_1\left(h - 3\frac{q}{h^2} - \zeta h\partial_x h + h\partial_{xxx}h\right) + \kappa_2\delta\frac{q}{h}\partial_x q + \kappa_3\delta\frac{q^2}{h^2}\partial_x h, \tag{6.53}$$

and by studying the convergence of the coefficients κ_i toward the corresponding values in (6.51) as the truncation level increases.

6.7.1 Method of Subdomains

This is a generalization of the averaging method leading to the Kapitza–Shkadov model (6.13a), (6.13b): Integrating (6.3) over the layer depth using just f_0 and a uniform weight and neglecting terms of $\mathcal{O}(\varepsilon^2)$ indeed yields (6.13b). The $\bar{y}$-interval $[0, 1]$ is split into $j_{\max} + 1$ equal adjacent subintervals by $j_{\max}$ break points $\bar{y}_i = (i/j_{\max} + 1)$. Equation (6.3) is integrated over each of these subintervals:

$$\int_{\bar{y}_i}^{\bar{y}_{i+1}} \big[3\varepsilon Re(\partial_t u + u\partial_x u + v\partial_y u) - \partial_{yy}u - 1 + \varepsilon Ct\partial_x h - \varepsilon^3 We\partial_{xxx}h\big]\,d\bar{y} = 0,$$

$$\tag{6.54}$$

Table 6.1 Method of subdomains

j_{max}	$h - 3\dfrac{q}{h^2} - \zeta h\partial_x h + h\partial_{xxx}h$	$\delta\dfrac{q}{h}\partial_x q$	$\delta\dfrac{q^2}{h^2}\partial_x h$	Re_c/Ct
0	1	$-\dfrac{12}{5} = -2.40$	$\dfrac{6}{5} = 1.20$	1
1	$\dfrac{16}{19} \approx 0.84$	$-\dfrac{1851}{760} \approx -2.44$	$\dfrac{993}{760} \approx 1.31$	$\dfrac{16}{19} \approx 0.84$
2	$\dfrac{5}{6}$ exact	$-\dfrac{175}{72} \approx -2.43$	$\dfrac{31}{24} \approx 1.29$	$\dfrac{5}{6}$ exact
3	$\dfrac{5}{6}$	$-\dfrac{2487}{1024} \approx -2.43$	$\dfrac{1317}{1024} \approx 1.29$	$\dfrac{5}{6}$
4	$\dfrac{5}{6}$	$-\dfrac{17}{7}$ exact	$\dfrac{9}{7}$ exact	$\dfrac{5}{6}$

which is equivalent to the definition of $j_{max} + 1$ "hat functions":

$$\bar{y} \rightarrow \begin{cases} 1 & \text{for } \bar{y} \in [\bar{y}_i, \bar{y}_{i+1}] \\ 0 & \text{for } \bar{y} \in [0, \bar{y}_i[\text{ and }]\bar{y}_{i+1}, 1]. \end{cases}$$

One recognizes easily the Kármán–Pohlhausen technique when $j_{max} = 0$. The linear system for the a_j resulting from (6.52) is then solved as discussed earlier. The corresponding coefficients κ_i appearing in (6.53) are given in Table 6.1. Linear properties are recovered for $j_{max} = 3$ ($\kappa_1 = 1$). Convergence is nearly achieved already for $j_{max} = 3$ but $j_{max} = 4$ is necessary for a complete nonlinear agreement.

6.7.2 Collocation Method

The weight functions w_j are now Dirac delta functions located at equally spaced collocation points in the interval $[0, 1]$. The vanishing residuals correspond to the exact fulfillment of the equation at the locations of the Dirac delta functions. When $j_{max} = 0$, the corresponding residual represents the evaluation of (6.3) at $\bar{y} = 1/2$. A solvability condition is next obtained in the form (6.53) with coefficients given in Table 6.2. Full convergence is observed at level $j_{max} = 4$.

6.7.3 Integral-Collocation Method

In this method, a simple averaging of (6.3) is augmented with additional conditions generally placed at the boundaries. As an example, one can choose to specify the derivatives of the momentum equation (6.3) at the wall:

$$\partial_{y^k}\left[3\varepsilon Re\left(\partial_t u + u\partial_x u + v\partial_y u\right) - \partial_{yy}u\right] = 0, \quad \text{at } y = 0, 1 \le k \le j_{max}. \tag{6.55}$$

Table 6.2 Collocation method

j_{max}	$h - 3\dfrac{q}{h^2} - \zeta h \partial_x h + h \partial_{xxx} h$	$\delta \dfrac{q}{h} \partial_x q$	$\delta \dfrac{q^2}{h^2} \partial_x h$	Re_c/Ct
0	$\dfrac{2}{3} \approx 0.67$	$-\dfrac{5}{2} = -2.50$	$\dfrac{3}{2} = 1.50$	$\dfrac{2}{3} \approx 0.67$
1	$\dfrac{32}{39} \approx 0.82$	$-\dfrac{251}{104} \approx -2.41$	$\dfrac{129}{104} \approx 1.24$	$\dfrac{32}{39} \approx 0.82$
2	$\dfrac{5}{6}$ exact	$-\dfrac{697}{288} \approx -2.42$	$\dfrac{121}{96} \approx 1.26$	$\dfrac{5}{6}$ exact
3	$\dfrac{5}{6}$	$-\dfrac{4973}{2048} \approx -2.43$	$\dfrac{2631}{2048} \approx 1.28$	$\dfrac{5}{6}$
4	$\dfrac{5}{6}$	$-\dfrac{17}{7}$ exact	$\dfrac{9}{7}$ exact	$\dfrac{5}{6}$

Table 6.3 Integral-collocation method

j_{max}	$h - 3\dfrac{q}{h^2} - \zeta h \partial_x h + h \partial_{xxx} h$	$\delta \dfrac{q}{h} \partial_x q$	$\delta \dfrac{q^2}{h^2} \partial_x h$	Re_c/Ct
0	1	$-\dfrac{12}{5} = -2.40$	$\dfrac{6}{5} = 1.20$	1
1	$\dfrac{8}{11} \approx 0.73$	$-\dfrac{126}{55} \approx -2.29$	$\dfrac{48}{55} \approx 0.87$	$\dfrac{8}{11} \approx 0.73$
2	$\dfrac{5}{6}$ exact	$-\dfrac{21}{8} \approx -2.62$	$\dfrac{15}{8} \approx 1.87$	$\dfrac{5}{6}$ exact
3	$\dfrac{5}{6}$	$-\dfrac{19}{8} \approx -2.37$	$\dfrac{9}{8} \approx 1.12$	$\dfrac{5}{6}$
4	$\dfrac{5}{6}$	$-\dfrac{17}{7}$ exact	$\dfrac{9}{7}$ exact	$\dfrac{5}{6}$

In fact, expanding the streamwise velocity u on a basis of polynomials in $\bar{y}$ is similar to performing a Taylor expansion of the solution at $y = 0$. Shkadov's averaging method is recovered at level 0. Results are given in Table 6.3. The hope is that the additional "regularity" demanded by the solution at the boundaries might accelerate the convergence process, as is the case in the treatment of the Blasius equation for the semi-infinite boundary layer along a flat plate (see [243], Chap. XII). However, despite our expectations, the integral-collocation method has rather poor convergence properties (cf. Table 6.3).

6.7.4 Method of Moments

The weights used at the projection step are monomials of increasing degree $w_k = \bar{y}^k$ so that the residuals correspond to the moments of $\mathcal{E}$ in increasing order ($\langle \bar{y} | \mathcal{E} \rangle$ is the average value, $\langle \bar{y}^2 | \mathcal{E} \rangle$ is a linear combination of the variance and the mean of $\mathcal{E}$,

Table 6.4 Method of moments

$j_{\max}$	$h - 3\dfrac{q}{h^2} - \zeta h\partial_x h + h\partial_{xxx}h$	$\delta\dfrac{q}{h}\partial_x q$	$\delta\dfrac{q^2}{h^2}\partial_x h$	Re_c/Ct
0	1	$-\dfrac{12}{5} = -2.40$	$\dfrac{6}{5} = 1.20$	1
1	$\dfrac{16}{19} \approx 0.84$	$-\dfrac{231}{95} \approx -2.43$	$\dfrac{123}{95} \approx 1.29$	$\dfrac{16}{19} \approx 0.84$
2	$\dfrac{5}{6}$ exact	$-\dfrac{17}{7}$ exact	$\dfrac{9}{7}$ exact	$\dfrac{5}{6}$ exact

etc.). The equation $\mathcal{E} = 0$ is fulfilled when its successive moments vanish. Level 0 with $w_0 \equiv 1$ again corresponds to simple averaging, thus leading to the Kapitza–Shkadov model (6.13a), (6.13b). The convergence of the method is rather fast, as seen from data in Table 6.4.

6.7.5 Galerkin Method

The Galerkin method is the most widely used of the weighted residuals methods. The test functions are now taken as the weight functions themselves, $w_j \equiv f_j$. The residuals $\mathcal{R}_j$ thus read

$$\int_0^h f_j(y/h)\big(3\varepsilon Re[\partial_t u + u\partial_x u + v\partial_y u] - \partial_{yy}u\big)\,dy$$

$$-\frac{2h}{(j+2)(j+3)}\big(1 - \varepsilon Ct\partial_x h + \varepsilon^3 We\partial_{xxx}h\big). \tag{6.56}$$

It can be shown that this method leads to the optimal equation (6.51) already at level 0. Let us consider the first residual $\mathcal{R}_0$ corresponding to the parabolic velocity profile f_0. Only the term $\int_0^h f_0(y/h)\partial_{yy}u\,dy$ of $\mathcal{R}_0$ is of special concern, since the first-order terms $\partial_t u + u\partial_x u + v\partial_y u$ involve a_0 and h only. Through a double integration by parts using the boundary conditions $f_j(0) = 0$ and $f_j'(1) = 0$, this term can be written as $\int_0^h u f_j''(\bar{y})\,dy$. In the case $j = 0$ for which $f_0''(\bar{y}) \equiv -1$ we get

$$\int_0^h f_0(y/h)\partial_{yy}u\,dy = -\frac{q}{h^2} \tag{6.57}$$

from the definition of $q = \int_0^h u\,dy$, which also contains the special combination of the a_i given by (6.42) we need to close the model. Residual $\mathcal{R}_0$ reads explicitly

$$3\varepsilon Re\left(\frac{2}{15}h\partial_t a_0 - \frac{7}{120}a_0\partial_t h + \frac{37}{840}ha_0\partial_x a_0 - \frac{11}{840}a_0^2\partial_x h\right)$$

$$+ \frac{q}{h^2} - \frac{1}{3}\big(h - \varepsilon Cth\partial_x h + \varepsilon^3 Weh\partial_{xxx}h\big). \tag{6.58}$$

Since at $\mathcal{O}(\varepsilon)$, a_0 and $3q/h$ are interchangeable in all terms containing derivatives in x or t, the cancellation of the residual (6.58) gives

$$3\varepsilon Re\left(\frac{2}{5}\partial_t q - \frac{23}{40}\frac{q}{h}\partial_t h - \frac{18}{35}\frac{q^2}{h^2}\partial_x h + \frac{111}{280}\frac{q}{h}\partial_x q\right)$$

$$= \frac{1}{3}h - \frac{q}{h^2} - \frac{\varepsilon Ct}{3}h\partial_x h + \frac{\varepsilon^3 We}{3}h\partial_{xxx}h. \tag{6.59}$$

Using the relationship $\partial_t h = -\partial_x q$ leads finally back to the optimal system (6.1) and (6.51). The Galerkin method is the most efficient one, requiring less algebra.

6.7.6 Remarks

Tables 6.1–6.4 show that the variations of the coefficients are not monotonic as the approximation level is increased, and that their limiting values can be reached from above as well as from below. This might have been expected since the full original problem has no underlying "variational structure." Even more interestingly, the subdomain method and the collocation method are seen to display slow convergence properties, which is quite likely connected with their "finite difference"-type of approximation. In contrast, the method of moments and the Galerkin method converge faster, very much like spectral methods. The latter, involve basis functions well adapted to the problem and hence it turns out to be the most efficient.

6.8   Second-Order Formulation

Having outlined the weighted residuals strategy at first order, let us now sketch the main steps leading to a model consistent at second order. The starting point is the second-order boundary layer equations derived in Chap. 4. For purposes of clarity we rewrite here the streamwise momentum balance,

$$3\varepsilon Re(\partial_t u + u\partial_x u + v\partial_y u) - \partial_{yy}u - 2\varepsilon^2\partial_{xx}u$$

$$= 1 + \varepsilon^2\partial_x(\partial_x u|_h) - \varepsilon Ct\partial_x h + \varepsilon^3 We\partial_{xxx}h, \tag{6.60}$$

and the tangential stress balance at the interface

$$\partial_y u|_h = \varepsilon^2(4\partial_x h\partial_x u|_h - \partial_x v|_h). \tag{6.61}$$

The continuity equation (6.2), the no-slip condition (6.4) and the kinematic boundary condition (6.5) (or equivalently the mass conservation equation (6.1)) remain unchanged. The cross-stream component of the velocity v is linked to the streamwise velocity component u through the continuity equation, $v = -\int_0^h \partial_x u\,dy$. Equations (6.60) and (6.61) differ from their first-order counterparts (6.3) and (6.6) by the retention of the viscous second-order terms, which as we shall demonstrate in Sect. 7.1.3 modify the dispersion of the waves.

Comparison of the different weighted residuals methods reveals that the Galerkin method is again well adapted to the problem even at second order and is the most efficient one, requiring less algebra.

6.8.1 Full Second-Order Model

At $\mathcal{O}(\varepsilon)$, test functions were chosen to satisfy boundary conditions (6.4) and (6.61), which leads to the definition (6.39) of the functions f_j. However, unlike the first-order problem where the tangential stress balance is homogeneous, the tangential stress balance (6.61) is nonhomogeneous, i.e., its right hand side is nonzero due to the presence of the second-order viscous terms and cannot be satisfied from the start by the test functions. This boundary condition then becomes a new constraint on the amplitudes a_j and is effectively satisfied through a *tau method*.[3] The set of test functions is therefore restricted to polynomials satisfying the no-slip condition $f_j(0) = 0$ only. The simplest possible basis then is the "canonical" one, y^i, $i \geq 1$. Unfortunately, the level of complexity of the system to solve increases dramatically with the required order of consistency in ε. In fact, by extending the argument leading to the conclusion that the first-order approximation to u, say u_{first}, is a polynomial of degree six, gives that the second-order approximation is a polynomial of degree ten: the inertia term, $\partial_t u + u \partial_x u + v \partial_y u$, is formally quadratic, hence of degree eight, the sum of the degrees of f_0 and u_{first}, and has to be compensated by a term originating from $\partial_{yy} u$, hence u, of degree ten. The general solution thus depends on h plus ten supplementary amplitudes (condition $u|_0 = 0$ suppresses one coefficient), of which five at most are first-order quantities, since the fields a_j, $0 \leq j \leq 4$, are nonzero at that order. Written as a polynomial in $\bar{y}$ equal to zero, the inversion of the resulting system by sequentially setting the coefficient of each monomial to zero, is still possible.

However, the resulting model is a complicated set of six equations with six unknowns! The choice of the canonical basis gives no indication on the way to reduce effectively the complexity of the model and to isolate significant amplitudes other than those that can be *adiabatically eliminated*, i.e., slaved to the evolutions of the former ones. Recall that our objective is to obtain the simplest formulation consistent at $\mathcal{O}(\varepsilon^2)$ that accounts for the observed physical phenomenon quantitatively, or at least qualitatively. As for the derivation process at first order, amplitudes can be eliminated easily if their derivatives are slowly varying, i.e., of order higher than ε^2.

---

[3] In weighted residuals terminology the term "tau method" typically refers to a variant of the Galerkin method invented by Lanczos in which a number of amplitudes of the projections of the unknown functions onto a set of test functions has been eliminated via substitution of the projection onto the boundary conditions. The resulting expansions of the unknown functions then satisfy the boundary conditions [Gottlieb & Orszag 1977]. In essence, the tau method "homogenizes" the nonhomogeneous boundary conditions.

This shows in turn that they are of order higher than ε, which means that the corresponding test functions are not needed to approximate the velocity field at that order. We are thus led to determine the minimal number of polynomials that are required to approximate u at $\mathcal{O}(\varepsilon)$ consistently, which corresponds also to the number of independent fields required to ensure consistency at second order (in addition to h, only one, namely q, is necessary at first order).

It can be shown that the minimum number of fields necessary to approximate u at $\mathcal{O}(\varepsilon)$ can be obtained directly from the first-order boundary layer system (6.2)–(6.6). Differentiating once (6.3) with respect to y and making use of the continuity equation (6.2) yields

$$3\varepsilon Re(\partial_{ty}u + u\partial_{xy}u + v\partial_{yy}u) - \partial_{yyy}u = 0, \tag{6.62}$$

that we next apply at $y = h$. The kinematic boundary condition (6.5) gives $v|_h\partial_{yy}u|_h = (\partial_t h + u|_h\partial_x h)\partial_{yy}u|_h$ so that,

$$3\varepsilon Re\big[\partial_t(\partial_y u|_h) + u|_h\partial_x(\partial_y u|_h)\big] - \partial_{yyy}u|_h = 0, \tag{6.63}$$

which in view of the tangential stress balance (6.6) reduces to

$$\partial_{yyy}u|_h = 0. \tag{6.64}$$

Similarly, differentiating now three times (6.3) with respect to y and making use of the continuity equation (6.2), one obtains

$$3\varepsilon Re(\partial_{ty^3}u + u\partial_{xy^3}u + v\partial_{y^4}u + 2\partial_y u\partial_{xyy}u + 2\partial_y v\partial_{yyy}u) - \partial_{y^5}u = 0. \tag{6.65}$$

Written at $y = h$ with the help of (6.6) and (6.64), (6.65) now reads

$$3\varepsilon Re(\partial_{ty^3}u|_h + u|_h\partial_{xy^3}u|_h + v|_h\partial_{y^4}u|_h) - \partial_{y^5}u|_h = 0, \tag{6.66}$$

or by making use of the kinematic boundary condition (6.5),

$$3\varepsilon Re\big[\partial_t(\partial_{yyy}u|_h) + u|_h\partial_x(\partial_{yyy}u|_h)\big] - \partial_{y^5}u|_h = 0. \tag{6.67}$$

Thus, we finally obtain

$$\partial_{y^5}u|_h = 0. \tag{6.68}$$

The argument now on the successive differentiation of (6.3) developed in Sect. 6.6 still applies and shows that for polynomials of degree higher or equal to seven, the associated amplitudes b_j are of order higher than ε.

Let us assume that at $\mathcal{O}(\varepsilon)$, u is given by $u = \sum_0^{j_{\max}-1} a_j f^{(j)}(\bar{y})$, where the fields a_j are functions of q, h and their derivatives, and where the a_i are not linearly independent (there is at last one linear combination of the a_i that is zero). Assume also that $f^{(i)}$ is a polynomial of degree i. Thus, $j_{\max} \leq 7$. Then (6.4), (6.6), (6.64) and (6.68) give, $f^{(i)}(0) = df^{(i)}/d\bar{y}(1) = d^3f^{(i)}/d\bar{y}^3(1) = d^5f^{(i)}/d\bar{y}^5(1) = 0$. Therefore,

$a_0 = a_1 = a_3 = a_5 = 0$, and hence $f^{(2)}$ must be proportional to f_0. Set $f^{(2)} \equiv f_0$. The polynomial $d^2 f^{(4)}/d\bar{y}^2$ is of degree two and satisfies (6.6). We thus have

$$\frac{d^2 f^{(4)}}{d\bar{y}^2} = c_1 f_0 + c_2, \tag{6.69}$$

where $c_{1,2}$ are constants. With similar arguments, we get $d^4 f^{(6)}/d\bar{y}^4 = c_3 f_0 + c_4$, where c_3, c_4 are constants, which upon integrating twice yields $d^2 f^{(6)}/d\bar{y}^2 = c_3(-\frac{1}{3}\bar{y} + \frac{1}{6}\bar{y}^3 - \frac{1}{24}\bar{y}^4) - c_4 f_0 + c_5$ or finally

$$\frac{d^2 f^{(6)}}{d\bar{y}^2} = \frac{1}{6}c_3\left(f_1 - \frac{1}{3}f_2\right) - \left(c_4 + \frac{1}{3}c_3\right)f_0 + c_5. \tag{6.70}$$

We have thus proved that only three fields are necessary to approximate the velocity field at $\mathcal{O}(\varepsilon)$ and that these fields correspond to polynomials of degree two, four and six, verifying relations (6.69) and (6.70).

Let us now turn to the explicit computation of the expression for u at $\mathcal{O}(\varepsilon)$ as a function of q, h and their derivatives. Substituting a_0 with q in the expressions (6.44b)–(6.44e) through $a_0 = 3q/h + \mathcal{O}(\varepsilon)$ we obtain

$$
\begin{aligned}
a_1 &= 3\varepsilon Re\left[-\frac{1}{2}h\partial_t q - \frac{3}{5}h\partial_x\left(\frac{q^2}{h}\right)\right], \\
a_2 &= 3\varepsilon Re\left[\frac{1}{6}h\partial_t q + \frac{2}{5}\frac{q^2\partial_x h}{h} + \frac{1}{5}q\partial_x q\right], \\
a_3 &= 3\varepsilon Re\left[\frac{3}{20}h^3 q\,\partial_x\left(\frac{q}{h^3}\right)\right], \\
a_4 &= 3\varepsilon Re\left[-\frac{3}{100}h^3 q\,\partial_x\left(\frac{q}{h^3}\right)\right].
\end{aligned}
\tag{6.71}
$$

As $a_1 + 3a_2 = -(3/5)h^3 q\,\partial_x(q/h^3) = -4a_3 = 20a_4$, and thus $a_2 = -(1/3)a_1 - (4/3)a_3$, the velocity field therefore at $\mathcal{O}(\varepsilon)$ reads

$$u = 3\frac{q}{h}f_0 + a_1\left(-\frac{2}{5}f_0 + f_1 - \frac{1}{3}f_2\right) + a_3\left(\frac{8}{35}f_0 - \frac{4}{3}f_2 + f_3 - \frac{1}{5}f_4\right). \tag{6.72}$$

It can be verified without difficulty that the polynomials appearing in (6.72) satisfy (6.69) and (6.70).

To take advantage of the specific form of u given by (6.72), it is best, instead of using the f_j, to turn to appropriate combinations of these functions dictated by the above expression for u. Let us denote them as F_j and the corresponding amplitudes as b_j. Again, $F_0 \equiv f_0$ is needed to ensure that b_0 (or equivalently q) is the only amplitude of $\mathcal{O}(\varepsilon^0)$; the other polynomials F_j are corrections to the parabolic profile F_0, and the associated b_j amplitudes are at least of $\mathcal{O}(\varepsilon)$. Two other polynomials, F_1 and F_2, must be defined to account for the departures of the velocity profile from its parabolic shape at $\mathcal{O}(\varepsilon)$ with the second one being of higher order. The difficulty here is to find the simplest formulation without having to invert the complicated system of seven linear equations for the expressions of b_j, $3 \le j \le 9$. Fortunately, a shortcut is possible by considering again the $\mathcal{O}(\varepsilon^0)$ terms of the residuals $\mathcal{R}_j$ of which $\int_0^h w_j \partial_{yy} u$ is the only term of special concern, since the unknowns b_j, $i \ge 3$, may enter into the evaluation of the residuals only through this term. Two integrations by parts give

$$\int_0^h w_j\left(\frac{y}{h}\right)\partial_{yy} u\, dy = \left[w_j\left(\frac{y}{h}\right)\partial_y u\right]_0^h - \frac{1}{h}\left[w_j'\left(\frac{y}{h}\right)u\right]_0^h + \frac{1}{h^2}\int_0^h w_j''\left(\frac{y}{h}\right)u\, dy.$$

$$(6.73)$$

As $\partial_y u|_h$ (6.61) is already of $\mathcal{O}(\varepsilon)$, it may only involve h, b_0 and their derivatives. By making also use of the no-slip condition on the plate, $u|_0 = 0$, only three terms are left, namely $w_j(0)\partial_y u|_0$, $w_j'(1)u|_h$ and $\int_0^h w_j''(y/h)u\, dy$. Considering relations (6.69) and (6.70), a complete set of orthogonal polynomials $\int_0^1 F_i F_j\, d\bar{y} \propto \delta_{ij}$ is the most appropriate. As a matter of fact, as the polynomials F_0, F_1 and F_2 are linear combinations of f_0, f_1 and f_2 all verify $w_j'(1) = 0$. More importantly, (6.69) and (6.70) ensure that F_0'', F_1'' and F_2'' are linear combinations of 1, F_0 and F_1. This dramatically simplifies the definitions of the seven polynomials and hence the approximation of the velocity field at $\mathcal{O}(\varepsilon^2)$ and the evaluation of the corresponding residuals.

For this reason, an orthogonal basis can be constructed through a "Gram–Schmidt orthogonalization" procedure. More specifically, F_1 is sought as a linear combination of F_0 and $f_1 - (1/3)f_2$. Next F_2 is sought as a linear combination of F_0, F_1 and $-(4/3)f_2 + f_3 - (1/5)f_4$. We then arrive at

$$F_0 = \bar{y} - \frac{1}{2}\bar{y}^2,$$

$$(6.74a)$$

$$F_1 = \bar{y} - \frac{17}{6}\bar{y}^2 + \frac{7}{3}\bar{y}^3 - \frac{7}{12}\bar{y}^4,$$

$$(6.74b)$$

$$F_2 = \bar{y} - \frac{13}{2}\bar{y}^2 + \frac{57}{4}\bar{y}^3 - \frac{111}{8}\bar{y}^4 + \frac{99}{16}\bar{y}^5 - \frac{33}{32}\bar{y}^6.$$

$$(6.74c)$$

The basis is next completed with other independent polynomials of increasing degree, whose expressions have no importance since the Galerkin procedure avoids the determination of their coefficients. As expected, F_0'', F_1'' and F_2'' are linear combinations of 1, F_0 and F_1. We have

$$[F_0]'' = -1, \qquad [F_1]'' = 14F_0 - \frac{17}{3}, \qquad \text{and} \qquad [F_2]'' = \frac{1485}{28}F_1 + \frac{909}{28}F_0 - 13,$$

and consequently

$$\int_0^h F_0(y/h)\partial_{yy} u\, dy = \frac{1}{2}\partial_y u|_h - \frac{q}{h^2},$$

$$(6.75a)$$

$$\int_0^h F_1(y/h)\partial_{yy}u\,dy = -\frac{1}{12}\partial_y u|_h - \frac{17}{3}\frac{q}{h^2} + \frac{14}{h^2}\int_0^h F_0(y/h)u\,dy, \quad (6.75b)$$

$$\int_0^h F_2(y/h)\partial_{yy}u\,dy = \frac{1}{32}\partial_y u|_h - 13\frac{q}{h^2} + \frac{909}{28h^2}\int_0^h F_0(y/h)u\,dy$$

$$+ \frac{1485}{28h^2}\int_0^h F_1(y/h)u\,dy. \quad (6.75c)$$

As for the first-order model, it is therefore appropriate to substitute b_0 with the flow rate q. Similarly, it is convenient to work with amplitudes homogeneous in q. We thus set

$$u = \frac{3}{h}(q - s_1 - s_2)F_0(\bar{y}) + 45\frac{s_1}{h}F_1(\bar{y}) + \frac{210}{h}\left(s_2 - \sum_{i=3}^{9} s_i\right)F_2(\bar{y})$$

$$+ \sum_{i=3}^{9} \frac{1}{\int_0^1 F_i(\bar{y})\,d\bar{y}}\frac{s_i}{h}F_i(\bar{y}), \quad (6.76)$$

where s_1, s_2 are at most first-order inertia corrections to the velocity distribution (they also contain terms of $\mathcal{O}(\varepsilon^2)$) and s_j, $j \geq 3$, are corrections at most of $\mathcal{O}(\varepsilon^2)$ (they also contain terms of $\mathcal{O}(\varepsilon^3)$). By noticing that $\int_0^1 F_0(\bar{y})\,d\bar{y} = 1/3$, $\int_0^1 F_1(\bar{y})\,d\bar{y} = 1/45$ and $\int_0^1 F_2(\bar{y})\,d\bar{y} = 1/210$, one can easily see that the flow rate definition, $q = \int_0^h u\,dy$, is still satisfied by the expansion (6.76). This also ensures that the projections of u on F_0 and F_1 only involve the corrections s_1, s_2 and not the s_j, $j \geq 3$, corrections,[4] which combined with (6.75a)–(6.75c) enables one to close the system of equations by obtaining the first three residuals only,

$$\int_0^h F_j(y/h)\big[3\varepsilon Re(\partial_t u + u\partial_x u + v\partial_y u) - \partial_{yy}u - 2\varepsilon^2\partial_{xx}u\big]dy$$

$$= h\big[1 + \varepsilon^2\partial_x(\partial_x u|_h) - \varepsilon Ct\partial_x h + \varepsilon^3 We\partial_{xxx}h\big]\int_0^1 F_j(\bar{y})\,d\bar{y}, \quad (6.77)$$

for $j = 0$, 1 and 2. Finally, we get a system of four evolution equations for the four unknowns h, q, s_1 and s_2 that we rewrite below using the Shkadov scaling:

$$\partial_t h = -\partial_x q, \quad (6.78a)$$

$$\delta\partial_t q = \frac{27}{28}h - \frac{81}{28}\frac{q}{h^2} - 33\frac{s_1}{h^2} - \frac{3069}{28}\frac{s_2}{h^2} - \frac{27}{28}\zeta h\partial_x h + \frac{27}{28}h\partial_{xxx}h$$

---

[4] Although s_j, $j \geq 3$, are at most of $\mathcal{O}(\varepsilon^2)$ (and they contain terms of $\mathcal{O}(\varepsilon^3)$), their contribution in the projection (6.75a)–(6.75c) and hence in the resulting system is of $\mathcal{O}(\varepsilon^3)$ and higher due to the orthogonality between F_j, $j \geq 2$, and F_0, F_1.

$$
+\delta\left(-\frac{12}{5}\frac{q s_1 \partial_x h}{h^2}-\frac{126}{65}\frac{q s_2 \partial_x h}{h^2}+\frac{12}{5}\frac{s_1 \partial_x q}{h}+\frac{171}{65}\frac{s_2 \partial_x q}{h}+\frac{12}{5}\frac{q \partial_x s_1}{h}\right.
$$

$$
\left.+\frac{1017}{455}\frac{q \partial_x s_2}{h}+\frac{6}{5}\frac{q^2 \partial_x h}{h^2}-\frac{12}{5}\frac{q \partial_x q}{h}\right)+\eta\left(\frac{5025}{896}\frac{q(\partial_x h)^2}{h^2}\right.
$$

$$
\left.-\frac{5055}{896}\frac{\partial_x q \partial_x h}{h}-\frac{10851}{1792}\frac{q \partial_{xx} h}{h}+\frac{2027}{448}\partial_{xx} q\right),
\tag{6.78b}
$$

$$
\delta\partial_t s_1 = \frac{1}{10}h-\frac{3}{10}\frac{q}{h^2}-\frac{126}{5}\frac{s_1}{h^2}-\frac{126}{5}\frac{s_2}{h^2}-\frac{1}{10}\zeta h\partial_x h+\frac{1}{10}h\partial_{xxx}h
$$

$$
+\delta\left(-\frac{3}{35}\frac{q^2 \partial_x h}{h^2}+\frac{1}{35}\frac{q \partial_x q}{h}+\frac{108}{55}\frac{q s_1 \partial_x h}{h^2}-\frac{5022}{5005}\frac{q s_2 \partial_x h}{h^2}\right.
$$

$$
\left.-\frac{103}{55}\frac{s_1 \partial_x q}{h}+\frac{9657}{5005}\frac{s_2 \partial_x q}{h}-\frac{39}{55}\frac{q \partial_x s_1}{h}+\frac{10557}{10010}\frac{q \partial_x s_2}{h}\right)
$$

$$
+\eta\left(\frac{93}{40}\frac{q(\partial_x h)^2}{h^2}-\frac{69}{40}\frac{\partial_x h \partial_x q}{h}+\frac{21}{80}\frac{q \partial_{xx} h}{h}-\frac{9}{40}\partial_{xx} q\right),
\tag{6.78c}
$$

$$
\delta\partial_t s_2 = \frac{13}{420}h-\frac{13}{140}\frac{q}{h^2}-\frac{39}{5}\frac{s_1}{h^2}-\frac{11817}{140}\frac{s_2}{h^2}-\frac{13}{420}\zeta h\partial_x h+\frac{13}{420}h\partial_{xxx}h
$$

$$
+\delta\left(-\frac{4}{11}\frac{q s_1 \partial_x h}{h^2}+\frac{18}{11}\frac{q s_2 \partial_x h}{h^2}-\frac{2}{33}\frac{s_1 \partial_x q}{h}-\frac{19}{11}\frac{s_2 \partial_x q}{h}+\frac{6}{55}\frac{q \partial_x s_1}{h}\right.
$$

$$
\left.-\frac{288}{385}\frac{q \partial_x s_2}{h}\right)+\eta\left(-\frac{3211}{4480}\frac{q(\partial_x h)^2}{h^2}+\frac{2613}{4480}\frac{\partial_x h \partial_x q}{h}-\frac{2847}{8960}\frac{q \partial_{xx} h}{h}\right.
$$

$$
\left.+\frac{559}{2240}\partial_{xx} q\right).
\tag{6.78d}
$$

As required, a gradient expansion $q = q^{(0)}+\varepsilon q^{(1)}+\varepsilon^2 q^{(2)}$, $s_{1,2}=\varepsilon s_{1,2}^{(1)}+\varepsilon^2 s_{1,2}^{(2)}$ (recall that s_1, s_2 contain terms of $\mathcal{O}(\varepsilon)$ and $\mathcal{O}(\varepsilon^2)$), recovers the second-order BE (5.13) (including the second-order surface tension terms in $q^{(2)}$). The above system will be referred to as the *full second-order model*.

6.8.2 Simplified Second-Order Model

The second order model (6.78a)–(6.78d) is complicated. But, it can be straightforwardly simplified by assuming s_1 and s_2 to be of order higher than second, i.e., that the dynamics of the flow is in effect governed by only two variables, the film thickness h and the flow rate q (a justification of this assumption will be given in the next section). Thus, the derivatives of s_1, s_2 or their products with h and q derivatives can be dropped so that they only enter into the calculation via the terms $\frac{1}{h^2}\int_0^h F_j'' u\, dy$

appearing in the evaluation of the residuals (6.77) as earlier noted. With this assumption and because $F_0'' = -1$, the quantities s_1 and s_2 do not appear into the evaluation of the first residual. Thus, applying the Galerkin method to the second-order problem but with a single function F_0 leads to the solvability condition:

$$\delta \partial_t q = \frac{5}{6} h - \frac{5}{2} \frac{q}{h^2} - \delta \frac{17}{7} \frac{q}{h} \partial_x q + \left(\delta \frac{9}{7} \frac{q^2}{h^2} - \frac{5}{6} \zeta h \right) \partial_x h + \frac{5}{6} h \partial_{xxx} h$$
$$+ \eta \left[4 \frac{q}{h^2} (\partial_x h)^2 - \frac{9}{2h} \partial_x q \partial_x h - 6 \frac{q}{h} \partial_{xx} h + \frac{9}{2} \partial_{xx} q \right]. \tag{6.79}$$

The terms within the square brackets are generated by the second-order contributions originating from $\eta [2 \partial_{xx} u + \partial_x (\partial_x u |_h)]$ in the momentum equation (6.60) and the tangential stress boundary condition (6.61). As such they include the effect of viscous dispersion that was lacking at first order. Hereinafter, (6.1) and (6.79) will be referred to as the *simplified second-order model*. As will be shown later, the system (6.1) and (6.79) is the simplest formulation that accurately accounts for the two-dimensional traveling wave evolutions of film flows at moderate Reynolds numbers (see Chap. 7).

Although taking s_1 and s_2 of order higher than second is a drastic assumption, it is not equivalent to using the self-similar parabolic profile (6.12). With this assumption, the system of the two residuals $\mathcal{R}_1$ and $\mathcal{R}_2$ that contain the weights F_1 and F_2 can be solved for s_1 and s_2. Consequently, the corrections of the velocity profile from its parabolic shape, which are assumed to be varying in time and length scales much slower than the film thickness, are not set to zero, but can still be computed from the solutions of the system (6.1), (6.79).

Yet this closure assumption and associated elimination of s_1 and s_2 is not satisfactory, since second-order inertia terms originating from the corrections to the velocity distribution are not taken into account in the simplified second-order model. Though these second-order corrections are small at onset, the gradient expansion of (6.79) fails to reproduce the exact expression for the flow rate at $\mathcal{O}(\varepsilon^2)$ given by the long wave theory. Actually, the flow rate expression obtained from a gradient expansion of (6.79) differs from the exact one only through the coefficient of the first inertia term whose value is $\frac{636}{175}$ instead of the correct value $\frac{127}{35}$, a very small deviation indeed of $\sim 0.2\%$. Still, small differences with the exact result, i.e., full Navier–Stokes, quite likely will be amplified as Re increases and especially when the flow becomes three-dimensional (in Chap. 8 we discuss the effects of second-order inertial terms on three-dimensional wave patterns). A different strategy for eliminating s_1 and s_2 is then required.

6.9 Reduction of the Full Second-Order Model

Our discussion above highlighted the need to include the second-order inertia corrections to the velocity distribution. At the same time of particular interest would be the derivation of accurate second-order models taking into account these corrections but involving only two equations for two independent variables such as h and q. The theoretical analysis and the numerical integration of models such as (6.1) and (6.78a)–(6.78d) are indeed simpler than full Navier–Stokes and the boundary layer formulation (6.60). However, dealing with the four fields of (6.1) and (6.78a)–(6.78d) remains a difficult task, and a definitive two-field formulation consistent at $\mathcal{O}(\varepsilon^2)$ seems desirable. Accordingly, we follow [238] and develop a reduction strategy aiming at obtaining a two-equation model which also yields with a gradient expansion the BE at $\mathcal{O}(\varepsilon^2)$.

A simple argument permits us to justify the elimination of s_1 and s_2. Since viscosity acts so as to ensure the coherence of the flow across the layer, velocity perturbations varying rapidly in the direction normal to the wall are efficiently damped by viscosity. Thus s_1 and s_2 corresponding to high degree polynomials should be also efficiently damped. This concept can be checked by linearizing system (6.78a)–(6.78d) around the Nusselt flat film flow in the zero wavenumber limit, that is, assuming no spatial variations. The flat film mass balance (6.1) thus suggests a constant thickness. By writing $h = 1 + \epsilon \tilde{h}$ and $q = 1/3 + \epsilon \tilde{q}$, $s_i = \varepsilon \tilde{s}_i$ where $\epsilon \ll 1$, we get,

$$\delta \frac{d\tilde{\mathbf{V}}}{dt} = \mathbf{M}\tilde{\mathbf{V}}, \tag{6.80}$$

where $\tilde{\mathbf{V}} = (\tilde{h}, \tilde{q}, \tilde{s}_1, \tilde{s}_2)$ and $\mathbf{M}$ is a matrix corresponding to the linear part of (6.78a)–(6.78d). The first equation of (6.80) is simply $d\tilde{h}/dt = 0$. The eigenvalues λ_i of $\tilde{\mathbf{V}}$ are 0, -2.47, -22.3, and -87.7. Because of the large gap between (λ_1, λ_2) and (λ_3, λ_4), it is evident that, at low Reynolds numbers and provided that the long wave assumption is valid, the evolution of the flow is governed by the neutral mode with $\lambda_1 = 0$ associated with the free surface elevation (the corresponding eigenvector is $(\tilde{h}, \tilde{q}, \tilde{s}_1, \tilde{s}_2) = (1, 0, 0, 0)$) and the mode corresponding to λ_2; the corresponding eigenvector is $(\tilde{h}, \tilde{q}, \tilde{s}_1, \tilde{s}_2) = (0, 1.00, -1.33 \times 10^{-2}, 1.38 \times 10^{-4})$ (this argument is similar to the presence of a sufficient spectral gap necessary for the center manifold approach followed by Roberts). Consequently, given that this eigenvector is nearly aligned with the null eigenvector, the quantities s_1 and s_2 are truly slaved to the evolution of the thickness and of the flow rate, at least close to the instability threshold.

6.9.1 Elimination of s_1 and s_2

Having justified the elimination of s_1 and s_2, let us attempt its practical implementation. The fields s_1 and s_2 are corrections to the Nusselt flat film parabolic profile corresponding to F_0. Hence, they are at least first-order terms produced by the deformation of the free surface. In the first residual $\mathcal{R}_0$ associated with the weight F_0, the amplitudes s_1 and s_2 appear through inertia terms involving their space and time derivatives or through products with derivatives of h and q, which are terms of $\mathcal{O}(\varepsilon^2)$. In fact, as already mentioned, the corrections to the velocity field cannot appear in $\mathcal{R}_0$ at lowest order since the viscous term $\int_0^h F_0(y/h)\partial_{yy}u\,dy$ yields $\frac{1}{2}\partial_y u|_{y=h} - q/h^2$, owing to the definition of the streamwise flow rate, $q = \int_0^h u\,dy$, and $\frac{1}{2}\partial_y u|_{y=h}$ is already of $\mathcal{O}(\varepsilon^2)$, as can be seen from the expression of the tangential stress balance at the free surface (6.61).

At this stage, we need to determine both quantities s_1 and s_2 as functions of h, q and their derivatives truncated at $\mathcal{O}(\varepsilon)$. Such relations can easily be obtained by dropping all second-order terms from residuals $\mathcal{R}_1$ and $\mathcal{R}_2$ and then solving for s_1 and s_2. One gets

$$s_1 = 3\varepsilon Re\left(\frac{1}{210}h^2\partial_t q - \frac{19}{1925}q^2\partial_x h + \frac{74}{5775}hq\,\partial_x q\right) + O(\varepsilon^2), \quad (6.81\text{a})$$

$$s_2 = 3\varepsilon Re\left(\frac{2}{5775}q^2\partial_x h - \frac{2}{17325}hq\,\partial_x q\right) + O(\varepsilon^2). \quad (6.81\text{b})$$

Finally, substitution of (6.81a), (6.81b) into $\mathcal{R}_0$ and introducing the Shkadov scaling gives

$$\delta\partial_t q = \frac{5}{6}h - \frac{5}{2}\frac{q}{h^2} - \delta\frac{17}{7}\frac{q}{h}\partial_x q + \left(\delta\frac{9}{7}\frac{q^2}{h^2} - \frac{5}{6}\zeta h\right)\partial_x h + \frac{5}{6}h\partial_{xxx}h$$

$$+ \delta^2 \mathcal{K}(h,q) + \eta\left(4\frac{q}{h^2}(\partial_x h)^2 - \frac{9}{2h}\partial_x q\,\partial_x h - 6\frac{q}{h}\partial_{xx}h + \frac{9}{2}\partial_{xx}q\right), \quad (6.82)$$

where the additional terms arising from the elimination of s_1 and s_2 are second-order inertia terms all gathered in $\mathcal{K}$:

$$\mathcal{K} = \frac{1}{210}h^2\partial_{tt}q - \frac{1}{105}q\,\partial_x h\partial_t q + \frac{1}{42}h\partial_x q\partial_t q + \frac{17}{630}hq\,\partial_{xt}q + \frac{653}{8085}q(\partial_x q)^2$$

$$- \frac{26}{231}\frac{q^2}{h}\partial_x h\partial_x q + \frac{386}{8085}q^2\partial_{xx}q + \frac{104}{2695}\frac{q^3}{h^2}(\partial_x h)^2 - \frac{78}{2695}\frac{q^3}{h}\partial_{xx}h. \quad (6.83)$$

These corrections contain nonlinearities up to seventh order. They also contain time derivatives that are difficult to handle in numerical computations. Fortunately, the Nusselt flat film/zeroth-order relationship between q and h, $q = h^3/3$, allows us to simplify the expression of $\mathcal{K}$. Using also $\partial_t h = -h^2\partial_x h + \mathcal{O}(\varepsilon)$, we obtain the more compact expression

$$\mathcal{K} = -\frac{1}{630}h^7(\partial_x h)^2. \quad (6.84)$$

The behavior of (6.82) with the inertia corrections $\mathcal{K}$ given either by (6.83) or (6.84) in the drag-inertia regime can be understood by computing the single-hump solitary wave solutions for a vertical wall and by neglecting second-order viscous effects ($\eta = 0$). The computations were performed using the software AUTO-07P with the HOMCONT subroutines for continuation of homoclinic orbits

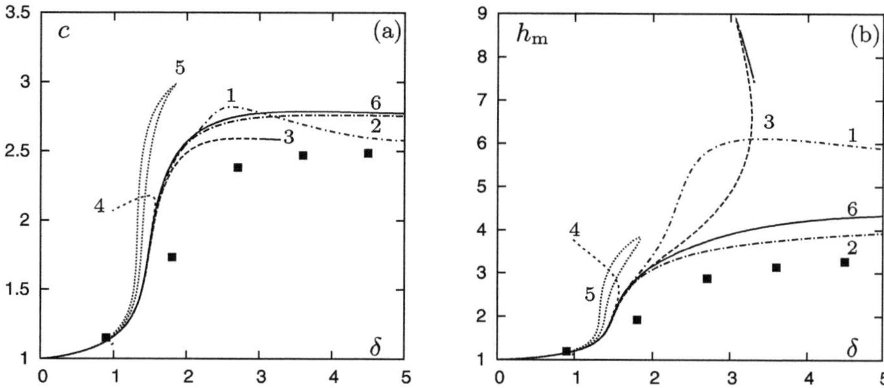

Fig. 6.2 Speed c (**a**) and amplitude h_{m} (**b**) of the single-hump solitary waves as functions of the reduced Reynolds number δ. The wall is vertical and viscous dispersion is omitted ($\zeta = \eta = 0$). 1: Full second-order model (6.78a)–(6.78d); 2: simplified second-order model (6.1, 6.79); 3: (6.1, 6.82) with $\mathcal{K}$ given by (6.83); 4: with $\mathcal{K}$ given by (6.84); 5: with $\mathcal{K}$ given by (6.85); 6: regularized model (6.1, 6.92); *solid squares*: solutions to the first-order boundary layer equations from [46]

(an introduction to AUTO can be found in Appendices F.1 and F.2). Figure 6.2 displays the speed and amplitude of the solitary waves as functions of the reduced Reynolds number δ from (6.82) with the inertia corrections $\mathcal{K}$ given either by (6.83) or (6.84), compared against the solutions of the full-second order model (6.78a)–(6.78d), the simplified second-order model (6.1), (6.79), the "regularized model," to be introduced in the next section, and the results obtained by Chang et al. [46] with the first-order boundary layer equations (6.2)–(6.6). The construction of the solitary wave solution branches is an important test for the validity of the different models obtained from the weighted residuals methods.

The simplified model and the full second-order model both exhibit single-hump solitary wave solutions for all δ and their speeds agree well with the results of Chang et al. [46]. Noteworthy is that with both expressions (6.83) and (6.84) of $\mathcal{K}$ the corresponding branch of single-hump solitary wave solutions has a turning point in the transition region between the drag-gravity and the drag-inertia regimes. This unorthodox behavior is similar to that encountered with the BE in Chap. 5 and is related to the high degree nonlinearities present in (6.83) and (6.84) resulting from the elimination of s_1, s_2. We therefore end up with basically the same difficulty as in the case of single-variable interface equations such as the BE. The fundamental problem then is how to obtain the inertia terms in a form that accounts accurately for the drag-inertia regime for the widest possible range of values of the reduced Reynolds number δ.

On the other hand, the full second-order, simplified second-order and regularized models seem to be performing well without any turning points and hence unphysical behavior. However, the full second-order model seems to "overshoot" the speed of the solitary waves at $\delta \simeq 2$ where the solution branch for the amplitude of the waves exhibits a slight "bend," which then makes the branch move toward higher amplitudes and eventually overestimates the amplitudes of the waves for $\delta \gtrsim 2.5$.

Interestingly, the full second-order model that has a relatively large number of fields (four), h, q, s_1 and s_2 performs worse for $\eta = 0$ than the simplified second-order model, where s_1 and s_2 are suppressed. This appears to be in contradiction with a basic property of the weighted residuals methods: As we have already pointed out, weighted residuals methods have good convergence characteristics and in fact their convergence improves as the number of test functions increases. However, in the approach adopted here, it is the truncation of the averaged equations resulting from a certain ordering of their terms implied by the long wave assumption—with the consequence that several terms are dropped out—that affects the convergence properties of the weighted residuals method. One then anticipates that were we to follow closely the spirit of the weighted residuals technique as a numerical tool for the solution of sets of partial differential equations, i.e., in a purely numerical implementation of the technique in our problem, we would find that indeed increasing the number of fields s_i improves the convergence of the technique. However, the drawback would be that the resulting system of equations would be substantially more complicated than (6.78a)–(6.78d).

Therefore, the problem with the full second-order model is not the number of fields but rather the simplification of the average equations using the long wave approximation. This then produces high-order nonlinearities, an effect similar to what is happening with the long wave theory where higher-order terms with high-order nonlinearities lead to poor convergence characteristics of the corresponding expansions. Hence, as for the BE, whose validity domain when traveling wave solutions are considered shrinks in the limit $k \to 0$, i.e., when traveling waves become increasingly localized leading to solitary pulses (Chap. 5), the range of parameters for which boundedness of the solutions and reasonable agreement with DNS and experiments can be expected for the different weighted residuals models can also be limited when solitary waves are considered. We note that the high-order nonlinearities produced by the long wave approximation, which are responsible for the overshoot of the full second-order model in Fig. 6.2, are due to inertia. These second-order inertia terms are gathered under the parameter δ in (6.78b)–(6.78d).

Setting $\eta = 0$, as in the construction of the solitary wave branches of solutions displayed in Fig. 6.2, is the most stringent test one can think of, since the stabilizing effects of the streamwise viscous dispersion are absent. Decreasing η has little effect on the main solitary hump but affects the amplitude and number of the oscillations at its front (Sect. 4.3) as predicted by the linear stability analysis of the flat film (Sect. 7.1.1). As far as the number of frontal oscillations are concerned, decreasing η increases the band of unstable modes as the cut-off wavenumber k_c increases. This follows from the analysis in Sect. 7.1.1. For example, for the simplified second-order model, rewrite the equation obtained by substituting the expression for c in (7.11a) into (7.11b) in terms of the Shkadov scaling. The resulting equation shows that as η decreases, k_c increases. The same conclusion can also be drawn from Fig. 7.2 for the cut-off frequency as a function of Re (the cut-off frequency is directly related to k_c). Simply compare the first-order model with the simplified and full second-order models: When $\eta \neq 0$, f_c (and similarly k_c) is reduced.

As η decreases, the number of oscillations at the front increases while the oscillations are close packed and with a smaller wavelength (because k_c increases). An example of a solitary pulse with many oscillations at the front is given in Fig. 7.31 for the large value $\delta = 5$, i.e., deeply in the drag-inertia regime. This pulse is difficult to construct numerically with AUTO-07P. Indeed, in the numerical implementation of AUTO-07P, we find that in the absence of viscous dispersion, the number of points in the domain must increase substantially in order to achieve convergence. But convergence has indeed been achieved in all cases in Fig. 6.2. The reason for the overshoot of the second-order model is not numerical but it is due to the second-order inertia terms as first noted above. The long wave approximation fails in the oscillatory region in front of a pulse due to the presence of rapid and sharp oscillations there: Because of the high-order nonlinearities in the second-order terms and large slopes in the oscillatory region, the second-order terms are no longer small compared to the first-order ones (the order of magnitude of nonlinearities is more sensitive to the local slope when nonlinearities are high, as in the BE, where indeed the "dangerous" terms originate from the high-order nonlinearities). On the other hand, we note that the behavior of the first-order model (not shown in the figure) is similar to the simplified one, i.e., there is no overshoot despite the rapid and sharp oscillations at the front. Simply, there are no second-order inertia terms in the first-order model.

We then expect the performance of the full second-order model to improve substantially in the drag-inertia regime when more realistic situations with $\eta \neq 0$ are considered (and still the plane remains vertical). The local slope in the oscillatory region in front of the pulse is now smaller while the oscillations have a larger wavelength. Indeed, this is the case in Fig. 6.3 where single hump solitary wave solutions of the different models are shown. The Kapitza number is fixed at the value $\Gamma = 529$ corresponding to the conditions of an experiment by Kapitza [141] ($\nu = 2 \times 10^{-6}$ m^2 s^{-1} and $\sigma/\rho = 29 \times 10^{-6}$ m^3 s^{-2}).

With the help of the definitions $\delta = 3Re/\kappa$, $\eta = \kappa^{-2}$ and $\kappa = We^{1/3} = \Gamma(3Re)^{-2/3}$, one obtains the dependence $\eta = \delta^{4/11}\Gamma^{-6/11}$ of the viscous dispersion number η in terms of Γ and δ. As a consequence, for given liquid, i.e., $\Gamma = $ const, the effect of streamwise viscous dispersion increases with the reduced Reynolds number δ so that, for $\Gamma = 529$, η reaches 0.059 at $\delta = 5$. Though "small," the stabilizing effect of viscous dispersion in reducing the amplitude and number of oscillations at the front of the pulses is noticeable and the questionable overshoot of the branch of solutions to the full second-order model observed in Fig. 6.2 is no longer present in Fig. 6.3. Figure 6.3 then represents a rather compelling test of the validity of the full second-order in realistic situations where $\eta \neq 0$.

Let us note the topological similitude of the solution branches obtained from the simplified second-order and regularized models (labeled 2 and 6 in Figs. 6.2 and 6.3) to those obtained from the first-order boundary layer equations (solutions to these equations are identical for the computations shown in Figs. 6.2 and 6.3 since second-order viscous terms are neglected). The simplified second-order model seems to be

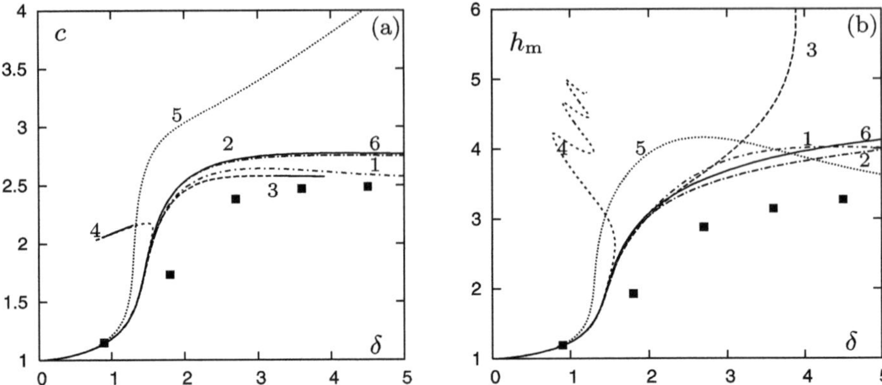

Fig. 6.3 Speed c (**a**) and amplitude h_{m} (**b**) of the single-hump solitary waves as functions of the reduced Reynolds number δ. The wall is vertical and the liquid is fixed ($\zeta = 0$ and $\Gamma = 529$). 1: Full second-order model (6.78a)–(6.78d); 2: second-order simplified model (6.1, 6.79); 3: (6.1, 6.82) with $\mathcal{K}$ given by (6.83); 4: with $\mathcal{K}$ given by (6.84); 5: with $\mathcal{K}$ given by (6.85); 6: regularized model (6.1), (6.92); *solid squares*: solutions to the first-order boundary layer equations from [46]

following more closely the first-order boundary layer equations compared to the regularized one. However, the numerical solution of the first-order boundary layer equations might not be accurate enough in the drag-inertia regime, where the steepness of the solitary wave front and the large number of capillary ripples preceding the main solitary hump call for refined numerical mesh grids in the region in front of the main hump. The precise details of the numerical scheme used by Chang et al. [46] are not given in their study. Yet, some details of their algorithm are given, i.e., a maximum of 7 mesh points in the cross-stream direction and a number of 70 complex Fourier modes in the streamwise direction, which in our view is insufficient to represent correctly one-hump solitary wave solutions deeply in the drag-inertia regime.

As far as the regularized model is concerned, our anticipation is that it is more accurate than the simplified second-order one. In Chap. 7 we shall demonstrate, however, that for two-dimensional traveling waves the regularized and the simplified models give similar results. In fact, the regularized model turns out to be more accurate than the simplified one when the wave dynamics of the film become three-dimensional at larger Reynolds numbers. Further, in Chap. 8 we shall see that the inertial terms included into the regularized model (via an appropriate regularization procedure) capture the *synchronous* three-dimensional patterns observed in the experiments by Liu et al. [170]. The same is true for the full second-order model but not for the simplified second-order one. A precise answer to the question of which model is more accurate in the drag-inertia regime can only be given by comparisons with exact solitary wave solutions from DNS, which are not available as of yet, unlike traveling wave solutions. DNS solutions for homoclinic orbits are not straightforward, especially in the region of moderate-to-large Re, where the number of capillary ripples at the front of the primary solitary hump increases substantially, thus necessitating long computational domains.

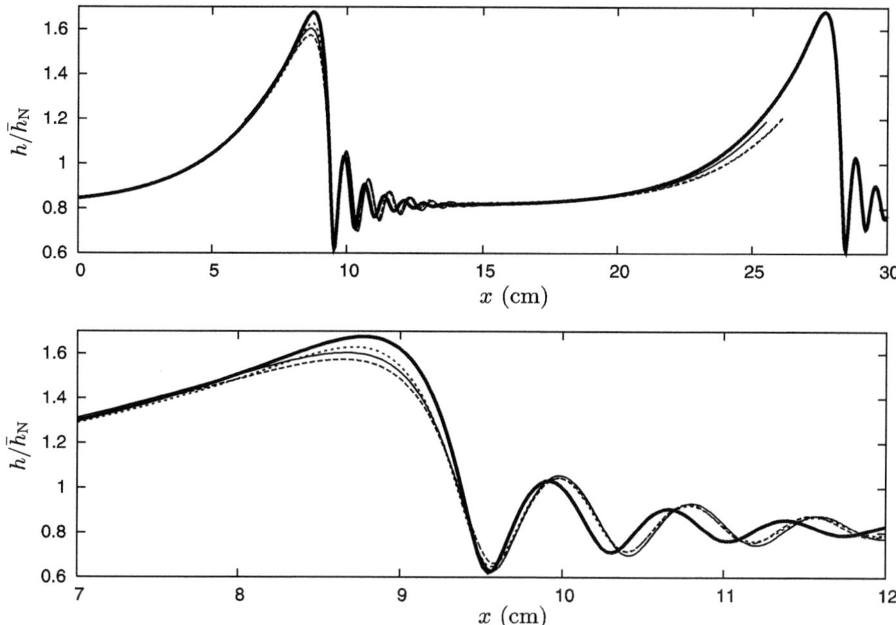

Fig. 6.4 Traveling wavetrain approaching solitary waves corresponding to an experiment by Liu and Gollub [168]. The forcing frequency is $f = 1.5$ Hz. Parameters are $\beta = 6.4°$, $Re = 19.33$ and $\Gamma = 526$ ($\delta = 17.7$, $\zeta = 2.72$ and $\eta = 0.093$). The *thick solid line* is obtained from the DNS study by Malamataris et al. [176]. Results of the regularized (simplified) model correspond to the *solid* (*dashed*) line. Solution to the full second-order model is the *dotted line*. The *bottom panel* is a zoom of the *top one* at the front of the main hump

Figure 6.4 compares the profiles of traveling wave solutions of the different models to the DNS study performed by Malamataris et al. [176]. The conditions correspond to an experiment by Liu and Gollub [168] ($\beta = 6.4°$, $Re = 19.33$, $\Gamma = 526$, and a time period of 0.67 s). The computed traveling waves are sufficiently long to start resembling isolated solitary waves. Results of the simplified second-order, regularized and full second-order models are in excellent agreement with the DNS results. The full second-order result is closest to the DNS one around the primary solitary hump, whereas the simplified model predicts a wave amplitude slightly lower than the DNS result. The regularized model is somewhere in between.

Other forms of the second-order inertia corrections $\mathcal{K}$ can be obtained by using the Nusselt flat film/zeroth-order relation $q = h^3/3$. For example, the center manifold analysis outlined in Sect. 6.3 yields the inertia corrections [221]

$$\mathcal{K} = \frac{1}{100}\left(-0.1961\frac{q^3}{h^2}(\partial_x h)^2 - 1.78\frac{q^2}{h}\partial_x h \partial_x q + 0.1226 q (\partial_x q)^2\right.$$

$$\left. - 1.792\frac{q^3}{h}\partial_{xx}h + 0.7778 q^2 \partial_{xx}q\right), \tag{6.85}$$

which effectively correspond to the terms gathered under δ^2 in (6.26).

The results obtained with this expression of $\mathcal{K}$ are also displayed in Figs. 6.2 and 6.3. For $\eta = 0$, a bending of the solution branch is observed at $\delta \approx 2$ and subsequently the branch turns back on itself in a very small area. Of course, the center manifold analysis is based on a perturbation from the Nusselt flat film solution, which is strictly valid only in the drag-gravity regime where inertia effects are small compared to both gravity and viscous drag. The same assumption has been used in the average models presented here. However, the center manifold analysis produces nonlinearities in the second-order inertia terms $\mathcal{K}$ higher than those in the second-order model (which in turn are responsible for its overshoot in Fig. 6.2). In the presence of viscous dispersion, no turning points are observed for the center manifold analysis (cf. Fig. 6.3). Yet, the presence of a significant overshoot for the speed c at large δ is a clear indication of failure.

Equation (6.82) with $\mathcal{K}$ given either by (6.83) or (6.84) also leads to failure for the solution branches. This is due to the high-order nonlinearities produced by the elimination of s_1, s_2, as noted earlier, in fact the same with those resulting in the center manifold analysis—as a general rule, the smaller the number of degrees of freedom in the system, the higher the nonlinearities are produced. Then, reducing the number of degrees of freedom, and thus the complexity of the system of equations to solve, does not necessarily lead to better convergence properties in the drag-inertia regime as all two-equation models consistent at second order presented up to this point behave worse than the four-equation second-order model. As a matter of fact, the contribution of the second-order inertia terms to the inability of the corresponding two-equation second-order models to accurately describe the drag-inertia regime can be proved by switching them off. Setting $\mathcal{K}$ to zero in (6.82) thus leads to the second-order simplified model (6.1), (6.79) which exhibits solitary wave solutions for all values of δ. It seems therefore possible to describe the drag-inertia regime with a relatively simple model involving only two degrees of freedom: h and q. Yet, as stressed before, the simplified model is not consistent at second order, precisely because the second-order inertia terms have been neglected, so that the reasonable behavior of the solitary wave solutions in the drag-inertia regime occurs at the price of accuracy (in terms of consistency with the long wave theory and also in terms of comparisons to the experiments for three-dimensional flows as will be shown in Chap. 8). This raises the question of a possible expression of $\mathcal{K}$ that yet ensures validity for a large range of values of the Reynolds number and also consistency at second order. This is the objective of the next section.

6.9.2 Padé-Like Regularization

As mentioned earlier, the second-order inertia corrections $\mathcal{K}$ contain high-order nonlinearities, which are responsible for the turning points of the single-hump solitary wave solution branches, which in turn can trigger unphysical blow ups in time-dependent computations in the drag-inertia regime, much like with the BE. To cure the singular behavior of these terms, a procedure has been formulated [238] that follows more closely the classical Padé approximants technique outlined in Appendix C.7, rather than Ooshida's regularization approach [196].

The starting point is the residual $\mathcal{R}_0$ obtained by averaging the momentum equation (6.60) with weight F_0 and written as a series in ε, $\mathcal{R}_0^{(0)} + \varepsilon \mathcal{R}_0^{(1)} + \varepsilon^2 \mathcal{R}_0^{(2),\eta} + \varepsilon^2 \mathcal{R}_0^{(2),\delta}$. In the second-order terms of this expansion, terms having viscous origin (superscript η) have been isolated from those accounting for the convective acceleration induced by the departures of the velocity profile from the parabolic shape (superscript δ). The simplified equation (6.79) is recovered by neglecting $\mathcal{R}_0^{(2),\delta}$. In line with the Padé approximants technique, $\mathcal{R}_0$ is then sought in the form $\mathcal{G}^{-1}\mathcal{F}$ where $\mathcal{G}$ is now simply a function of h, q and their derivatives (an "algebraic preconditioner" instead of "differential one," like in Ooshida's approach), and $\mathcal{F}$ is reduced to $\mathcal{R}_0^{(0)} + \varepsilon \mathcal{R}_0^{(1)} + \varepsilon^2 \mathcal{R}_0^{(2),\eta}$. This is the residual obtained by assuming a parabolic velocity profile corresponding to the simplified model ($\mathcal{F} = 0$ is the residual for the simplified model, the momentum equation of this model). We then form the residual equation $\mathcal{R}_0 = 0$, or more precisely $\mathcal{G}\mathcal{R}_0 = 0$ (since solutions to $\mathcal{R}_0 = 0$ should also verify $\mathcal{G}\mathcal{R}_0 = 0$),

$$3\varepsilon Re\mathcal{G}(h, q) \int_0^h F_0(\bar{y})(\partial_t u + u\partial_x u + v\partial_y u)\, dy$$

$$= \mathcal{G} \int_0^h F_0(\bar{y})\left\{1 + \partial_{yy}u - \varepsilon Ct\partial_x h + \varepsilon^3 We\partial_{xxx}h + \varepsilon^2[2\partial_{xx}u + \partial_x(\partial_x u|_h)]\right\}dy,$$

which must be supplemented with the mass equation (6.1) and where the inertia terms isolated on the left hand side read

$$3\varepsilon Re\mathcal{G} \int_0^h F_0(y/h)[\partial_t u + u\partial_x u + v\partial_y u]\, dy$$

$$= 3\varepsilon Re\mathcal{G}\left[\left(\frac{2}{5}\partial_t q - \frac{18}{35}\frac{q^2}{h^2}\partial_x h + \frac{34}{35}\frac{q}{h}\partial_x q\right) - \frac{6}{5}\varepsilon Re\mathcal{K}\right]$$

$$\equiv \mathcal{G}\left(\varepsilon \mathcal{R}_0^{(1),\delta} + \varepsilon^2 \mathcal{R}_0^{(2),\delta}\right). \tag{6.86}$$

"Matching" (6.86) with

$$3Re\left(\frac{2}{5}\partial_t q - \frac{18}{35}\frac{q^2}{h^2}\partial_x h + \frac{34}{35}\frac{q}{h}\partial_x q\right) \equiv \mathcal{R}_0^{(1),\delta} \tag{6.87}$$

leads to the regularization factor

$$\mathcal{G} = \left(1 + \varepsilon \frac{\mathcal{R}_0^{(2),\delta}}{\mathcal{R}_0^{(1),\delta}}\right)^{-1}. \tag{6.88}$$

An asymptotically equivalent expression of $\mathcal{G}$ can be found using

$$q = \frac{h^3}{3} + O(\varepsilon), \tag{6.89}$$

and $\partial_t h = -h^2 \partial_x h + \mathcal{O}(\varepsilon)$. We then obtain

$$\mathcal{R}_0^{(1),\delta} = -\frac{6}{15} Re h^4 \partial_x h + \mathcal{O}(\varepsilon) \quad \text{and} \quad \mathcal{R}_0^{(2),\delta} = \frac{9 Re^2}{1575} h^7 (\partial_x h)^2 + \mathcal{O}(\varepsilon),$$

which, when substituted into (6.88), yields

$$\mathcal{G} = \left(1 - \varepsilon \frac{3Re}{210} h^3 \partial_x h\right)^{-1} + \mathcal{O}(\varepsilon^2). \tag{6.90}$$

In order to keep the order of nonlinearities as small as possible, $\mathcal{G}$ is rewritten in terms of the local slope $\partial_x h$ and the "local Reynolds number" $3Req$ (defined in Sect. 5.4)

$$\mathcal{G} = \left(1 - \varepsilon \frac{3Re}{70} q \partial_x h\right)^{-1}. \tag{6.91}$$

Finally, the resulting equation in terms of the Shkadov scaling is

$$\delta \partial_t q = \delta \left(\frac{9}{7} \frac{q^2}{h^2} \partial_x h - \frac{17}{7} \frac{q}{h} \partial_x q\right)$$
$$+ \left[\frac{5}{6} h - \frac{5}{2} \frac{q}{h^2} + \eta \left(4 \frac{q}{h^2} (\partial_x h)^2 - \frac{9}{2h} \partial_x q \partial_x h - 6 \frac{q}{h} \partial_{xx} h + \frac{9}{2} \partial_{xx} q\right)\right.$$
$$\left. - \frac{5}{6} \zeta h \partial_x h + \frac{5}{6} h \partial_{xxx} h\right] \times \left(1 - \frac{\delta}{70} q \partial_x h\right)^{-1}, \tag{6.92}$$

together with the mass balance equation (6.1).

Hereinafter, the system (6.1) and (6.92) will be referred to as the *regularized model*. Homoclinic orbits corresponding to single-hump solitary wave solutions to (6.1), (6.92) have been computed and are displayed as curves labeled 6 in Figs. 6.2 and 6.3. Unphysical turning points resulting in bending of the solution branches have never been observed for all values of δ we examined. Further, the system (6.1), (6.92) is consistent at second-order with the BE long wave theory but at the same time it takes into account modifications of the momentum balance of the film induced by the departures of the velocity profile from the parabolic Nusselt flat film solution, which become crucial deeply into the drag-inertia regime.

Of course to be consistent with the perturbation approach for small ε, the regularization factor $\mathcal{G}$ in (6.91) should be expanded for small ε, i.e., it should have the form $1 + \varepsilon, \dots$. A natural question that might be asked here is. Why is $\mathcal{G}$ not expanded? First of all, many different forms of the regularization factor $\mathcal{G}$ have been tried and only the one that works has been kept. In fact, expanding $\mathcal{G} = 1/(1 - \varepsilon, \dots)$

as $\sim 1 + \varepsilon, \dots$ leads to failure of the model, i.e., occurrence of unphysical turning points for the speed c of the single-hump solitary wave solutions as a function of δ. The reason for this failure can be seen from the dynamical system for the traveling wave solutions of (6.92): We immediately get that h''' is equal to a fraction with a denominator equal to $1 + \varepsilon 3Req\partial_x h/70$, which can go to zero. In other words, this form of $\mathcal{G}$ creates a region in the phase space that is forbidden for the homoclinic trajectories (a "singular surface" in the phase plane that we cannot cross).

Different forms (but asymptotically equivalent) of the regularization factor not leading to the same result is a sign of the gradient approach starting to fail.[5] In the drag-inertia regime, inertia effects cannot be treated as corrections to gravity and viscous drag, and the hypothesis sustaining the gradient expansion approach is violated (see also the discussion in Sect. 6.1). Regularization is a way to extend the two-equation second-order model to a region in the parameter space where the gradient expansion starts to fail. In fact, Ooshida followed a similar approach. But his procedure, also based on a gradient expansion, cannot overcome the BE inherent limitation that inertia is a small correction. Yet, it provides a way to extend the BE approach to a regime where the gradient expansion should not work.

The main idea of the regularization procedure followed here is to manipulate the inertia terms in $\mathcal{K}$ to obtain asymptotically equivalent expressions that allow us to move into the drag-inertia regime. It is the regularization factor $\mathcal{G}$ that allows us to obtain $\mathcal{K}$ asymptotically from, e.g., (6.92); recall that both the regularized model and the models with $\mathcal{K}$ are consistent at second order. After the elimination of the fields s_1 and s_2 to obtain (6.83) and the utilization of the equivalence $\partial_t h = -h^2 \partial_x h + \mathcal{O}(\varepsilon)$ to obtain the more compact expression in (6.84), we produce high-order nonlinearities. The regularization procedure then attempts to reduce the order of the nonlinearities so as to avoid nonphysical blow ups. For instance, we reduce the order of the nonlinearities when we go from (6.90) to (6.91). Of course, much like with the BE, there is no guarantee beforehand that the regularization procedure will work.

But, as it turns out, the regularization procedure does work, and the reasons for that can be summarized as follows:

  (i) it reduces the order of the nonlinearities associated with second-order inertia corrections;
 (ii) it avoids the presence of a denominator that can vanish in the dynamical system for traveling waves obtained from the resulting regularized model;
(iii) the numerical factor $3/70$ in front of $\varepsilon Req\partial_x h$ in (6.91) is small.

Finally, as mentioned earlier both regularized and second-order models capture the synchronous three-dimensional patterns observed in the experiments by Liu et

---

[5] Similarly, Oron and Gottlieb [199] in their study of the subcritical or supercritical nature of the primary bifurcation from the Nusselt flow found dramatically different results when using the first-order and second-order BE: The bifurcation is supercritical for the first-order BE but it can change to subcritical when the second-order BE is used. They attributed this to the poor convergence characteristics of the gradient expansion used to obtain the BE.

al. [170]. However, the simplicity of the regularized model compared to the full second-order model, with two dynamic variables as opposed to four, makes it a useful prototype for numerical and mathematical scrutiny for both two-dimensional and three-dimensional effects in falling film flows.

6.10 Contrasting the Center Manifold Analysis and the Method of Weighted Residuals

To conclude this chapter, let us underline the similarities and differences of the center manifold analysis by Roberts detailed in Sect. 6.3 and of the weighted residuals formulation developed later on in this chapter.

The center manifold analysis does not rely on an expansion in series of polynomials for the velocity field, but nevertheless it is based on similar hypotheses utilized in the averaging methodology: coherence of the flow in the cross-stream direction and the long wave assumption, $\varepsilon \ll 1$. In addition, the center manifold analysis requires that inertia effects are weak corrections to the viscous drag-gravity balance. The momentum equation in the Roberts model (6.26) contains all the terms of the Kapitza–Shkadov averaged momentum balance (6.13b) or the momentum equation (6.51) of the first-order model but with different coefficients. This agreement originates in the fact that the velocity profile, $u_{\mathrm{rob}} \propto \sin(\pi \bar{y}/2)$ (first nonzero mode for the velocity with $l = \pi/2$ for $\gamma = 1$ in (6.22)), is very close to the parabolic profile since $\langle u_{\mathrm{rob}}, F_0 \rangle / \sqrt{\langle u_{\mathrm{rob}}, u_{\mathrm{rob}} \rangle \langle F_0, F_0 \rangle} \approx 0.999$, where $F_0 = \bar{y} - (1/2)\bar{y}^2$. However, the Roberts model also contains high-order additional terms including high-order nonlinearities, which then necessarily restrict the applicability of the model in the drag-gravity regime.

Similarly to Roberts' analysis, the weighted residuals formulation developed in this chapter presumes that inertia effects are weak corrections to the balance of viscous drag and gravity. Once again, this, strictly speaking, holds only in the drag-gravity regime as we have stated several times. Nevertheless, our hope was that the averaged models we obtained can be accurate outside their region of validity and are thus capable of describing the drag-inertia regime. As a matter of fact, we have already demonstrated (Figs. 6.2 and 6.3) that the full second-order, simplified second-order and regularized averaged models contain nonlinearities that do not lead to the unphysical loss of the solitary wave branch of solutions at $\delta > 1$, and hence they cure the deficiencies of the BE/long wave theory in the drag-inertia regime.

Still, solutions to the averaged models must be checked against both experiments and DNS. This was done here with DNS for traveling waves (Fig. 6.4) and in the next chapter where we demonstrate good agreement with both experiments and DNS for traveling waves. Thus, the method of weighted residuals is the definitive low-dimensional modeling of film flows leading to a small number of coupled evolution equations which are accurate in the drag-inertia regime.

Chapter 7
Isothermal Case: Two-Dimensional Flow

A falling liquid film can serve as a paradigm for the study of open flow hydrodynamic systems. This is because: (i) the flow is nearly parallel, unlike other hydrodynamic systems, e.g., jets that break up into drops; (ii) it can be studied experimentally relatively easily; (iii) the Reynolds number is small-to-moderate, which makes the problem amenable to theoretical analysis.

In fact, falling liquid films have several similarities with many other hydrodynamic systems. The analogy with boundary layer flows has already been emphasized in the modeling approaches described in Chaps. 4 and 6, while similarities with the three-dimensional instabilities developed in boundary layer flows will be discussed in Chap. 8. Similarities can also be found with the propagation of *bores* in rivers (in the "torrential regime"), a topic that will be discussed in this chapter.

More importantly, falling film flows offer an excellent opportunity for the theoretical study of the route toward spatio-temporal disorder and the specific events characterizing its development, not only in open flow hydrodynamic systems but other nonlinear systems as well. The wide variety of phenomena that can be investigated with falling film flows are: (i) development of convective instabilities; (ii) spatial response to external perturbations; (iii) development of traveling waves; (iv) competition/interplay between different instability mechanisms, e.g., for the problem of a heated film; (v) "condensation" phenomena such as formation of *bound states*, i.e., well-defined and robust groups of *coherent structures*.

Figure 7.1 shows a snapshot of the thickness of the film at the end of a simulation of a naturally excited wavy motion. The initial growth of the waves at the inlet is rapidly followed by a wavy regime where localized structures are separated by relatively large portions of nearly flat films. Subsequently, the dynamics on the film is dominated by these *dissipative structures*, which seem to organize the flow. The dynamics is therefore "weakly disordered" and the spatial evolution of the film is an example of *weak/dissipative turbulence* in the Manneville sense [177]. Isolated waves look like tear drops made of a large-amplitude hump preceded by small capillary ripples, also referred to as radiation, as noted in previous chapters. These waves resemble the infinite-domain solitary waves we have already encountered at several places in this monograph. The evolution of the film is therefore dominated

S. Kalliadasis et al., *Falling Liquid Films*, Applied Mathematical Sciences 176, DOI 10.1007/978-1-84882-367-9_7, © Springer-Verlag London Limited 2012

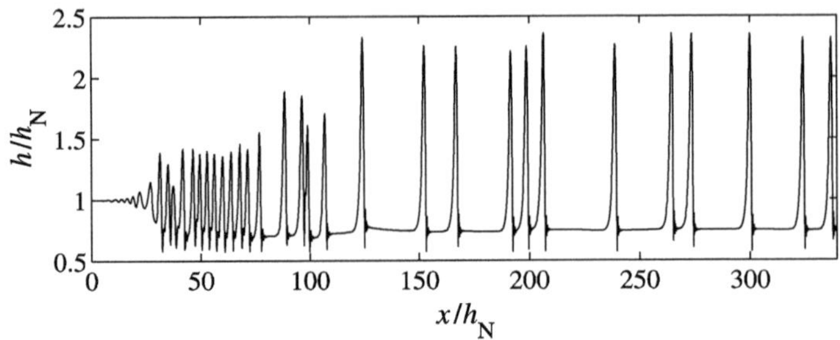

Fig. 7.1 Simulation of a noise-driven film flow down a vertical wall (see also Sect. 7.3.2). Parameter values correspond to an experiment by Kapitza [141] with alcohol ($v = 2.02 \times 10^{-2}\ \mathrm{cm^2\,s^{-1}}$, $\gamma/\rho = 29\ \mathrm{cm^3\,s^{-2}}$). The flow rate is $0.123\ \mathrm{cm^2\,s^{-1}}$ ($Re = 6.07$, $We = 76.4$)

by solitary-like *coherent structures*, which are stable and robust and interact indefinitely with each other as "quasi-particles."

Increasing the Reynolds number in the simulation of Fig. 7.1 makes the interface appear more complicated, but despite the apparent complexity one can still identify solitary-like coherent structures in what appears to be a randomly disturbed surface. It is then essential that in order to understand the spatio-temporal evolution of the film, we fully understand the properties of individual solitary waves, which in turn can help us understand the way they interact with each other. In fact, these coherent structures are truly elementary processes so that the dynamics of the film can be described by their superposition. Hence, the falling liquid film can serve as a canonical reference system for the study of weak/dissipative turbulence. This is further facilitated by the substantial reduction of the complexity of the governing equations offered by the long wave nature of the instability.

The "self-organization" of the flow into interacting solitary waves is characteristic of *active dispersive-dissipative* nonlinear media, where typically an "instability" mechanism generates waves that are subsequently "synchronized" through a process involving dispersion. The instability mechanism acts at large scales and pumps energy from the basic state ("main flow" in the case of hydrodynamic systems) to the perturbations, while at small scales a "stability" mechanism is effective. This mechanism is also responsible for energy transfer from the large scales to the small ones, known as "dissipation".

In the case of falling films, dissipation is triggered by surface tension and instability is due to inertia (H mode). As far as the dispersion of the waves is concerned, it is due to second-order viscous effects, often underestimated in the literature, which we have already referred to in this monograph as "viscous dispersion." The primary instability considered in Chap. 3 as well as the nontrivial dispersive effect of the viscosity will be examined in this chapter within the framework of the wave hierarchy concept. We shall further scrutinize the models derived in Chap. 6 with the tools of linear stability analysis, dynamical systems theory and numerical simulations. We shall demonstrate that these models satisfactorily account for all features of the two-dimensional wavy regime and are in quantitative agreement with both experiments and DNS.

7.1 Linear Stability Analysis

7.1.1 Dispersion Relations and Neutral Stability

Let us now consider infinitesimal perturbations around the Nusselt flat film solution,

$$h = 1, \qquad q = \frac{1}{3},$$

and in the case of (6.78), $s_i = 0$, to obtain conditions for instability.

By substituting $w = w_0 + \tilde{w}$ into the Kapitza–Shkadov model (6.13a), (6.13b), where w_0 and $\tilde{w}$ refer to the values of the different variables for the Nusselt flat film solution and their deviations from these values, respectively, and linearizing for $\tilde{w} \ll 1$ leads to

$$\partial_t \tilde{h} = -\partial_x \tilde{q}, \tag{7.1a}$$

$$\delta \partial_t \tilde{q} = -\frac{4}{5}\delta \partial_x \tilde{q} + \left(\frac{2}{15}\delta - \zeta\right)\partial_x \tilde{h} + 3\tilde{h} - 3\tilde{q} + \partial_{xxx}\tilde{h}. \tag{7.1b}$$

The solutions of these equations can be sought in the form of normal modes,

$$\tilde{w} = w' \exp\{i(kx - \omega t)\}, \tag{7.2}$$

where k and ω are the complex wavenumber and complex angular frequency, respectively. Inserting (7.2) into (7.1a), (7.1b) leads to the dispersion relation

$$\delta \omega^2 + \left(-\frac{4}{5}k\delta + 3i\right)\omega - k^4 + k^2\left(\frac{2}{15}\delta - \zeta\right) - 3ik = 0. \tag{7.3}$$

To simplify comparisons with the linear stability analysis presented in Chap. 3, it is convenient to rewrite (7.3) using the Nusselt scaling. Recall that this is a necessary step whenever a comparison is needed either with full Navier–Stokes (including Orr–Sommerfeld), or with experiments. The change of scales does not affect the speed and does not introduce any modifications to the coefficients of the equations except for a coefficient of 3 along with the Reynolds number (which retains its standard definition (2.35)). With the transformation $\omega \to \kappa h_N$, $k \to \kappa k/h_N$ and using the definition of the reduced variables $\delta = 3Re/\kappa$, $\zeta = Ct/\kappa$, $\eta = 1/\kappa^2$ and $\kappa^3 = We$,

$$3i\omega + 3Re\omega^2 + \left(-3i - \frac{12}{5}\omega Re\right)k + \left(-Ct + \frac{2}{5}Re\right)k^2 - Wek^4 = 0, \tag{7.4}$$

the dispersion relation of the Kapitza–Shkadov model in terms of the Nusselt scaling.

Controlled experiments devoted to the detection of neutral stability conditions are generally performed by forcing either the film thickness or its flow rate at some frequency and by detecting the *cut-off frequency* f_c beyond which the film remains flat. The cut-off frequency f_c is thus determined from the dispersion relation by imposing that $k_i = \omega_i = 0$ in this relation (otherwise disturbances grow in both space and time). This then yields the Reynolds number as a function of the cut-off frequency f_c ($\equiv \omega_r/(2\pi)$).

By considering then both ω and k to be real in the dispersion relation (7.4) and by separating real and imaginary parts, we obtain

$$\omega = k_c \quad \text{and} \quad Re = Ct + k_c^2 We, \tag{7.5}$$

or

$$k_c = \sqrt{\frac{1}{We}(Re - Ct)}, \tag{7.6}$$

where the second relation is the neutral stability curve and k_c is the cut-off wavenumber, which can be directly related to the cut-off frequency, $k_c = 2\pi f_c$. Hence the phase velocity is $c = \omega/k_c \equiv 1$ and is equal to the speed of the *kinematic waves* (we shall discuss these waves in detail later on in this chapter) described by the mass conservation equation $\partial_t h + \partial_x q = 0$ with $q \equiv h^3/3$. The absence of any wavenumber dependence for the phase velocity is due to the absence of viscous dispersion in the Kapitza–Shkadov model. Moreover, the minimum of Re at the neutral curve is Ct (alternatively, simply set $k_c = 0$; the maximum growing linear mode at criticality has a vanishing growth rate and wavenumber) and hence the critical value of Re is $Re_c = Ct$, deviating 20% when compared to the correct value $Re_c = \frac{5}{6}Ct$ obtained from Orr–Sommerfeld.

The dispersion relation of the first-order model (6.1), (6.51) reads,

$$3i\omega + \frac{18}{5}Re\omega^2 + \left(-3i - \frac{102}{35}\omega Re\right)k + \left(-Ct + \frac{18}{35}Re\right)k^2 - Wek^4 = 0, \tag{7.7}$$

which differs from (7.4) only in the numerical coefficients. The neutral stability conditions corresponding to (7.7) are now,

$$\omega = k_c \quad \text{and} \quad Re = \frac{5}{6}(Ct + k_c^2 We), \tag{7.8}$$

or

$$k_c = \sqrt{\frac{1}{We}\left(\frac{6}{5}Re - Ct\right)}. \tag{7.9}$$

The minimum of Re at the neutral curve then is the correct critical Reynolds number, $Re_c = \frac{5}{6}Ct$, obtained from Orr–Sommerfeld.

The simplified and the regularized models, (6.1), (6.79) and (6.1), (6.92), respectively, yield the same dispersion relation:

$$
3i\omega + \frac{18}{5}Re\omega^2 + \left(-3i - \frac{102}{35}\omega Re\right)k
$$
$$
+ \left(-Ct + \frac{18}{35}Re + i\frac{27}{5}\omega\right)k^2 - \frac{12}{5}ik^3 - Wek^4 = 0. \qquad (7.10)
$$

A comparison of (7.10) to (7.7) shows that two new terms appear in (7.10): they account for viscous effects. Taking both ω and k as real and separating real and imaginary parts in (7.10) yields:

$$
\omega = k_c\frac{1 + 4k_c^2/5}{1 + 9k_c^2/5} \quad \text{and} \quad c = \frac{\omega}{k_c} = \frac{1 + 4k_c^2/5}{1 + 9k_c^2/5}, \qquad (7.11a)
$$
$$
\frac{6}{35}(21c^2 - 17c + 3)Re = Ct + k_c^2 We. \qquad (7.11b)
$$

k_c for the simplified and regularized models is not the same with that in (7.9). To obtain k_c for these models, substitute the expression for c in (7.11a) into (7.11b) to obtain a single equation for k_c.

It is not difficult to see that for the simplified and regularized models the minimum value of Re at the neutral curve, i.e., the critical value of Re, occurs at $k_c = 0$ and $c = 1$ or $Re_c = \frac{5}{6}Ct$, the correct answer.

Notice the presence of the dispersive terms $4k_c^2/5$ and $9k_c^2/5$ in the expression for the phase velocity, unlike the Kapitza–Shkadov and first-order models. Therefore, on the neutral curve the phase velocity is a function of wavenumber and in fact smaller to the speed of the kinematic waves, i.e., unity. Hence, viscous effects make the falling film dispersive at the instability onset. In other words, viscous effects modify the phase velocity at onset by introducing a wavenumber dependence and we may therefore refer to this effect as *viscous dispersion*. In fact, in terms of the Shkadov scaling, i.e., with the transformation $k \to (1/\kappa)k$, the above expression for the phase velocity becomes,

$$
c_k = \frac{1 + 4\eta k_c^2/5}{1 + 9\eta k_c^2/5},
$$

thus justifying the term "viscous dispersion number" for η, introduced in Sect. 4.6.

The derivation of the linear dispersion relation corresponding to the full second-order model (6.1), (6.78) is more involved and the relation more complicated than the ones we've just given. The result is:

$$
A + Bk + Ck^2 + Dk^3 + Ek^4 + Fk^5 + Gk^6 = 0, \qquad (7.12)
$$

with

$$
A = 3i\omega + \frac{54}{13}Re\omega^2 - \frac{90}{143}iRe^2\omega^3 - \frac{12}{715}Re^3\omega^4, \qquad (7.13a)
$$

$$B = -3i - \frac{522}{143}Re\omega + \frac{98}{143}iRe^2\omega^2 + \frac{108}{5005}Re^3\omega^3, \tag{7.13b}$$

$$C = -Ct + \frac{498}{715}Re + \omega\left(\frac{27}{5}i + \frac{12}{65}iCtRe - \frac{26424}{117117}iRe^2\right)$$
$$+ \omega^2\left(\frac{3231}{3640}Re + \frac{27}{5005}CtRe^2 - \frac{612}{65065}Re^3\right) - \frac{2027}{80080}iRe^2\omega^3, \tag{7.13c}$$

$$D = -\frac{12}{5}i - \frac{304}{5005}iCtRe + \frac{1368}{65065}iRe^2 + \omega\left(-\frac{2441}{20020}Re - \frac{16}{5005}CtRe^2\right.$$
$$\left.+ \frac{1104}{715715}Re^3\right) + \frac{3439}{145600}iRe^2\omega^2, \tag{7.13d}$$

$$E = \frac{30993}{320320}Re + \frac{148}{325325}CtRe^2 - \frac{48}{715715}Re^3 - We$$
$$+ \omega\left(-\frac{4591}{650650}iRe^2 + \frac{12}{65}iWeRe\right) + \frac{27}{5005}WeRe^2\omega^2, \tag{7.13e}$$

$$F = \frac{1773}{2602600}iRe^2 - \frac{304}{5005}iWeRe - \frac{16}{5005}WeRe^2\omega, \tag{7.13f}$$

$$G = \frac{148}{325325}WeRe^2. \tag{7.13g}$$

Comparison of (7.10) and (7.12) reveals that the terms independent of Re are identical. The terms linear in Re are recovered but with slightly different coefficients. The remaining terms in (7.12) have Re at some power. These terms are due the inclusion of inertia corrections in the second-order model not present in the simplified and regularized models.

Although complicated, the dispersion relation (7.12) is still much simpler to solve than the Orr–Sommerfeld eigenvalue problem, for which the system of ordinary differential equations (3.22a)–(3.22i) needs to be solved. In fact, solving (7.12) amounts to finding the roots of a polynomial in k and ω. The neutral stability conditions corresponding to (7.12) have been obtained by continuation using the software AUTO-07P [79] starting from the instability threshold $Re = Re_c$, $k = 0$ and $c = 1$. The numerical procedure is similar to that outlined in Appendix F.1.

The neutral stability curves obtained from the different models are compared in Fig. 7.2 with the experiments carried out Liu et al. [170]. The curves are obtained by fixing the liquid properties and the inclination angle (hence the Kapitza number) to the values used in the experiments. The only free parameter then is the film thickness h_N or equivalently the Reynolds number. The simplified and full second-order dispersion relations (7.10) and (7.12), are displayed in the figure as a thick dashed and thin solid line, respectively. Recall that the regularized model (6.1), (6.92) and the simplified model share the same dispersion relation and so we no longer refer to the regularized model for the remainder of this section.

Both results agree equally well with the experimental data by Liu et al. [170]. However, comparison with the Orr–Sommerfeld analysis (thick solid line) shows that, as expected, the full second-order is more accurate than the simplified one. At the same time we notice that the simplified model is closer to the experiments than

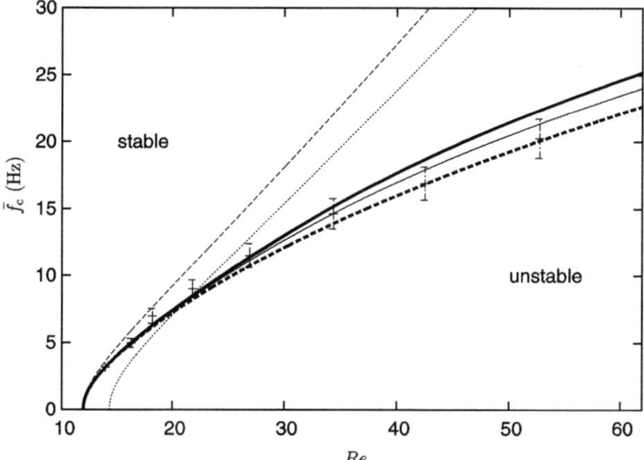

Fig. 7.2 Cut-off dimensional frequency $\bar{f}_c$ in Hz as a function of the Reynolds number Re for the conditions of the experiments in [170] (glycerin–water mixture, $\beta = 4°$, $\nu = 2.3 \times 10^{-6}$ m^2 s^{-1}, $\sigma/\rho = 62.6 \times 10^{-6}$ m^3 s^{-2}, i.e., $\Gamma = 2341$). Orr–Sommerfeld analysis (*thick solid line*) and experimental results (*crosses*) compared to predictions from the Kapitza–Shkadov model (*dotted line*), first-order model (*dashed line*), simplified second-order model (*thick dashed line*), and full second-order model (*thin solid line*)

both Orr–Sommerfeld analysis and full second-order model, but the departure of the Orr–Sommerfeld and full second-order model results from the experimental points is small. This small departure can be accounted for by noting the sensitivity of the results with respect to the inclination angle β at small β: a small error in the measurement of β can cause a visible shift of the whole neutral curve in the plane (Re, f). We also note the increasing discrepancy between the prediction of the first-order model represented by the dashed line in Fig. 7.2—the Kapitza–Shkadov model (dotted line) does even worse as it predicts erroneously the instability threshold—and the experimental data. This is due to the neglect of viscous effects, which influence the change of the phase velocity with Re [226].

From Chap. 2, the time scale for the Nusselt scaling is $t_\nu l_\nu / \bar{h}_N (= \bar{h}_N / 3\bar{u}_N) = t_\nu / h_N = [(\nu / (g \sin \beta)^2]^{1/3} / h_N$. Fixing the liquid and the inclination angle, i.e., the Kapitza number, then means that the time scale is $\sim 1/h_N$. Hence, the relation between the dimensional and dimensionless frequency is $\bar{f}_c \sim h_N f_c \sim Re^{1/3} f_c$, which with $f_c \sim k_c$ becomes $\bar{f}_c \sim Re^{1/3} k_c$. But for large Re, (7.5) and (7.8) give $k_c \sim (Re/We)^{1/2} \sim h_N^{5/2} \sim Re^{5/6}$. Hence, $\bar{f}_c \sim Re^{7/6}$ for the Kapitza–Shkadov and first-order models, which explains the seemingly linear behavior seen in Fig. 7.2 for the corresponding neutral curves.

7.1.2 Absolute and Convective Instabilities

A more stringent test of the accuracy of the different models can be obtained from the study of the linear dynamics of *wave packets*, for which numerical results from

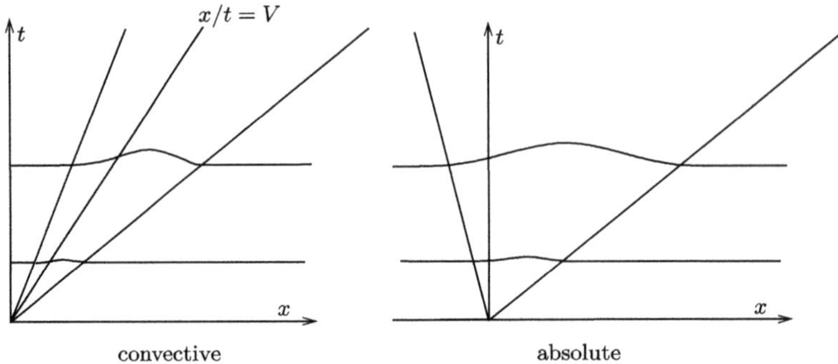

Fig. 7.3 Schematic of the development of a wave packet in the laboratory frame for convective/absolute instabilities

Orr–Sommerfeld are available [31]. We start by writing formally the dispersion relation as, $D(k, \omega) = 0$. Following Huerre and Monkewitz [120] we can associate a differential operator $D(-i\partial_x, i\partial_t)$ in the physical space (x, t) to the dispersion relation $D(k, \omega) = 0$ in the (k, ω) space such that the perturbations $\tilde{w}$ satisfy

$$D(-i\partial_x, i\partial_t)\tilde{w} = 0.$$

Let us now consider the fate of an infinitesimal perturbation initiated at position $x = 0$ and time $t = 0$ (see Fig. 7.3). Mathematically, the linear response to a localized perturbation in space and time can be obtained by the solution to

$$D(-i\partial_x, i\partial_t)G(x, t) = \delta(x)\delta(t), \tag{7.14}$$

where δ denotes the Dirac delta function and $G(x, t)$ is the "Green's function"

$$G(x, t) = \frac{1}{(2\pi)^2} \int_{L_\omega} \int_{F_k} \frac{\exp[i(kx - \omega t)]}{D(k, \omega)} \, dk \, d\omega, \tag{7.15}$$

where the integration paths L_ω and F_k are appropriately chosen in the complex ω- and k-planes so as to ensure the convergence of the two integrals. A natural choice for F_k is the real axis, $k \in \mathbb{R}$, so that a necessary condition for the convergence of the integration in time is that L_ω lies above all zeroes of the dispersion relation for $k \in F_k = \mathbb{R}$, i.e., above all *temporal modes*. *Spatial modes* are conversely obtained by considering ω real and k complex, whereas the integration path L_ω defines *generalized spatial modes* as the solutions to the dispersion relation (7.16a) for an angular frequency ω belonging to the contour L_ω in the complex plane (ω_r, ω_i).

The asymptotic behavior of the Green's function as $t \to \infty$ in a frame of reference moving at a speed V with respect to the laboratory frame of reference, i.e., on the ray $x/t = V$, is determined by the fastest growing part of the wave packet which travels with a *group velocity* $v_g = d\omega/dk$ equal to the ray velocity V. By writing $\omega(k) = \omega'(k) + Vk$, this is equivalent to finding the solution, denoted $\omega_{V,i}$, of largest imaginary part of the angular frequency ω'_i of the system:

$$D(k, \omega' + Vk) = 0, \qquad\qquad (7.16a)$$

$$d\omega'/dk = 0. \qquad\qquad (7.16b)$$

The solutions of system (7.16a), (7.16b) correspond to double roots of the dispersion relation (7.16a). Considering the generalized spatial branches $k(\omega')$ defined by the contour $L_{\omega'} = \{\omega' \in \mathbb{C}, \omega'_i = \mathrm{const}\}$, system (7.16a), (7.16b) defines *saddle points* in the complex plane (k_r, k_i), where two generalized spatial branches $k(\omega')$ collide. A saddle point must satisfy the so-called *collision criterion* established by Briggs [32, 121]. This criterion, which follows from "causality" (for $t < 0$, the film is at rest), states that, in order to be physically acceptable, the saddle point has to arise from the "pinching" of two $k(\omega')$-branches coming from different sides of the real axis $k_i = 0$ as the contour $L_{\omega'}$ is displaced downward in the complex plane (ω'_r, ω'_i). Practically, one proceeds by lowering the integration path $L_{\omega'}$ lying initially far above all temporal modes [120]. As $L_{\omega'}$ approaches one of the temporal branches, i.e., when $\omega'_i = \max\{\omega'_i, D(k, \omega' + Vk) = 0, k \in \mathbb{R}\}$, the integration paths F_k must be deformed to ensure the convergence of the integrals in (7.15). The deformation of the integration path F_k from the real axis then defines *generalized temporal modes* as the solutions $(k, \omega'(k))$ to the dispersion relation (7.16a) for k lying in $F_k \neq \mathbb{R}$. The process ends with the finding of the first saddle point verifying the Briggs criterion, as it is no more possible to deform the integration path F_k in the k-plane. However, in practice the construction of the generalized temporal modes is unnecessary and the monitoring of the generalized spatial modes in the k-plane is sufficient.[1]

The basic flow is stable whenever the initial perturbation is asymptotically damped along all rays, i.e., $\omega_{V,i} < 0$ for all ray velocities V. When instability occurs, there exists a range of values of V for which the perturbation grows along the ray $x/t = V$, i.e., $\omega_{V,i} > 0$. The front and the rear of the linear wave packet are then defined by $\omega_{V,i} = 0$ and are illustrated by straight lines in Fig. 7.3.

Considering the ray $x/t = 0$, i.e., the laboratory frame, the saddle point given by (7.16a), (7.16b) defines the *absolute frequency* ω_0 and *absolute wavenumber* k_0. Further, the instability is termed *convective* if the perturbation vanishes on the spot of initiation ($\omega_{0,i} < 0$) and *absolute* in the opposite case ($\omega_{0,i} > 0$). In the connectively unstable case the perturbations originating from the noise upstream are convected downstream by the flow: The flow responds to the upstream perturbations as a *noise amplifier*. In the absolutely unstable case, the perturbations grow, i.e., they are able to move upstream, and the flow behaves as an oscillator having its own intrinsic dynamics. The convective or absolute nature of the instability can be determined

---

[1]In most cases, the task of identifying the relevant saddle point can be reduced to monitoring of only one generalized spatial mode, which is identified as follows: After the temporal mode corresponding to the instability is found, which is very easy in our case since $\omega = 0$ must belong to this mode, the intersection of this temporal mode and the real axis $\omega \in \mathbb{R}$ defines a point also belonging to a spatial mode ($w \in \mathbb{R}, k \in \mathbb{R}$). Increasing the growth rate ω_i above the maximum growth rate then defines a point on the generalized spatial mode whose deformations in the k-plane are then monitored as ω_i is lowered up to the finding of a pinching point.

from the sign of the maximum of ω_i for all the k roots of the dispersion relation $D(k, \omega) = 0$, verifying the collision criterion. From a spatial Orr–Sommerfeld analysis, it has been shown that the flow over inclined plates is connectively unstable at least up to very high Reynolds numbers [31].

Let us now compare the linear stability analysis of the models introduced in Chap. 6 to the exact results from the Orr–Sommerfeld analysis, restricting our attention to the full second-order model (6.78) and its simplified version (6.1), (6.79), since the first-order model (6.1), (6.51) fails to reproduce the neutral stability conditions correctly. Whereas for full Navier–Stokes, the dispersion relation is obtained from the numerical solution of the Orr–Sommerfeld eigenvalue problem, a differential eigenvalue problem in the cross-stream coordinate, for the models introduced in Chap. 6 the dispersion relation is just a polynomial equation in k and ω whose numerical solution is easier than that of Orr–Sommerfeld.

We first consider the case $Re = 26.7$, $\Gamma = 769.8$, $\beta = 4.6°$, corresponding to an experiment by Liu and Gollub [167] and to the Orr–Sommerfeld analysis performed by Brevdo et al. [31], reproduced here by using a numerical scheme similar to the one described in Appendix F.1. For this purpose an additional variable k_i and an additional constraint $\omega_i = \mathrm{const}$ must be added in subroutine ICND to define the integration path L_ω. Generalized spatial branches k_n in the complex plane (k_r, k_i) are displayed in Figs. 7.4 and 7.5 for the dispersion relations (7.12) and (7.10), respectively. The k roots are computed as ω_r is varied for different values of ω_i. Upon decreasing the imaginary part ω_i of the angular frequency from positive to negative values, no pinching of the generalized spatial branches k_n is observed before ω_i becomes negative, which is a clear indication of the convective nature of the instability. The agreement between the full second-order dispersion relation (7.12) and the exact results obtained from Orr–Sommerfeld is remarkable. All the branches observed by Brevdo et al. as well as their change as ω_i is varied from 0.01 (Fig. 7.4, top) to 0 (bottom), are recovered. Small departures from the exact solutions are only significant far from the origin (Fig. 7.4, left). Approaching the origin $k = 0$ (right) makes the predictions indistinguishable from the exact results obtained from Orr–Sommerfeld. The agreement turns out to be excellent when approaching the origin $k = 0$ (right), in line with the expectations from the long wave assumption underlying the models. Hence, (7.12) may be seen as an expansion of the exact dispersion relation for $k, \omega \ll 1$.

The behavior of the generalized spatial branches corresponding to the simplified second-order dispersion relation (7.10) is displayed in Fig. 7.5. The dispersion relation now is a polynomial of degree four in k so that not all the branches obtained with the Orr–Sommerfeld analysis can be recovered. However, branch 2 in Fig. 7.5 seems to result from a hybridization of branches 2 and 3 in Fig. 7.4. Branch 1, which is more physically important since it crosses the real axis ($k_i = 0$) from above (compare panels b and c in Fig. 7.5) and ultimately pinches with branches 4 and 5 as ω_i is further lowered, approximates well those obtained using either the full second-order model or the Orr–Sommerfeld analysis.

The agreement close to the origin between the models and the exact results found for the case $V = 0$, extends to the case $V \neq 0$. Figures 7.6 and 7.7 summarize the

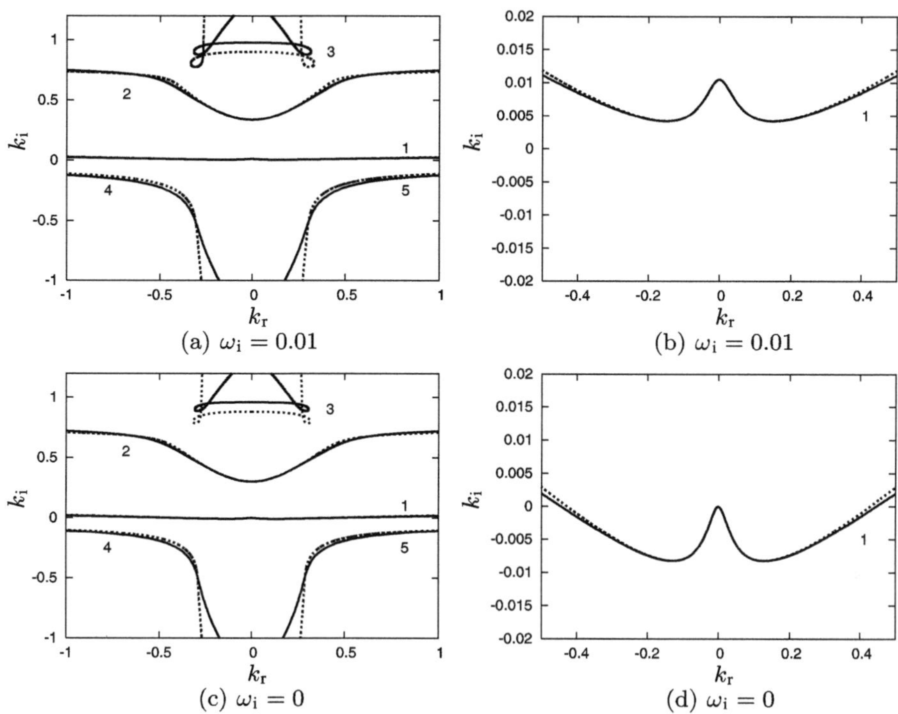

Fig. 7.4 Generalized spatial branches k_n in the complex (k_r, k_i)-plane with $\omega_i = 0.01$ (*top*) and $\omega_i = 0.0$ (*bottom*). $Re = 26.7$, $\beta = 4.6°$, $We = 41.46$ ($\Gamma = 769.8$). Solutions of the full second-order dispersion relation (7.12) and the Orr–Sommerfeld equations (3.22a)–(3.22i) are represented by the *dotted* and *solid lines*, respectively. *Left*: Overall view. *Right*: Enlarged view of the neighborhood of the origin in the complex k-plane

characteristics of the saddle points, verifying the collision criterion as a function of the speed V of the moving frame considered and for different values of the Reynolds number, from $Re = 13.3$ to $Re = 133$. The maximum of the temporal growth rate $\omega_i(V)$ governing the long time evolution of the perturbations on the ray $x/t = V$ is found to be positive only for $V > 0$ (see panel (a) in Fig. 7.6). Therefore, the instability is always convective as observed in experiments [169]. Presumably, this is a consequence of the high speed of the waves, approximately three times the average velocity of the Nusselt flat film flow.

In all cases an excellent agreement is observed between the results obtained from the full second-order dispersion relation (7.12) and the solution of the Orr–Sommerfeld equations, including the very peculiar change of the saddle point verifying the Briggs criterion and corresponding to the largest growth rate ω_i', or equivalently the "dominant" saddle point (see Table 7.1). The simplified model appears accurate only up to about $Re = 67$, a sufficiently large value nevertheless. Beyond $Re = 67$ the latter model does not succeed in reproducing the two branches. How-

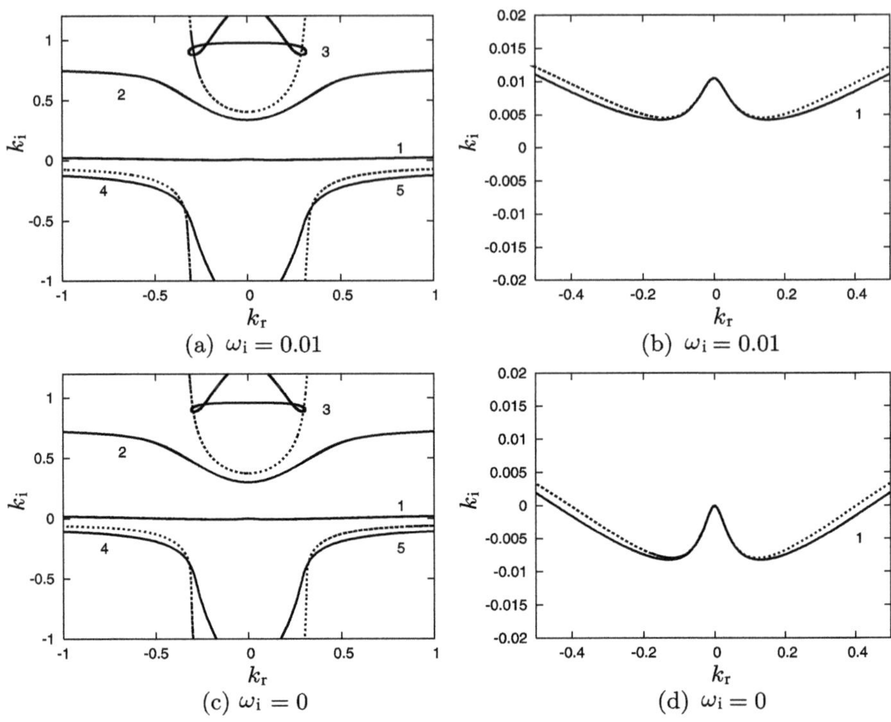

Fig. 7.5 Generalized spatial branches k_n in the complex (k_r, k_i)-plane for dispersion relation (7.10) (*dotted lines*) and the Orr–Sommerfeld equations (3.22a)–(3.22i) (*solid lines*). See also the caption of Fig. 7.4

ever, it seems to interpolate smoothly between them, predicting the total V-width of the unstable band and all other features of the instability satisfactorily as a function of V—compare Fig. 7.7 to Fig. 7.6.

The displacement of the spatial branches in the complex plane (k_r, k_i) is depicted in Fig. 7.8 as the temporal growth rate ω_i' decreases. The ray $x/t = 0.58$ is chosen to closely correspond to the exchange of the dominant saddle points as the ray velocity V is varied. Two successive collisions of the branches are clearly observable and should be contrasted to the results obtained with the simplified dispersion relation shown in Fig. 7.9. For the latter, the absence of the change of dominant saddle points is clearly related to the absence of one generalized spatial branch of solutions to (7.16a) due to the lower degree of the polynomial of the dispersion relation (7.10). A possible reason for this behavior is that, while the number of basis functions used for the projection in the Galerkin method is not sufficiently large to fully account for all the flow properties, a projection onto the first parabolic test function only recovers most of the main linear stability results, including the positions of the dominant saddle points.

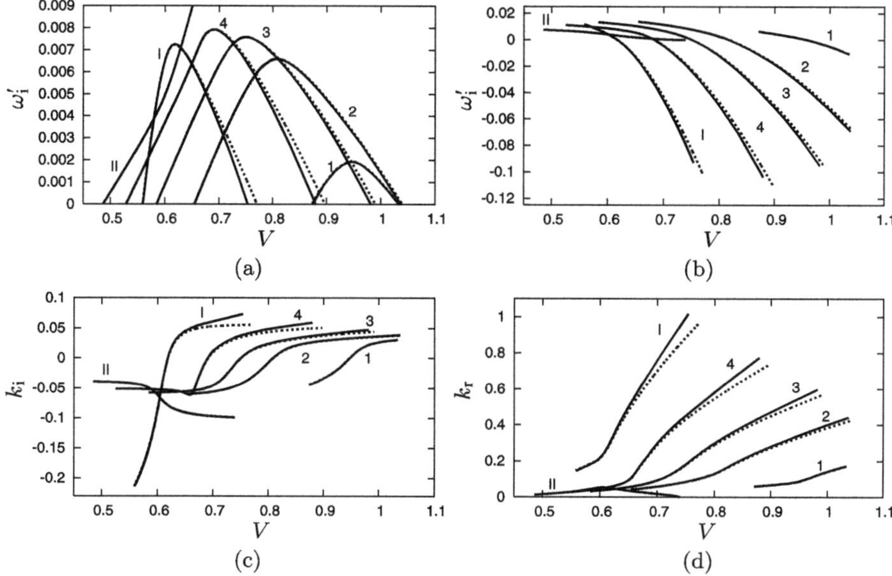

Fig. 7.6 Characteristics of the saddle point solutions to (7.16a), (7.16b) as function of the speed V of the moving frame for the Orr–Sommerfeld equations (3.22a)–(3.22i) (*solid lines*) and the full second-order dispersion relation (7.12) (*dotted lines*). Parameter values correspond to a glycerin–water mixture and the inclination angle considered in [167] ($\Gamma = 769.8$, $\beta = 4.6°$). $Re = 13.3$ (curve 1), $Re = 26.7$ (curve 2), $Re = 40$ (curve 3), $Re = 66.7$ (curve 4) and $Re = 133$ (curves I and II)

7.1.3 Wave Hierarchy

The origin of the primary instability can be understood within the framework of the *wave hierarchy* theory proposed initially by Whitham [299]. The theory has found applications in many different settings, from traffic flows and gas dynamics to two-phase flows, shallow water waves and even crystal growth. An account of the Nusselt flow stability in terms of wave hierarchy has been offered by Alekseenko et al. [4] and Ooshida [196]. In this section we provide a brief outline of the concept of wave hierarchy restricted to linear waves, we extend both Alekseenko's and Ooshida's works to include viscous effects and connect with the shallow water theory.

To illustrate the main ideas of the wave hierarchy approach with a minimum of algebra, let us first consider the linearized Kapitza–Shkadov model, the simplest of all averaged models. For the sake of simplicity we neglect surface tension. The simplified model leading to more complicated algebra and including both surface tension and second-order dispersion effects will be considered later on.

The linearized equations (7.1a), (7.1b) are rewritten here in terms of the Nusselt scaling

$$\partial_t \tilde{h} = -\partial_x \tilde{q}, \tag{7.17a}$$

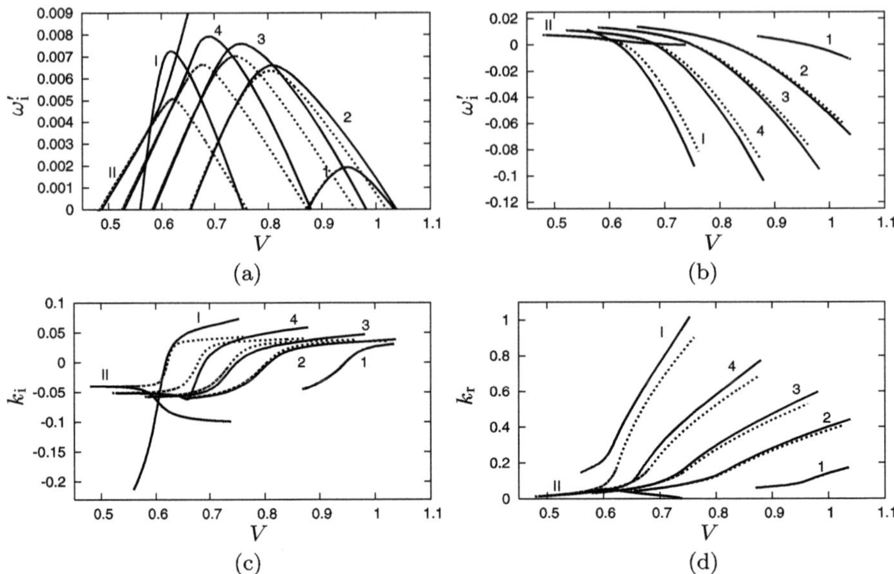

Fig. 7.7 Characteristics of the saddle point solutions to (7.16a), (7.16b) as function of the speed V of the moving frame for the Orr–Sommerfeld equations (3.22a)–(3.22i) (*solid lines*) and the simplified second-order dispersion relation (7.10) (*dotted lines*). $Re = 13.3$ (curve 1), $Re = 26.7$ (curve 2), $Re = 40$ (curve 3), $Re = 66.7$ (curve 4) and $Re = 133$ (curve 5). See also the caption of Fig. 7.6

$$Re\partial_t \tilde{q} = -\frac{4}{5}Re\partial_x \tilde{q} + \left(\frac{2}{15}Re - \frac{1}{3}Ct\right)\partial_x \tilde{h} + \tilde{h} - \tilde{q}. \tag{7.17b}$$

Differentiating (7.17b) with respect to x and replacing $\partial_x \tilde{q}$ with $-\partial_t \tilde{h}$ gives the following wave equation:

$$(\partial_t + \partial_x)\tilde{h} + Re\left(\partial_{tt} + \frac{4}{5}\partial_{xt} + \frac{2}{15}\partial_{xx}\right)\tilde{h} - \frac{1}{3}Ct\partial_{xx}\tilde{h} = 0. \tag{7.18}$$

To consider the behavior of the solutions to (7.18) when $x, t \to \infty$, let us rescale the time and space coordinates according to

$$\Delta x = X, \qquad \Delta t = T, \tag{7.19}$$

so that $\Delta \to 0$ corresponds to $x, t \to \infty$. Then, the limit $\Delta \to 0$ gives to leading order $(\partial_T + \partial_X)\tilde{h} = 0$, which when rewritten with the original scales becomes

$$\partial_t \tilde{h} + \partial_x \tilde{h} = 0. \tag{7.20}$$

Conversely, the initial stage of the development of a localized wave packet $x, t \to 0$ corresponds to $\Delta \to \infty$. This limit gives to leading order

$$Re\left(\partial_{tt} + \frac{4}{5}\partial_{xt} + \frac{2}{15}\partial_{xx}\right)\tilde{h} - \frac{1}{3}Ct\,\partial_{xx}\tilde{h} = 0, \tag{7.21}$$

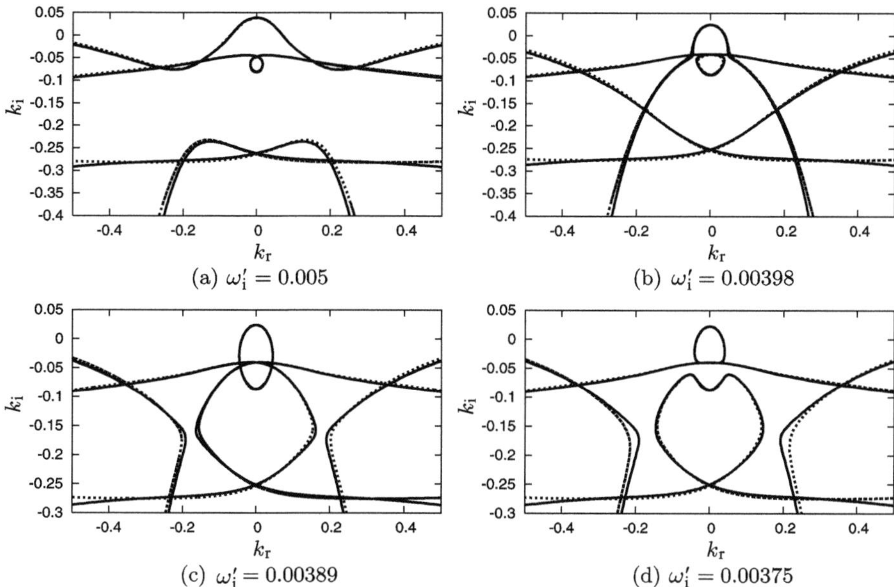

Fig. 7.8 Displacement of the spatial branches k_n in the complex (k_r, k_i)-plane as ω_i decreases for dispersion relation (7.12) and $Re = 133$, $V = 0.58$, $\beta = 4.6°$, $\Gamma = 769.8$. *Solid and dotted lines* refer to the solutions to the Orr–Sommerfeld equations (3.22a)–(3.22i) and to the full second-order dispersion relation (7.12), respectively

which can be rewritten as

$$(\partial_t + c_{d-}\partial_x)(\partial_t + c_{d+}\partial_x)\tilde{h} = 0, \tag{7.22a}$$

where

$$c_{d\pm} = \frac{2}{5} \pm \sqrt{\frac{2}{75} + \frac{1}{Fr^2}}. \tag{7.22b}$$

The *Froude number Fr*, which compares the advection by the flow at speed $3\bar{u}_N$, i.e., the speed of the kinematic waves and the speed of *gravity waves* $\sqrt{g\bar{h}_N \cos\beta}$ (see below), is given by, $Fr^2 = 3Re/Ct = (3\bar{u}_N)^2/(g\bar{h}_N \cos\beta)$. The above discussion shows that the wave equation (7.18) is effectively the combination of two levels of description corresponding to the first-order wave equation (7.20) and second-order wave equation (7.21). The series of wave equations of different order obtained from the original wave equation is precisely what Whitham referred to as "wave hierarchy." The early stages of a localized perturbation are governed by the higher-order wave equation (7.22a) and consequently the wavefronts at the front and back of the produced wave packet must travel at speeds c_{d+} and c_{d-}, respectively. The effects of lower-order waves moving at speed $c_k = 1$ on the higher-order ones traveling at speed c_{d-} can be approximated by substituting the time derivatives ∂_t with $-c_{d-}\partial_x$ in (7.18), except for the operator $\partial_t + c_{d-}\partial_x$, which vanishes with this substitution.

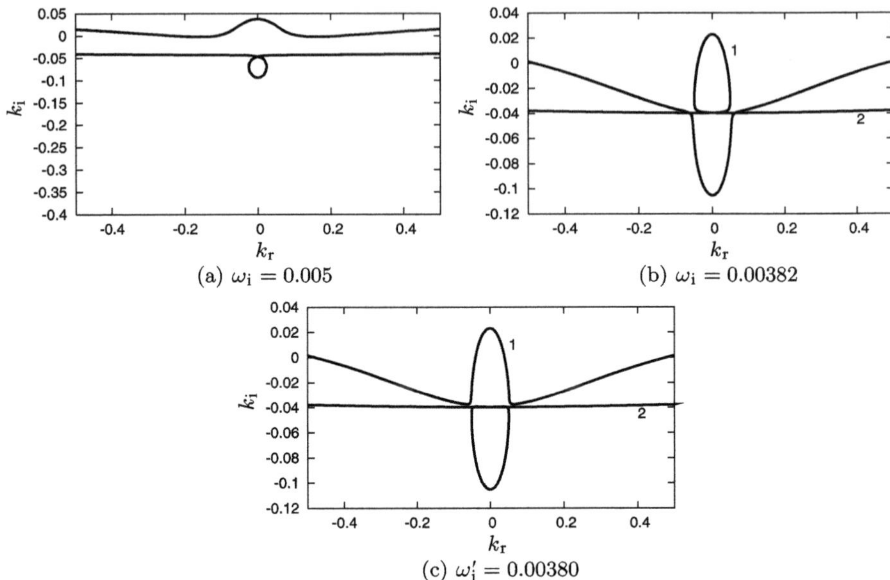

(a) $\omega_i = 0.005$

(b) $\omega_i = 0.00382$

(c) $\omega_i' = 0.00380$

Fig. 7.9 Displacement of the generalized spatial branches k_n in the complex (k_r, k_i)-plane as ω_i decreases for dispersion relation (7.10). See also the caption of Fig. 7.8

Table 7.1 Comparison of the saddle point positions between solutions of the dispersion relation of the second-order model, (7.12), of the simplified model, (7.10) and Orr–Sommerfeld [31] ($Re = 133$, $\beta = 4.6°$ and $\Gamma = 769.8$)

	Branch	k_r	k_i	ω_r'	ω_i'
$V = 0.575$					
Orr–Sommerfeld	I	0.170	−0.179	0.0092	0.0031
Equation (7.12)	I	0.172	−0.176	0.0091	0.0031
Orr–Sommerfeld	II	0.040	−0.046	0.0053	0.0037
Equation (7.12)	II	0.040	−0.046	0.0054	0.0037
Equation (7.10)		0.0478	−0.0387	0.0050	0.0036
$V = 0.580$					
Orr–Sommerfeld	I	0.177	−0.165	0.0083	0.0039
Equation (7.12)	I	0.181	−0.163	0.0083	0.0040
Orr–Sommerfeld	II	0.0431	−0.0475	0.0051	0.0039
Equation (7.12)	II	0.0430	−0.0475	0.0051	0.0039
Equation (7.10)		0.0516	−0.0385	0.0049	0.0038

We thus have after one integration in space:

$$(\partial_t + c_{d-}\partial_x)\tilde{h} \approx \frac{1}{Re}\frac{c_{d-} - c_k}{c_{d+} - c_{d-}}\tilde{h}$$

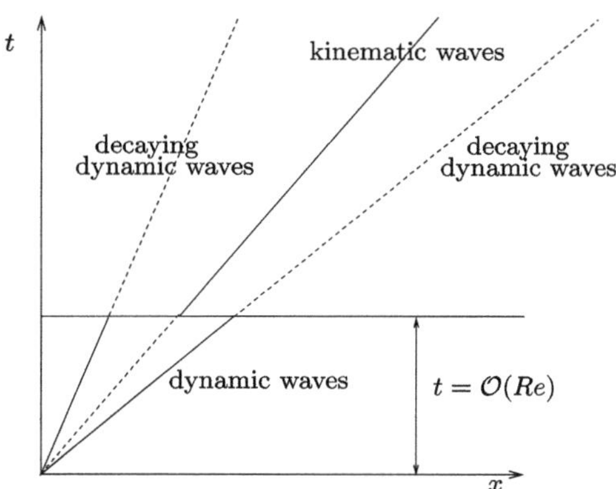

Fig. 7.10 Sketch of the spatio-temporal diagram of wave solutions to (7.18) when stability condition (7.23) holds (after [125, 299])

($c_k = 1$). The long time evolution of the signal, i.e., for $t = \mathcal{O}(Re)$, thus leads to an exponential growth of the higher-order waves if $c_{d-} > c_k = 1$, and conversely, an exponential decay if $c_{d-} < c_k = 1$. Since the same argument applies for waves traveling at speed close to c_{d+}, we conclude that the inequality

$$c_{d-} \leq c_k \leq c_{d+} \tag{7.23}$$

must hold. When (7.23) is not satisfied, the long time evolution of the wave packet cannot be described with the linear hyperbolic equation (7.18). When (7.23) holds, higher-order waves decay for $t = \mathcal{O}(Re^{-1})$ or longer, and the phase speed of a wave solution to (7.18) must be close to the speed of the lower-order waves, $c \approx c_k$. Substituting the time derivative ∂_t in the highest order terms of (7.18) with $-c_k\partial_x$ gives

$$\partial_t \tilde{h} + \partial_x \tilde{h} \approx -Re(c_{d-} - c_k)(c_{d+} - c_k)\partial_{xx}\tilde{h}. \tag{7.24}$$

Therefore, the effect of higher-order waves on the lower-order ones is "diffusive." When the hierarchy condition (7.23) is not satisfied, the "diffusion coefficient," $-Re(c_{d-} - c_k)(c_{d+} - c_k)$ is negative, which signals again the occurrence of an instability (see Fig. 7.10).

   Physically, wave solutions to (7.20) and (7.22a) have different characteristics. Solutions to (7.20) are *kinematic waves*. They correspond to the linearized conservation equation for the mass (7.17a) for which the flux $\tilde{q}$ is written as an explicit function of $\tilde{h}$, $\tilde{q} = \tilde{h}$, which corresponds to the vertical wall and inertia-less limit of (7.17b) ($Re \rightarrow 0$ and $Ct \rightarrow 0$). The term "kinematic" has its origin in the formal equivalence between the mass conservation, $\partial_t h + \partial_x q = 0$, and the kinematic condition at the free surface, $\partial_t h + u|_h \partial_x h = v|_h$. These waves are fast. They propagate at speed $c_k = 1$, which is three times the average speed of the base flow.

   Wave solutions to (7.22a) have a more subtle interpretation. This second-order wave equation is obtained from (7.17b) by dropping out the drag and gravity contribution $\tilde{q}$ and $-\tilde{h}$, respectively, which corresponds to $Re \rightarrow \infty$ and $Ct \rightarrow \infty$ with $Fr^2 = 3Re/Ct = \mathcal{O}(1)$. This limit corresponds to the propagation of free-surface long waves in hydraulics when we deal with problems such as "flood on rivers," "tidal waves" and "hydraulic jumps." For large Reynolds numbers, the flow can be assumed to be potential, except for at a viscous boundary layer at the bottom of the channel. The velocity is therefore assumed to be independent of the cross-stream co-ordinate y and one is led to the *shallow-water equations*, or *Saint-Venant equations*, written in dimensional form as [215, 299]:

$$\partial_t h + \partial_x q = 0, \tag{7.25a}$$

$$\partial_t q + \partial_x \left(\frac{q^2}{h} \right) = gh \sin \beta - \tau_{\mathrm{w}} - gh \cos \beta \partial_x h. \tag{7.25b}$$

The reader can recognize in (7.25a) the mass conservation equation and in (7.25b) the averaged momentum balance. The wall drag τ_{w} is generally modeled using the so-called "Chézy law", $\tau_{\mathrm{w}} = C_f q^2 / h^2$, where the dimensionless "friction coefficient" C_f is assumed to be constant [84]. The Chézy law is an empirical friction law obtained from experimental data [33]. Apart from the wall drag, and a different coefficient in front of the convective term $\partial_x (q^2 / h)$, the Saint-Venant equations are very similar to the Kapitza–Shkadov model when surface tension is neglected. The balance of gravity acceleration and of wall friction gives $q = \sqrt{gh^3 \sin \beta / C_f}$ and the corresponding linear kinematic waves travel at speed $\frac{3}{2} q / h$, which is 1.5 times the speed of the flow. Using the Nusselt scaling based on the uniform film thickness $\bar{h}_{\mathrm{N}}$ and the speed $\frac{3}{2} \bar{q}_{\mathrm{N}} / \bar{h}_{\mathrm{N}}$ (so that kinematic waves again propagate at a dimensionless speed equal to unity), the Saint-Venant equations in dimensionless form are

$$\partial_t h + \partial_x q = 0, \tag{7.26a}$$

$$\partial_t q + \partial_x \left(\frac{q^2}{h} \right) = \frac{1}{Re} \left(\frac{2}{3} h - \frac{q^2}{h^2} \right) - \frac{1}{Fr^2} h \partial_x h, \tag{7.26b}$$

where $Re = \frac{2}{3} C_f^{-1}$ and $Fr^2 = \frac{3}{2} Re / Ct$. The Reynolds number $Re = \frac{2}{3} C_f^{-1}$ compares the wall friction time $\bar{h}_{\mathrm{N}}^2 \bar{q}_{\mathrm{N}}^{-1} C_f^{-1}$ and the advection time $\frac{2}{3} \bar{h}_{\mathrm{N}}^2 \bar{q}_{\mathrm{N}}^{-1}$, whereas the Froude number $Fr^2 = \frac{3}{2} Re / Ct$ compares the speed of the kinematic waves, $\frac{3}{2} \sqrt{g \bar{h}_{\mathrm{N}} \sin \beta / C_f}$, to the speed, $\sqrt{g \bar{h}_{\mathrm{N}} \cos \beta}$, of the gravity waves created by the action of gravity on perturbations of the free surface elevation. (Fr^2 is again proportional to Re / Ct, but with a factor 3/2 instead of the factor 3 earlier, a difference that arises from a different definition of the speed of the kinematic waves, $3 \bar{u}_{\mathrm{N}}$ in the earlier case, $(3/2) \bar{u}_{\mathrm{N}}$ now. But in both cases, the definition of the Froude number as the ratio of the speed of the kinematic waves to that of "gravity waves" remains the same.)

   Linearizing the Saint-Venant equations leads to a single multispeed equation similar to (7.18), with lower- and higher-order waves traveling at speed unity and

$$c_{d\pm} = \frac{2}{3} \pm \frac{1}{Fr}, \tag{7.27}$$

respectively. Higher-order linear wave solutions to the Saint-Venant equations thus correspond to the advection by the flow, whose velocity is $2/3$, of two waves propagating in opposite directions at speed Fr^{-1}, that is, in dimensional form $\sqrt{g\bar{h}_{\mathrm{N}}\cos\beta}$. Physically, perturbations of the free surface elevation induce perturbations of hydrostatic pressure, which are transported in the moving frame of the flow at speed, $\pm\sqrt{g\bar{h}_{\mathrm{N}}\cos\beta}$. For this reason, these waves are generally referred to as *surface gravity waves*. However, when a vertical wall is considered ($Fr \to \infty$), these two different wave solutions do not vanish but reduce to a single solution that remains stationary in the frame moving at the speed $2/3$ of the flow. Yet, in that case the hydrostatic pressure is not affected by perturbations of the free surface elevation. In fact, considering a vertical wall and negligible wall drag ($Fr, Re \to \infty$), the momentum balance (7.26b) reduces to a single wave equation $\partial_t u + u\partial_x u = 0$ for the averaged speed of the flow $u = q/h$. For this reason, higher-order linear wave solutions to (7.26a), (7.26b) are also referred to as *dynamic waves* [126], in contrast with the kinematic waves induced by the mass conservation equation (7.26a). Since film flows are generally considered in nearly vertical wall geometries, we prefer the latter terminology. The differences between expressions (7.27) and (7.22b) of the speed of the dynamic waves in the case of the Kapitza–Shkadov model and the Saint-Venant equations originate from the assumption of different velocity profiles.

The propagation of long wave small perturbations on a general inviscid shear flow was examined by Burns [35]. The pertinent equations are

$$\partial_t \tilde{u} + U\partial_x \tilde{u} + DU\tilde{v} + \partial_x \tilde{p} = 0, \qquad \partial_y \tilde{p} = 0, \qquad \partial_x \tilde{u} + \partial_y \tilde{u} = 0, \qquad (7.28a)$$

where tildes refer to the perturbations, $D \equiv d/dy$ and $U(y)$ is the streamwise velocity of the base flow. The system of equations is completed with the boundary conditions

$$\tilde{v} = 0, \quad \text{on } y = 0, \tag{7.28b}$$

$$\tilde{v} = \partial_t \tilde{h} + U\partial_x \tilde{h}, \qquad \tilde{p} = -\tilde{h}DP, \quad \text{and}$$
$$\partial_y \tilde{u} = D^2 U\tilde{h} = 0, \quad \text{all on } y = 1, \tag{7.28c}$$

where $P(y) = -Fr^{-2}y$ is the hydrostatic pressure distribution of the base flow. By looking for solutions to (7.28a)–(7.28c) that remain stationary in the moving frame, $\xi = x - ct$, we have

$$(U - c)^2 D\left[\frac{\tilde{v}}{U - c}\right] = -DP\tilde{h}',$$

where the prime refers to differentiation with respect to ξ. Integration with respect to y then gives

$$\tilde{v} = \frac{1}{Fr^2}(U - c)\tilde{h}' \int_0^y \frac{dy}{(U - c)^2}.$$

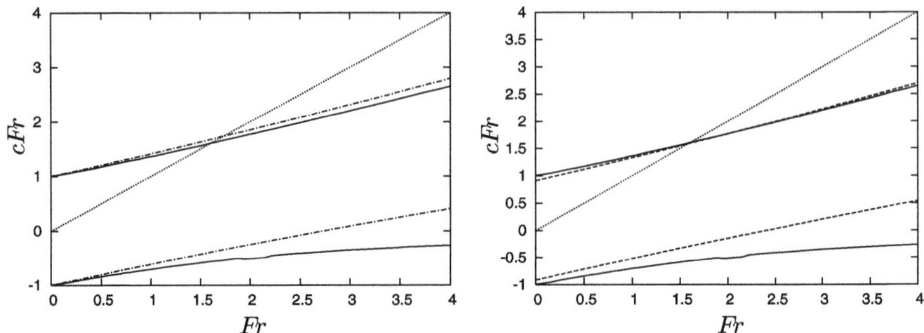

Fig. 7.11 Speed times Froude number *cFr* as function of the Froude number *Fr*. *Solid lines* correspond to the solutions to the Burns condition (7.29). *Dotted lines* correspond to the kinematic waves. *Dashed-dotted lines* (*left*) and *dashed lines* (*right*) refer to the dynamic waves with speed given by (7.22b) and (7.33) with $k \to 0$, respectively

The kinematic condition at the free surface, $\tilde{v}|_1 = (U(1) - c)\tilde{h}'$, thus leads to an equation for the phase speed c,

$$Fr^2 = \int_0^1 \frac{dy}{(U - c)^2}, \tag{7.29}$$

which admits two solutions for a given Froude number. The numerical solution to the *Burns condition* (7.29) for a parabolic velocity profile $U = y - \frac{1}{2}y^2$ is compared to the expression (7.22b) in Fig. 7.11. The Burns condition admits a backward moving ($c < 0$) and a forward moving ($c > 0$) solution, whereas both dynamic waves corresponding to (7.22b) are forward moving for $Fr > 2.74$. However forward moving Burns dynamic waves have a speed close to the corresponding dynamic waves of the Kapitza–Shkadov model.

The stability condition (7.23) with the speed of the dynamic waves given by (7.22b) gives $Fr < \sqrt{3}$ or, by making use of the definition of the Froude number, $Re < Ct$, with Ct corresponding to the value of the critical Reynolds number for the onset of the primary instability for the Kapitza–Shkadov model (7.5).

Let us now turn to the analysis of the primary instability based on the simplified second-order model (6.1), (6.79). As with the Kapitza–Shkadov model (6.13a), (6.13b), a multispeed wave equation governing the evolution of infinitesimal perturbations around the Nusselt flat film solution can similarly be obtained:

$$(\partial_t + \partial_x)\tilde{h} + \frac{6}{5}Re\left(\partial_{tt} + \frac{17}{21}\partial_{xt} + \frac{1}{7}\partial_{xx}\right)\tilde{h} - \frac{Ct}{3}\partial_{xx}\tilde{h}$$

$$+ \frac{We}{3}\partial_{xxxx}\tilde{h} - \frac{9}{5}\left(\partial_{xxt} + \frac{4}{9}\partial_{xxx}\right)\tilde{h} = 0. \tag{7.30}$$

This equation is also obtained from the regularized model (6.1), (6.92), as expected, since the simplified and regularized models are equivalent at the linear stage.

Compared to (7.18), new terms enter into (7.30) corresponding to viscous effects (third-order terms) and surface tension (fourth-order terms), such that (7.30) is no more hyperbolic. Yet, we may still take advantage of the idea of wave hierarchy to give a physical interpretation of the different terms in the dispersion relation. Apart from third-order viscous terms missing in the formulation by Alekseenko et al. [4], (7.30) is similar to the linear part of the equation obtained by these authors. The decomposition of the infinitesimal perturbations into normal modes of wavenumber k and angular frequency ω then leads back to the dispersion relation (7.10). By recognizing that (7.10) can be split into two parts having a $\pi/2$ phase difference due to the parity of differentiation, (7.30) after separating odd and even derivatives becomes [196]

$$(\partial_t + \partial_x)\tilde{h} - \frac{9}{5}\left(\partial_{xxt} + \frac{4}{9}\partial_{xxx}\right)\tilde{h} = \lambda, \tag{7.31a}$$

$$\frac{6}{5}Re\left(\partial_{tt} + \frac{17}{21}\partial_{xt} + \frac{1}{7}\partial_{xx}\right)\tilde{h} - \frac{Ct}{3}\partial_{xx}\tilde{h} + \frac{We}{3}\partial_{xxxx}\tilde{h} = -\lambda. \tag{7.31b}$$

Hence, viscous effects modify the speed c_k of the kinematic wave solutions to (7.31a), which is now dependent on the wavenumber k,

$$c_k = \frac{1 + 4k^2/5}{1 + 9k^2/5}, \tag{7.32}$$

an expression identical to (7.11a). Dynamic wave solutions to (7.31b) travel at speeds

$$c_{d\pm} = \frac{17}{42} \pm \sqrt{\frac{37}{1764} + \frac{5}{6}\left(\frac{1}{Fr^2} + 3k^2\frac{We}{Re}\right)}. \tag{7.33}$$

Therefore, surface tension has a dispersive effect on dynamic waves similar to the effect of second-order viscous terms on kinematic waves. Since the phase speed of the slower dynamic waves is always lower than the speed of kinematic waves, the stability condition (7.23) reduces to

$$\frac{1 + 4k^2/5}{1 + 9k^2/5} \leq \frac{17}{42} + \sqrt{\frac{37}{1764} + \frac{5}{6}\left(\frac{1}{Fr^2} + 3k^2\frac{We}{Re}\right)}. \tag{7.34}$$

One can easily confirm that equality of the two sides in (7.34) is equivalent to the neutral stability condition (7.11a), (7.11b). Notice that the stability condition $c_k = c_{d+}$ implies that neutral waves propagate at the speed of the kinematic waves thus justifying the equality of the expressions for the phase speed (7.11a) and (7.32).

The stabilizing effects of viscous dispersion and surface tension can now be explained within the framework of wave hierarchy. Viscous dispersion reduces the speed of the kinematic waves whereas surface tension accelerates the dynamic waves, both effects being obviously stabilizing by shifting upward the critical Froude number at which the kinematic waves move faster than the dynamic ones.

As the onset of the primary instability for falling film flows occurs at $k = 0$, it is always possible to neglect the dispersive effects of viscosity and surface tension on the kinematic and dynamic waves, at least for a Reynolds number very close to the critical value, $Re - Re_c \ll 1$. As already discussed earlier in our analysis of the Kapitza–Shkadov model, there is an analogy between the onset of long wave instability in film flows and the torrential regime of rivers characterized by the presence of turbulence. In steep channels, such as spillways from dams, run-off channels or open aqueducts, a uniform and steady (on average) flow evolves eventually to a series of "breaking waves," or "bores," separated by regions of gradually varying flows arranged in a staircase manner.

These bores are generally referred to as *roll waves*. The breaking of bores cannot be described in principle by the Saint-Venant equations as their derivation requires a slowly varying free surface, an assumption that is violated at the sharp fronts of the waves, i.e., when they form shocks prior to breaking. Yet the instability leading to bores is well captured by these equations. In fact, for not too high Froude numbers, the steep fronts of the shocks can be neglected and the roll wave properties are satisfactorily predicted by (7.26a), (7.26b) [33]. As for film flows, kinematic waves are controlled by the balance of the gravity acceleration along the slope and the wall drag. The speed of the kinematic waves exceeds the speed of the fastest dynamic waves (7.27) when $Fr > 3$ (actually, in most studies dealing with roll waves the velocity scale corresponds to the average velocity of the uniform flow giving a critical Froude number equal to 2). The instability is here possible because kinematic waves move at a different speed than the fluid velocity. In fact, for large inclinations or flow rates, that is $Fr \to \infty$, dynamic waves, which transport the kinetic and potential energies of the perturbations, are simply advected by the flow.

When an inviscid free surface flow with a semiparabolic velocity profile is considered,[2] the speed of the fastest (Burns) dynamic waves tends to the velocity of the fluid moving at the interface as $Fr \to \infty$, which is again slower than the kinematic waves and the flow becomes unstable for $Fr > \sqrt{1 + \pi/2} \approx 1.62$ [125], a value that is quite close to the onset of the film flow instability ($R_c = \frac{5}{6} Ct$ gives a constant value of the Froude number $Fr = \sqrt{5/2} \approx 1.58$). In Fig. 7.11, the speed of the fastest linear dynamic wave solutions to the model (6.1), (6.79) is compared to the corresponding results from the Burns condition (7.29). The agreement is excellent, which explains why the two predictions give a similar value for the critical Froude number. The speed c_{d+} given by (7.33) tends to ≈ 0.55 as $k \to 0$ and $Fr \to \infty$, a value close to the velocity of the fluid at the interface. Due to the normal component of the gravity acceleration, dynamic waves are much faster than the fluid when the channel is slightly inclined ($Fr \ll 1$) and then tend to the largest possible velocity of the flow for a vertical wall ($Fr \to \infty$). After all, in the first case dynamic waves essentially transport gravitational potential energy of the perturbations, whereas in

---

[2]The roll wave instability is a shear-driven instability produced by inertia and with negligible contribution from viscous effects. But the basic laminar flow does result from viscous effects, e.g., Johnson [126] neglected viscous effects except in the definition of the basic laminar flow. Considering a shear-driven instability by assuming that viscous forces are negligible, but at the same time having a basic velocity profile that results from viscous effects, is quite common in the analysis, e.g., of a shear-driven Kelvin–Helmholtz instability of a mixing layer.

the second case they mostly transport the kinetic energy of the perturbations. Consequently, the instability can occur only if the speed of the kinematic waves is not in the range of admissible velocities for the undisturbed flow. As for the onset of roll waves in channel flows, the ability of the kinematic waves to move faster than any fluid particle is therefore a crucial ingredient of the instability mechanism [256].

7.2 Traveling Waves

With periodic forcing at the inlet, waves rapidly reach a constant shape and speed after the inception region; Since the film acts as a spatial noise amplifier, regularly spaced waves are observed downstream without further change in the film texture. Although these waves remain almost stationary in their moving frame, they are commonly referred to as *traveling waves*. Traveling waves have been the subject of numerous theoretical and experimental studies for more than 60 years since Kapitza's pioneering efforts and they have already been encountered in earlier chapters. A comprehensive study of all possible traveling wave solutions is not done here. Our principal aims are to analyze traveling waves within the framework of dynamical systems theory, to examine the influence of the physical parameters on the wave characteristics and to assess the validity domain of the weighted residuals models developed in Chap. 6 (for solitary waves, a particular class of traveling waves with an infinite period, this was already done in Chap. 6). As already noted, these models capture most of the features of the nonlinear dynamics of film flows.

7.2.1 Dynamical Systems Approach

7.2.1.1 General Settings

Let us consider periodic waves steady in their moving frame $\xi = x - ct$, where c is the speed of the waves. Time dependence can then be eliminated via a suitable Galilean transformation. The initial system of partial differential equations can thus be reduced to a system of ordinary differential equations. For example, the isothermal BE in terms of the Shkadov scaling ((5.55) with $\mathcal{M} = 0$) gives a fourth-order ordinary differential equation which can be integrated once, yielding a third-order differential equation (see also Appendix F.2),

$$\frac{1}{3}h^3 h''' - \zeta \frac{1}{3}h^3 h' + \frac{2}{15}\delta h^6 h' + \frac{1}{3}h^3 - ch - q_0 = 0, \tag{7.35}$$

where the prime denotes differentiation with respect to the moving coordinate ξ.

For the two-equation formulations, such as the Kapitza–Shkadov model in (6.13a), (6.13b) or the first-order model in (6.1) and (6.51), a similar equation can be obtained. First, the mass conservation equation (6.1), $-ch' + q' = 0$, can be integrated once to yield

$$q = ch + q_0, \tag{7.36}$$

where $q_0 = \int_0^h (u - c)\,dy$ is an integration constant corresponding to the rate at which the fluid moves under the wave in its moving frame. This constant is negative, as the waves move faster than the flow. In fact, surface equations (such as the BE or the Ooshida equation) or two-equation models in the absence of second-order viscous dispersion effects, all lead to the following generic equation:

$$\frac{1}{3}h^3 h''' - \zeta\frac{1}{3}h^3 h' + \delta\mathcal{N}(h,c)h' + \frac{1}{3}h^3 - ch - q_0 = 0. \tag{7.37}$$

The functional $\mathcal{N}$ contains all inertia effects as is evident by the presence of the multiplicative factor δ. The third-order derivative arises from surface tension effects, the term $\frac{1}{3}h^3$ corresponds to the gravity acceleration and the two last terms account for the viscous drag.

In the case of the two second-order two-equation models—simplified model in (6.1), (6.79) and regularized model in (6.1), (6.92)—we get,

$$\frac{1}{3}h^3 h''' - \zeta\frac{1}{3}h^3 h' + \delta\mathcal{N}(h,c)h' + \frac{1}{3}h^3 - ch - q_0$$
$$+ \eta\big[\mathcal{I}(h,c)[h']^2 + \mathcal{J}(h,c)h''\big] = 0, \tag{7.38}$$

where

$$\mathcal{I}(h,c) = \frac{8}{5}q_0 - \frac{1}{5}ch \quad \text{and} \quad \mathcal{J}(h,c) = -\frac{3}{5}ch^2 - \frac{12}{5}q_0 h$$

account for viscous dispersion effects.

The integration constant q_0 can be fixed by demanding $h = 1$ as a solution to (7.37), corresponding to the unperturbed Nusselt film thickness h_N which leads to

$$q_0 = 1/3 - c. \tag{7.39}$$

Notice that there is actually a one-parameter infinite family of solutions with constant film thickness. Making the transformation

$$h \to Hh, \qquad c \to Cc, \qquad q \to Qq, \tag{7.40}$$

preserves the structure of equations (7.35), (7.37) and (7.38) provided that ξ is also rescaled as $\xi \to \Xi\xi$ and the control parameters as $\delta \to \Delta\delta$, $\zeta \to \Lambda\zeta$ and $\eta \to \Upsilon\eta$. By substitution then one is led to $\Xi = H^{1/3}$, $\Delta = H^{-11/3}$, $\Lambda = H^{-2/3}$ and $\Upsilon = H^{-4/3}$, whereas $C = H^2$ and $Q = H^3$.

Our starting point will thus be the general equation

$$\tfrac{1}{3}h^3 h''' - \zeta\tfrac{1}{3}h^3 h' + \delta\mathcal{N}(h,c)\,h' + \eta\big[\mathcal{I}(h,c)[h']^2 + \mathcal{J}(h,c)\,h''\big] + \mathcal{H}(h,c) = 0,$$
$$\tag{7.41a}$$

which contains all previous equations, surface equations and two-equation models as special cases. The functional $\mathcal{H}$ is given by

$$\mathcal{H}(h,c) \equiv \frac{1}{3}h^3 - ch - q_0 = \frac{1}{3}(h-1)\big(h^2 + h + 1 - 3c\big), \tag{7.41b}$$

and the functionals $\mathcal{N}(h, c)$ corresponding to the different cases are detailed below:

BE (5.55): $\qquad\qquad\qquad\qquad \dfrac{2}{15}h^6$

Ooshida equation (5.62): $\quad \dfrac{10}{21}ch^4 - \dfrac{12}{35}h^6$

Kapitza–Shkadov

model (6.13a), (6.13b): $\quad \dfrac{2}{5}q^2 - \dfrac{4}{5}cqh + \dfrac{1}{3}c^2h^2 = \dfrac{2}{5}c^2 - \dfrac{4}{15}c + \dfrac{2}{45} - \dfrac{1}{15}c^2h^2$

Models (6.1), (6.51)

and (6.1), (6.79): $\qquad \dfrac{18}{35}q^2 - \dfrac{34}{35}cqh + \dfrac{2}{5}c^2h^2$

$$= \dfrac{1}{35}\left[18c^2 + \dfrac{2}{3}ch - 12c - 2c^2h(h+1) + 2\right]$$

Model (6.1), (6.92): $\qquad \left\{\dfrac{18}{35}q^2 - \dfrac{34}{35}cqh + \dfrac{2}{5}c^2h^2\right\}\left[1 - \dfrac{1}{70}\delta qh'\right].$

$$\tag{7.41c}$$

Notice that the expression of q given by (7.36) and (7.39) has been utilized in the above expressions for $\mathcal{N}$ to eliminate q_0. Notice also that due to a misprint in [229] the formula for $\mathcal{N}$ corresponding to the Kapitza–Shkadov model was given as $\mathcal{N} = \frac{2}{5}c^2 - \frac{4}{15}c + \frac{2}{45} - \frac{2}{15}c^2h^2$ in that study.

Equation (7.41a) can be recast as a three-dimensional dynamical system:

$$U_1' = U_2, \qquad U_2' = U_3,$$

$$U_3' = -3\left\{\left[\delta\mathcal{N}(U_1, c) - \dfrac{1}{3}\zeta U_1^3\right]U_2 + \mathcal{H}(U_1, c)\right.$$

$$\left. + \eta\left[\mathcal{I}(U_1, c)[U_2]^2 + \mathcal{J}(U_1, c)U_3\right]\right\}U_1^{-3}, \tag{7.42}$$

in the phase space spanned by $\mathbf{U} = (U_1, U_2, U_3)$ where $U_1 = h$, $U_2 = h'$, $U_3 = h''$. The traveling wave solutions of the full second-order model (6.78) are governed by a five-dimensional dynamical system—see Appendix E.1. The solutions to (7.41a) are trajectories in the phase space, also referred to as "phase curves"; "level curves" is another term. The reader should consult some of the numerous texts on dynamical systems theory, e.g., [111, 292, 301], but only basic elements of the theory are required.

The vector field $\mathbf{U}$ generates a "phase flow" or simply "flow." A common notation for flows is $\mathbf{\Phi}(\mathbf{U}, t)$; in our case the moving coordinate ξ plays the role of "time" t. A flow satisfies the associated dynamical system, say $d\mathbf{U}/dt = \mathbf{F}(\mathbf{U})$. Formally we write: $\frac{d}{dt}(\mathbf{\Phi}(\mathbf{U}, t))|_{t=\tau} = \mathbf{F}(\mathbf{\Phi}(\mathbf{U}, \tau)$ [111]. It then looks like flow is just the solution $\mathbf{U}$ of the dynamical system and so the natural question is: Why is there a need for a different symbol? Let us represent the solution of a dynamical system as $\mathbf{U}(t, t_0, \mathbf{U}_0)$ [301]: This is the solution through point $\mathbf{U}_0$ at $t = t_0$ with $\mathbf{U}(t_0, t_0, \mathbf{U}_0) = \mathbf{U}_0$. In the solution $\mathbf{U}(t, t_0, \mathbf{U}_0)$ we can now think of t and t_0 as fixed and then study how the "map" $\mathbf{U}(t, t_0, \mathbf{U}_0)$ moves sets of points around in phase space. For a set $\mathbf{S}$ we could denote its image under this map by $\mathbf{U}(t, t_0, \mathbf{S})$. Since points in phase space are also labeled by the symbol $\mathbf{U}$, it is often less confusing to change the notation for the solution, and hence the use of the symbol $\boldsymbol{\Phi}$. From the physical point of view one can think of the flow as simply the motion along a phase curve.

We now turn to the evolution of phase-space volumes as governed by the dynamical system $d\mathbf{U}/dt = \mathbf{F}(\mathbf{U})$ (a "volume" in phase space can be defined by a set of points in the phase space such as a "hypercube," $\Delta U_1 \Delta U_2 \Delta U_3$). Consider a domain D in the phase space which is supposed to have volume V. We denote by $D(t)$ the image of the region D under the action of the phase flow $\boldsymbol{\Phi}$, and by $V(t)$ we denote the volume of the region $D(t)$. From "Liouville's theorem" [292] we have

$$\frac{dV}{dt} = \int_{D(t)} \boldsymbol{\nabla}_U \cdot \mathbf{F} \, dU_1 \, dU_2 \, dU_2,$$

where the gradient operator $\boldsymbol{\nabla}_U$ is defined as $\boldsymbol{\nabla}_U = (\partial_{U_1}, \partial_{U_2}, \partial_{U_3})$. In other words, the variation with respect to time of the volume of a region in the phase space is given by the volume integral of the divergence of $\mathbf{F}$ in this region. Therefore, if $\boldsymbol{\nabla}_U \cdot \mathbf{F} = 0$, the phase flow preserves the volume of any region of the phase space. Such a flow can be viewed as the flow of an "incompressible phase fluid" in the phase space. On the other hand, for the special case where the divergence of $\mathbf{F}$ is a negative constant, say $\boldsymbol{\nabla}_U \cdot \mathbf{F} = -\lambda$, as is the case with the simple dynamical system $dU/dt = -U$, where we have attraction toward $U = 0$, Liouville's theorem yields $dV/dt = -\lambda V$, so that $V(t) = V(0) \exp(-\lambda t)$. Thus, phase space volumes shrink exponentially in time.

For (7.42), the divergence of $\mathbf{F}$ is equal to $-3\eta \mathcal{J}(U_1, c) U_1^{-3}$ and it is directly proportional to the viscous dispersion parameter η. For $\eta = 0$ the dynamical system is volume-conserving. For $\eta > 0$, an initial volume in the phase space will expand or contract depending on the sign of $\mathcal{J}(U_1, c)$. From the expression of the functional $\mathcal{J}$ in (7.38) and (7.39), the zeroes of $\boldsymbol{\nabla}_U \cdot \mathbf{F}$ correspond to $U_1 = h_{\mathrm{div0}}$ given by

$$h_{\mathrm{div0}} = 4\left(1 - \frac{1}{3c}\right), \tag{7.43}$$

where the subscript "div0" is used to denote zero divergence. For most of the solutions to (7.42) considered here, when $\eta > 0$, $\mathcal{J}(U_1, c)$ is found to be positive and a contraction of the phase volume surrounding an orbit is observed for all positions of the vector $\mathbf{U}$ belonging to this orbit (this point will be discussed further in Sect. 7.2.1.5).

Fixed points of (7.42) are its equilibrium points corresponding to $\mathbf{F} = \mathbf{0}$ or $U_2 = U_3 = 0$ and

$$3\mathcal{H}(U_1, c) = (U_1 - 1)\left(U_1^2 + U_1 + 1 - 3c\right) = 0, \tag{7.44}$$

from which it is seen that $U_1 = 1$ is a solution corresponding to the Nusselt flat film solution $h = 1$. Additional roots are given by

$$U_1^2 + U_1 + 1 - 3c = 0. \tag{7.45}$$

Accordingly, for $c > 1/3$, i.e., for waves traveling faster than the average speed of the Nusselt flat film flow, there is an additional positive solution

$$h_{\mathrm{II}} \equiv -1/2 + \sqrt{3(c - 1/4)}. \tag{7.46}$$

Hence (7.42) admits two fixed points, $\mathbf{U}_{\mathrm{I}} = (1, 0, 0)$ and $\mathbf{U}_{\mathrm{II}} = (h_{\mathrm{II}}, 0, 0)$, with the uniform film of constant thickness corresponding to one of the two. This multiplicity of solutions of (7.44) indicates that for a given flow rate two different flat film solutions are possible so that hydraulic jumps can be constructed as trajectories in the phase space connecting one flat film solution to the other. To obtain the behavior of a dynamical system close to a fixed point we must linearize the dynamical system about this fixed point. This will yield a Jacobian whose eigenvalue spectrum determines the stability of the fixed point.

Before proceeding further we need to define the *saddle-focus fixed point* and *saddle fixed point*. A saddle-focus point has either a real positive eigenvalue and a pair of complex conjugate ones with negative real part or a real negative eigenvalue and a pair of complex conjugate ones with positive real part. Each of the two cases can be obtained from the other by reflection with respect to the imaginary axis. Further, when the fixed point admits a real eigenvalue, say λ_1, and two complex conjugate eigenvalues, say λ_2, λ_3, that are equidistant from the imaginary axis, such as $\Re(\lambda_{2,3}) = -\lambda_1$, the fixed point is a saddle-focus that is said to be "neutral" or "resonant." On the other hand, a saddle point has three real eigenvalues, either a positive and two negative ones, or a negative and two positive ones. Each of the two cases can be obtained from the other by reflection with respect to the imaginary axis. Neither saddle-focus nor saddle points are "stable" fixed points (the real parts of the eigenvalues do not have the same sign; a stable fixed point is one for which all eigenvalues have negative real parts).

Spatially, periodic waves generated experimentally by forcing at the inlet (provided there is synchronization between the forcing and the flow at any location in space as was emphasized in Sect. 5.3.1) correspond to periodic orbits in the phase space or *limit cycles*. They arise through a change of the nature of the fixed point from *focus* (all eigenvalues have a negative real part) to saddle-focus when a pair of complex conjugate eigenvalues, say λ_2, λ_3, cross the imaginary axis, $\Re(\lambda_2) = \Re(\lambda_3) = 0$, corresponding to a Hopf bifurcation. Instead, solitary waves in real space correspond to *homoclinic orbits* in phase space: They leave the fixed point for an excursion in the phase space and eventually return back to it. More precisely, for a positive-hump solitary wave, a homoclinic orbit leaves the fixed point at $\xi = -\infty$ on the *unstable manifold* $\mathcal{W}^{\mathrm{u}}$ and gets back to it at $\xi = +\infty$ on the *stable manifold* $\mathcal{W}^{\mathrm{s}}$ (as the flow of (7.42) is "steady" on the fixed points, trajectories start or return to a fixed point infinitely slowly).

The approach of setting $h = 1$ to obtain the value of q_0 in (7.39) corresponding to the flow rate at which the fluid flows under the wave in its moving frame is used for both limit cycles, around $\mathbf{U}_\mathrm{I}$ or $\mathbf{U}_\mathrm{II}$—limit cycles in the latter case are constructed in Sect. 7.2.1.3—and solitary waves. After all, $h = 1$ corresponds to the fixed point $\mathbf{U}_\mathrm{I} = (1, 0, 0)$ of the associated dynamical system and the aim is to use the position of one fixed point as a reference solution for this dynamical system. As far as the condition $h \to 1$ as $x \to \pm\infty$ for solitary waves is concerned, in practice solitary waves are numerically constructed with periodic boundary conditions in a finite domain. The closed flow condition $\langle h \rangle_\xi = 1$ is then recovered asymptotically as the domain tends to infinity. However, the condition $h \to 1$ as $x \to \pm\infty$ cannot be related to time-dependent computations as it cannot be related to $\langle h \rangle_x = 1$ or $\langle q \rangle_t = 1/3$ (see Sect. 5.3.1; of course, the solitary pulses obtained in time-dependent computations do resemble the infinite domain computations).

We note that our analysis of homoclinicity is restricted to positive-hump solitary waves with corresponding homoclinic orbits in a three-dimensional phase space. The associated dynamical system admits a real positive eigenvalue, say λ_1, corresponding to the unstable manifold, and either a complex conjugate pair with negative real parts, or two negative real eigenvalues, say λ_2, λ_3, corresponding to the stable manifold. This situation arises explicitly when homoclinic orbits connecting $\mathbf{U}_\mathrm{I}$ to itself are considered. In the case of homoclinic orbits around $\mathbf{U}_\mathrm{II}$, the situation is reversed: One eigenvalue is negative and the two others have positive real parts so that the unstable manifold is two-dimensional, whereas the stable manifold is one dimensional. The extension to negative-hump waves would simply require exchanging the role of the unstable and stable manifolds and is thus straightforward.

Actually, a homoclinic orbit can only exist for specific values of c, as most of the time, an orbit leaving the fixed point along its unstable manifold $\mathcal{W}^\mathrm{u}$ never comes back. Equivalently, the construction of homoclinic orbits in a three-dimensional phase space is a nonlinear eigenvalue problem for the speed c of the corresponding solitary wave as function of the other parameters of the associated dynamical system. Depending on the value for c, the homoclinic orbit may come back to the vicinity of the fixed point and eventually onto its stable manifold, either through a single loop in the phase space, or it can be repelled from the fixed point several times and perform several loops in the phase space before it eventually returns onto the stable manifold $\mathcal{W}^\mathrm{s}$. In dynamical systems theory the former homoclinic orbit has been christened *principal/primary homoclinic orbit* and the latter *subsidiary homoclinic orbit* [103, 301]. The precise conditions under which subsidiary homoclinic orbits are possible are given below. Principal/single-loop homoclinic orbits correspond to single-hump solitary pulses in real space, and subsidiary/multi-loop ones to "multi-hump solitary waves" or "trains of solitary waves" ("multi-pulse waves").

By now the reader must have understood that a solitary *hump* is the (clearly identifiable) large amplitude bump of a solitary wave as, e.g., Fig. 4.1 nicely demonstrates. A hump corresponds to a loop in the phase space. A series of humps, as is the case with multi-hump waves, resembles a large-amplitude oscillation and corresponds to several loops in the phase space. Hence, a solitary wave can be either single-hump or a multi-hump. For a multi-hump wave, the extra loops of the corresponding homoclinic orbit about a fixed point do not pass close to it but revolve around a second fixed point, i.e., they require two fixed points as is the case with the falling film. For a dissipative wave the first fixed point is a saddle-focus and has one real and positive eigenvalue and a complex conjugate eigenvalue pair with negative real parts.

Solitary pulses on a falling film are dissipative waves (the second fixed point in the falling film problem is also a saddle-focus but with one negative real eigenvalue and a complex conjugate pair with positive real parts as noted earlier). The profile of a solitary pulse consists of a monotonic increase of the interface from the flat film up to the first hump, corresponding to the monotonic departure of the homoclinic orbit from the fixed point in the direction of the unstable manifold, while the final hump is always preceded by an oscillatory radiation structure corresponding to the spiraling return of the homoclinic orbit to the fixed point on the stable manifold. For nondissipative solitary waves, the fixed point is a saddle with three real eigenvalues from which only one is positive. When a homoclinic orbit is connected to a saddle point the corresponding solitary wave exhibits a smooth entry in the positive ξ direction without any oscillations.

Even though, strictly speaking, a solitary wave exists in an infinite domain, quite frequently we use this term loosely to denote a localized structure in a finite domain that resembles an infinite-domain solitary pulse, or to denote trains of localized structures in an infinite domain each of which resembles an infinite-domain solitary pulse. Such structures are typically separated by portions of nearly flat film corresponding to the returns of the phase space trajectory in the neighborhood of the fixed point associated with the flat film solution. Such trains are what we have referred to as "trains of solitary waves" or "multi-pulse waves"; their separation distance has to be sufficiently large for each wave in the train to resemble the infinite-domain solitary pulse—see, e.g., Fig. 6.4.

From the above discussion it is clear that a homoclinic orbit is necessarily associated with a saddle-focus point or a saddle point. Stable and unstable manifolds $\mathcal{W}^{s,u}$ are locally tangent to the linear eigenspaces $E^{s,u}$ spanned by the eigenvectors associated with the eigenvalues λ_i with negative and positive real parts, respectively. It is then clear that a homoclinic orbit departs from the fixed point following a direction in the subspace of the tangent eigenspace spanned by the eigenvectors corresponding to eigenvalues whose positive real parts are closest to zero. Conversely, by reversing the time direction the same argument applies also to the stable manifold. It can be concluded that trajectories in the vicinity of a fixed point are mainly influenced by the eigenvalues that are the closest to the imaginary axis, known as *leading eigenvalues*. The closest to the imaginary axis of all is sometimes called the *determining eigenvalue*.

As far as the study of the homoclinic orbit solution to the five-dimensional dynamical system corresponding to the full second-order model (6.78) is concerned, it requires the continuation of only three eigenvalues in the parameter space, as is the case with the three-dimensional flows in (7.42). There are in fact three leading eigenvalues for the dynamical system corresponding to the full second-order model

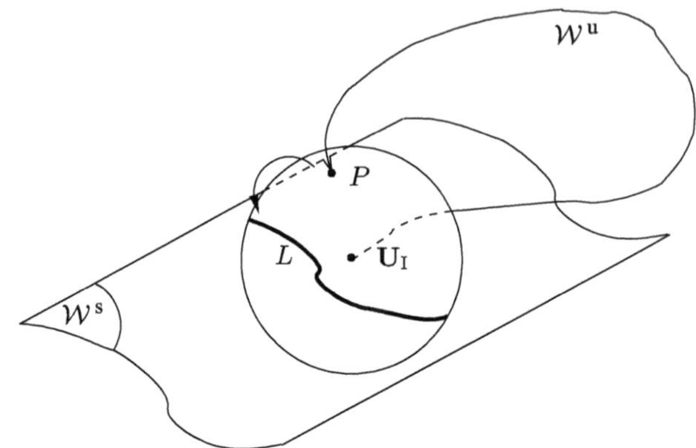

Fig. 7.12 Intersections P and L of the unstable and stable manifolds $\mathcal{W}^{u,s}$ of the fixed point $\mathbf{U}_{I}$ with a sphere. The *arrow* indicates that when P is in L, a homoclinic orbit is created. In the case of fixed point $\mathbf{U}_{II}$, the roles of $\mathcal{W}^{s}$ and $\mathcal{W}^{u}$ are simply exchanged

as the other two eigenvalues are real and positive and located further away from the imaginary axis. As a consequence, the study of homoclinic orbits for the full second-order model is qualitatively similar to the much simpler case of the three-dimensional flows in (7.42) and we restrict ourselves to the latter.

Let us see now how a homoclinic orbit can be created in a three-dimensional phase space. Assume that a fixed point of a three-dimensional dynamical system admits a real positive eigenvalue, say λ_1, corresponding to the unstable manifold, and either a complex conjugate pair with negative real parts, or two negative real eigenvalues, say λ_2, λ_3, corresponding to the stable manifold. (Once again this situation arises explicitly when homoclinic orbits connecting $\mathbf{U}_{I}$ to itself are considered. In the case of homoclinic orbits around $\mathbf{U}_{II}$, the situation is reversed. However, apart from exchanging the role of the unstable and stable manifolds the demonstration remains unchanged.) The homoclinic orbit corresponds to an intersection of the unstable manifold with the stable one. Imagine a sphere in the phase space centered at the fixed point as shown in Fig. 7.12. Since the unstable manifold is one-dimensional and the stable manifold is two-dimensional, the intersections of the sphere with the unstable and stable manifolds, $\mathcal{W}^{u}$ and $\mathcal{W}^{s}$, are a point P and a curve L, respectively. A trajectory starting along the unstable manifold most often never comes back to the neighborhood of the fixed point, as already emphasized. However, it might do so for a specific value of c and we might assume that the corresponding trajectory intersects the sphere centered on the fixed point at point P. If we suppose now that the sphere is sufficiently small and that the considered trajectory is homoclinic, this trajectory would return to the fixed point along the stable manifold so that its intersection with the sphere would belong to $\mathcal{W}^{s}$ and thus to L. The idea is then to place P on L. If P does not belong to L, the trajectory is only approximately homoclinic and the point P will be repelled from the vicinity of the fixed point in the direction of the unstable manifold. Notice that a point movement on a curve in a two-dimensional manifold is a codim 1 phenomenon, i.e., it requires the monitoring of only one parameter. In our case we can simply adjust the speed c to ensure that $P \in L$, thus obtaining the homoclinic trajectory. The condition $P \in L$ gives a unique relation for the speed c as a function of the other parameters.

A "shooting method" to numerically obtain homoclinic orbits follows directly from the above idea of adjusting P onto L. The phase speed c is adjusted by monitoring the function $d(c)$ of first comeback of a trajectory starting close to the fixed point in the direction of the unstable manifold, i.e., the first local minimum of the distance of the trajectory to the fixed point in the phase space. For appropriately chosen values of c, a loop is observed in the phase space before the trajectory is repelled from the vicinity of the fixed point, which enables us to define $d(c)$ unambiguously. Then c can be refined by dichotomy to make $d(c)$ as small as possible [216]. Another numerical procedure to obtain homoclinic orbits is to look for the *homoclinic bifurcation* of a limit cycle approaching a fixed point. This procedure is easy to implement using the continuation software AUTO07P and its toolbox HOM-CONT and the homoclinic orbits constructed throughout the monograph have been obtained using the second method. If the determining eigenvalue is real, homoclinic orbits correspond to clearly separated codim 1 surfaces $c(\delta, \zeta, \eta)$ in the parameter space or *solution branches* (also referred to as *families of solutions*) in the four-dimensional parameter space. The phase portrait becomes much more complicated when the determining eigenvalues are complex while the corresponding solution branches/codim 1 surfaces are no longer well separated. In this case, the so-called *Shil'nikov theorem* states in its most common form that if a primary homoclinic orbit exists in the parameter space, e.g., in our case we have a homoclinic solution branch $c = c^*(\delta, \zeta, e)$, then a countable infinite number of limit cycles/periodic orbits and subsidiary/multi-loop homoclinic orbits also exist in a local neighborhood of c^* provided that the *Shil'nikov criterion*

$$-\Re(\lambda_{2,3})/\lambda_1 < 1 \qquad (7.47)$$

is satisfied or, alternatively, when the sign of the *Shil'nikov number* $\lambda_1 + \Re(\lambda_{2,3})$ is > 0 [103]. If the Shil'nikov criterion is not satisfied, we either have a finite number of subsidiary/multi-loop orbits or none.

We note that the subsidiary/multi-loop orbits about a fixed point of the Shil'nikov criterion are of the type for which the extra loops pass close enough to the fixed point. The Shil'nikov criterion does not ensure the existence of the other type in which the extra loops of a homoclinic orbit about a fixed point do not pass close enough to this point. However, the criterion can be extended to other orbits, for example a double-loop subsidiary homoclinic orbit with a loop that does not pass close to the fixed point; after all it is based on a local analysis near the fixed point (construction of a "Poincaré section") on a plane close to the fixed point [103]. Hence, in its most general form, the Shil'nikov criterion ensures the existence of subsidiary homoclinic orbits for both primary and multi-loop subsidiary orbits of the type with extra loops that do not pass close to the fixed point.

The existence of an infinite number of subsidiary homoclinic orbits in the neighborhood of a primary one corresponds to a situation that is usually called *homoclinic chaos*. This term actually refers to the existence of a "chaotic attractor" in the vicinity of a homoclinic orbit in phase space and does not imply the construct of chaos in time and space; i.e., complex spatio-temporal dynamics of the corresponding partial differential equations. The presence of resonant saddle-focus points signals the onset of homoclinic chaos: In general, by changing appropriately the governing parameters of a system one can go from a case with $\lambda_1 + \Re(\lambda_{2,3}) < 0$ to one with $\lambda_1 + \Re(\lambda_{2,3}) > 0$.

We finally note that formation of subsidiary homoclinic orbits in the neighborhood of a principal homoclinic orbit (such as the one we have been discussing due to the intersection of a one-dimensional manifold with a two-dimensional one) is connected with the fact that homoclinic orbits are "structurally unstable": An arbitrarily small variation of one of the parameters can break this intersection and can lead to subsidiary homoclinic orbits in the vicinity of the primary homoclinic loop [301].[3]

7.2.1.2 Stability of Fixed Points

To complete the study of the traveling wave solutions to the different models we need to examine the stability characteristics of the fixed points of the corresponding dynamical systems. We shall focus on the Hopf bifurcation and resonant saddle-focus points that signal the onset of homoclinic chaos.

Let us first consider the stability analysis of the fixed points of the models described by (7.37), corresponding to the BE, Ooshida equation, Kapitza–Shkadov and first-order model, which neglect viscous dispersion ($\eta = 0$). Linearization of the associated dynamical systems around the fixed point $\mathbf{U}_\mathrm{I}$ yields

$$\mathbf{u}' = \mathbf{L}_\mathrm{I}\mathbf{u}, \tag{7.48}$$

where $\mathbf{u}$ represents a small perturbation around $\mathbf{U}_\mathrm{I}$ and $\mathbf{L}_\mathrm{I}$ is the Jacobian matrix $[\mathbf{L}]_{ij} = \partial_{U_j} F_i|_{\mathbf{U}_\mathrm{I}}, i, j = 1, 2, 3$, where $\mathbf{F}$ is the right hand side of (7.42) with $\eta = 0$. The eigenvalues of $\mathbf{L}_\mathrm{I}$ are given by the characteristic equation $|\mathbf{L}_1 - \lambda\mathbf{I}| = 0$, where $\mathbf{I}$ is the 3×3 identity matrix, or

$$\lambda^3 + \left[3\delta\mathcal{N}(1, c) - \zeta\right]\lambda - 3(c - 1) = 0. \tag{7.49}$$

As viscous dispersion is neglected, volume is conserved in the phase space, as pointed out earlier: $\nabla_U \cdot \mathbf{F} = 0$. This property implies that $\lambda_1 + \lambda_2 + \lambda_3 = 0$ where $\lambda_i, i = 1, 2, 3$, are the roots of (7.49). Indeed, with $\mathbf{F} = \mathbf{L}_\mathrm{I}\mathbf{u}$,

$$\nabla_U \cdot \mathbf{F} = \nabla_U \cdot (\mathbf{L}_\mathrm{I}\mathbf{u}) = \nabla_u \cdot (\mathbf{L}_\mathrm{I}\mathbf{u}) = \mathrm{tr}(\mathbf{L}_\mathrm{I}) = \sum_i \lambda_i = 0, \tag{7.50}$$

as the trace of a matrix is equal to the sum of its eigenvalues. Hence, a zero trace for the Jacobian is linked directly to conservation of volume. Further, one of the roots is

---

[3]This concept characterizes the response of a dynamical system to a weak perturbation, i.e., a weak vector field added to the dynamical system yielding a perturbed dynamical system. If the phase portraits of the perturbed and unperturbed dynamical systems are topologically equivalent ("homeomorphism"), i.e., if it is possible to go from one to the other by a continuous deformation, then the dynamical system is structurally stable.

real, λ_1, and has the sign of the product $\lambda_1\lambda_2\lambda_3 = 3(c-1)$, which is positive when $c > 1$ and negative when $c < 1$. The two others roots are complex conjugate (real) when

$$\Delta_{\mathrm{I}} = 4\big[3\delta\mathcal{N}(1,c) - \zeta\big]^3 + 243(c-1)^2 \tag{7.51}$$

is positive (negative).

When $c = 1$, since the sum of the roots is equal to zero and their product is also equal to zero, the fixed point $\mathbf{U}_{\mathrm{I}}$ undergoes a Hopf bifurcation and a *transcritical bifurcation* (an eigenvalue crosses the imaginary axis) at the same time: The real eigenvalue λ_1 and the pair of complex conjugate ones λ_2, λ_3 cross the imaginary axis simultaneously. This is the case of a "codim 2 bifurcation point" in the parameter space, or *Gavrilov–Guckenheimer point* [39], which can be thought of as a conjunction of a stationary instability and an oscillatory instability, $\lambda_1 = 0$ and $\lambda_2, \lambda_3 = \pm i\omega$, respectively, with respect to "time" ξ. In the vicinity of this point, homoclinic chaos is expected [101]. Recall that the two types of instability, "stationary" and "oscillatory," have been defined in Sect. 3.3. As we pointed out there, there is something very subtle about long wave instabilities that have $\omega_r = \omega_i = 0$ at the critical wavenumber $k_0 = 0$. Since $\omega_r = 0$ at criticality, one might think of them as stationary instabilities, but actually they are oscillatory—otherwise we would not get traveling waves. But here when $\lambda_1 = 0$, we do not have the same problem as with traveling waves; the issue with $k_0 = 0$ does not appear here; in the discussion of Sect. 3.3 we considered both time and space dependence, as opposed to "time" ξ only here.

Needless to say, there is a close relation between the linear stability analysis of the flat film and the stability analysis of the fixed points. In fact, the changes $\partial_x \to i\,k$ and $\partial_t \to -i\,\omega$ associated with the linear stability of the flat film are formally equivalent to the sequence $\partial_x \to \partial_\xi$, $\partial_t \to -c\,\partial_\xi$ and $\partial_\xi \to \lambda$ where c and λ are complex. Consequently, the Hopf bifurcation occurs precisely at the neutral conditions of the linear stability analysis of the flat film ($c = 1$). At the Hopf bifurcation point, the complex eigenvalues are

$$\lambda_2 \text{ and } \lambda_3 = \pm i\sqrt{3\delta\mathcal{N}(1,1) - \zeta}. \tag{7.52}$$

Therefore, the onset of limit cycles demands also that $3\delta\mathcal{N}(1,1) \geq \zeta$. This condition coincides with the instability threshold of the Nusselt flat film. From the linear stability analysis of the Nusselt flat film, in the case of the BE and the first-order model, the critical reduced Reynolds number is $\delta_c = \frac{5}{2}\zeta$, whereas in the case of the Kapitza–Shkadov model, $\delta_c = 3\zeta$. The instability of the Nusselt flat film, which occurs for $\delta \geq \delta_c$, is therefore a necessary condition for the onset of limit cycles.

At the second fixed point $\mathbf{U}_{\mathrm{II}}$, we get

$$h_{\mathrm{II}}^3\lambda^3 + \big[3\delta\mathcal{N}(h_{\mathrm{II}},c) - h_{\mathrm{II}}^3\zeta\big]\lambda - 3\big(c - h_{\mathrm{II}}^2\big) = 0, \tag{7.53}$$

and the sum of the roots is again zero. One of the roots is real and has the sign of $c - h_{\mathrm{II}}^2$, hence negative when $c > 1$, since $c > h_{\mathrm{II}}$ implies $c - h_{\mathrm{II}}^2 = c - (3c - 1 - h_{\mathrm{II}}) = (1-c) + (h_{\mathrm{II}} - c) < 0$. By performing the transformation (7.40) that allowed

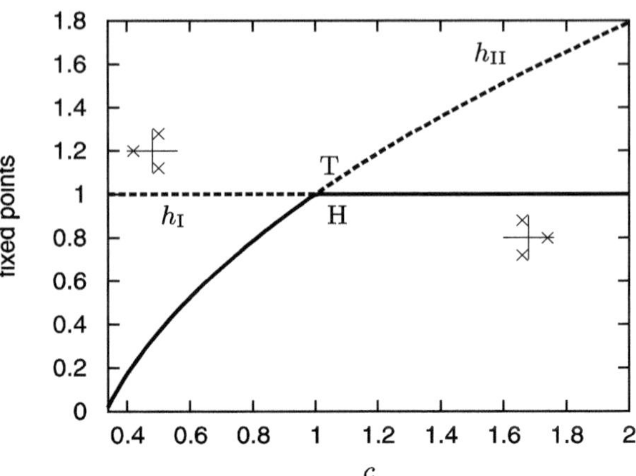

Fig. 7.13 Location of the fixed points h_I and h_II and their stability diagrams as function of the wave speed c in the case of the BE (5.55). The positions of the eigenvalues in the complex plane ($\Re(\lambda)$, $\Im(\lambda)$) are indicated by crosses. A Hopf (H) and transcritical (T) bifurcations are simultaneously observed at $c = 1$

us to rescale equations (7.41a)–(7.41c) in order to reset h_II to unity, it can be seen that the sign of

$$\Delta_\mathrm{II} = 4\big[3\delta\mathcal{N}(h_\mathrm{II}, c) - h_\mathrm{II}^3\zeta\big]^3 + 243 h_\mathrm{II}^3\big(c - h_\mathrm{II}^2\big)^2 \tag{7.54}$$

is the same with that of Δ_I. Hence the stability properties of $\mathbf{U}_\mathrm{II}$ can be obtained from those of $\mathbf{U}_\mathrm{I}$ by just exchanging the dimensions of their stable and unstable manifolds.

The case of the BE (5.55) [216] is the easiest one, thanks to the simplicity of the corresponding expression of $\mathcal{N}$, $\mathcal{N}(h, c) = \frac{2}{15}h^6$, independent of c. Since $\mathcal{N}$ is always positive, both fixed points have one real root and one complex conjugate pair for all c. As shown in Fig. 7.13, they are both saddle-foci. As $\mathcal{N}(1, c)$ is positive, when $c > 1$ we have $\lambda_1 > 0$, while λ_2, λ_3 are complex conjugate with negative real parts. As discussed earlier, in this case the homoclinic orbit leaves the fixed point $\mathbf{U}_\mathrm{I}$ monotonically along the one-dimensional unstable manifold $\mathcal{W}^\mathrm{u}$ and returns to the fixed point by spiraling on the two-dimensional stable manifold $\mathcal{W}^\mathrm{s}$. Because $h_\mathrm{II} > h_\mathrm{I} = 1$, the corresponding wave profile is a hump preceded by radiation corresponding to the spiraling return to $\mathbf{U}_\mathrm{I}$. One can check that $\mathcal{N}(1, c) > 0$ for the Ooshida equation, as well as for the two-equation models (6.13a), (6.13b) and (6.1), (6.51). Accordingly, Δ_I is positive and the fixed point $\mathbf{U}_\mathrm{I}$ is of the same type with the BE: a saddle-focus with a one-dimensional unstable manifold and a two-dimensional stable manifold. Note that for all first-order models and surface equations, i.e., (7.41a)–(7.41c) with $\eta = 0$, the sum of the roots of (7.49) is equal to zero so that $\Re(\lambda_2) = \Re(\lambda_3) = -\frac{1}{2}\lambda_1$ and hence the Shil'nikov criterion (7.47) is always satisfied whenever a homoclinic orbit exists.

Let us now turn to the second-order models, i.e., the simplified (6.1), (6.79) and regularized models (6.1), (6.92) for which viscous dispersion is taken into account ($\eta \neq 0$). The characteristic equation of the associated dynamical system (7.41a)–

(7.41c) is

$$\lambda^3 + \lambda^2 \eta\left(-\frac{12}{5} + \frac{27}{5}c\right) + \lambda\left[\delta\left(\frac{6}{5}c^2 - \frac{34}{35}c + \frac{6}{35}\right) - \zeta\right] - 3(c-1) = 0, \quad (7.55a)$$

which can be written in compact form as

$$\lambda^3 + A\lambda^2 + B\lambda + C = 0, \qquad (7.55b)$$

where $A = \eta(-\frac{12}{5} + \frac{27}{5}c)$, $B = \delta(\frac{6}{5}c^2 - \frac{34}{35}c + \frac{6}{35}) - \zeta$ and $C = 3 - 3c$.

For $\eta \neq 0$, the sum of the roots is no longer equal to zero but to $-A \propto \eta$. Equating $h_{\text{div}0}$ given by (7.43) with 1 gives $c = \frac{4}{9} \approx 0.44$. When $c \geq \frac{4}{9}$ the sum of the roots is negative and $\nabla_U \cdot \mathbf{F} < 0$, i.e., the flow is locally contracting and the local "amount of phase flow" entering a small neighborhood around $\mathbf{U}_\mathrm{I}$ is less than the amount leaving it. From $h_{\text{div}0} = h_\mathrm{II}$ we similarly obtain an upper bound, $c \approx 3.8$, above which the phase volume is locally expanding in the vicinity of the second fixed point h_II. It should be emphasized, however, that for all the solutions to the dynamical system (7.42) considered in this chapter, the speed of the corresponding periodic or solitary waves is in the interval $(0.44, 3.8)$ and the two fixed points are located in contracting regions of the phase space. A transcritical bifurcation at $c = 1$ corresponding to the coalescence of the two fixed points $\mathbf{U}_\mathrm{I}$ and $\mathbf{U}_\mathrm{II}$ appears in addition to the Hopf bifurcation. The simultaneous occurrence of an oscillatory instability ($\lambda_2, \lambda_3 = \pm i\omega$) and a stationary instability ($\lambda_1 = 0$) is encountered with all first-order models including the BE. This degeneracy is removed by the introduction of the viscous second-order effects, i.e. the Hopf bifurcation and the transcritical bifurcation cease to occur simultaneously.

Setting $\lambda = \lambda_r + i\lambda_i$ for λ and separating the real and imaginary parts of (7.55a), (7.55b) leads to

$$\lambda_i^2 = B \quad \text{and} \quad C = AB. \qquad (7.56)$$

The behavior of the imaginary part λ_i and speed c is presented in Fig. 7.14 as function of the reduced Reynolds number δ when the Hopf bifurcation occurs, i.e., when the complex eigenvalues cross the imaginary axis $\lambda_r = 0$. This corresponds exactly to the neutral stability of the Nusselt flat film (see Sect. 7.1.1). The imaginary part λ_i decreases with viscous dispersion η corresponding to the reduction of the range of unstable wavenumbers for the flat film ($\lambda_i \sim k$). The decrease of λ_i as function of the distance from the threshold $\delta - \frac{5}{2}\zeta$ is faster when the wall is less inclined (ζ large) compared to the vertical case ($\zeta = 0$). However, the effect of viscosity (η) on the speed of the waves is more visible when the wall is vertical than when it is slightly inclined.

Double eigenvalues of (7.55b) written as a polynomial $\mathcal{P} = \lambda^3 + A\lambda^2 + B\lambda + C = 0$ can be found by demanding that λ be a root of both $\mathcal{P}$ and its derivative $\mathcal{P}'$, or that $\mathcal{P}'$ divides $\mathcal{P}$, i.e., that λ is a root of the polynomial of second degree $\mathcal{P}(\lambda) - \lambda\mathcal{P}'(\lambda)$:

$$\left(C - \frac{AB}{9}\right)\left(-\frac{4}{9}A^3 + \frac{5}{3}AB - 3C\right) - B\left(-\frac{2}{9}A^2 + \frac{2}{3}B\right)^2 = 0. \qquad (7.57)$$

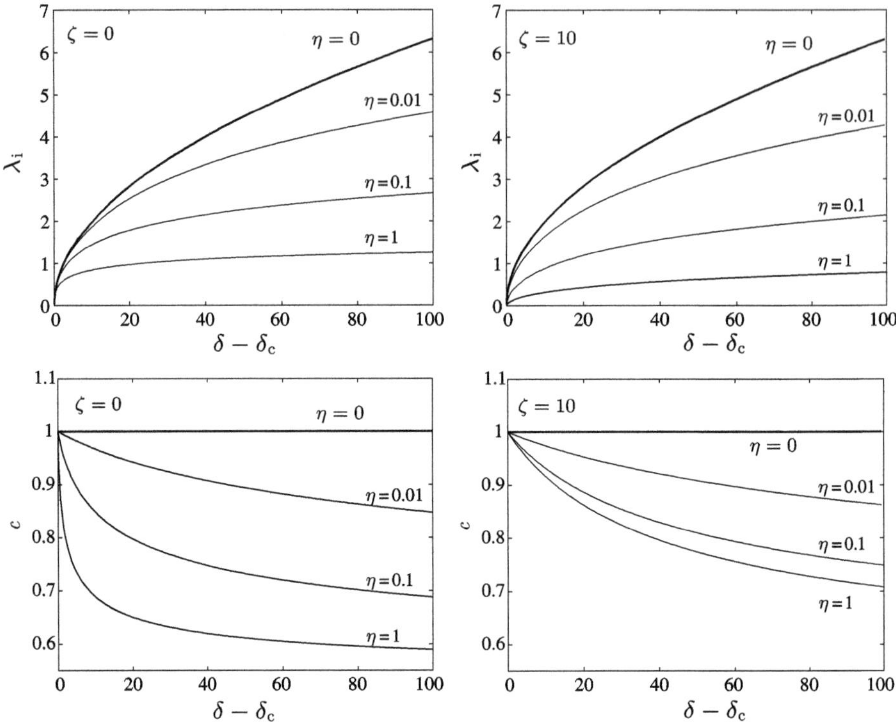

Fig. 7.14 Location of the Hopf bifurcation for different values of the parameters ζ and η for the characteristic equations (7.55a), (7.55b) corresponding to the simplified model (6.1), (6.79) and regularized model (6.1), (6.92). *Upper and lower figures* correspond to the imaginary part of the eigenvalue λ_i and phase speed c as functions of the distance from threshold $\delta_c = 0$, respectively

In a point of the parameter space where (7.57) is satisfied, the fixed point $\mathbf{U_I}$ changes from a saddle-focus to a saddle or from a stable focus to a "node" [177] (a node fixed point has real eigenvalues of the same sign). In the first case and as pointed out earlier, the shapes of the homoclinic orbits change as they cease to spiral back to the fixed point, which in turn corresponds to the disappearance of the radiation preceding the hump of a solitary wave and the appearance of a smooth monotonic entry in the positive ξ direction. In the second case, the unstable manifold of $\mathbf{U_I}$ is empty, which then disallows the existence of solitary waves.

We have already seen that when $\eta = 0$ the sum of the eigenvalues vanishes, so that $\Re(\lambda_2) = \Re(\lambda_3) = -\frac{1}{2}\lambda_1$ and the Shil'nikov criterion (7.47) is satisfied whenever a homoclinic orbit is found. For $\eta > 0$, this is not automatically the case. At the onset of homoclinic chaos, the eigenvalues are equally distant from the imaginary axis, $\lambda_1 = -\Re(\lambda_2) = -\Re(\lambda_3)$. The sum of the eigenvalues is thus $-\lambda_1 = -A$ and the monomial $\lambda - A$ divides $\mathcal{P}$. We then have:

$$2A^3 + AB + C = 0. \tag{7.58}$$

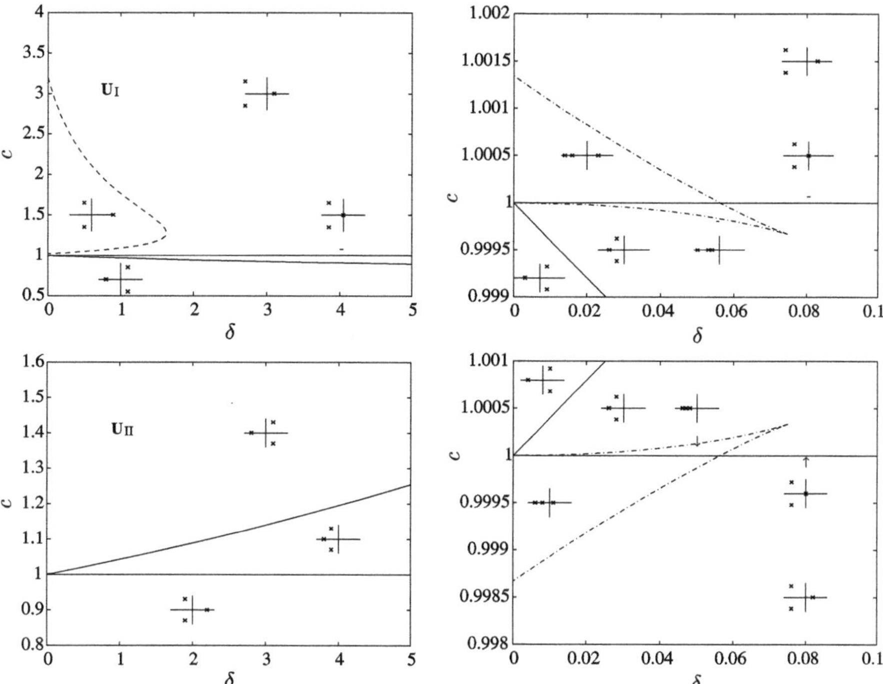

Fig. 7.15 Behavior of the eigenvalues in the complex plane as functions of δ and c for $\zeta = 0$ (vertical wall) and $\eta = 0.1$. *Upper and lower figures* correspond to the fixed points $\mathbf{U}_\mathrm{I}$ and $\mathbf{U}_\mathrm{II}$, respectively. The *diagrams on the right* are blow ups of the *left figures* at $c = 1$ and $\delta = 0$. The locus of the transcritical bifurcation $c = 1$ and of the Hopf bifurcation are indicated by *solid lines. Dashed line* (*upper left panel*) and *dashed-dotted lines* (*right panels*) refer to the onset of the Shil'nikov homoclinic chaos (7.58) and to the transition from saddle to saddle-focus (7.57), respectively

Since for $\eta = 0$, $\Re(\lambda_2) = \Re(\lambda_3) = -\frac{1}{2}\lambda_1$, which combined with the equidistance of the eigenvalues from the imaginary axis leads to $\lambda_1 = \Re(\lambda_2) = \Re(\lambda_3) = 0$, the product of the eigenvalues vanishes and (7.58) reduces to $c = 1$.

Additional pieces of information can be extracted by investigating the behavior of the eigenvalues in the complex plane as the parameter space is explored. We restrict our attention to situations where the flat film is unstable ($\delta > \frac{5}{2}\zeta$). Since surface tension is predominant, viscous dispersion is small, but we shall assume here that it can be as large as unity. The location of the eigenvalues in the complex plane is schematically indicated in Figs. 7.15, 7.16, 7.17, 7.18, and 7.19 for $\zeta = 0$, $\zeta = 10$, $\eta = 0.01$, $\eta = 0.1$ and $\eta = 1$. Again the figures are obtained with the characteristic equations (7.55a), (7.55b) corresponding to the simplified model (6.1), (6.79) and regularized model (6.1), (6.92).

Let us first discuss the behavior of the eigenvalues of the fixed point $\mathbf{U}_\mathrm{I}$ by varying δ and c and for given values of ζ and η. When the wall is vertical ($\zeta = 0$), and in the presence of moderate viscous dispersion, $\eta = 0.1$ (see Fig. 7.15), the location of the Hopf bifurcation remains close to the axis $c = 1$. The Shil'nikov criterion (7.47) is satisfied in a rather small region of the (δ, c) plane limited by the axis $\delta = 0$ and

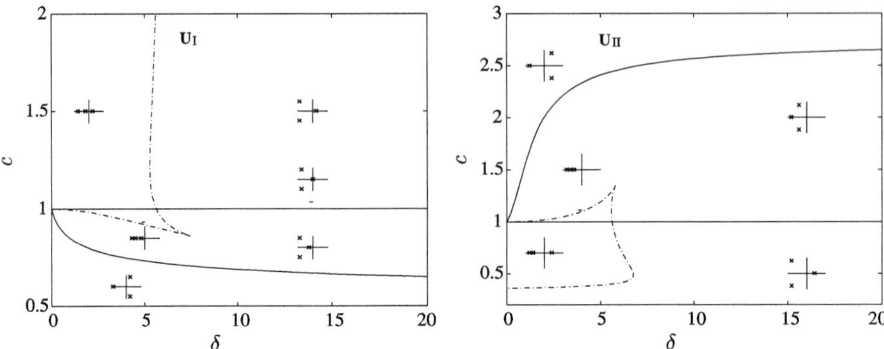

Fig. 7.16 Behavior of the eigenvalues in the complex plane as functions of δ and c for $\zeta = 0$ (vertical wall) and $\eta = 1$. *Left and right figures* correspond to fixed points $\mathbf{U}_\mathrm{I}$ and $\mathbf{U}_\mathrm{II}$, respectively. *Dashed-dotted lines* refer to the transition from saddle to saddle-focus (7.57)

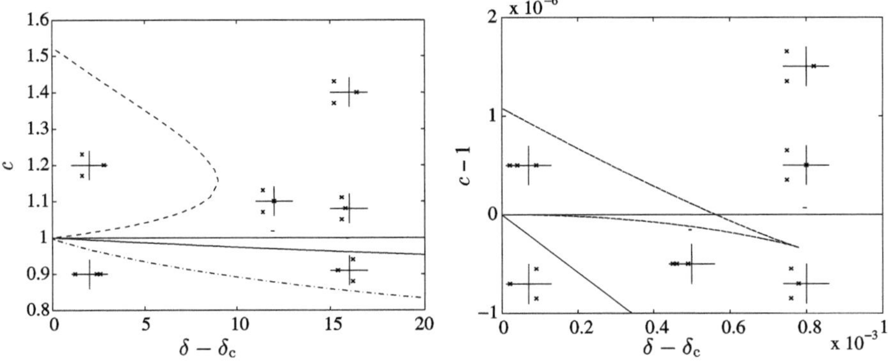

Fig. 7.17 Behavior of the eigenvalues in the complex plane as functions of δ and c for the fixed point $\mathbf{U}_\mathrm{I}$. Parameter values are $\zeta = 10$ and $\eta = 0.01$. The *figure on the right* is an enlarged view of the region around $c = 1$, $\delta = 25$. *Left panel*: *dashed line* $(c > 1)$ refers to the onset of the Shil'nikov homoclinic chaos (7.58) whereas *dashed-dotted line* $(c < 1)$ corresponds to the transition from saddle to saddle-focus (7.57). *Right panel*: *dashed lines* refer to the onset of the Shil'nikov homoclinic chaos (7.58)

the dotted line and below the solid line indicating the Hopf bifurcation. Moreover, $\mathbf{U}_\mathrm{I}$ is a focus except for a small neighborhood of the point $(\delta = 0,\ c = 1)$ corresponding to the instability threshold of the film. For a large value of the dispersion parameter $\eta = 1$, the region where the Shil'nikov criterion is satisfied reduces to the portion of the plane below the location of the Hopf bifurcation.

The situation is again modified for a slightly inclined wall $(\zeta = 10)$. For $\eta = 0.01$ (see Fig. 7.17), the regions where homoclinic chaos are possible, i.e., where the Shil'nikov criterion applies, form a small strip limited by the locations of the Hopf bifurcation (solid line) and double eigenvalues (dashed-dotted line) and a region limited by the axis $\delta = \frac{5}{2}\zeta$ and the dotted line. This last region disappears when $\eta = 0.1$ and the domain where the Shil'nikov criterion is satisfied diminishes again for $\eta = 1$ (cf. Fig. 7.19).

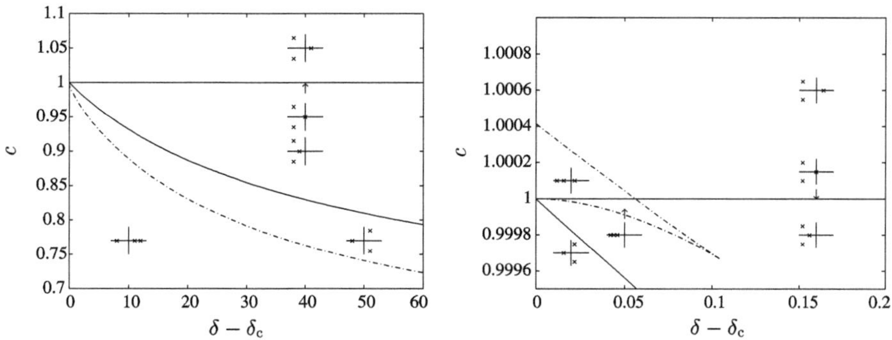

Fig. 7.18 Behavior of the eigenvalues in the complex plane as functions δ and c for the fixed point $\mathbf{U}_\mathrm{I}$. Parameter values are $\zeta = 10$ and $\eta = 0.1$. The *figure on the right* is an enlarged view of the region around $c = 1$, $\delta = 25$. *Dashed-dotted lines* refer to the transition from saddle to saddle-focus (7.57)

Fig. 7.19 Behavior of the eigenvalues in the complex plane as functions δ and c for the fixed point $\mathbf{U}_\mathrm{I}$. Parameter values are $\zeta = 10$ and $\eta = 1$. *Dashed-dotted lines* refer to the transition from saddle to saddle-focus (7.57)

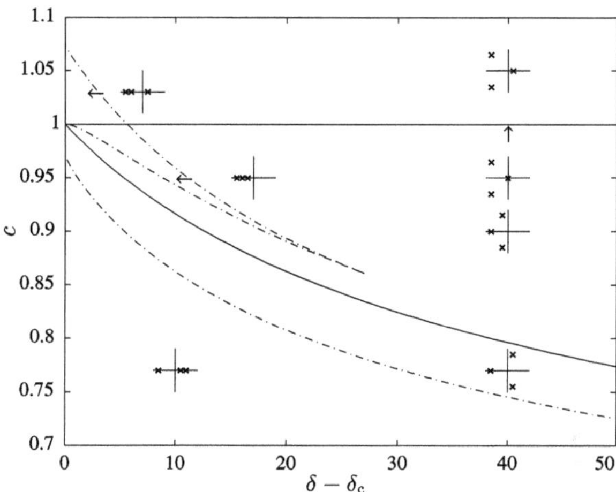

The stability diagrams of the fixed point $\mathbf{U}_\mathrm{II}$ can be obtained from those of $\mathbf{U}_\mathrm{I}$ by performing a transformation similar to (7.40),

$$U_1 \to h_\mathrm{II}^{-1} U_1, \qquad c \to h_\mathrm{II}^{-2} c, \qquad \delta \to h_\mathrm{II}^{-11/3} \delta, \qquad \zeta \to h_\mathrm{II}^{2/3} \zeta, \qquad \eta \to h_\mathrm{II}^{4/3} \eta,$$

$$(7.59)$$

which allows us to reset h_II to unity. As for the BE, the Hopf bifurcation now occurs at values of c larger than unity.

7.2.1.3 Limit Cycles and Homoclinic Bifurcations

To fix ideas we limit ourselves to the dynamical system (7.42) corresponding to the simplified model (6.1), (6.79). The results are quantitatively similar when another dynamical system is considered.

As noted earlier, limit cycles/periodic orbits can emerge from a Hopf bifurcation. The resulting family of limit cycles is referred to as the *principal/primary periodic orbits* in contrast to the *subsidiary periodic orbits* that may further branch off the principal family [103]. As a control parameter changes and we pass the Hopf bifurcation point, the period of these solutions, say $2\pi/k$, with k the wavenumber of the corresponding periodic wave, increases. Consequently, their size in phase space increases and they start approaching the fixed point until they "collide" with this point, i.e., they intersect this point exactly (topologically, when the coalescence with the fixed point occurs, the homoclinic orbits approach the eigenvectors of this point tangentially), yielding an orbit of infinite period ($k \to 0$), i.e., a homoclinic orbit. After the creation of a homoclinic orbit there is no longer a periodic orbit.

This type of bifurcation is a *homoclinic bifurcation*. It is a "global bifurcation" as the corresponding topological change affects the entire phase space and its study cannot be reduced to the study of the neighborhood of the fixed point. In general, a global bifurcation corresponds to a modification of the phase space that cannot be reduced to the study of a neighborhood of a fixed point [111]. On the contrary, a "local bifurcation," e.g., Hopf, occurs when a parameter change causes the stability of the fixed point to change.

A global bifurcation offers a relatively simple way to numerically construct homoclinic orbits. This is the method used by the continuation software AUTO07P as noted earlier. Limit cycles emerging from both fixed points $\mathbf{U}_I$ (branch labeled 1, $c < 1$) and $\mathbf{U}_{II}$ (branch labeled 2, $c > 1$) are displayed in Fig. 7.20. In this example the limit cycle emerging from $\mathbf{U}_{II}$ is seen to approach the vicinity of $\mathbf{U}_I$ and finally ends up as a homoclinic orbit connecting $\mathbf{U}_I$ to itself within only one loop. It is a typical example of the principal homoclinic orbit solutions that correspond to single-hump solitary waves. Since $\mathbf{U}_I$ is a saddle-focus with a one-dimensional unstable manifold and a two-dimensional stable one, the wave profile consists of a large hump preceded by radiation (see Fig. 7.20, bottom). But quite frequently the waves observed in both experiments and computations are trains of solitary waves, which, as pointed out earlier, are sometimes referred to as "multi-pulse waves"; recall that both multi-pulse and multi-hump waves correspond to subsidiary/multi-loop homoclinic orbits in the phase space.

Primary homoclinic orbits can arise form the homoclinic bifurcation of the primary limit cycle branch that appears through a Hopf bifurcation. On the other hand, subsidiary homoclinic orbits can emerge through homoclinic bifurcations of subsidiary limit cycles, which have bifurcated in turn from the primary limit cycle branch (through a *period-doubling bifurcation*, as we shall see below). Clearly, a point in the phase space, say $M(\xi)$, of a limit cycle is invariant under the transformation $\xi \to \xi + nP$ where P is the period in the phase space and n an integer. A countable infinite set of new limit cycles can then emerge by breaking this symmetry.

An example is given in Figs. 7.21 and 7.22, showing that the corresponding bifurcation diagram and phase space can become complicated. Figure 7.21 depicts a bifurcation diagram in the plane (c, k). The Hopf bifurcation is indicated by label 1, the principal limit cycle branch emerging in this example from fixed point $\mathbf{U}_{II}$. As

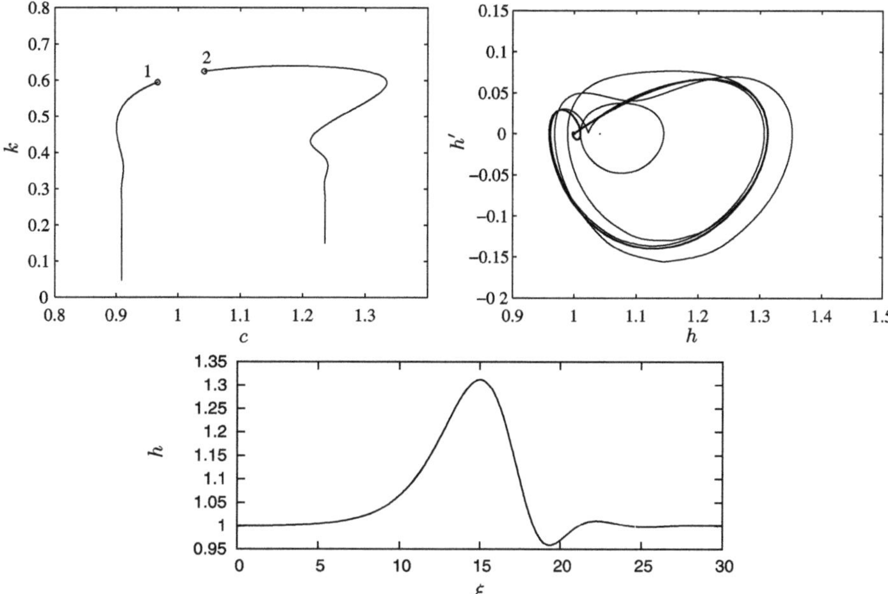

Fig. 7.20 Limit cycles emerging from Hopf bifurcations for $\delta = 1$ and $\eta = 0.1$ from fixed points $\mathbf{U_I}$ (label 1) and $\mathbf{U_{II}}$ (label 2). The control parameter is the speed c and the *upper left diagram* shows the wavenumber k of the corresponding periodic wave as function of its speed c. The *upper right diagram* corresponds to projections of the limit cycles emerging from $\mathbf{U_{II}}$ in the three-dimensional phase plane onto the plane ($U_1 \equiv h$, $U_2 \equiv h'$) for different values of the wave speed c. For $k \ll 1$, the limit cycle approaches a homoclinic orbit around $\mathbf{U_I}$ (*thick solid line* in the *upper right panel*). The wave profile corresponding to this homoclinic orbit is shown in the *lower diagram*

the amplitude of the limit cycle is increased, a secondary branch emerges through period doubling, i.e., breaking of the symmetry $\xi \rightarrow \xi + P$ (label 2). This branch of solutions then terminates at a homoclinic orbit connecting $\mathbf{U_I}$ to itself and made of two loops around $\mathbf{U_{II}}$ (Fig. 7.22, left). At point 4a, this secondary branch undergoes another period doubling giving rise to a tertiary branch of limit cycles (the terms "secondary" and "tertiary" here are used to simply count the branches of solutions) that merges back to the secondary one through a reverse period-doubling bifurcation (point 4b). Finally, the tertiary branch shows up again at point 4c of the bifurcation diagram through period doubling and terminates at a homoclinic orbit made of four loops around fixed point $\mathbf{U_{II}}$ (Fig.7.22, right).

7.2.1.4 Bifurcation Diagrams of Limit Cycles

It is now pertinent to discuss the stability in the phase space of the different limit cycles obtained from Hopf bifurcations and subsequent period-doubling bifurcations. With the term "stability," we refer here to the sensitivity of trajectories in the phase space to small perturbations of the initial conditions at $\xi = 0$, i.e., stability in the

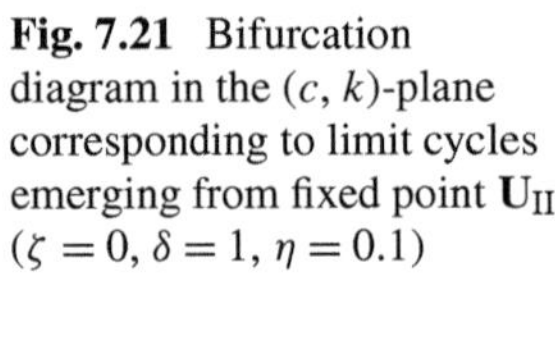

Fig. 7.21 Bifurcation diagram in the (c, k)-plane corresponding to limit cycles emerging from fixed point $\mathbf{U}_{\mathrm{II}}$ ($\zeta = 0$, $\delta = 1$, $\eta = 0.1$)

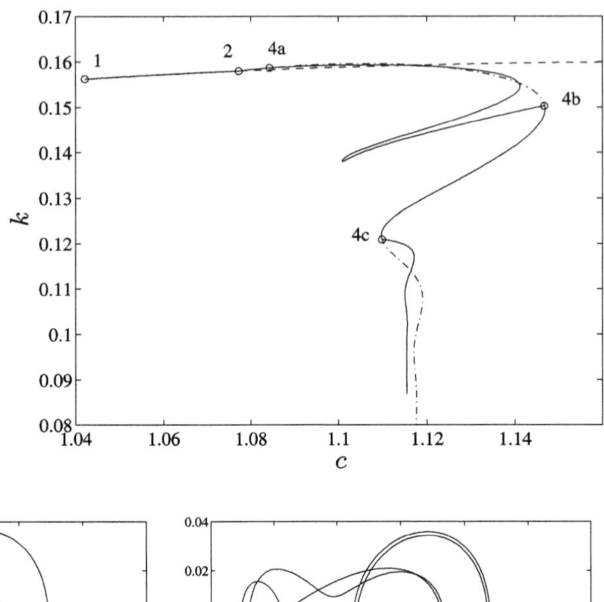

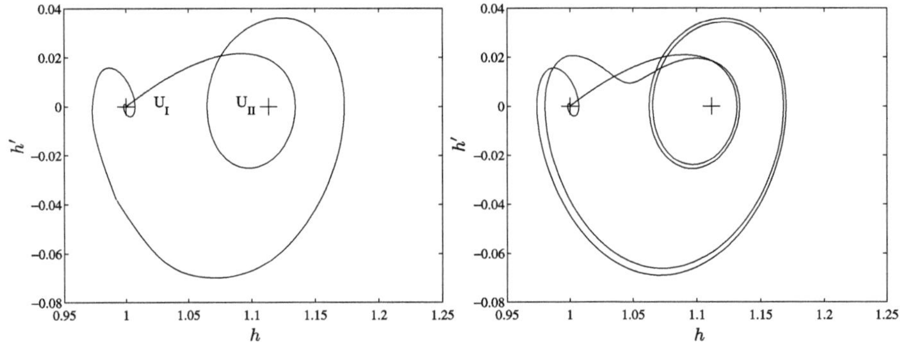

Fig. 7.22 Projections on the plane ($U_1 \equiv h$, $U_2 \equiv h'$) of the subsidiary homoclinic orbits where the limit cycle branches of solutions shown in Fig. 7.21 terminate for $k \to 0$. Locations of the fixed points are indicated by *crosses*

Lyapunov sense.[4] A relatively easy way to address this question is to compute trajectories in the phase space starting from points in the vicinity of the considered limit cycle. If the limit cycle is recovered, it is a stable orbit, and moreover a "local attractor."

Local attractors in the vicinity of limit cycle branches of solutions have been obtained starting from a trajectory in the vicinity of the fixed point $\mathbf{U}_{\mathrm{II}}$ and increasing the wave speed c gradually. For each value of c, the trajectory is computed for a long time, leading to one local attractor or diverging to infinity when no attractors can be found. Parameter values are initially chosen so as to correspond to the limit

---

[4]This concept is different from the "structural stability" defined earlier in this chapter which deals with the topological properties of the complete phase portrait of a dynamical system when a perturbation is applied, for example when one parameter is slightly modified. Here we consider only the stability of a single trajectory in the phase space.

cycles displayed in Figs. 7.20 and 7.22 ($\delta = 1$, $\zeta = 0$, $\eta = 0.1$). Trajectories systematically return to $\mathbf{U}_{\mathrm{II}}$ before the principal limit cycle branch of solutions arises, which signals an "exchange of stability" between the fixed point and the limit cycle at the Hopf bifurcation point. Similar exchanges of stability are next observed when period-doubling bifurcations occur.

To visualize the sequence of bifurcations observed when c is increased, a Poincaré section in the phase plane can be useful (e.g., [301]). A simple example of such sections can be obtained by searching for local maxima h_{m} of the wave height ($h \equiv U_1$). Typical examples of the corresponding bifurcation diagrams for these maxima as functions of the bifurcation parameter c are offered in Fig. 7.23 for a moderate reduced Reynolds number and a vertical wall ($\delta = 1$ and $\zeta = 0$). The bifurcation diagram obtained for weak viscous dispersion ($\eta = 0.01$) and displayed in panel (a) exhibits a well-defined period-doubling route to chaos [177]. As the bifurcation parameter c is increased, the initial period doubling is followed by a rapid sequence of period doublings, ending in a disordered trajectory in the phase space, for which the local maxima are no longer isolated points of the Poincaré section but now cover finite-size bands. The trajectory is at this stage weakly chaotic and is nothing more than a periodic trajectory superimposed with some noise. As c is further increased, a reverse cascade is next observed with bands merging with one another so that local maxima occupy an interval of increasing size. If trajectories are chaotic, windows of periodic stable attractors exist in the bifurcation diagram as predicted by the theory [177]. At the last stage of our computations, the chaotic trajectory starts to approach the vicinity of the first fixed point $\mathbf{U}_{\mathrm{I}}$, as indicated by the presence of local maxima close to unity, until finally the trajectory diverges to infinity and no attractors are detected. The corresponding wavetrain ($h = U_1$ versus ξ) is shown in Fig. 7.24(a).

When η is increased to $\eta = 0.15$ (Fig. 7.23(b)), chaotic attractors are still obtained through a period-doubling route, but the range of values reached by the local maxima of h is smaller. For the largest values of c, periodic attractors are systematically obtained and a homoclinic bifurcation leading to a complicated homoclinic orbit is finally observed. More precisely, the wavetrain shown in Fig. 7.24(b) reveals the formation of a group of four two-hump waves.

At a larger value of the viscous dispersion parameter, $\eta = 0.17$, the range of the bifurcation parameter c for which chaos is observed is reduced to only two intervals separated by a window of stable limit cycles. Curiously, two cascades of period doublings are observed: One by decreasing the value of c and one by increasing it. For the largest values of c, the trajectory is again periodic and a homoclinic bifurcation is observed again, leading this time to a two-hump subsidiary homoclinic orbit (see Fig. 7.24(c)).

At even stronger viscous dispersion ($\eta = 0.2$), no chaotic attractors are found and a single period doubling is directly followed by a homoclinic bifurcation leading to a two-hump solitary wave (Figs. 7.23(d) and 7.24(d)). Increasing η further, the period-doubling bifurcation observed for $\eta = 0.2$ disappears and the bifurcation diagram reduces to a Hopf bifurcation, followed by a homoclinic bifurcation leading to the principal homoclinic orbit.

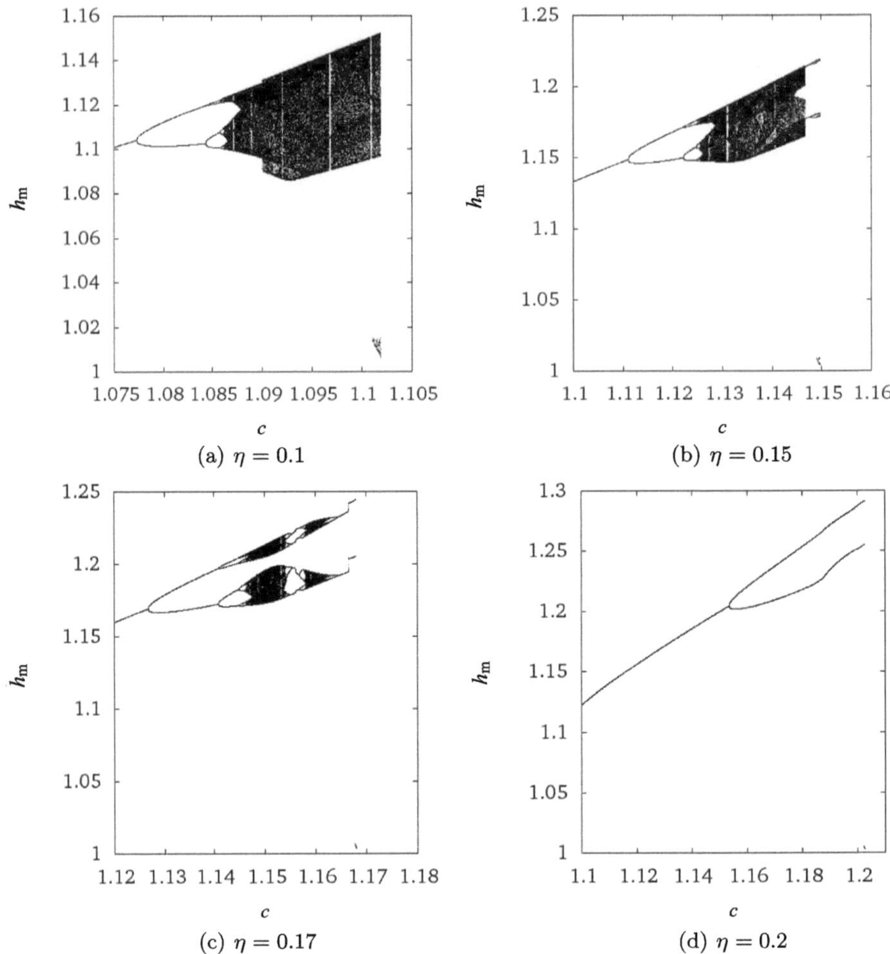

Fig. 7.23 Bifurcation diagrams of limit cycles emanating from $\mathbf{U}_{\mathrm{II}}$ for increasing values of the viscous dispersion parameter η with $\delta = 1$ and $\zeta = 0$ (vertical wall)

The above discussion is limited to the bifurcation diagrams starting from the Hopf bifurcation of fixed point $\mathbf{U}_{\mathrm{II}}$ as c is increased. However, limit cycles arising from $\mathbf{U}_{\mathrm{I}}$ can be recovered by our making use of the transformation (7.59). The corresponding bifurcation diagrams are therefore similar to those we have just discussed except that the bifurcation parameter c decreases from its value at the Hopf bifurcation and that the corresponding bifurcating homoclinic orbits connect $\mathbf{U}_{\mathrm{II}}$ to itself (examples of such diagrams can be found in the literature, e.g., [161, 288]).

Finally, let us underline the link between the presence of homoclinic chaos and the chaotic trajectories observed in the vicinity of limit cycle branches of solutions emanating from $\mathbf{U}_{\mathrm{II}}$. In the case of the three bifurcation diagrams corresponding to $\eta = 0.15$, $\eta = 0.17$ and $\eta = 0.2$, the Shil'nikov numbers $\lambda_1 + \Re(\lambda_{2,3})$, computed for

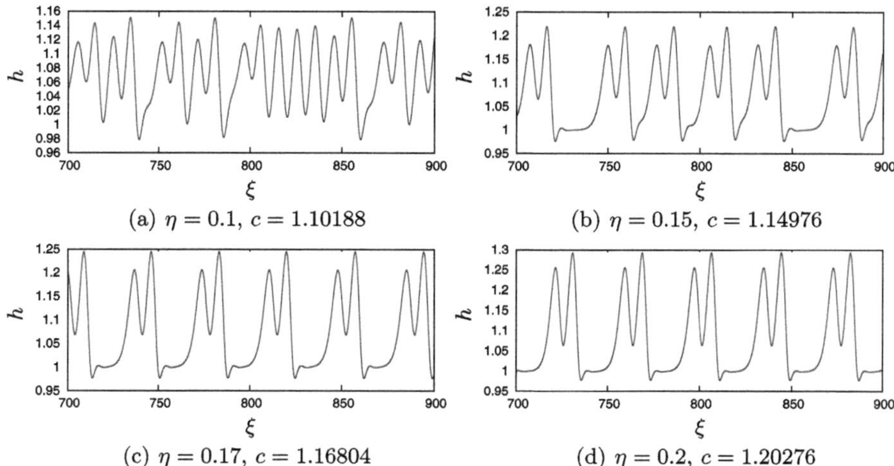

Fig. 7.24 caption panels:
(a) $\eta = 0.1$, $c = 1.10188$
(b) $\eta = 0.15$, $c = 1.14976$
(c) $\eta = 0.17$, $c = 1.16804$
(d) $\eta = 0.2$, $c = 1.20276$

Fig. 7.24 Film thickness h versus ξ corresponding to the orbits obtained at the last stage of the bifurcation diagrams displayed in Fig. 7.23

the homoclinic orbits around $\mathbf{U}_{\mathrm{I}}$ are negative and homoclinic chaos is not present, which explains that stable homoclinic orbits (in the Lyapunov sense) are observed (panels b, c and d in Fig. 7.24). Conversely, the Shil'nikov number is positive for $\eta = 0.1$, in agreement with the observation of a chaotic trajectory at the final stage of the computation. However, the wavetrain illustrated in Fig. 7.24(a) is rather far from the succession of nearly identical pulses separated by portions of flat films of various lengths, which is expected when homoclinic chaos is observed. This discrepancy can be explained by noting that homoclinic chaos is observed for values of the wave speed c that are very close to the speed $c^{\star}$ of the homoclinic orbit giving birth to it. Unfortunately, whenever the speed c exceeds $c^{\star}$ the computed trajectory diverges to infinity, which makes the computation of trajectories for values of c close to $c^{\star}$ through the numerical procedure followed here rather difficult.

To conclude, the procedure adopted here, i.e., following long-time trajectories starting close to a fixed point of the associated dynamical system, has enabled us to detect structural instabilities of the dynamical system when the speed is varied. For instance, the transition from the phase portrait observed when η is raised from 0.1 to 0.15 is an example of structural instability. Indeed, the phase portrait is suddenly drastically modified.

7.2.1.5 Principal Homoclinic Orbits: Single-Hump Solitary Waves

Needless to say, the autonomous dynamical system (7.42) has a rich dynamics in the phase space and exhibits a large variety of limit cycles and homoclinic orbits, of which only a few representative examples have been singled out. Even when viscous dispersion is neglected ($\eta = 0$), as is the case, e.g., with the simple Kapitza–Shkadov model (6.13a), (6.13b), giving a thorough account of all possible traveling wave solutions and a detailed unfolding of their bifurcation scenarios is an almost impossible task. In fact, many branches/families of solutions can be obtained showing complex interconnections and forming many-folded and many-sheeted surfaces in the parameter space [253, 281, 287].

In spite of this complex bifurcation picture, only a few families of solutions are pertinent, the most important being the principal homoclinic orbits. In order to fix ideas, let us consider the principal homoclinic orbits around the fixed point $\mathbf{U}_I$ of the simplified model (6.1), (6.79). As already noted, such orbits correspond to positive-hump solitary waves in real space. The corresponding speed c and wavetrains are shown in Figs. 7.25 and 7.26 for a vertical wall $\zeta = 0$, or a slightly inclined plate $\zeta = 10$, respectively. The viscous dispersion parameter η ranges from $\eta = 0.01$ to $\eta = 1$. At threshold ($\delta = \delta_c = \frac{5}{2}\zeta$), solitary waves have an infinitesimal amplitude and thus correspond to the neutral linear waves in the limit $k \to 0$, i.e., kinematic waves with speed $c = 1$ (cf. Sect. 7.1.3). The speed and amplitudes of the waves increase with the distance from threshold, $\delta - \frac{5}{2}\zeta$. The back tail of the waves is also increasing as the reduced Reynolds number δ is increased. Unlike inclination (ζ), which affects the maximum amplitude and speed of the waves, viscous dispersion (η) has a small effect on these quantities. However, the amplitude and frequency/number of ripples preceding the hump is very sensitive to the value of η to the point that they disappear for $\delta < 5.8$, $\eta = 1$ and a vertical wall. The disappearance of radiation corresponds to the transition of the fixed point $\mathbf{U}_I$ from a saddle-focus to a saddle, with the pair of complex eigenvalues λ_2, λ_3 reducing to a double real eigenvalue ($\Im(\lambda_2)$ and $\Im(\lambda_3) = 0$ indicated by the label "DE" in Figs. 7.25 and 7.26). Hence, for sufficiently large viscous dispersion the falling film can have nondissipative solitary waves.

From the stability diagrams of Figs. 7.15–7.19 and the locations of the principal homoclinic orbits in Figs. 7.25, 7.26, the portions of hypersurfaces (codim 1 sets of points) in the four-dimensional parameter space (δ, ζ, η, c) where the flow exhibits homoclinic chaos can be readily determined. Recall that when viscous dispersion is negligible (η small), the Shil'nikov criterion (7.47) is satisfied and these hypersurfaces then correspond entirely to the locations of the homoclinic orbits. However, as soon as viscous dispersion is taken into account, the size of the hypersurfaces where homoclinic chaos is found shrinks dramatically. When the wall is vertical ($\zeta = 0$) and $\eta = 0.01$, homoclinic chaos disappears for $\delta > 7$ (the transition point from complex to real determining eigenvalues is labeled "HC" in Figs. 7.25 and 7.26). At $\eta = 0.1$, the part of the principal homoclinic branch in the space (c, δ) where homoclinic chaos is possible has already been reduced to a small portion. At $\eta = 1$, homoclinic chaos is never observed. The effect of viscous dispersion on the existence of homoclinic chaos is even more dramatic for slightly inclined plates ($\zeta = 10$). The portion of the principal homoclinic curve is already small at $\eta = 0.01$ and homoclinic chaos cannot be observed above $\eta = 0.1$.

Noteworthy is that homoclinic chaos is independent of the form of the homoclinic orbit (e.g., primary around the fixed point $\mathbf{U}_I$ or double-loop subsidiary one with one loop not passing close to the fixed point $\mathbf{U}_I$ but revolving around $\mathbf{U}_{II}$). From our previous discussion, it is clear that homoclinic chaos requires: (i) a homoclinic orbit; (ii) that the Shil'nikov criterion is satisfied.

Here we only checked for the principal homoclinic orbits, the regions in the parameter space where the Shil'nikov criterion is satisfied. These regions are located with the help of the stability diagrams in Figs. 7.15–7.19, which pinpoint where the Shil'nikov criterion is satisfied (for $\eta = 0$ it is satisfied everywhere; for $\eta \neq 0$ it is satisfied only in certain regions that shrink rapidly as η increases). By checking the curves in the bifurcation diagrams of Figs. 7.25 and 7.26, which give us the locations of the homoclinic orbits in the parameter space, with the diagrams in Figs. 7.15–7.19, we can mark in Figs. 7.25 and 7.26 the upper boundary of homoclinic chaos. In this way, we determine the regions in the parameter space where the above two criteria for the existence of homoclinic chaos are satisfied.

But, clearly, since we only examined the principal homoclinic orbits, the diagram in Fig. 7.23 is incomplete. In fact, we could have double-loop subsidiary homoclinic orbits with one loop that does not pass close to the fixed point $\mathbf{U}_I$, located sufficiently far from the primary orbits (the Shil'nikov criterion depends on the speed; double loop orbits could have speeds sufficiently different to those of the principal ones and hence located far from the principal ones in the parameter space). Then, if the Shil'nikov criterion is satisfied for these double-loop orbits, we could also have homoclinic chaos in the regions of the parameter space where such orbits exist. On the other hand, if these double-loop orbits have speeds sufficiently close to the primary ones, there is a chance that the Shil'nikov criterion is satisfied for these orbits as well so that the conditions given here (in terms of the parameters δ, ζ and η) for the existence of homoclinic chaos associated with primary orbits are also conditions for homoclinic chaos associated with the double-loop orbits.

An additional piece of information can be extracted by considering the contracting or expanding nature of the right hand side $\mathbf{F}$ of the dynamical system (7.42) in the vicinity of the principal homoclinic orbits obtained so far. From Liouville's theorem, expansion or contraction is locally observed in the phase space depending on the sign of $\nabla_U \cdot \mathbf{F}$, which is proportional to the viscous dispersion parameter η and whose zeroes are given by (7.43). Therefore, when $\eta > 0$ and U_1 is smaller (respectively, larger) than $h_{\mathrm{div}0}(c)$, $\mathbf{F}$ is contracting (expanding). In Fig. 7.27, we compare the thickness limit $h_{\mathrm{div}0}(c)$ (7.43) corresponding to the speed c of the solitary waves with the maximum height $h_{\mathrm{m}} = \max(U_1)$ achieved by the homoclinic orbits. Above $U_1 = h_{\mathrm{div}0}$ the volume in the phase space is expanded by the dynamical system (i.e., $\nabla_U \cdot \mathbf{F} > 0$), whereas whenever $U_1 < h_{\mathrm{div}0}$ the volume is contracted. For $\eta = 0.1$ and a vertical wall, $\zeta = 0$, $h_{\mathrm{m}} < h_{\mathrm{div}0}$ when $\delta < \delta_{\mathrm{lim}} \approx 2.9$ and the volume in phase space is always contracted by $\mathbf{F}$ in the vicinity of the corresponding principal homoclinic orbits. However, when $\delta > \delta_{\mathrm{lim}}$, $h_{\mathrm{m}} > h_{\mathrm{div}0}$, the loop in phase space corresponding to the main hump of the solitary waves, crosses regions of the phase space where the volume is expanded by the dynamical system. The curves obtained for $\eta = 0.01$ and $\eta = 1$ (not shown) are qualitatively similar with a threshold δ_{lim}, above which the homoclinic orbits start to visit expanding regions in phase space, equal to 2.7 and 3.5, respectively. The same phenomenon is observed for a slightly inclined plane (cf. Fig. 7.27(b)) but the crossing of the curves is now observed for much larger values of the distance from threshold $\delta - \delta_{\mathrm{c}}$.

The occurrence of volume expansion in phase space in the vicinity of the homoclinic orbits for $\delta > \delta_{\mathrm{lim}}$ is questionable because of the viscous origin of a

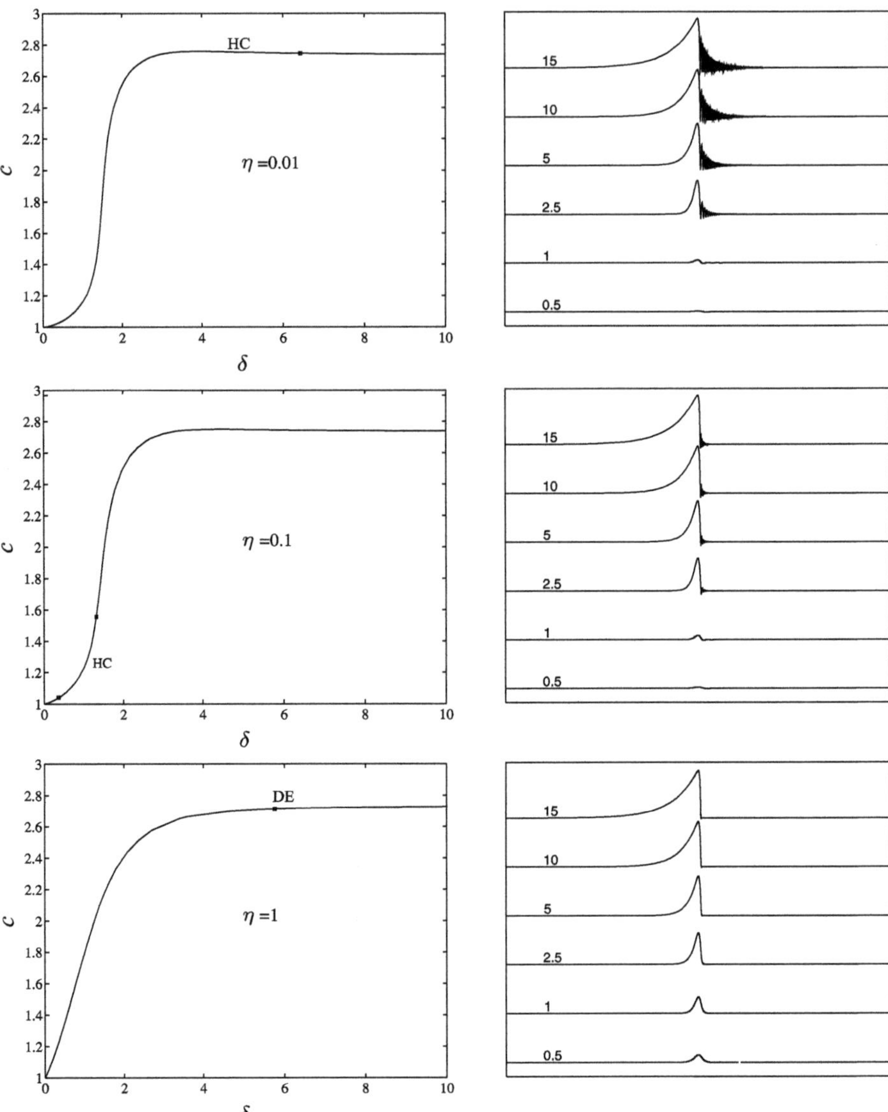

Fig. 7.25 *Left*: Speed of single-hump solitary waves (principal homoclinic orbits) as function of the distance δ from threshold $\delta_c = 0$. Homoclinic chaos is indicated with the label HC. The transition from a saddle-focus to a saddle is labeled DE. *Right*: Wave profile $h(\xi) = U_1$ corresponding to increasing values of δ (0.5, 1, 2.5, 5, 10 and 15). Other parameter values are $\zeta = 0$ (vertical wall) and *from the top panel to the bottom one* $\eta = 0.01$, $\eta = 0.1$ and $\eta = 1$

nonzero divergence of $\mathbf{F}$ (see also our earlier discussion where we showed that $\nabla_U \cdot \mathbf{F} = -3\eta \mathcal{J}(U_1, c)U_1^{-3}$). Hence, viscous dispersion is the ingredient that makes our dynamical system "nonconservative," a property that is followed by a contrac-

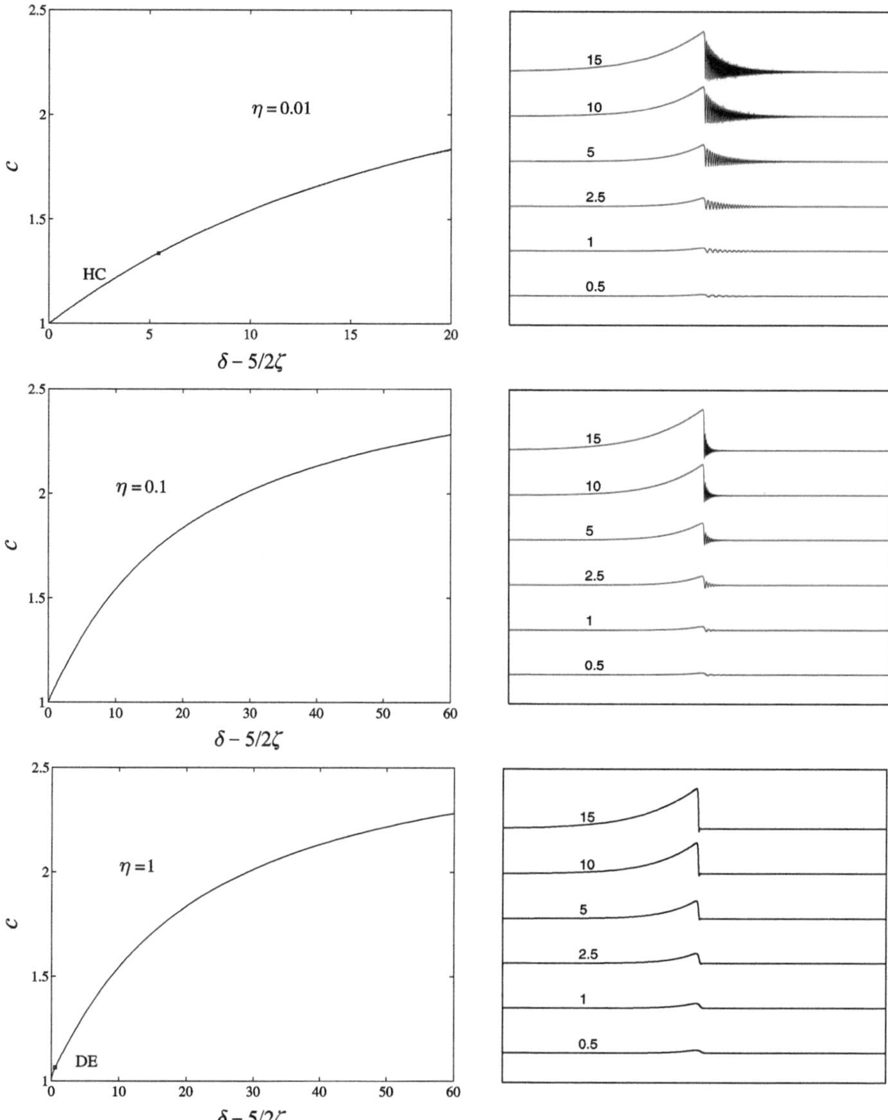

Fig. 7.26 See caption of Fig. 7.25. Parameter values are identical except for $\zeta = 10$, so that the threshold of instability is given by $\delta_c = \frac{5}{2}\zeta = 25$

tion of volume in phase space. (Volume conservation in the phase space is not related to the possibility of writing the original dynamical system in conservative form— given in Sect. 5.1.1—e.g., the first-order model does conserve phase volume but cannot be written in conservative form.) However, we note that volume expansion in phase space in the vicinity of the maximum amplitude is not problematic and has

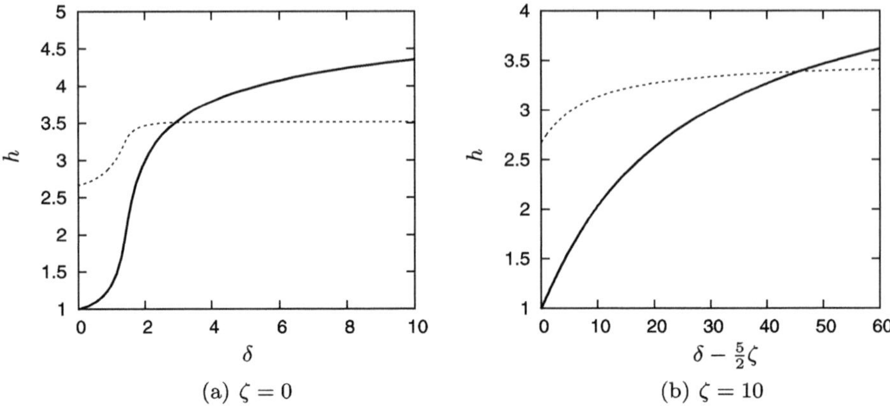

$$\text{(a) } \zeta = 0 \qquad\qquad\qquad\qquad\qquad \text{(b) } \zeta = 10$$

Fig. 7.27 Maximum height $h_{\mathrm{m}} = \max(U_1)$ as function of the distance from threshold $\delta - \delta_{\mathrm{c}}$, for the principal homoclinic orbits (*solid lines*) and limit height h_{div0}, given by (7.43), above which the volume in phase space starts to expand in the vicinity of the orbits (*dashed lines*)

no consequences on the observed dynamics in the phase space. On the other hand, volume contraction in the vicinity of the fixed points—which is always observed in the presence of viscous dispersion ($\eta \neq 0$) and for all orbits of interest—can dramatically modify the dynamics in the phase space (consider, e.g., the bifurcation diagrams in Fig. 7.23).

Further, recall a basic underlying assumption of the derivation process of low-dimensional models is weak inertia effects. Volume expansion is observed only for large solitary waves with a maximum thickness that is at least 3.5 times the thickness of the uniform film at infinity. For such large waves inertia effects become dominant. However, we are unaware of any experimental data that show the existence of solitary waves with such large amplitude. In fact, recent experimental studies for a film on an inclined plane [273] report waves for which h_{m} does not exceed ≈ 1.8. Therefore, the occurrence of volume expansion for the dynamical system (7.42) does not necessarily inhibit its ability to account for the dynamics of film flows, as we further demonstrate in this chapter.

7.2.1.6 Subsidiary Homoclinic Orbits: Bound States and Multi-hump Solitary Waves

Up to now we have focused on single-hump solitary waves corresponding to principal homoclinic orbits of the dynamical system (7.42). We are now ready to consider the subsidiary homoclinic orbit solutions to (7.42). We have already seen that some of them can be found by looking for homoclinic bifurcations of subsidiary limit cycles emerging from the primary branch through period doublings (cf. Figs. 7.21 and 7.22).

By comparing the subsidiary orbits in Fig. 7.22 with the principal homoclinic orbit represented by a thick solid line in the upper right plot of Fig. 7.20, it is clear that the extra loops of the orbits in Fig. 7.22 do not pass close to the fixed point $\mathbf{U}_I$. The corresponding waves are multi-hump. When the extra loops pass close enough to $\mathbf{U}_I$, the corresponding waves are trains of solitary waves (multi-pulse waves). The subsidiary homoclinic orbits in the latter case can be found numerically using Lin's method as described in [195] and implemented in the AUTO-07P software. Examples of switching from principal to subsidiary branches of homoclinic orbits are detailed in AUTO-07P user's manual, Chap. 27 [79].

The general idea of Lin's method is to construct an n-pulse solitary wave through the addition of n "truncated" single-hump solitary waves, in fact traveling waves approaching homoclinicity in a periodic domain of period, say P. One therefore sets up a set of $(n-1)$ boundary value problems corresponding to the connections of the different waves with one another. This method introduces $(2n-2)$ parameters corresponding to the $(n-1)$ "gaps" separating the trajectories at the extremities of the subdomains and the $(n-1)$ lengths of the subdomains in ξ, all equal initially to P. A rigorous and detailed exposure of the method can be found in [195]. Using the AUTO-07P implementation of Lin's method, we have computed the branches of subsidiary homoclinic orbits emerging from the principal homoclinic orbit for $\delta = 1$, a vertical wall $\zeta = 0$ in the presence of viscous dispersion, $\eta = 0.1$, and in the absence of viscous dispersion, $\eta = 0$ when two or three pulses ($n = 2$ and 3) are concatenated.

Figure 7.28 shows the loci of the numerically constructed subsidiary branches in the (δ, c)-plane. When second-order viscous terms are neglected, the loci of the subsidiary orbits are relatively close to the primary homoclinic branch, indicating all solitary waves (single-pulse and multi-pulse) have similar speeds for the given parameters, despite the large variety of possible solutions. Conversely, when second-order viscous effects are accounted for, the subsidiary homoclinic orbits explore a wide interval corresponding to a significant variation of the speed of all solitary waves. Therefore, the second-order viscous effects influence not only the linear waves as demonstrated in Sect. 7.1.3, but also the nonlinear waves associated with subsidiary orbits, precisely because they influence the tails through which nearly homoclinic loops are connected. However, recall that speed and amplitude of the principal homoclinic orbits are hardly affected by viscous dispersion.

Some of the wave profiles obtained for $\delta = 1.5$, $\zeta = 0$ and $\eta = 0$ are presented in Fig. 7.29. As the Shil'nikov criterion is satisfied, we expect a countable infinite family of subsidiary homoclinic orbits whose corresponding wave speeds are in a neighborhood of the speed $c^\star$ associated with the principal homoclinic orbit. Some of these subsidiary orbits are examples of *bound states*, in which several solitary waves are locked into constant relative positions in a moving frame (a "bound state" is a generic term in physics used to denote a composite of two or more particles or bodies that behave as a single body). Such orbits are shown, e.g., in Figs. 7.29(a)–7.29(c); they belong to the same solution branch.

The bound states in these figures are made of two single-hump solitary waves propagating steadily with a speed less than that of an infinite-domain solitary wave and separated by a constant distance as they propagate. Notice that the separation distance between the two pulses in Figs. 7.29(a)–7.29(c) increases as $|c - c^*|$ decreases, which is to be expected as each of the two waves starts approaching the

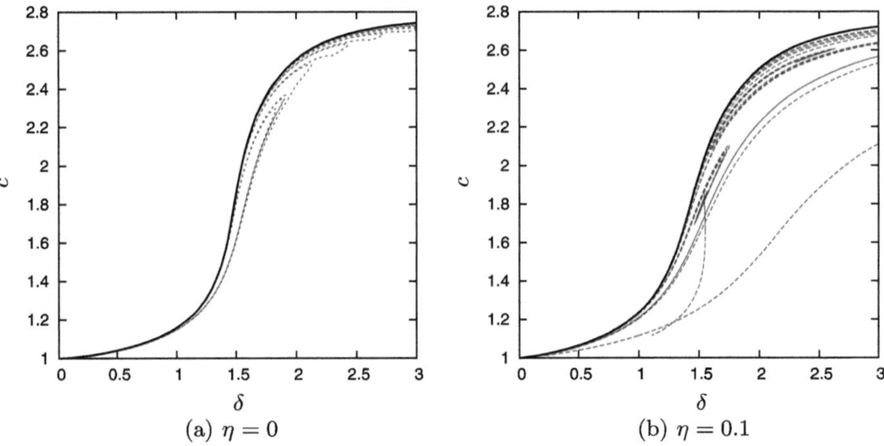

Fig. 7.28 Speed c of solitary waves as a function of the reduced Reynolds number δ. *Dashed lines* refer to the loci of some branches of subsidiary homoclinic orbits corresponding to multi-pulse waves. Primary homoclinic orbits are indicated by *solid lines*. The plane is vertical ($\zeta = 0$). *Left*: $\eta = 0$ and *right*: $\eta = 0.1$

infinite-domain solitary wave. Of course, Shil'nikov's criterion predicts a countable infinite number of multi-pulse waves, only three of which are shown in the figure, but numerically they are difficult to construct, e.g., a fourth member of the family of these waves would correspond to a $|c - c^*| \lesssim 10^{-8}$ which is beyond the precision of AUTO-07P. It is also possible to obtain bound states consisting of three single-hump pulses (with a separation distance roughly the same as that between the two pulses in a two-pulse bound state) or several single-hump waves so that the whole of the computational domain is occupied by these waves ("periodic wavetrains").

The computations reveal a wide variety of wave shapes. In most cases, the homoclinic orbit performs one or several loops in the phase space, each repelled from the vicinity of the fixed point $\mathbf{U}_{\mathrm{I}}$ after several oscillations corresponding to radiation waves (Fig. 7.29(a)–7.29(h)). But, interestingly, in some cases, the homoclinic orbits approach the vicinity of the second fixed point $\mathbf{U}_{\mathrm{II}}$ (cf. Figs. 7.29(g) and 7.29(h)). Figures 7.29(i) and 7.29(j) in particular are examples of three-hump waves found relatively far from $c^\star$. Recall that the Shil'nikov criterion applies not only for a primary homoclinic orbit about a fixed point but also for a subsidiary homoclinic orbit about a fixed point of the type in which the extra loops do not pass close to the fixed point. Therefore, we expect a countable infinite number of subsidiary homoclinic orbits associated with Figs. 7.29(i) and 7.29(j). The corresponding waves are trains of three-hump solitary waves.

Figure 7.29 demonstrates clearly the two types of subsidiary homoclinic orbits that we have introduced: those whose extra loops are repelled from the vicinity of the fixed point $\mathbf{U}_{\mathrm{I}}$ (cf. Fig. 7.29(a)) and those whose extra loops do not pass close enough to $\mathbf{U}_{\mathrm{I}}$ (cf. Figs. 7.29(i), 7.29(j)).

We close this section with comments on the study by Glendinning and Sparrow [103]. These authors introduce a clear distinction between the two types of

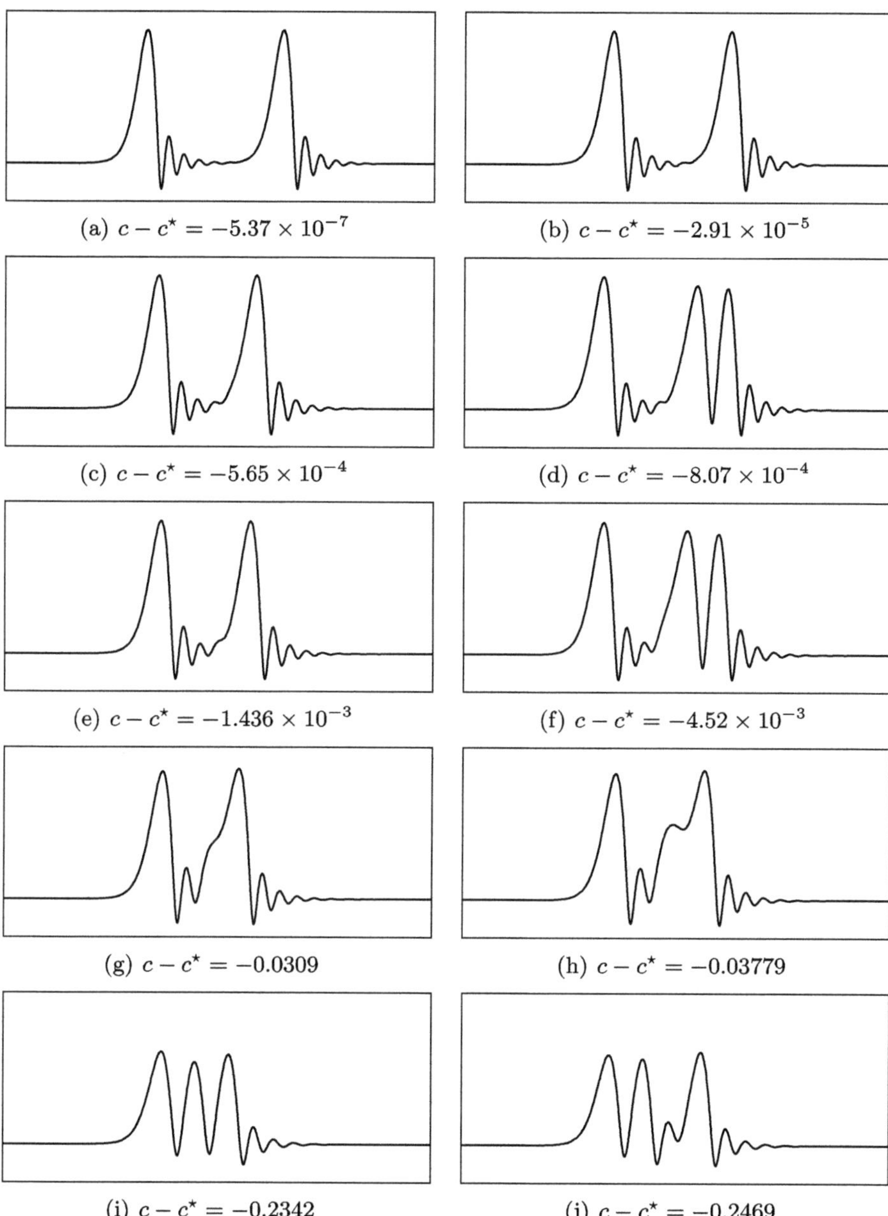

(a) $c - c^\star = -5.37 \times 10^{-7}$

(b) $c - c^\star = -2.91 \times 10^{-5}$

(c) $c - c^\star = -5.65 \times 10^{-4}$

(d) $c - c^\star = -8.07 \times 10^{-4}$

(e) $c - c^\star = -1.436 \times 10^{-3}$

(f) $c - c^\star = -4.52 \times 10^{-3}$

(g) $c - c^\star = -0.0309$

(h) $c - c^\star = -0.03779$

(i) $c - c^\star = -0.2342$

(j) $c - c^\star = -0.2469$

Fig. 7.29 Wave profiles of some subsidiary homoclinic orbits. Parameters are $\delta = 1.5$, $\zeta = 0$ and $\eta = 0$. $c^\star$ refers to the speed of the primary homoclinic orbit (infinite-domain single-hump solitary wave)

orbits by adopting the term "subsidiary" homoclinic orbits for the former and "secondary" for the latter. Nevertheless, we have found that such subsidiary and secondary orbits may belong to the same family of solutions. For example, one can move continuously from Fig. 7.29(c) to a two-hump wave (not shown), by following two parameters, i.e., by changing both c and δ (a codim 2 phenomenon). On the other hand, the subsidiary orbit shown in Fig. 7.29(e) belongs to the same branch of solutions as the secondary orbit displayed in Fig. 7.29(f), but of course the orbit (f) is not a straightforward secondary one as it consists of a single-hump wave together with a two-hump one ((g) and (h) also belong to the same branch; (d), (i) and (j) belong to branches which are not connected in the (δ,c)-plane).

Of course in general there is no guarantee that subsidiary orbits can be continued to secondary ones, i.e., that the two families of solutions are connected in the parameter space (if that were the case, the existence of a subsidiary orbit, i.e., if the Shil'nikov criterion were satisfied, would ensure the existence of a secondary one as well). But in our case we refrain from making a clear distinction between subsidiary and secondary orbits, precisely because the two families of solutions are connected with one another through parameter continuation. It is for this reason, that all multi-loop orbits are referred to as subsidiary in this monograph.

Finally, it is worth noting that Glendinning and Sparrow [103] also introduce a clear distinction between subsidiary and secondary limit cycles. In our case, subsidiary ones consist of two loops around $\mathbf{U_I}$, while secondary ones of two loops around $\mathbf{U_{II}}$. When these limit cycles grow and collide with $\mathbf{U_I}$ we have the formation of double-loop subsidiary and secondary homoclinic orbits, respectively. In Sect. 7.2.1.3 we focused on the secondary limit cycles only. From our discussion there it is clear that secondary homoclinic orbits originate from the homoclinic bifurcation of the secondary limit cycle branches that come up through period-doubling bifurcations of the primary limit cycle branch (on the other hand, recall that primary homoclinic orbits arise from the homoclinic bifurcation of the primary limit cycle branch that appears through a Hopf bifurcation). But again, much like with homoclinic orbits, in our case we can go from one type of limit cycle to the other, and hence the distinction between subsidiary and secondary limit cycles in the Glendinning and Sparrow [103] terminology is not clear. As a consequence we also adopt the term subsidiary limit cycles for all multi-loop limit cycles.

7.2.2 Solitary Wave Characteristics for $\delta \ll 1$ and $\delta \gg 1$

We now focus on the behavior of single-hump solitary waves as the reduced Reynolds number δ approaches zero or infinity. For simplicity consider a vertical wall ($\zeta = 0$) and without loss of generality assume negligible dispersion ($\eta = 0$). Even though strictly speaking δ must be at most of $\mathcal{O}(1)$, the question of an asymptotic behavior for the solitary waves for large δ is still a valid one within the context of the boundary layer equations as model equations. This is similar to what we have done with η: again, even though, strictly speaking, it is a small parameter, at times we have taken it as large as $\mathcal{O}(1)$. Besides, Fig. 7.25 indicates that δ does not really have to be very large in order for the solitary waves to reach their asymptotic state:

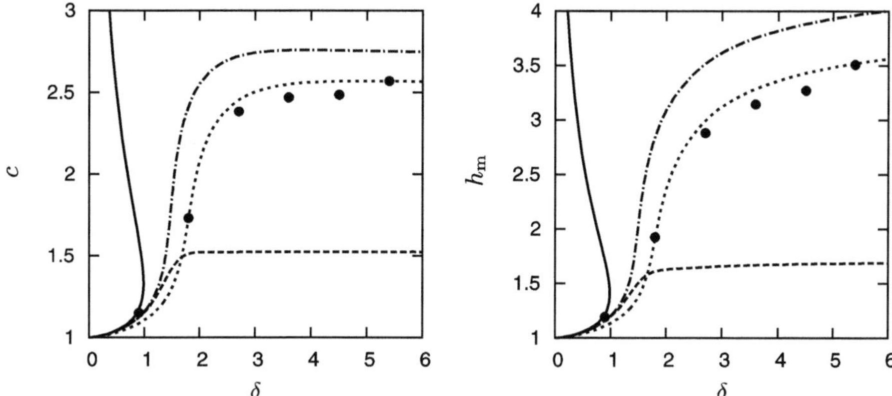

Fig. 7.30 Speed c (*left*) and maximum height h_m (*right*) as functions of the reduced Reynolds number δ for the single-hump solitary wave solutions of the different models: *solid and dashed lines* correspond to the BE and the Ooshida equation, respectively. *Dotted and dash-dotted lines* correspond to the Kapitza–Shkadov (6.13a), (6.13b) and to the first-order models (6.1), (6.51), respectively. *Solid circles* correspond to the first-order boundary layer equations (6.2)–(6.6) after [46]

$\delta \simeq 3$ is large enough to be in the asymptotic regime. The results given here can be easily generalized to include viscous dispersion ($\eta \neq 0$).

Whenever possible we include the more general situation of a fixed Froude number ($Fr^2 = 3Re/Ct = \delta/\zeta$, see Sect. 7.1.3). This is relevant for the study of the analogy of solitary waves with roll waves in the torrential regime of river flows for which we took the distinguished limit $Re \to \infty$, $Ct \to \infty$ and $Fr = \mathcal{O}(1)$. The speed c and maximum height h_m of the single-hump solitary waves (the principal homoclinic orbits) of the dynamical system (7.42) corresponding to the BE (5.55), the Ooshida equation (5.62) and two-equation models (6.13a), (6.13b) and (6.1), (6.51) are displayed in Fig. 7.30 as functions of δ. We have encountered similar diagrams comparing the speeds and amplitudes of solitary wave solutions of different models as function of δ in previous chapters: Figs. 4.8, 5.10, 6.2 and 6.3.

Figure 7.30 shows that all models are in reasonable agreement in the small Reynolds number limit, $\delta \ll 1$ (which underlines the slaving of the velocity field to the free-surface evolution—see Chaps. 5 and 6), except for the Kapitza–Shkadov model (6.13a), (6.13b) whose gradient expansion does not lead to the correct result. Yet, for a vertical wall ($\zeta = 0$), the prediction of the instability threshold corresponding to the Kapitza–Shkadov model, $\delta_\mathrm{c} = 3\zeta = 0$, agrees with the correct answer, $\delta_\mathrm{c} = \frac{5}{2}\zeta = 0$, from Orr–Sommerfeld analysis, which explains that all curves in Fig. 7.30 converge to the point $(\delta, c) = (0, 1)$.

The turning point of the BE solution branch signals a loss of solutions for δ greater than $\delta^\star \approx 0.986$, a value that closely corresponds to the occurrence of blow-ups of unsteady solutions for the BE examined in detail in Chap. 5. By contrast, the Ooshida equation (5.62) and models (6.13a), (6.13b) and (6.1), (6.51) possess single-hump solitary wave solutions for all values of δ. The same is true of the numerical results by Chang et al. [46] for the first-order boundary layer equations (6.2)–(6.6) (recall, however, that the accuracy of the numerical results of the Chang et al. study can be questioned). With the exception of the BE, for all other equations the speed of solitary waves saturates rapidly after $\delta \simeq 1.5$–2. But the outcome of the Ooshida regularized equation is clearly the least satisfactory in comparison with the more reliable results from the two-equation models and the first-order boundary layer equations. As a matter of fact, the Ooshida equation predicts saturated speeds and amplitudes that are by a factor of more than 2 smaller than those obtained from models (6.13a), (6.13b) and (6.1), (6.51) and the first-order boundary layer equations (6.2)–(6.6). Moreover, the amplitudes of the two-equation models and the first-order boundary layer equations continue to increase beyond $\delta = 2$. This trend is not reproduced by Ooshida's solutions, suggesting that the shape of the waves is also not properly approximated by (5.62) at large δ.

Figure 7.25 also shows the two different behaviors in the drag-gravity and drag-inertia regime of the speed and amplitude of the waves as a function of the reduced Reynolds number δ. The speed and amplitude increase slowly first, followed by a rapid increase at $\delta \simeq 1$, and then they reach an asymptote. This transition at $\delta \simeq 1$ demarcates the two different regimes—the drag-gravity and drag-inertia regimes. It is then pertinent to analyze the characteristics of the solitary waves in the two regimes, i.e., to consider the limits $\delta \ll 1$ and $\delta \gg 1$.

7.2.2.1 Drag-Gravity Regime, $\delta \ll 1$

For small reduced Reynolds number, the amplitudes of the waves are small which allows for a weakly nonlinear expansion by imposing $h = 1 + \alpha H$ where $\alpha \ll 1$. The characteristic equation (7.49) gives $\lambda = 0$ at $\delta = 0$ and $c = 1$. Therefore, homoclinic orbits connecting $\mathbf{U_I}$ to itself have lengths with tails $\sim 1/|\Re(\lambda_i)|$ going to infinity in the limit $\delta \to 0$, which justifies the scaling, $\xi = \beta^{-1}X$ with $\beta \ll 1$. After substitution in (7.37), one gets to leading order in β:

$$\frac{1}{3}\beta^3 \frac{d^3}{dX^3}H + \beta\delta\left[\mathcal{N}(1,1) - \frac{1}{3Fr^2}\right]\frac{d}{dX}H + (1-c)H + \alpha H^2 = 0. \qquad (7.60)$$

To have all terms in (7.60) of the same order, $\delta \sim \beta^2$, $\alpha \sim \beta^3 \sim c - 1$. By setting $\frac{1}{3}\beta^3 = 2\alpha = \beta\delta[\mathcal{N}(1,1) - (3Fr^2)^{-1}]$, (7.60) reads

$$\frac{d^3}{dX^3}H + \frac{d}{dX}H + \frac{1}{2}H^2 - \mu H = 0, \qquad (7.61)$$

where $\mu = 3(c - 1)/[\delta(3\mathcal{N}(1,1) - Fr^{-2})]^{3/2}$. Hence, in the limit $\delta \to 0$, branches of homoclinic orbits correspond to particular values of the relative speed μ and of the relative maximum amplitude, say μ_n and H_n. Further, $\beta \sim \sqrt{\delta}$, $\alpha H_n \sim \delta^{3/2}$ and $c - 1 \propto \mu_n \delta^{3/2}$. Consequently, as δ tends to zero, the speed c and maximum amplitude h_m of the solitary waves corresponding to the homoclinic orbits satisfy

$$h_\mathrm{m} - 1 \sim c - 1 \sim \delta^{3/2}, \qquad (7.62)$$

which agrees with the results for the case of the BE (5.55) for a vertical wall
($Fr \to \infty$) [216] (see also Fig. 5.10 in Chap. 5).

Notice that (7.61) corresponds precisely to the traveling wave solutions of the KS
equation (5.25),

$$\partial_T H + H \partial_X H + \partial_{XX} H + \partial_{XXXX} H = 0 \tag{7.63}$$

(see also Sect. 4.7). Looking for stationary solutions of (7.63) in the moving frame
$X = X - CT$ gives after integration

$$\frac{d^3}{dX^3} H + \frac{d}{dX} H + \frac{1}{2} H^2 - CH = Q, \tag{7.64}$$

where Q is an integration constant. Setting Q to zero, which corresponds to demanding that $H = 0$ be a solution, and identifying C with μ, finally leads back to
(7.61).

7.2.2.2   Drag-Inertia Regime with $\delta \gg 1$

We now investigate the limit $\delta \gg 1$. The speeds and amplitudes of the single-hump
solitary wave solutions to the Ooshida equation (5.65) and to the two-equation models (6.13a), (6.13b) and (6.1), (6.51) saturate as δ increases. As seen in Fig. 7.31, the
large-δ trajectories have three different parts: two of them extending the linearized
dynamics around $\mathbf{U}_I$ to the weakly nonlinear regime; the third one, in-between, accounts for the strongly nonlinear region away from $\mathbf{U}_I$ where they bend back. To
proceed we will need only two assumptions supported by numerical evidence: (i)
smooth single-hump solitary waves (with a monotonic rear and an oscillatory front),
i.e., corresponding to a surface elevation $h(\xi)$ where $h(\xi)$ and all its derivatives are
continuous, do exist in the limit $\delta \to \infty$; and (ii) their speeds are larger than unity.

The linearized dynamics around $\mathbf{U}_I$ are controlled by (7.49) so that $\sum_i \lambda_i = 0$.
By setting $\lambda_1 = 2\sigma$ and $\lambda_2, \lambda_3 = -\sigma \pm i\omega$, we get

$$\omega^2 - 3\sigma^2 = \delta\left[3\mathcal{N}(1, c) - Fr^{-2}\right] \quad \text{and} \quad 2\sigma\left(\sigma^2 + \omega^2\right) = 3(c - 1),$$

since $Fr^2 = \delta/\zeta$. Assuming $c > 1$ and $Fr > Fr_c$, and thus $3\mathcal{N}(1, c) - Fr^{-2} > 0$, we
have

$$\omega \sim \delta^{1/2}\sqrt{3\mathcal{N}(1, c) - Fr^{-2}} \quad \text{and} \quad \sigma \sim \frac{3}{2}\delta^{-1}(c - 1)/\left[3\mathcal{N}(1, c) - Fr^{-2}\right].$$

The above estimates agree with the numerical result shown in Fig. 4.5. Accordingly,
the escape from $\mathbf{U}_I$ along the one-dimensional unstable manifold $\mathcal{W}^u$ is slow and
monotonic while the convergence toward $\mathbf{U}_I$ along its two-dimensional stable manifold $\mathcal{W}^s$ should lead to a slow relaxation of fast oscillations, in agreement with what
is observed.

When the trajectory has left the immediate vicinity of $\mathbf{U}_I$, one must return to the
nonlinear equation (7.41a) with $\mathcal{H}(h, c)$ given by (7.41b) (recall that second-order

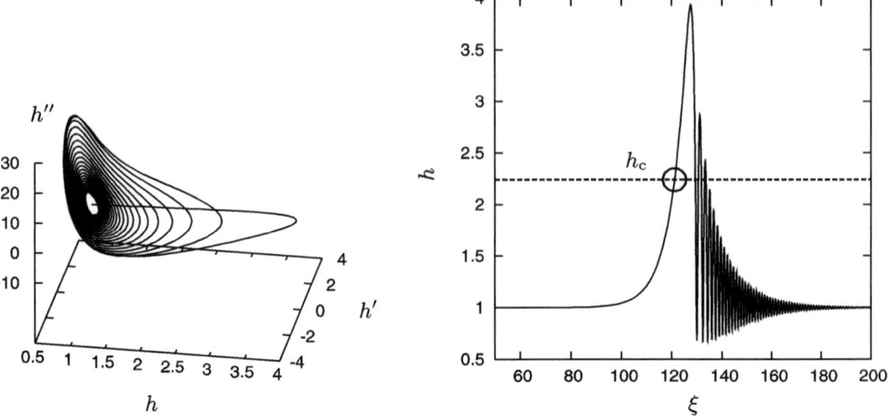

Fig. 7.31 Principal homoclinic orbit (single-hump solitary waves) solutions to model equations (6.1), (6.51) at large δ, $\delta = 5$. *Left*: Trajectory in the phase space. The spiraling behavior toward the fixed point $\mathbf{U_I}$ has been truncated in order to illustrate the escape of the trajectory from the vicinity of $\mathbf{U_I}$ along the unstable manifold $\mathcal{W}^{\mathrm{u}}$. *Right*: Profile of the wave $h = h(\xi)$ in the frame moving at speed c. The critical level h_{c} is indicated by a *dotted line*

corrections $\mathcal{I}$ and $\mathcal{J}$ are set to zero). The homoclinic trajectory starts along the one-dimensional unstable manifold of $\mathbf{U_I}$, whose tangent eigenspace corresponds to the eigenvalue $2\sigma \sim \delta^{-1} \ll 1$. The initial escape from the fixed point of the trajectory $h - 1 \propto \exp(2\sigma\xi)$ is therefore slow, which suggests the introduction of the slow variable $\tilde{\xi} = \xi/\delta$. Equation (7.41a) then reads

$$\left[\mathcal{N}(h, c) - \frac{h^3}{3Fr^2}\right]h' = -\mathcal{H}(h, c) - \frac{1}{3}\delta^{-3}h^3 h''', \tag{7.65}$$

where prime now denotes differentiation with respect to $\tilde{\xi}$. The last term in (7.65) is negligible all along the first part of the trajectory corresponding to ξ (or $\tilde{\xi}$) coming from $-\infty$, i.e., at the rear of the wave, so that (7.65) reduces to

$$\left[\mathcal{N}(h, c) - \frac{h^3}{3Fr^2}\right]h' = -\mathcal{H}(h, c). \tag{7.66}$$

For *Fr* larger than its critical value Fr_{c} corresponding to the onset of the primary instability, $\mathcal{N}(1, c) - 1/(3Fr^2) > 0$, and, at a given c, $\mathcal{N}(h, c) - h^3/(3Fr^2)$ decreases as h increases, which is readily seen from the expressions given earlier in the three cases of interest. This follows immediately for the Kapitza–Shkadov model and for model (6.1), (6.51) since $c > 1$ is assumed. For the Ooshida equation the decrease only occurs for $h^2 > \frac{25}{27}c$ but this does not change the argument. On the other hand, h increases with $\tilde{\xi}$ as long as $h' > 0$. Since $\mathcal{H}(h, c) < 0$ for $1 < h < h_{\mathrm{II}}$, where h_{II} is given by (7.46), h' is positive as long as h is below h_{II} and a "critical level" h_{c}

defined as the root for h of

$$\mathcal{N}(h_\mathrm{c}, c) - \frac{h_\mathrm{c}^3}{3Fr^2} = 0. \tag{7.67}$$

If, for the considered value of c, $h_\mathrm{II} < h_\mathrm{c}$, h' does not diverge and h generally goes through a maximum so that it cannot reach h_c, at least in the rear part of the trajectory, which contradicts the assumption of single-hump solitary waves. On the other hand, if h_c is reached first, then a singularity takes place with h' diverging at $\tilde{\xi} = \tilde{\xi}_\mathrm{c}$, which now contradicts the assumption of nonbreaking solitary waves (continuous free surface elevation $h(\xi)$ and continuity of its derivatives) originating from empirical evidence. The only possibility then to remove the divergence of h' is when $h_\mathrm{II} = h_\mathrm{c}$, in which case $\mathcal{N}(h, c) - h^3/(3Fr^2)$ and $\mathcal{H}$ are both zero for the same value of h. Solving (7.67) with $h_\mathrm{c} = h_\mathrm{II}$ given by (7.46) yields the asymptotic values c_∞ reached by c in the limit $\delta \to \infty$. For a vertical wall ($Fr \to \infty$), the values obtained for (5.65), (6.13a), (6.13b) and (6.1), (6.51) are [229]

Ooshida equation (5.65): $\qquad c_\infty = \dfrac{9}{841}\left(83 + 5\sqrt{141}\right) \approx 1.524$

Kapitza–Shkadov

model (6.13a), (6.13b): $\qquad c_\infty = 1 + 1/\sqrt{6} + \sqrt{1/2 + \sqrt{2/3}} \approx 2.556$

First-order model (6.1), (6.51): $\quad c_\infty = \dfrac{1}{6}\left(9 + \sqrt{43 + 2\sqrt{37}}\right) \approx 2.738.$

The values are in good agreement with the value obtained from computations of the corresponding equations (Fig. 7.30, left). For the BE (5.55), (7.67) admits no nonzero solutions, which explains the lack of solitary wave solutions at large Reynolds numbers.

For δ large but finite, the singularity in (7.65) remains at $h = h_\mathrm{c}$ defined by (7.67). This singularity will be avoided again if the right hand side is zero when $h = h_\mathrm{c}$. In the region $h \sim h_\infty$ where $h_\infty = h_\mathrm{II}(\delta \to \infty)$ as determined above, the shape of the solution has no reason to change rapidly as δ increases since $\mathcal{N} - h^3/(3Fr^2) \neq 0$ at $h = h_\mathrm{II}(\delta \to \infty)$. Therefore, one can generally expect $h''' \sim h'''_\infty$, where $h'''_\infty \neq 0$ is the asymptotic value of the third derivative of h in $\tilde{\xi}$ (the slow variable). We thus have

$$\mathcal{N}\left(h_\mathrm{c}(\delta), c(\delta)\right) - \frac{[h_\mathrm{c}(\delta)]^3}{3Fr^2} = K\delta^{-3}, \tag{7.68}$$

where K is a numerical constant depending on h_∞ and h'''_∞. The solution to (7.67) and (7.68) through their expansion around (h_∞, c_∞) yields

$$c - c_\infty \propto \delta^{-3}, \tag{7.69}$$

a convergence rate complemented by direct numerical integration of the different models as shown in Fig. 7.32.

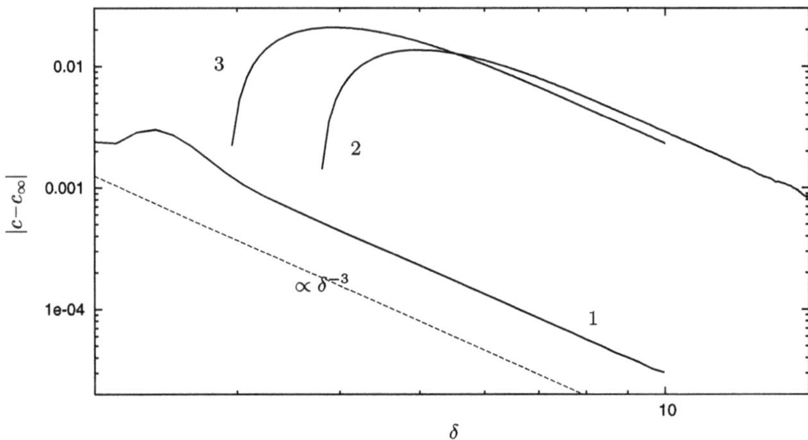

Fig. 7.32 Convergence of the speeds of solitary waves toward their asymptotic values c_∞ as function of δ and for a vertical wall ($Fr \to \infty$). 1: Ooshida equation (5.65). 2: Kapitza–Shkadov model (6.13a), (6.13b). 3: Model (6.1), (6.51)

While the above analysis explains the asymptotic behavior of the speed of single-hump solitary waves, their existence has not been justified and was taken for granted. This existence property is likely to be more difficult to prove than in the small-δ limit, where analysis of the vicinity to a codim 2 bifurcation point corresponding to the simultaneous onset of a stationary and an oscillatory instability ($\lambda_1 = 0$, λ_2, $\lambda_3 = \pm i\omega$) enables one to show the existence of homoclinic orbits [101].

It should be noted that the critical value h_c introduced in the above derivation, makes sense only on the slow rear part of the wave, i.e., for the value of $\tilde{\xi}$ where the critical condition is achieved for the first time when $\tilde{\xi}$ increases from $-\infty$, since $h = h_c$ also happens at least once in the fast oscillating front part when h decreases from its maximum value, $h_m > h_c$. However, the dominant term (7.41a) is then the surface tension term $\frac{1}{3}h^3 h'''$ and to deal with it one has to follow the dynamics in terms of a fast variable, say $\hat{\xi} = \xi\sqrt{\delta}$. The presence of the surface tension term prevents the divergence of h' when $h \sim h_c$. The third derivative does not diverge as long as h is not close to zero.

7.2.2.3 Analogy to Roll Waves in Open Channels

The distinguished limit $\delta \to \infty$ with $Fr = \mathcal{O}(1)$ is equivalent to the limit $Re \to \infty$ and $Fr = \mathcal{O}(1)$ already discussed in the wave hierarchy context. There is again analogy between the features of solitary waves of film flows and the roll waves observed on steep water channels when the flow is "supercritical," i.e., when the Froude number is above the instability threshold of the uniform thickness solution (we recall from our discussion in Sect. 7.1.3 that the instability threshold $Re_c = \frac{5}{6}Ct$ corresponds to a fixed value of the Froude number $Fr_c = \sqrt{5/2}$).

This analogy follows from the presence of the Poiseuille solution to the full Navier–Stokes equations in an open channel with a fluid layer of arbitrary uniform thickness. The Poiseuille solution is identical to the Nusselt flat film solution at small thicknesses and links the film flow problem, for which surface tension is a dominant effect, i.e., for $We \gtrsim 1$, when the fluid layer thickness $\bar{h}_N$ is at most comparable to the capillary length $l_\sigma = (\sigma/\rho g \sin \beta)^{1/2}$, to the problem of a hydraulic flow in an inclined open channel, for which surface tension can be neglected (i.e., $We \ll 1$ or $\bar{h}_N \gg l_\sigma$). Indeed, to leading order in δ^{-1}, the escape of the homoclinic trajectory from the fixed point $\mathbf{U}_I$, hence the rear of the solitary wave, is governed by (7.66), where the effects of surface tension are negligible. Equation (7.66) is similar to the equation that governs traveling wave solutions to the Saint-Venant equations (7.26a), (7.26b)

$$\left[(u - c)^2 - \frac{h}{Fr^2} \right] h' = u^2 - \frac{4}{9}h, \tag{7.70}$$

where $u = q/h$ is the local mean velocity of the flow. Integration of the mass conservation equation (7.26a) gives (7.36) and thus, $u = c + q_0/h$. Notice that the coordinate ξ has been compressed by the transformation $\xi \to Re\xi$. Equation (7.70) can be recast as

$$\left[c_{d+}(u, h) - c \right]\left[c_{d-}(u, h) - c \right] h' = u^2 - \frac{4}{9}h, \tag{7.71}$$

where $c_{d\pm}(u, h) = u \pm \sqrt{h}/Fr$ are the speeds of the infinitesimal dynamic waves propagating on the uniform flow of depth h and velocity u.

Roll waves consist of *hydraulic jumps* connected by sections of gradually varying flows. In a roll wave flow, the crests of the waves are connected to their troughs by discontinuities, that is, hydraulic jumps or bores. The flow crosses a hydraulic jump from the "supercritical region," where dynamic waves travel slower than the speed of the wave ($c_{d\pm} < c$), to the "subcritical region," where some dynamic waves can propagate upstream toward the hydraulic jump ($c_{d+} > c$) (see Fig. 7.33). One may therefore define a reduced Froude number Fr_r as the ratio of the speed of the flow relative to the roll wave and the speed of surface gravitational waves $\sqrt{g\bar{h}\cos\beta}$:

$$Fr_r = \frac{|u - c|}{\sqrt{h}/Fr} = Fr\frac{c - u}{\sqrt{h}},$$

since the wave moves faster than the flow ($u - c < 0$). The supercritical region at the front of the moving hydraulic jump therefore corresponds to $Fr_r > 1$ and, conversely, the subcritical region at the back corresponds to $Fr_r < 1$. In the sections of slowly varying flows separating the hydraulic jumps, the fluid must return from subcritical to supercritical regions. As a consequence, between two hydraulic jumps there must be at least one critical point ($Fr_r = 1$). Since u is a function of h, this defines a critical level h_c. As the flow is continuous between the hydraulic jumps, h' is finite so that at $h = h_c$ the left hand side of (7.71) vanishes. This demands that the right hand side of (7.71) vanishes as well [270]. Therefore, at the critical level

Fig. 7.33 Sketches of roll waves propagating down a steep water channel, i.e., sufficiently steep for the flow to be supercritical, $Fr > Fr_c$. Directions of propagation of the dynamic waves with respect to the roll waves are indicated by *arrows*. In the frame moving at the speed of the roll waves, the flow is moving upstream from the supercritical to subcritical regions ($u - c < 0$) (after [33, 172])

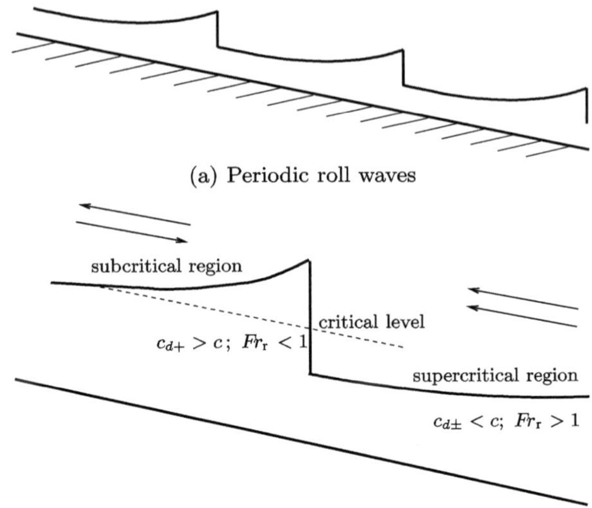

(b) Sketch of a roll wave moving at speed c. The sections of gradually varying flows are connected by a discontinuity (hydraulic jump or bore)

where dynamic waves travel at the speed of the roll wave, the wall friction must compensate exactly the gravitational acceleration.

For a film flow, this is precisely what condition $h_{\mathrm{II}} = h_c$ stipulates—by demanding that at the critical level the right hand side of (7.66), $-\mathcal{H}(h, c)$, must vanish. The dynamics of the flow at the rear of the wave are governed by (7.66), which can be recast in a form similar to (7.71) with a right hand side given by $3u/h - h$ and $c_{\pm d}(u, h) = \frac{6}{5}u \pm \sqrt{\frac{6}{25}u^2 + hFr^{-2}}$ in the case of the Kapitza–Shkadov model (6.13a), (6.13b), and $c_{\pm d}(u, h) = \frac{17}{14}u \pm \sqrt{\frac{37}{196}u^2 + \frac{5}{6}hFr^{-2}}$ in the case of the first-order model (6.1), (6.51) and the simplified model (6.1), (6.79). Similarly to roll waves, solitary waves traveling on film flows can be divided into two parts. For $h > h_c$, in the subcritical region, infinitesimal dynamic waves can travel faster than the solitary waves, whereas for $h < h_c$, in the supercritical region, infinitesimal dynamic waves are slower than the solitary wave.

One is therefore tempted to view solitary waves on film flows at large reduced Reynolds numbers δ as roll waves whose breaking is arrested by surface tension. Let us then compare the features of roll waves traveling on film flows when surface tension is negligible to the characteristics of the corresponding solitary waves. Infinitely long roll waves must verify condition $h_{\mathrm{II}} = h_c$ such that their speed is exactly c_∞. Their amplitude can be determined through the conservation of mass and momentum across the "shock." Considering the Ooshida equation, shock conditions can be obtained by splitting (5.62) into

$$\partial_t h + \partial_x q = 0$$

and

$$3q = h^3 - \delta\left[\frac{2}{7}\partial_t(h^5) + \frac{36}{245}\partial_x(h^7) - \frac{1}{4Fr^2}\partial_x(h^4)\right] + h^3\partial_{xxx}h.$$

By neglecting then surface tension and the thickness of the shock, the jump conditions across the discontinuity are obtained through the formal substitutions $\partial_x \to [\,]$ and $\partial_t \to -c_\infty[\,]$, where the brackets indicate a jump in the corresponding quantities [299]:

$$-c_\infty[h] + [q] = 0 \quad \text{and} \quad -c_\infty\frac{2}{7}[h^5] + \frac{36}{245}[h^7] - \frac{1}{4Fr^2}[h^4] = 0. \qquad (7.72)$$

Across the shock, the thickness of the film goes from its maximum, h_m, to its minimum, 1. For $c_\infty \approx 1.524$ one obtains $h_\mathrm{m} \approx 1.68$. Computations of the solitary wave solutions to (5.62) seem to indicate that the maximum of the thickness approaches an asymptotic value ≈ 1.70. Written in conservative form (see Sect. 5.1.1), the Kapitza–Shkadov equations read

$$\partial_t h + \partial_x q = 0 \quad \text{and} \quad \delta\left[\partial_t q + \partial_x\left(\frac{6}{5}\frac{q^2}{h} + \frac{1}{Fr^2}\frac{h^2}{2}\right)\right] = h - 3\frac{q}{h^2} + h\partial_{xxx}h, \qquad (7.73)$$

and the corresponding shock conditions are

$$-c_\infty[h] + [q] = 0 \quad \text{and} \quad -c_\infty[q] + \left[\frac{6}{5}\frac{q^2}{h} + \frac{1}{Fr^2}\frac{h^2}{2}\right] = 0. \qquad (7.74)$$

$c_\infty \approx 2.556$ and q is given by (7.36) and (7.39) so that one obtains from (7.74), $h_\mathrm{m} \approx 4.54$ while the maximum thickness of the solitary wave solution to (6.13a), (6.13b) $h_\mathrm{m} \approx 3.7$ at $\delta = 10$.

Of course replacing the radiation preceding the hump of a solitary wave with a shock is a crude assumption for at least two reasons. First, it does not give any information on the way the actual solution approaches its shocklike characteristics. From Sect. 7.2.2.2 we know that the convergence of the wave speed to its limit c_∞ depends on the balance of inertia and surface tension in the oscillatory region at the front of the main solitary hump. Second, the derivation of the jump conditions (7.74) is based on the formulation of the Kapitza–Shkadov equations in the form (7.73), a conservative form following Sect. 5.1.1: $\mathbf{H} = (h, q)$, the flux $\mathbf{Q}$ is a function of $\mathbf{H}$ and the right hand side $\mathbf{R}$ contains the influence of gravity acceleration, viscosity and surface tension terms, all of them being neglected across the shock. As a consequence, the estimates obtained with the Kapitza–Shkadov model for the amplitude of the waves cannot be extended to the first-order model (6.1), (6.51) and to the simplified second-order one (6.1), (6.79). (Recall that the first-order model does not have a conservative form; the simplified second-order model is essentially the first-order one with additional second-order terms due to dispersion, and hence it also does not have a conservative form.)

Nevertheless, the estimates of the wave amplitudes from the shock conditions (7.72) and (7.74) are quite reasonable. This supports the idea that solitary waves at large δ are in essence roll waves whose breaking is arrested by surface tension.

7.2.3 Closed Flow Conditions

We have already seen that computation of traveling wave solutions requires assigning a value to the rate q_0 at which the fluid flows under the wave in its moving frame. This was done by demanding that the constant thickness $h = 1$ be a solution. Observed solitary waves are positive-hump waves (see, e.g., Fig. 7.1) under which the fluid is locally accelerated so that the local flow rate q can be several times larger under the wave humps than in the portions of the flat film (also known as "substrate"; see Sect. 5.4) separating the solitary waves. Consider now the flow rate at a given location x. Since the time average $T^{-1} \int_0^T q \, dt$ of the flow rate over a sufficiently long time T is conserved (see Sect. 5.3.1) and corresponds to the Nusselt flat film at the inlet, i.e., $T^{-1} \int_0^T q \, dt = 1/3$, the flow rate in the substrate flat films separating the solitary waves is lower than the Nusselt flow rate at the inlet. Thus, the corresponding substrate thickness is also smaller than the inlet thickness of the film, i.e., unity (for a waveless flow, the film thickness equals the inlet thickness throughout).

The above observations imply that the relationship (7.39) between q_0 and the speed c of the waves cannot be used to account for experimental results. As already discussed in Sect. 5.3.1, where we defined the open and closed flow conditions, periodic-wave regimes observed in experiments with forcing at the inlet correspond to a constant temporal average flow rate $\langle q \rangle_t = \tau^{-1} \int_0^\tau q \, dt = 1/3$, with τ the period of the oscillation, or "open flow condition," provided of course that the flow synchronizes all along the plate with the inlet forcing. Yet, as was first pointed out in Sect. 5.3.1, several studies (e.g., [128, 198, 232, 239]) rely on periodic boundary conditions in space and hence implicitly prescribe a constant space average film thickness, $\langle h \rangle_x = 1$, the "closed flow condition." This is inherent to time-dependent numerical simulations that use a spectral method in which periodic boundary conditions are enforced. Thus, the amount of liquid leaving the domain downstream is reinjected upstream and the average film thickness is constant. A simulation of the actual film flow dynamics using periodic boundary conditions is still possible though, but requires sufficiently extended computational integration domains (an example of such simulations can be found in [218]).

Many studies have been devoted to obtaining a detailed picture of the different branches of traveling wave solutions of the Kapitza–Shkadov model (6.13a), (6.13b) or the first-order boundary layer equations (6.2)–(6.6) (hence $\eta = 0$) [50, 250, 253]. These studies have been restricted to the case of a vertical wall ($\zeta = 0$) and the closed flow condition $\langle h \rangle_\xi = 1$. The family of γ_1 waves bifurcating from the neutral stability curve was computed first by Shkadov [248] (the γ_1, γ_2 wave families were first introduced in Sect. 5.3.2 and they correspond to slow and fast waves, respectively). The slow waves have a speed smaller than that of spatially amplified infinitesimal waves at the same frequency. The corresponding solution branch terminates at small frequency as a (infinite-domain) negative-hump solitary wave with a deep trough and capillary ripples at the back. However, the experimentally observed waves excited by low frequency forcing are fast waves, i.e., trains of positive-hump solitary waves having speeds larger than that of infinitesimal waves (at large frequencies we have trains of negative-hump waves—we shall examine this

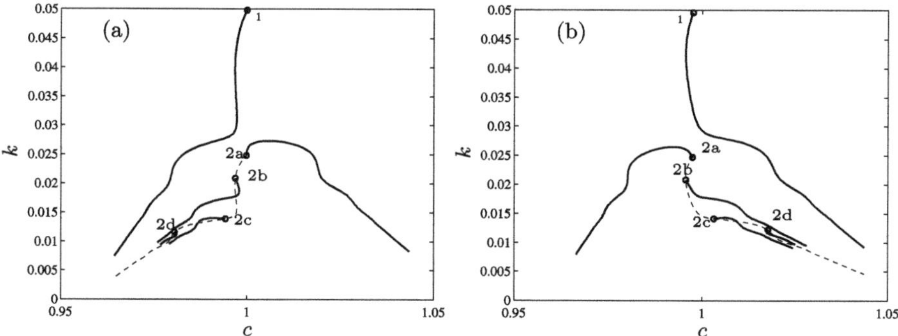

Fig. 7.34 Speed of periodic wavetrains (limit cycles) as functions of their wavenumber k for a vertical wall. Parameter values are $\delta = 0.844$, $\zeta = 0$ and $\eta = 0.0112$ ($Ct = 0$, $Re = 2.66$ and $\Gamma = 3375$). The closed flow condition $\langle h \rangle_\xi = 1$ is enforced. *Left*: First-order optimal model (6.1), (6.51); *Right*: Simplified second-order (6.1), (6.79). To simplify comparisons with Figs. 12 and 14 in [232], the wavenumber k is made dimensionless using the film thickness $\bar{h}_{\mathrm{N}}$. Branch 1 bifurcates from the flat film solution through a Hopf bifurcation at $k = k_{\mathrm{c}}$. Branches 2 bifurcate from the solutions made of two branch 1 waves (*dashed lines*) through period-doubling bifurcations indicated by *solid circles*

point in Sect. 7.3.1). This second family of waves γ_2 was first obtained by Bunov et al. [34], who showed that it emerges from the γ_1 branch through a period-doubling bifurcation. Due to the symmetry-breaking of a wave made of n identical γ_1 waves (i.e., the $\gamma_1^{(n)}$ waves introduced in Sect. 5.3.2), many more branches of solutions, denoted here as γ_1^n, γ_2^n, exist. γ_1^n, γ_2^n waves bifurcate from $\gamma_1^{(n)}$ and $\gamma_2^{(n)}$ through period-doubling bifurcations at low values of δ in pairs of slow and fast waves that terminate at low wavenumber as (infinite-domain) solitary waves having different numbers of troughs and humps [253]. Increasing δ, a series of pinchings of the γ_2 branch of solutions with the γ_1^n waves is observed, [50]. New branches of solutions originate from these pinchings. At large δ, all branches of solutions correspond to slow waves except for the single γ_2 family [250].

Several studies have also been devoted to the computation of traveling wave solutions from DNS [13, 116, 218, 232]. They are based on spectral/finite-element methods and have enforced the closed flow condition $\langle h \rangle_\xi = 1$. They provide reliable data to which the results of the low-dimensional models developed in Chap. 6 can be compared. Computations by Salamon et al. [232] have notably shown that drastically different bifurcation scenarios take place for constant δ and ζ but viscous dispersion effects vary, i.e., when η is modified. Bifurcation diagrams for the wavenumber versus velocity corresponding to the first-order model (6.1), (6.51) and the simplified second-order model (6.1), (6.79) are shown Fig. 7.34. In agreement with the DNS results in [232] (to be discussed later together with Fig. 7.36), the structure of the bifurcation diagram is drastically changed when viscous dispersion effects are taken into account, i.e., when we move from the first-order model (a) to the second-order one (b). Indeed, in the first case with no viscous dispersion, branch 1, which arises from a Hopf bifurcation at a wavenumber k equal to the cut-off wavenumber k_{c}, connects the primary solution to slow γ_1 waves ($c < 1$) whereas branch 2a, which bifurcates off branch 1 by a first period doubling, corresponds to fast γ_2 waves ($c > 1$). When viscous dispersion effects are taken into account (see Fig. 7.34b), the connections of the branches are reversed. The other curves, labeled 2b, 2c, 2d in Fig. 7.34(a, b), correspond to different branches of solutions arising by secondary period-doubling bifurcations off branch 1 (dashed lines indicate the loci of solutions made of two branch 1 waves) and approaching at small wavenumber multi-hump solitary waves.

The bifurcation diagrams formed by branch 1 and branch 2a in Fig. 7.34 are typical examples of an imperfect pitchfork bifurcation with k as the bifurcation parameter. Its origin close to the instability onset ($\delta - \delta_c \ll 1$) can be understood within the framework of weakly nonlinear analysis by considering traveling wave solutions to the KS equation (7.63) and its extension when dispersion is considered, hence using the Kawahara equation (5.31) [49, 73]. Indeed, the KS equation has been derived in Sect. 5.2 from the first- and second-order BE while the Kawahara equation has been derived from the second-order BE. The first- and second-order BE can be in turn obtained from a gradient expansion of the first-order and full second-order models (6.1), (6.51) and (6.78), respectively (see Sects. 6.6 and 6.8.1). Dispersion, characterized by the parameter δ_K in (5.31), originates from second-order viscous effects.[5] This origin can be made explicit by recasting the expression of δ_K in (5.31) in terms of the reduced parameters δ and η:

$$\delta_K = 3 \left[\frac{5}{2(\delta - \delta_c)} \right]^{1/2} \eta. \tag{7.75}$$

Traveling wave solutions to the ordinary differential equation (7.64) when the closed flow condition $\langle H \rangle_X = 0$ is enforced are displayed in Fig. 7.35. The wavenumber K is defined with respect to the variable X introduced in the derivation of the KS equation (7.63). Standing waves ($C = 0$) evolve from neutral wavenumbers $K = 1$ and $K = 1/2$, where two identical waves are put in the same computational domain. A period-doubling bifurcation occurs at $K = 0.497783$, from which another branch of standing waves emerges. These standing waves correspond to kinematic waves traveling at speed $c = 1$ since $C \propto c - 1$ is in fact a deviation speed. The pitchfork bifurcation occurs at $K = 0.5462$ leading to the two branches of slow γ_1 and fast γ_2 waves.

Since (7.64) is invariant under the transformation (see also Sect. 5.3.2),

$$H \to -H, \qquad X \to -X, \qquad C \to -C, \qquad Q \to Q, \tag{7.76}$$

the wave profiles shown in Fig. 7.35 are symmetric. Hence, for any γ_2 wave traveling to the right there exists a corresponding symmetric γ_1 wave traveling to the left. As emphasized in Sect. 5.3.2, these negative waves do not actually propagate backward. In fact, turning back to the laboratory frame, γ_1 waves effectively correspond to right-moving waves traveling slower than the kinematic waves.

---

[5]Contributions from inertia to the dispersive term $\delta_K \partial_{XXX} H$ are of higher order. To include inertia in the dispersive term, the Kawahara equation would have to be modified to include higher-order nonlinearities—see Sect. 5.2.

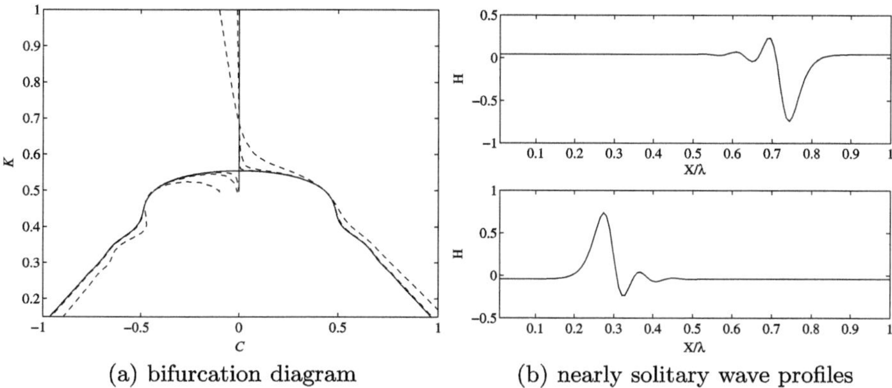

(a) bifurcation diagram   (b) nearly solitary wave profiles

Fig. 7.35 (**a**) Wavenumber K of primary traveling waves versus deviation speed C. Solutions to the KS equation (7.63) correspond to *solid lines*. Solutions to the Kawahara equation (5.31) are displayed for increasing dispersion $\delta_K = 0.001$, $\delta_K = 0.01$ and $\delta_K = 0.1$ as *dashed lines*. The branches of waves with $\delta_K < 0$ can be obtained by symmetry around the axis $c = 0$. (**b**) Profiles of periodic waves close to the solitary wave limit at $K = 0.1$ as function of the reduced coordinate X/λ where λ is the period of the waves

Noteworthy is that the invariance of the KS equation under the transformation (7.76) is broken when positive dispersion is included and the KS equation becomes the Kawahara equation, i.e., when the term $\delta_K d^2 H/dX^2$ with $\delta_K > 0$ is added to (7.64) as shown in Fig. 7.35 (see also Sect. 5.3.2). The original branch of standing waves separates into two branches, one being connected to the slow γ_1 waves, and another one being connected to the fast γ_2 waves. Notice also from Fig. 7.35 that the curves for $\delta_K < 0$ are simply mirror images of those with $\delta_K > 0$ with respect to the $c = 0$ axis—recall from Sect. 5.3.2 the reversible symmetry of the Kawahara equation. Of course, as was pointed out there, for both isothermal and heated falling film problems, $\delta_K > 0$.

It is then clear that the imperfection in Fig. 7.34(a) is due to the absence of symmetry, $h - 1 \to -(h - 1)$, $x \to -x$ and $c \to -c$, for the first-order models, much like the first-order BE (see Sect. 5.3.2 and Fig. 5.2). In Fig. 7.34(b), the symmetry-breaking is due to the same reason as that in Fig. 7.34(a), but the presence of dispersion changes the connectivity of the branches.

In Fig. 7.34(a), branch 1 of traveling waves bifurcating from $k = k_c$ is connected to γ_1 slow waves as $k \to 0$, whereas branch 2a, which emerges through a period-doubling bifurcation, terminates into γ_2 fast waves as $k \to 0$. The situation is reversed in Fig. 7.34(b) due to the effect of viscous dispersion. The exchange of connections between branches 1, 2a, hence γ_1 and γ_2, is still observable at a larger value of the reduced Reynolds number, $\delta = 2.79$, as the viscous dispersion parameter η is increased from 0.015 to 0.075, as demonstrated in Fig. 7.36. The phase velocity c_r of the temporally most amplified linear waves ($\omega \in \mathbb{C}$ and $k \in \mathbb{R}$) is here compared to the speed c of the traveling waves. Since the speed of the waves is now significantly different from unity (the reference speed of kinematic waves in the limit $k \to 0$), a rigorous demarcation between slow and fast waves is achieved by their

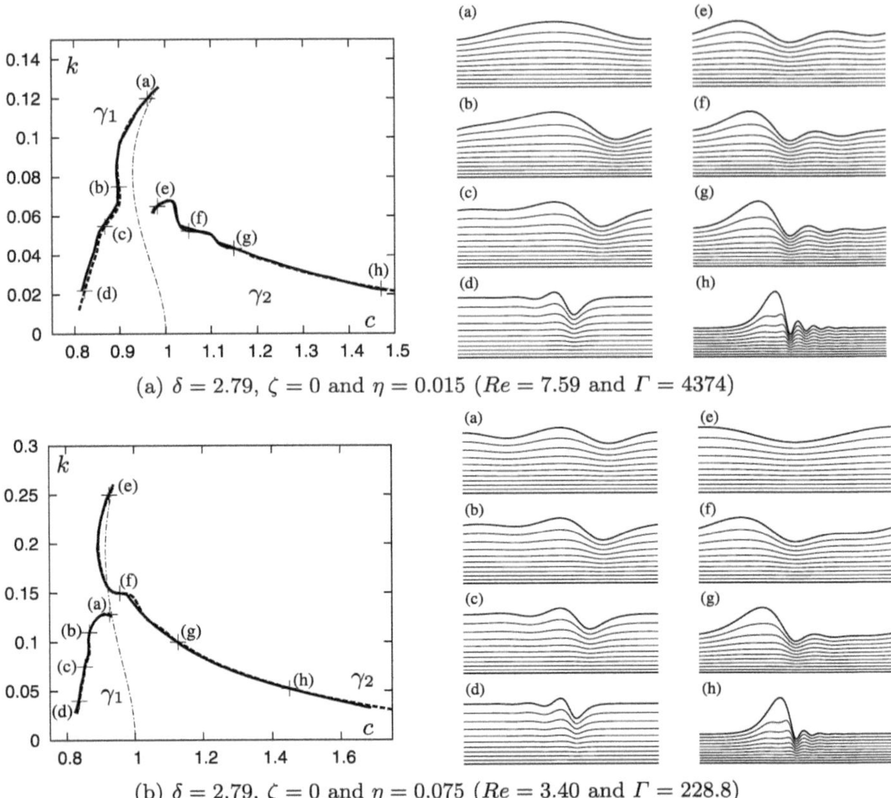

(a) $\delta = 2.79$, $\zeta = 0$ and $\eta = 0.015$ ($Re = 7.59$ and $\Gamma = 4374$)

(b) $\delta = 2.79$, $\zeta = 0$ and $\eta = 0.075$ ($Re = 3.40$ and $\Gamma = 228.8$)

Fig. 7.36 Periodic wavetrains (limit cycles). *Left*: Wavenumber k versus speed c. *Dashed lines* refer to solutions of model (6.1), (6.79), whereas *solid lines* refer to the DNS results in [232]. The phase velocity c_r of the temporally most amplified linear waves is indicated by *dashed-dotted lines*. *Right*: Wave profiles and streamlines in the moving frame of reference for the wave families γ_1 (*panels a to d*) and γ_2 (*panels e to h*). As in Fig. 7.34, k is made dimensionless using the film thickness $\bar{h}_N$

comparison to the most amplified linear waves with the same wavelength. The bifurcation diagrams and solution profiles obtained with the second-order simplified model (6.1), (6.79) are in remarkable agreement with the DNS study in [232].

It is important to note at this stage that unlike the models considered here, e.g., the simplified second-order model (6.1), (6.79), where viscous dispersion can be suppressed by simply setting $\eta = 0$, the same is not true with a DNS study. But one can approach an equivalent limit of small η with large Γ or large We (indeed η decreases when We increases). This then means that in order to keep δ the same, Re must be increased when We is increased. This is precisely how the topological change of the solution structure from Fig. 7.36(a) to Fig. 7.36(b) was found by Salamon et al. [232].

Comparisons of the profiles of solitary-like waves (panels h in Fig. 7.36) show that viscous dispersion affects significantly the number and amplitude of the cap-

Table 7.2 Phase speed c ($\mathrm{cm\,s^{-1}}$) of the traveling waves from the experimental data by Kapitza and Kapitza [141], from DNS and from the second-order models. Parameters are $Re = 6.07$, $We = 76.4$ (inlet flow rate $0.123\ \mathrm{cm^2\,s^{-1}}$, kinematic surface tension $\sigma/\rho = 29 \times 10^{-6}\ \mathrm{m^3\,s^{-2}}$, kinematic viscosity $2.02 \times 10^{-6}\ \mathrm{m^2\,s^{-1}}$ and wavelength $\lambda = 1.77$ cm). The *first column* corresponds to the closed flow condition and the second to the open flow condition

	$\langle h \rangle_\xi = 1$	$\langle q \rangle_\xi = 1/3$
Full second-order model (6.78)	23.5	20.4
Simplified model (6.1), (6.79)	23.5	20.3
Regularized model (6.1), (6.92)	23.5	20.3
Kapitza and Kapitza [141]	–	19.5
Ho and Patera [116]	24.7	–
Salamon et al. [232]	23.5	–
Ramaswamy et al. [218]	23.1	–

illary ripples preceding the humps, as already noted several times in this monograph. An excellent agreement (not shown) is also obtained with the full second-order model (6.78) and the regularized model (6.1), (6.92) derived in Chap. 6. Table 7.2 shows a comparison of the wavespeed of an experimental γ_2 traveling wave reported by Kapitza [141] to the solutions obtained by DNS and to the solutions of the different second-order models derived in Chap. 6. Some corresponding wave profiles have already been shown in Chap. 4 (see Figs. 4.1 and 4.3). The agreement of the results obtained with the low-dimensional second-order models to DNS is very convincing indeed. The discrepancy between the wavespeed of the DNS solutions and Kapitza's experimental result is due to the use of the closed flow condition $\langle h \rangle_\xi = 1$, instead of the open flow condition $\langle q \rangle_\xi = 1/3$, which corresponds to the experiments (but once again, this requires synchronization between the flow and the inlet forcing—see Sect. 5.3.1).

7.2.4 Open Flow Conditions

As already emphasized, most numerical studies on falling liquid films enforce the closed flow condition, $\langle h \rangle_\xi = 1$. Here we impose the open flow condition $\langle q \rangle_\xi = 1/3$, which fits experimental settings where typically the film is forced periodically at its inlet. In fact, we shall demonstrate in Sect. 7.3, where we examine the spatio-temporal evolution of the film, that by imposing the open flow condition we are able to capture the traveling waves observed experimentally at the final stage of the spatio-temporal evolution of the film. We examine the case of an inclined plate when viscous dispersion is taken into account. Different branches of traveling waves obtained for the conditions of the Liu and Gollub's experiments [170] are depicted in Fig. 7.37 at increasing values of the Reynolds number for a fixed inclination $\beta = 4°$ and Kapitza number $\Gamma = 2340$. For illustration purposes we present the solutions to the regularized model (6.1), (6.92). The bifurcation diagrams obtained with the simplified second-order model (6.1), (6.79) (not shown) are nearly identical. (Discrepancies between the results from the simplified and regularized models are observed further from the threshold of the instability, when the wave dynamics become three-dimensional; in particular, the inertial terms included in the regularized model capture the synchronous three-dimensional patterns observed in the experiments by Liu et al. [170]—see Chap. 8.) We chose to display the wavespeed as a function of the dimensional frequency in order to simplify comparison with the experiments. Among a wide variety of solution branches we chose to show only those that relate to the branch of the fast γ_2 waves. In particular, only the single-hump nearly solitary γ_2 fast waves are displayed in Fig. 7.37, whereas n-hump nearly solitary fast waves can also be found. (Thus, we will not discuss in this section the families $\gamma_1^{(n)}, \gamma_2^{(n)}$ corresponding to trains of n-negative or n-positive identical traveling wave solutions discussed in Sect. 5.3.2—except $\gamma_1^{(2)}, \gamma_2^{(2)}$.) However, single-hump nearly solitary γ_2 waves are always the fastest at a given frequency.

Very close to criticality ($Re_c = 11.9$), the γ_2 family emerges at the cut-off wavenumber k_c whereas the γ_1 waves appear through a period-doubling bifurcation from the family of $n = 2$ harmonic solutions, i.e., the $\gamma_2^{(2)}$ branch. This situation is different from the description given by Chang et al. [50], where a Hopf bifurcation at the cut-off wavenumber gives rise to the slow γ_1 branch of solutions and a period-doubling bifurcation from the branch $\gamma_1^{(2)}$ leads to the formation of the fast γ_2 waves. A similar discrepancy is observed with the DNS study performed by [232] for a vertical film but smaller Reynolds number and small η (we discussed earlier how the limit of small η can be approached in a DNS study) in which case the DNS study gives a bifurcation diagram similar to that in Fig. 7.36(a). Hence, the different bifurcation diagram obtained by Chang et al. is a direct consequence of neglecting viscous dispersion, which is not taken into account by the simple Kapitza–Shkadov model used by Chang et al. (see also Sect. 7.2.3).

Figure 7.37 indicates that close to criticality, the γ_1 and γ_2 families form an imperfect bifurcation similar to what is observed for the Kawahara equation (compare the top panel of the figure to Fig. 7.35). As the Reynolds number is increased, the fast-wave γ_2 branch experiences several collisions with other branches. In Fig. 7.37, two successive pinchings are displayed, for $14.3 < Re < 14.4$ (middle panels) and $16.3 < Re < 16.4$ (bottom panels). Each of these pinching events is reminiscent of the imperfect bifurcation affecting the γ_1 and γ_2 wave branches. They occur at frequencies close to $1/3$ and $1/4$ of the cut-off frequency, respectively, and give rise to the secondary γ_1' and γ_1'' branches of slow-wave solutions—at low frequency, these waves are slower than the most amplified linear waves and thus belong to the γ_1 type of waves. Close to the pinching points in the speed-frequency diagram, the waves resemble a train of three and four γ_1 waves. Accordingly, we refer to these branches as γ_1^3 and γ_1^4, respectively (γ_1^3 and γ_1^4 branches must not be confused with the $\gamma_1^{(3)}$ and $\gamma_1^{(4)}$ solutions consisting of trains of three and four identical γ_1 waves—see Sect. 5.3.2).

However, no bifurcation has been found at $Re \approx 14.3$ and $Re \approx 16.3$ when the emergence of new wave branches has been sought from the γ_1 family by placing

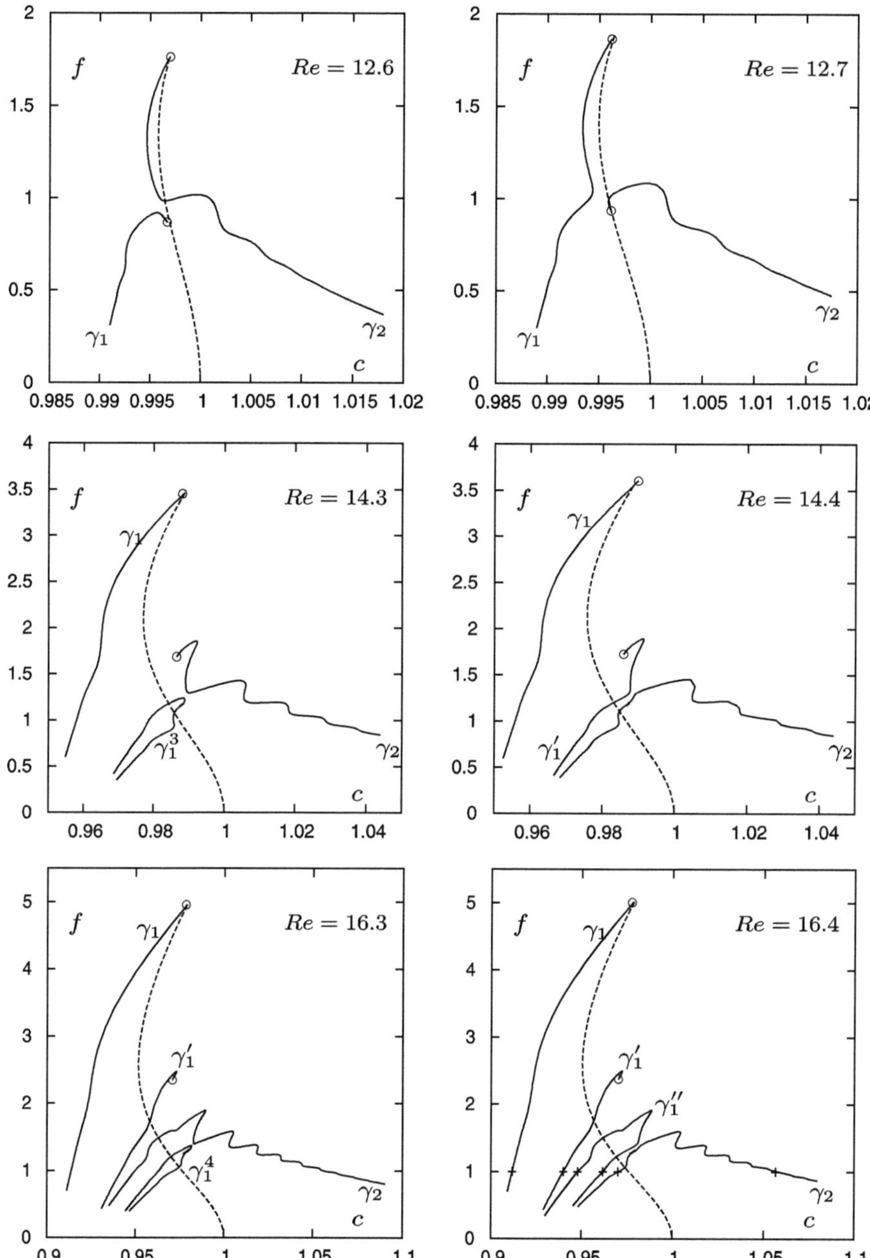

Fig. 7.37 Wavespeed c versus dimensional forcing frequency f in Hz at increasing Reynolds numbers. Parameter values are $\beta = 4°$ and $\Gamma = 2340$ corresponding to the experimental conditions in [170]. *Solid lines* are computed with the regularized model (6.1), (6.92). *Dashed lines* refer to the spatially most amplified solutions of the corresponding linear dispersion relation. The initial Hopf bifurcations and the period-doubling bifurcation giving birth to the γ_2 family and also the γ_1' family for $Re > 14.4$ are indicated by *open circles*

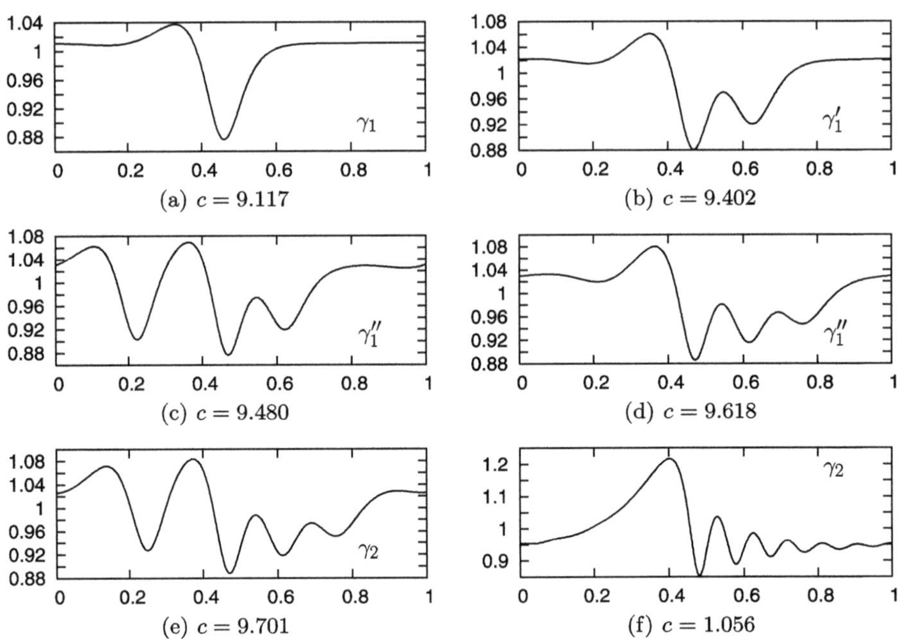

Fig. 7.38 Shape of different wave families at $Re = 16.4$ and frequency $f = 1$ Hz. Parameter values are $\beta = 4°$ and $\Gamma = 2340$ corresponding to the experimental conditions in [170]

three or four identical waves close to each other in the computational domain. In fact, the γ_1^3 and γ_1^4 families have been found through continuation of the γ_2 branch from larger values of Re after pinching of the branches has occurred. The profiles of the waves corresponding to the different branches shown in Fig. 7.37 are given in Fig. 7.38 for $Re = 16.4$ and $f = 1$ Hz. Figures 7.38(a–e) correspond to slow waves made of several troughs followed by radiation before the troughs return to the level of the flat film. The shape of the fast-speed end of the γ_2 branch is different from the shape of slow waves γ_1^n with a main hump preceded by ripples (compare Fig. 7.38(f) to Figs. 7.38(a–e)). The γ_1 waves correspond to a unique trough whereas γ_1', γ_1'' and the low-speed end of the γ_2 branch have several troughs. Notice the similarities between the shapes of the waves shown in Figs. 7.38(b) and 7.38(c), due to the fact that they both originate from the same γ_1^3 family. Figures 7.38(d) and 7.38(e) are even more similar since $Re = 16.4$ corresponds more closely to the value at which the γ_2 and γ_1^4 branches collide, giving rise to the γ_1'' waves. γ_2^n and γ_2' wave families also exist but are not shown in Fig. 7.37, to limit the complexity of the figure.

As it can be seen from Fig. 7.37, the bifurcation diagram becomes increasingly complicated as the Reynolds number is increased, and at a given frequency several traveling wave solutions exist. This raises the question of the attractiveness of these solutions, which is in turn related to their relevance and the way they attract initial conditions. In other words, we ask "Which traveling wave solutions will be selected by the flow following the inception region of linear growth of the inlet periodic perturbations?" An answer to this question can be given through numerical solution in time and space of the model used to describe the film dynamics. Shkadov and Sisoev [250] have performed a thorough analysis of the attractiveness of traveling wave solutions by considering the evolution in time of solutions to the Kapitza–Shkadov model when periodic boundary conditions are enforced. In addition to traveling wave solutions, they found "invariant tori," i.e., oscillatory modes made of the superpositions of two irrationally related periodic oscillations. The presence of these quasi-periodic attractors was also detected by the DNS study in [218]. Oscillatory modes generally correspond to successions of two different traveling wave solutions. When traveling wave solutions were found as the final stage of the unsteady computations, which is by far the most common situation, the selected waves are the fastest. We shall examine this point in Sect. 7.3.1. They also correspond to the largest maximum height among all traveling waves with the same wavelength and were referred to by Shkadov and Sisoev as "dominant waves" [250].

Finally, in line with experimental evidence, slow γ_1 waves are expected to follow the linear inception region at high frequency (i.e., close but below the cut-off), whereas at low frequency, γ_2 waves are expected [3, 168, 170]. Oscillatory modes, obtained by integration in time and space with periodic boundary conditions, are generally not observed in experiments. When they are observed, a *secondary instability* leads further downstream to a regular wavetrain of traveling waves [37, 168] (see Fig. 7.43 and its discussion below).

7.3 Spatio-temporal Evolution of Two-Dimensional Waves

The problem of the spatio-temporal evolution of two-dimensional waves on film flows has attracted great interest, both experimental [4, 37, 167–169, 294, 305] and computational [45, 176, 218]. These studies have revealed a rich variety of phenomena: traveling waves, quasi-periodic modulated waves, secondary instabilities and the complexity of wave interactions. It is not our strategy here to track all possible phenomena. Instead, we aim to capture the more pertinent ones by making use of time-dependent numerical simulations of the simplified model (6.1), (6.79) and the regularized model (6.1), (6.92) for parameter values corresponding to the experiments conducted by Liu and Gollub [168]. The finite-differences numerical scheme for the time-dependent computations is given in Appendix F.3.

7.3.1 Periodic Forcing

In his seminal work more than 60 years ago, Kapitza did not take long to realize that a falling liquid film behaves as an amplifier of the inlet noise [140, 141]. This observation is now clearly established theoretically as demonstrated in Sect. 7.1.2. Kapitza consequently applied well-controlled periodic perturbations at the inlet, and thus he was able to observe traveling waves with the same or related time periodicity. The same procedure was followed by almost all experimentalists studying falling

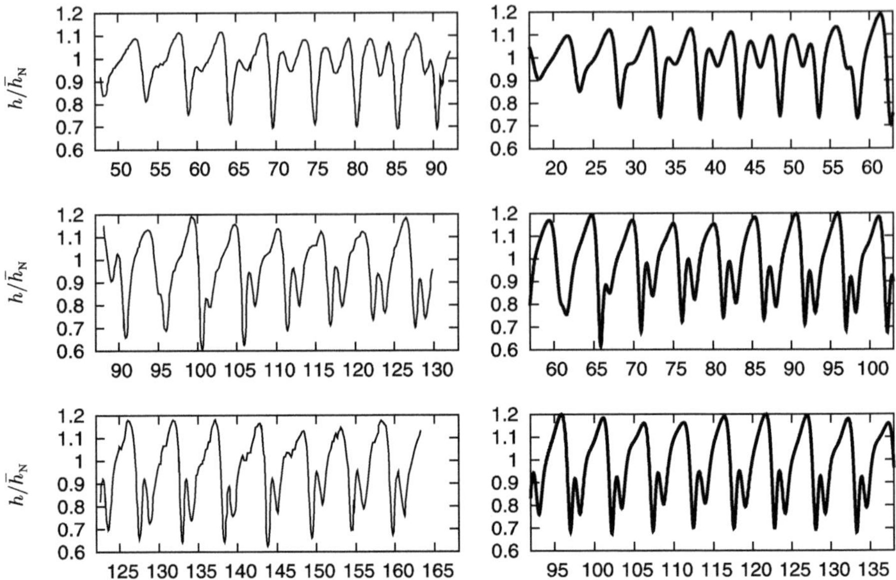

Fig. 7.39 Comparison between experiments [168] (*left*) and simulation of the simplified model (6.1), (6.79) (*right*). The plate is inclined at $\beta = 6.4°$ from the horizontal, the liquid is a glycerin–water mixture and the Reynolds and Kapitza numbers are $Re = 19.3$ and $\Gamma = 524.4$, respectively. The figure shows three snapshots of the film thickness at three different locations along the plate from upstream (*top*) to downstream (*bottom*) at forcing frequency $f = 4.5$ Hz and forcing amplitude, $A = 0.03$

film flows after Kapitza. Possibly the most comprehensive and definitive account of the spatial evolution of waves produced by a periodic forcing at the inlet can be found in the work by Gollub's group. Their observations can be summarized as follows. At high frequency, close but below the cut-off frequency f_c (above the cut-off frequency the film is stable—see Fig. 7.2), traveling waves of the slow γ_1 type are observed close to the inlet. At low frequency, in comparison to f_c, the exponential growth of the waves is directly followed by the formation of fast solitary-like wavetrains of γ_2 type.

The two situations are illustrated in Figs. 7.39 and 7.40, where results from the simplified model (6.1), (6.79) are compared to the experimental findings corresponding to a glycerin–water mixture film flowing down a plate inclined at an angle $\beta = 6.4°$ from the horizontal and at a Reynolds number $Re = 19.3$. The periodic forcing on the entrance flow rate has been simulated by setting

$$q(0, t) = \frac{1}{3}\left(1 + A\cos(2\pi f t)\right). \tag{7.77}$$

At lower—but not too low—frequencies, waves are multi-peaked (see Fig. 7.39). Both the experiments and our numerical simulations show in this case a complex nonlinear process leading to multi-peaked waves. These saturated multi-peaked

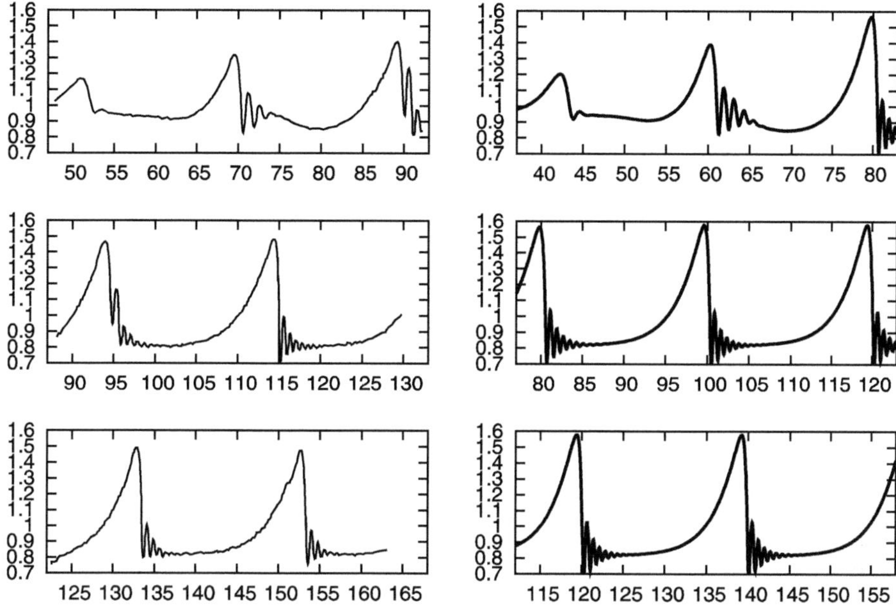

Fig. 7.40 Comparison between experiments [168] (*left*) and simulations of the simplified model (*right*) for $f = 1.5$ Hz. Values of the other parameters are given in the caption of Fig. 7.39

waves move slower than corresponding growing linear waves of infinitesimal amplitude at the inlet and thus belong to the γ_1 type of waves. After the growth of a subsidiary peak, the time shift separating the primary and subsidiary peaks in a time series at a given location x approaches an asymptote further downstream, a phenomenon that we refer to as *phase-locking*. A modulated two-peak wavetrain is then observed. This phase-locking can also be interpreted as a sequence of splittings and mergings of the primary and subsidiary peaks. If the details of the phase-locking process are modified by the amplitude A (see Fig. 7.41), the resulting multi-peaked wavetrain is not affected by the level of the inlet forcing. Computations of the corresponding traveling wave families by continuation using AUTO07P reveal that these multi-peaked waves are slow γ_1' waves coming out from a period-doubling bifurcation of the γ_1 branch that emerges at the cut-off frequency; the bifurcation diagram is then similar to those displayed in the lower panels of Fig. 7.37. When the inlet forcing is applied to the film thickness h rather than to the flow rate q, similar results are obtained. These observations also apply to the formation of the γ_2 solitary-like waves at low frequency (see Fig. 7.42).

The experiments by Gollub's group reveal a transition at $Re \approx 30$ between only spatially modulated wavetrains and wavetrains modulated in both space and time [37]. For $Re \lesssim 30$, the signal remains periodic in time at any location and the wavetrain modulation is only spatial: the splitting and merging events occur always at the same positions on the plate. At larger Reynolds numbers, the time periodicity of the signal is lost, and the splitting and merging of the waves occur at random

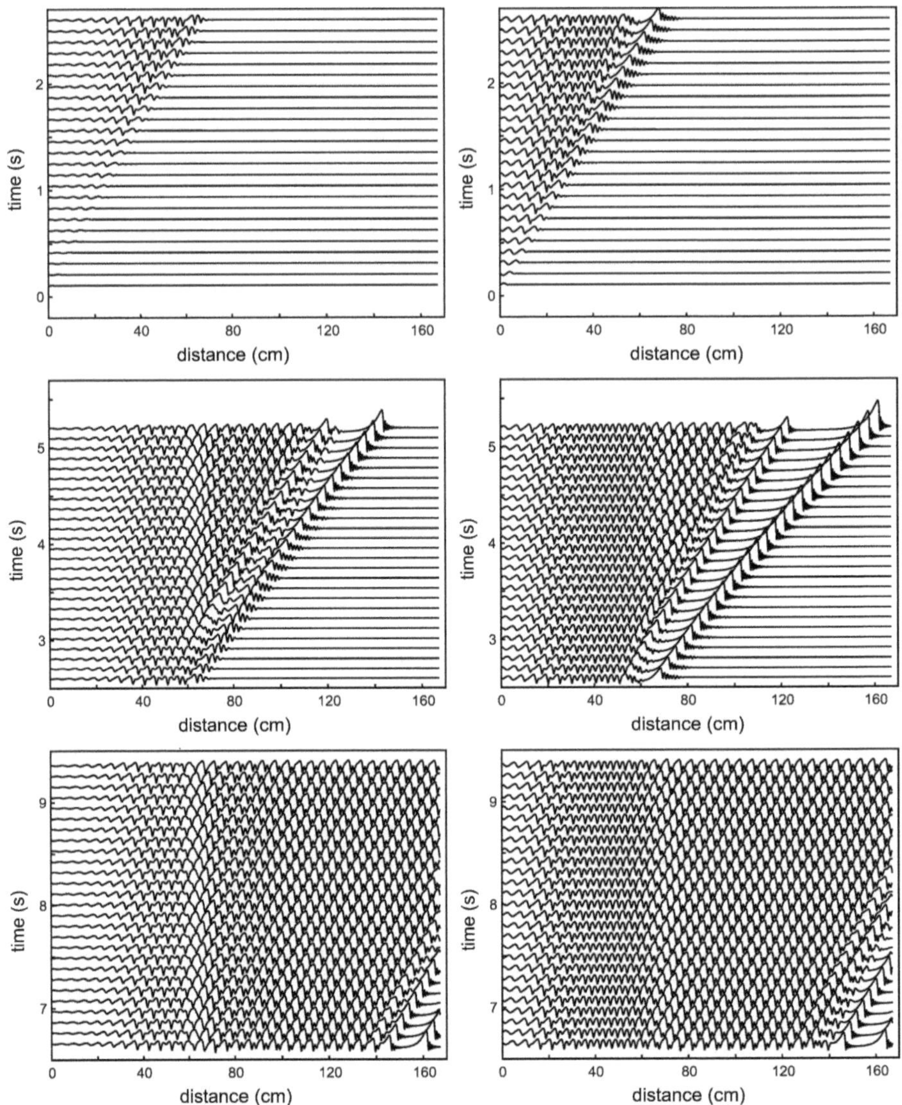

Fig. 7.41 Film thickness evolution for the experimental conditions [168] and in good agreement with the DNS study in [218]. The parameter values are given in the caption of Fig. 7.39 ($f = 4.5$ Hz). The film thickness is plotted at regular intervals of 0.104 s. *Left*: Forcing amplitude $A = 0.03$; *right* $A = 0.15$

locations. Numerical simulations of the spatial evolution of the flow based on the simplified model (6.1), (6.79) reveal that modulations in time of the primary wave-train can be obtained only when noise is added to the periodic forcing at the inlet (the numerical implementation of natural noise is detailed in Sect. 7.3.2). This suggests that the onset of space-time modulations observed in experiments is triggered

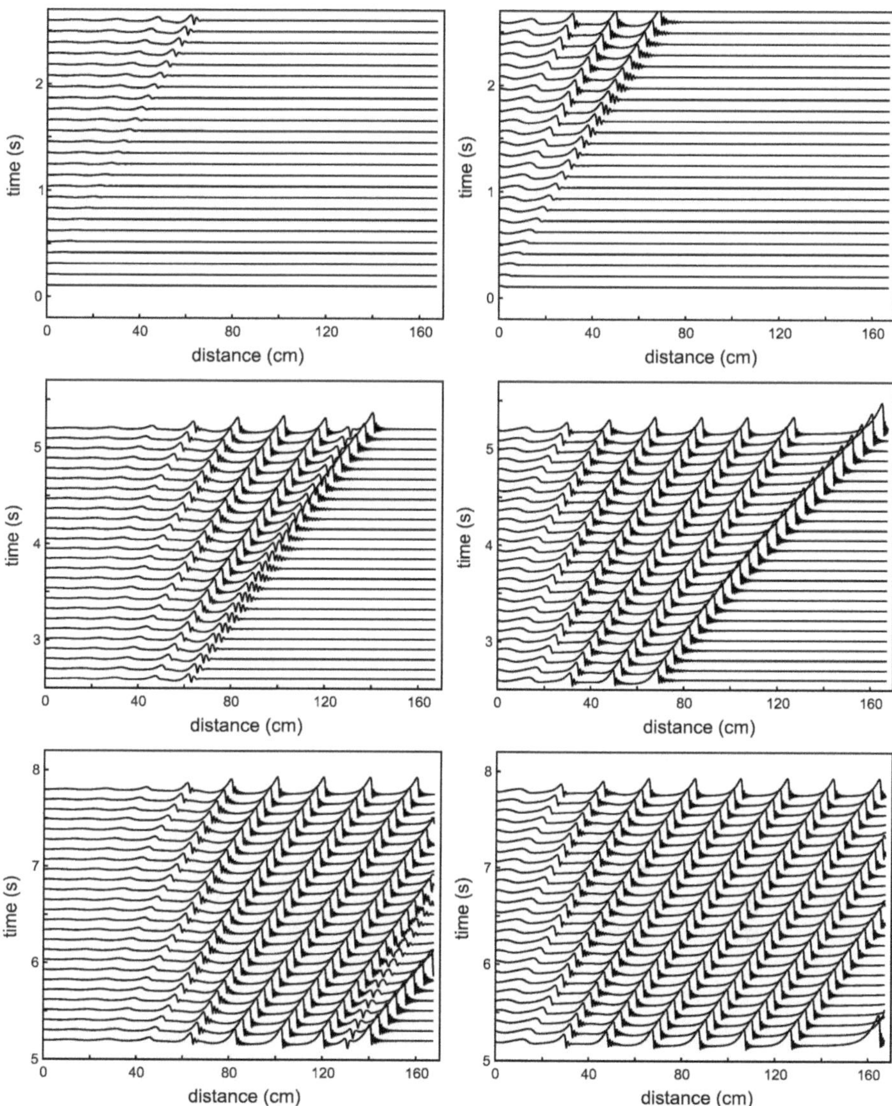

Fig. 7.42 Same caption as for Fig. 7.41 except for the forcing frequency $f = 1.5$ Hz and the forcing amplitude (*left*) $A = 0.03$; (*right*) $A = 0.15$

by the unavoidable ambient noise. However, our simulations, for which a small but noticeable noise was added to the periodic forcing, did not show a sharp transition at $Re \approx 30$ but rather a smooth one with the onset of space-time modulations depending on the level of noise added to the forcing at the inlet. Though our simulations unequivocally demonstrate the significant influence of inlet noise, the precise mechanism of the transition from spatial modulations to both spatial and temporal modulations remains an open question.

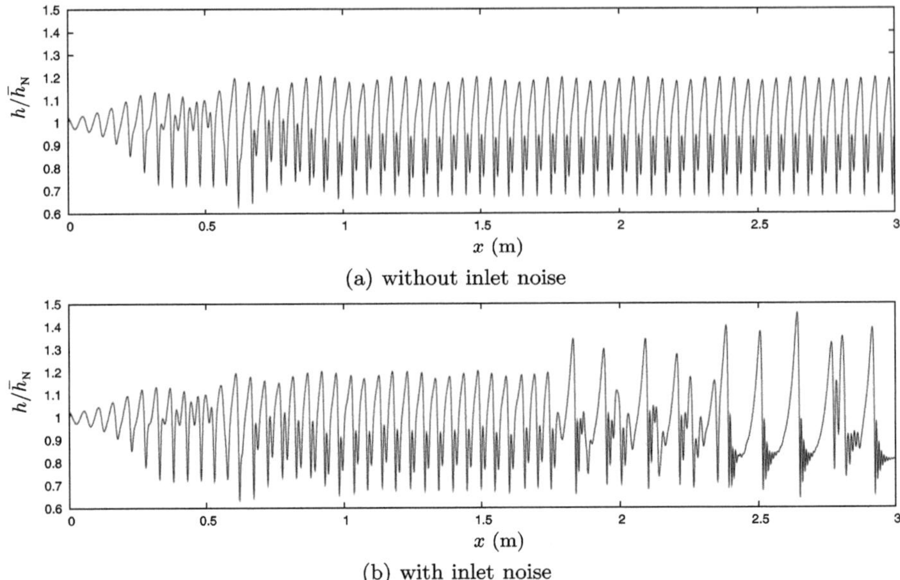

(a) without inlet noise

(b) with inlet noise

Fig. 7.43 Snapshots of the film thickness from a simulation using the regularized model (6.1), (6.92). Parameters correspond to an experiment in [168] ($\beta = 6.4°$, $R = 19.3$ and $\Gamma = 524.4$, $f = 4.5$ Hz). Forcing amplitude is $A = 0.05$. Contrast with Figs. 7.39 and 7.41 presenting simulations corresponding to the same set of parameters

Ultimately, the addition of a small noise to the inlet forcing leads to the disorganization of the modulated primary wavetrain, as illustrated in Fig. 7.43, which depicts snapshots of the film thickness at the end of simulations with an extended computational domain (corresponding to a 3-m long plane). The figure was computed with the regularized model (6.1), (6.92). The solution to the simplified model is again very similar. Our simulation shows that in the absence of inlet noise there is a spatial modulation of the primary wavetrain. This modulation is easily visible on the wavetrain envelope in panel a of Fig. 7.43: it is an oscillation in space of the envelope in the region ~ 0.7–2.5 m, but it seems to be damped further downstream. On the other hand, when a small amount of noise is added to the inlet forcing (panel b of Fig. 7.43), the spatial modulation is still visible close to the inlet but now the primary wavetrain is rapidly replaced by a train of solitary-like waves similar to the periodic wavetrain observed at low frequency (see Fig. 7.40). This is an example of a secondary instability of the primary wavetrain. Secondary instabilities will be further discussed in Chap. 8, where we shall also make the distinction between two-dimensional and three-dimensional secondary instabilities. The precise mechanism by which the spatial modulation triggers a secondary instability leading to a train of solitary-like waves is not known. However, it seems that without noise the system prefers to remain periodic in time; it synchronizes to the period of the forcing. To produce solitary waves, we need to "break" this synchronization by introducing noise at the inlet.

Since the presence of a small amount of noise cannot be avoided in experiments, we conclude that the two-dimensional dynamics of the flow is governed at the final stage by trains of solitary-like waves in interaction. The noise-driven dynamics of film flow is further investigated in the next section.

The waves at $x = 3$ m in Fig. 7.43(a) are of the γ_1-type (more precisely, they belong to the γ_1' branch bifurcating from the γ_1 branch through period doubling). On the other hand, in Fig. 7.43(b) γ_1 waves are observed in the interval $\simeq [0.6\,\text{m}, 1.5\,\text{m}]$ but they eventually give rise to γ_2 waves toward the end of the domain. This raises the question of stability of γ_1 waves. After all, as noted at the end of Sect. 7.2.4, previous studies show that in unsteady computations, the selected waves are the fastest [250]. But actually in the numerical implementation of the finite-difference numerical scheme used to discretize the regularized model in time and space, there is always some numerical noise, albeit small, i.e., the level of noise is not sufficient to excite the instability within 3 m from the inlet (that the instability of γ_1 waves is not excited does not make them stable; instead they are "metastable"-like states), and with a much longer plane, we could still observe the same disorganization as in Fig. 7.43(b), leading eventually to the formation of γ_2 waves.

Hence, γ_1 waves are unstable, giving rise to γ_2 waves. As a matter of fact, they are unstable to all disturbances, both streamwise and spanwise as will be shown with the Floquet analysis in Sect. 8.3. This is in agreement with the experimental observations by Gollub and collaborators, who observed γ_1 waves with high frequency (close but below the cut-off) near the inlet but which afterward become γ_2 waves or three-dimensional negative solitary waves, which in turn reorganize themselves into three-dimensional horseshoe waves. Without forcing γ_1 waves are not even observed.

Most interestingly, the middle left and bottom right panels of Fig. 7.41 and the middle left and bottom left panels of Fig. 7.42 indicate that an "excited" solitary wave, i.e., a soliton of larger amplitude and speed to those of an equilibrium solution for the same conditions and hence carrying additional mass compared to the equilibrium soliton, releases this mass through the formation of a γ_1 solitary wave, with a trough-like shape similar to the wave displayed in Fig. 7.38(a). But the formation of this wave is just a transient effect: an upcoming fast γ_2 wave will coalesce inelastically with the γ_1 one and absorb it. γ_1 waves are in fact too slow to be seen: If we excite both slow and fast waves, the fast ones will eventually win (hence the term "dominant waves" for the γ_2 waves of Shkadov and Sisoev [250]).

7.3.2 Noise-Driven Flows

The interaction between nonlinear waves in the evolution of noise-driven falling liquid films has received considerable attention over the past few years. Central to this quest has been the aim to describe through a coherent structure theory the dynamics of the film and hence provide a systematic description of spatio-temporal disorder or, equivalently, weak/dissipative turbulence as defined by Manneville [177, 189].

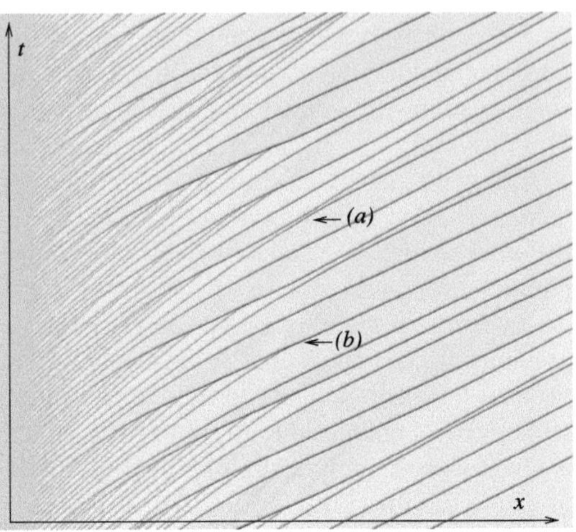

Fig. 7.44 Spatio-temporal diagram of a noise-driven film flow. The final stage of the simulation is shown in Fig. 7.1. Parameter values are $Re = 6.07$, $\beta = 90°$ and $We = 76.4$. The simulated duration is 50 s and the simulated spatial domain is 0.83 m. *Bright* (*dark*) *regions* correspond to depressions (elevations). *Arrows* labeled (*a*) and (*b*) indicate a repulsion event and a coalescence event, respectively

For example, for the conditions in Fig. 7.39 the final stage of the two-dimensional evolution of a falling liquid film seems to be a fairly regular train of coherent structures, each of which resembles an infinite-domain solitary pulse, or, equivalently, a periodic wavetrain close to the solitary wave limit (further downstream these structures suffer a transverse instability, which eventually gives rise to a three-dimensional wave regime; some features of this regime will be explored in Chap. 8). In the presence of random noise with sufficiently large amplitude, the dynamics become more complicated, e.g., dynamic interaction with continuously varying separation seems to persist indefinitely. This interaction seems to be of the type that makes neighboring coherent structures attract or repel each other continuously. Nevertheless, one can still identify the generic solitary wave shape in what appears to be a random interface; see for example Fig. 7.43(b). Solitary waves then become elementary processes so that the dynamics of the film can be described as their superposition.

An example of a computed noise-driven two-dimensional film flow is given in Fig. 7.44. Inlet noise is simulated by the inlet boundary condition [46]

$$q(0, t) = \frac{1}{3}\big(1 + F(t)\big),$$

$$F(t) = A \sum_{m=1}^{M} \cos\left(\frac{m}{M} 2\pi f_\star t - \theta_m\right), \tag{7.78}$$

where A is the noise amplitude and $f_\star$ is a multiple of the cut-off frequency f_c, i.e., $f_\star = n_f f_c$, with n_f an integer. The phases θ_m are generated randomly in the range $[0, 2\pi]$ using a generator of pseudo-random numbers with a uniform distribution between zero and unity. This procedure ensures that the inlet noise does not contain high frequency modes. The resulting noise is said to be "colored" to differentiate it from "white" noise that covers all possible frequencies.

Perturbations are introduced in time only and at the first node of the computational domain; the frequency content of the excitation is controlled by the time step. High frequency perturbations may trigger spurious numerical instabilities at the inlet by generating unrealistic gradients: a high frequency signal gives large time derivatives which in turn result in large spatial gradients (space derivatives are connected to time derivatives through the speed of the resulting wave). Such large spatial gradients might not be resolved accurately if the space step is not sufficiently small, because of the coarse sampling in the discretization of the computational domain, which in turn leads to numerical instabilities. Of course, this instability may be damped further downstream by surface tension and viscous dispersion but not when it is sufficiently strong. If there were no problem with numerical instabilities, white noise would give the same results as colored noise precisely because the instability is selective, i.e., at the inception region all disturbances with wavenumbers above the cut-off one k_c are damped (since the film behaves as a low frequency amplifier—Sect. 7.1.2—high frequency-modes are damped in the inception region). Similarly, with some very careful experiments in which we impose colored noise at the inlet (to carefully filter out high frequencies), the result downstream would be the same as that obtained in the presence of white noise.

After the linear inception region, nonlinear wavetrains undergo a series of several coalescence events. Faster waves, having also larger amplitudes, catch up with smaller ones and absorb them giving rise to waves of even larger amplitude. Further downstream, the dynamics are dominated by trains of large-amplitude nearly-solitary waves separated by portions of nearly flat films. These waves correspond to the principal homoclinic orbits constructed in Sect. 7.2, having the shape of isolated humps preceded by capillary ripples/radiation. They interact with each other through their (exponentially decaying) tails. As mentioned earlier, the interaction between neighboring coherent structures is either repulsive (see label a in the upper diagram of Fig. 7.45) or attractive, giving rising to coalescence (label b) and hence birth of even larger waves. This is a dynamic process with continuous coalescence events, but the number of these events decreases downstream.

Dispersion plays a crucial role in the formation of coherent structures. Through the corresponding nonlinearities, it enables the transfer of energy from the large scales, where the instability mechanism pumps energy from the mean flow to the perturbations, to the smaller scales where dissipation is triggered by surface tension. The simplest equation including "negative diffusion" (i.e., a diffusion process with a negative diffusion coefficient thus concentrating energy in the small-scale structures), associated here with the pumping of energy to the perturbations by inertia effects, as well as surface tension effects on small scales and dispersion (whose origin is viscosity), is the Kawahara equation (5.31).

Integration in time of the Kawahara equation with periodic boundary conditions has revealed a transition from the turbulent-like regime that is typical of the KS equation (7.63) to a regime dominated by localized coherent structures resembling periodic trains of nearly solitary waves as the dispersion parameter increases (see also Sect. 5.2.1). In the case of strong dispersion these coherent structures repel each other, whereas, at moderate dispersion, bound states of steady humps separated by

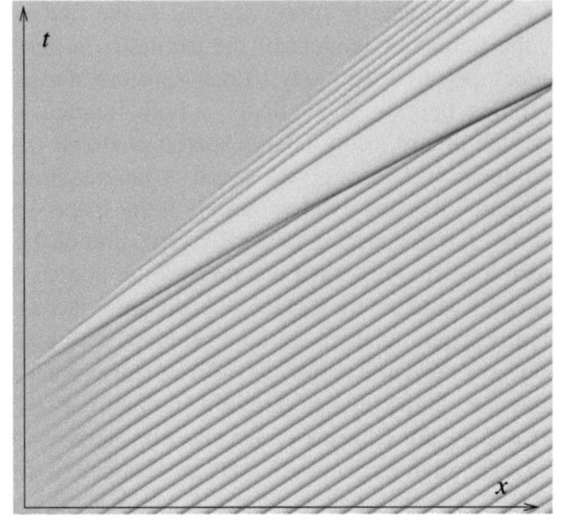

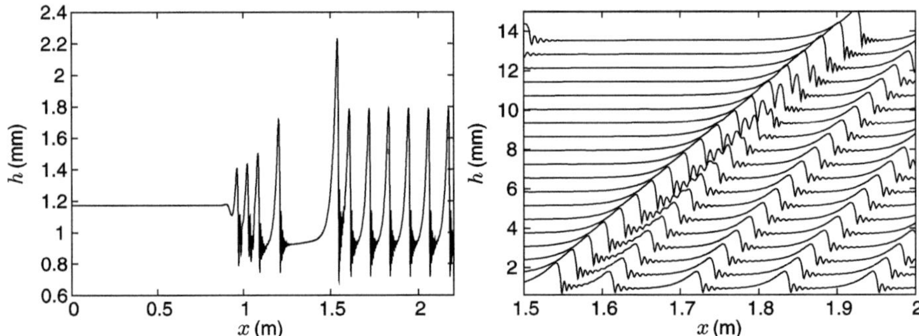

Fig. 7.45 Absorption of a train of traveling waves by a solitary wave of larger amplitude. Parameter values are $Re = 26$, $\beta = 8°$, $We = 35$ and $\nu = 6.28$ cS. After a periodic forcing of frequency $f = 2.5$ Hz has been applied, a localized perturbation is introduced at the inlet followed by the suppression of the forcing. The simulated duration is 12 s for a plate of 2.2 m length. *Upper panel* is a spatio-temporal diagram. *Dark (bright) regions* correspond to elevations (depressions). *Lower left panel* is a snapshot of the film thickness at $t = 8$ s. The absorption of a wave by a larger one is illustrated in the *lower left diagram*

fixed distances have been found [87, 90, 144, 284, 286]. As we already demonstrated in Sect. 7.2.1, the presence of bound states is connected with the behavior of the corresponding homoclinic orbits in the neighborhood of the fixed point in the phase space of the associated dynamical system.

As far as coherent structure theories are concerned, previous efforts include [15, 44, 45, 90], which, however, are either incomplete or which overlooked some serious details and subtleties, which are crucial for a complete and rigorous description of coherent structures interaction; these are resolved in the most recent studies in [87, 212, 284, 286]. Extensions to three-dimensional problems are given in [233–235, 283].

Most coherent structure theories are based upon the assumption of weak interaction of two successive humps through their (weakly) overlapping tails. This type of interaction can be described by the translational mode of the linearized operator that accounts for the perturbations introduced by a solitary wave to its neighbors. On the other hand, a key element of the approach developed by Chang and coworkers (e.g., [44]) based on the Kapitza–Shkadov model is the presence of an additional invariance that allows one to describe the process of mass exchange between neighboring structures. This type of interaction is a "strong" one and can be observed in Fig. 7.1: The nearly flat portions of the film separating each solitary pulse are not of equal thicknesses; hence the solitary waves move on substrates of different thicknesses. The presence of a continuous family of solitary wave solutions parameterized by the substrate thickness therefore introduces a second invariance to the system, which is effectively the Goldstone mode.

We shall not describe the various coherent structure approaches in this monograph. It should be emphasized, however, that the coherent structure theory for falling films developed by Chang and coworkers was based on the Kapitza–Shkadov model and as such it ignores the effect of viscous dispersion. Yet, as noted earlier, several times, viscous dispersion modifies the shape of the oscillatory structure at the front of solitary waves: it reduces the number and amplitude of radiation preceding the humps and therefore modifies the interaction between successive waves. Correct description of the radiation is crucial for an accurate prediction of the average separation distance between the coherent structures and hence of the stationary wave selection in the spatio-temporal evolution of the film. As a consequence, a comprehensive and accurate description of the noise-driven dynamics of film flows based on a coherent structure approach must account for viscous dispersion effects. This calls for a revision of the theory by Chang and coworkers through use of the more refined models developed in Chap. 6, such as the simplified second-order model in (6.1), (6.79). The theory must also be corrected in view of our comment above that previous studies overlooked some serious details. A decisive first effort in this direction is the recent study by Pradas et al. [212] which scrutinized the effects of viscous dispersion on coherent structures interaction and formation of bound states.

Chapter 8
Isothermal Case: Three-Dimensional Flow

In Chaps. 6 and 7 we focused on stationary two-dimensional periodic and solitary waves and their dynamics. We now examine the three-dimensional wave dynamics. Experiments show the development of three-dimensional wave patterns for moderate Reynolds numbers resulting from the instability of two-dimensional waves [3, 44, 170]. Since two-dimensional waves result from the primary instability of the Nusselt flat film solution, the transition from two-dimensional to three-dimensional waves is a type of a *secondary instability*.[1] The final state of wave evolution on a falling film corresponds to a weakly disordered dynamics where the interface is randomly covered by three-dimensional *coherent structures*, which are stable and robust and continuously interact with each other as quasi-particles, like their two-dimensional counterparts in the two-dimensional wave regime (see also discussion in the introduction of Chap. 7). These three-dimensional coherent structures resemble three-dimensional solitary pulses. Therefore, like their two-dimensional counterparts in the two-dimensional wave regime, three-dimensional pulses are also elementary processes so that the three-dimensional wave dynamics can be described as their superposition. This stage of the evolution is often referred to as *interfacial turbulence* or *soliton gas* [264], and much like the weakly disordered two-dimensional dynamics of the film, it is also an example of *weak/dissipative turbulence*.

Figure 8.1 is a replica of typical three-dimensional wave patterns observed on a falling film [3]. The rather rich phenomena of three-dimensional wave dynamics in falling films are still an open subject of research. Here we review the main known experimental results on the three-dimensional regime and explore it theoretically and numerically within the framework of the low-dimensional averaging approach developed in Chap. 6.

---

[1]Two-dimensional waves resulting from the primary instability may also undergo a spanwise-independent instability, a streamwise modulation leading to a secondary two-dimensional wavetrain. This instability is also referred to as a "two-dimensional secondary" instability. An example is provided in Chap. 7, Sect. 7.3.1, Fig. 7.43.

S. Kalliadasis et al., *Falling Liquid Films*, Applied Mathematical Sciences 176, DOI 10.1007/978-1-84882-367-9_8, © Springer-Verlag London Limited 2012

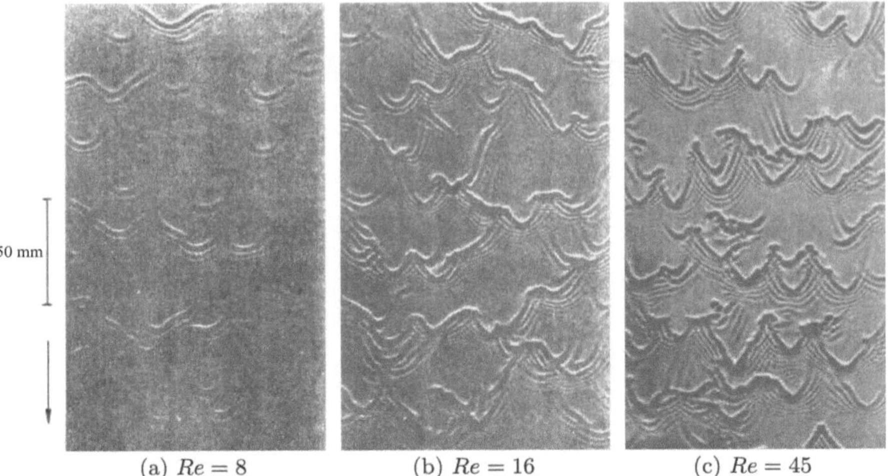

(a) $Re = 8$ (b) $Re = 16$ (c) $Re = 45$

Fig. 8.1 Three-dimensional wave patterns obtained experimentally (water–ethanol solution, inclination angle $\beta = 75°$). Reprinted from *Wave Flow of Liquid Films*, S.V. Alekseenko, V.E. Nakoryakov and B.G. Pokusaev, Copyright 1994, with permission from Begell House, Inc.

8.1 Phenomena

Experimental results on three-dimensional waves in falling films can be found in [1, 3, 170, 192, 193, 203]. The majority of these studies focused on vertical or near-vertical configurations. In contrast, Gollub and collaborators examined experimentally the three-dimensional dynamics of film flows for a moderately inclined plate [37, 167–170]. This configuration enabled them to consider the wave dynamics relatively close to the onset of the instability, $Re_c = \frac{5}{6}Ct$, where the sequence of the primary instability of the Nusselt flat film leading to a primary saturated wavetrain followed by secondary instabilities of the primary wavetrain, is more easily identified than in the vertical case for which the flow is always unstable ($Re_c = 0$) and therefore, by definition, in the experiments one is already far from onset. The picture by Gollub's group of the transition from a two-dimensional wave regime to a three-dimensional one has been recently completed by Nosoko and collaborators for a vertical wall [192]. Noteworthy is that Gollub and collaborators imposed spanwise-uniform perturbations of the inlet flow rate, whereas Nosoko and collaborators controlled the development of the three-dimensional waves with the help of an array of needles.

Based on these experimental observations, the different secondary instabilities leading to seemingly irregular patterns are schematically summarized in Fig. 8.2. Four stages, each corresponding to a different region, can be broadly distinguished by following the flow along the inclined plate.

The inception region is the domain close to the inlet where the primary linear instability of the flat film develops in space. Squire's theorem stipulates that the most dangerous (fastest growing) perturbations are spanwise-independent, so that

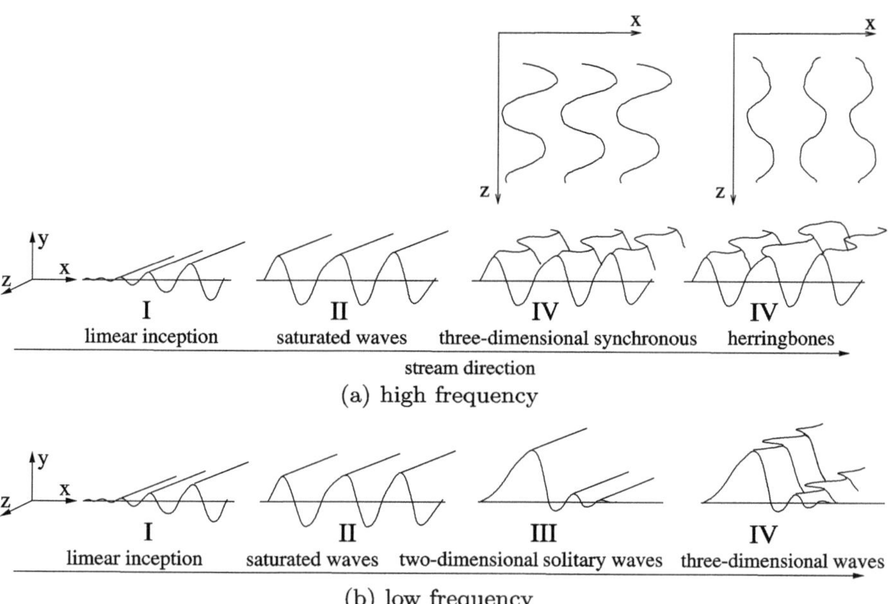

Fig. 8.2 Schematic description of the spatial evolution of film flows

the observed primary waves are two-dimensional [303] (see also Sect. 3.5.1). The amplitude of the waves next saturates and their shape remains practically unchanged over distances corresponding to a few wavelengths (region II). Subsequent events depend on the forcing frequency.

At large frequencies but below the cut-off frequency (Fig. 8.2(a)—after all, above the cut-off frequency the system is linearly stable), the observed primary waves are slow (here, as in previous chapters, slow and fast waves refer to waves propagating at rates slower or faster than the linear waves) and are characterized by wide bumpy crests and deep thin troughs. They belong to the γ_1 family following the terminology introduced in [48, 50] (see also Sects. 5.3.2, 7.2.3 and 7.2.4). Two different scenarios are possible for the subsequent evolution and, most interestingly, they are strongly reminiscent of what happens in boundary layers [244]. The first one, referred to as a *synchronous instability*, is characterized by in-phase deformation of neighboring troughs in the spanwise direction, whereas the crests remain—at least for a while—undisturbed (see Fig. 8.3). The second one, less commonly observed, is characterized by a phase shift of π between two successive crests. This leads to *herringbone patterns*, characteristic of a spanwise modulation combined with a streamwise *subharmonic instability*, corresponding to a resonance between the frequency f of the two-dimensional traveling waves and its subharmonic $f/2$ (see Fig. 8.3). This instability triggers a doubling of the wavelength of the primary waves. Herringbone patterns resemble checkerboards, which justifies the alternative term "checkerboard patterns" used sometimes in falling film studies. Both modes, the synchronous instability and herringbone pattern instability, are reminiscent of

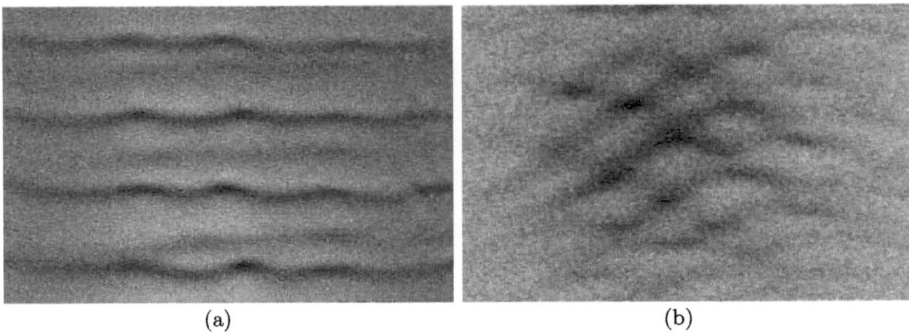

(a)							(b)

Fig. 8.3 (**a**) Synchronous instability; (**b**) herringbone pattern. From Émery and Brosse [91]. Courtesy of Prof. P. Manneville

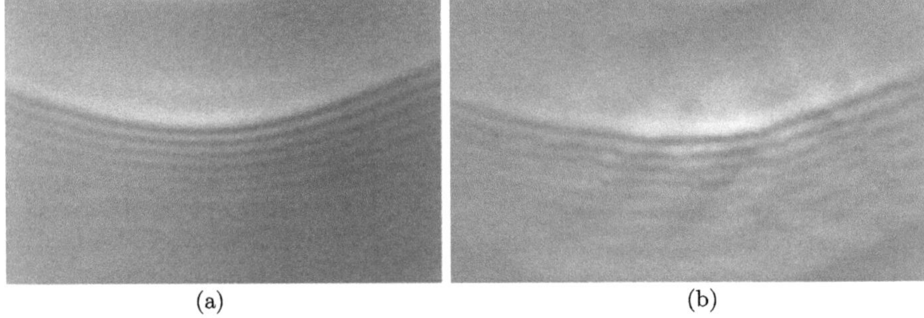

(a)							(b)

Fig. 8.4 (**a**) Quasi-two-dimensional solitary wave; (**b**) spanwise instability of a quasi-two-dimensional solitary wave. From Émery and Brosse [91]. Courtesy of Prof. P. Manneville

aligned and staggered Λ-vortices developing in transitional boundary layers, and they are analogous to the type-K and type-H transitions, respectively [113]. At high enough forcing frequency, the flow becomes disordered before the two-dimensional solitary waves have a chance to appear because three-dimensional instabilities are stronger than two-dimensional, which explains the absence of region III in the corresponding picture (Fig. 8.2(a)).

At low frequencies (Fig. 8.2(b)) saturated waves triggered by the flat film primary instability experience a secondary instability which eventually leads to large-amplitude two-dimensional solitary waves in the form of fast humps preceded by capillary ripples (region III) (see Fig. 7.43 in Chap. 7). Such waves belong to the γ_2 family, following the terminology introduced in [48, 50] (see also Sects. 5.3.2 and 7.2.3). They are generally unstable to transverse perturbations, which leads to the last stage of secondary three-dimensional instabilities (region IV, see Fig. 8.4). For films of water on a vertical wall, the spanwise modulations of the γ_2 waves seem to saturate for Re below approximately 40 [75]. By increasing Re, the waves tend to break into horseshoe-like solitary waves (their shape strongly resembles a horseshoe) having pointed fronts and long oblique legs. For the three-dimensional Kawahara equation, such waves were examined in [233–235] and for the three-dimensional Kapitza–Shkadov model in [76] (the model will be discussed shortly). These waves are reminiscent of the Λ vortices in boundary layers (e.g., [2]; hence the christening "Λ solitons" in [76]). At even higher values of Re the flow becomes increasingly disorganized (cf. Fig. 8.1), but one can still recognize three-dimensional pulses in what appears to be a randomly disturbed surface.

Finally, at very low forcing frequencies, saturated γ_1 waves (region II) do not show up. Inlet forcing then directly generates solitary-like wavetrains of the γ_2 family.

8.2 Modeling of Three-Dimensional Film Flows

We now turn to the three-dimensional formulation of the problem. the aim is to obtain equations in the streamwise (x) and spanwise (z) coordinates by averaging the governing equations over the cross-stream coordinate y. This is the approach followed in [238] and is an extension of the averaging procedure developed in Chap. 6 for two-dimensional flows. Therefore, it is based on the long-wave approximation, which ensures slow time and space modulations of the Nusselt flat film solution, formally expressed as $\partial_t, \partial_x, \partial_z = \mathcal{O}(\varepsilon)$ with $\varepsilon \ll 1$ the gradient expansion parameter.

The first step in reducing the Navier–Stokes equations and associated wall and free-surface boundary conditions to simpler equations consists of the elimination of the pressure in the Navier–Stokes equations truncated at $\mathcal{O}(\varepsilon^3)$. This leads to the three-dimensional second-order boundary layer equations, developed in Sect. 4.1 and rewritten below for purposes of clarity:

$$3\varepsilon Re\big[\partial_t u + \partial_x\left(u^2\right) + \partial_y(uv) + \partial_z(uw)\big]$$

$$= 1 + \partial_{yy} u - \varepsilon Ct\partial_x h + \varepsilon^2\big[2\partial_{xx} u + \partial_{zz} u + \partial_{xz} w - \partial_x(\partial_y v|_h)\big]$$

$$+ \varepsilon^3 We(\partial_{xxx} h + \partial_{xzz} h), \tag{8.1a}$$

$$3\varepsilon Re\big[\partial_t w + \partial_x(uw) + \partial_y(vw) + \partial_z\left(w^2\right)\big]$$

$$= \partial_{yy} w - \varepsilon Ct\partial_z h + \varepsilon^2\big[2\partial_{zz} w + \partial_{xx} w + \partial_{xz} u - \partial_z(\partial_y v|_h)\big]$$

$$+ \varepsilon^3 We(\partial_{xxz} h + \partial_{zzz} h), \tag{8.1b}$$

$$\partial_x u + \partial_y v + \partial_z w = 0, \tag{8.1c}$$

together with the no-slip and no-penetration condition at the wall,

$$u = v = w = 0 \quad \text{at } y = 0, \tag{8.1d}$$

and the projections of the tangential stress balance at the free surface along the x and z directions, respectively:

$$\partial_y u = \varepsilon^2\big[\partial_z h(\partial_z u + \partial_x w) + 2\partial_x h(2\partial_x u + \partial_z w) - \partial_x v\big] \quad \text{at } y = h, \tag{8.1e}$$

$$\partial_y w = \varepsilon^2 \left[\partial_x h (\partial_z u + \partial_x w) + 2 \partial_z h (2 \partial_z w + \partial_x u) - \partial_z v \right] \quad \text{at } y = h. \quad (8.1\text{f})$$

These equations are invariant under the exchange $\{u \leftrightarrow w, \; x \leftrightarrow z\}$, except for the gravity term, equal to unity, in (8.1a).

The Nusselt flat film solution is a parallel flow with no spanwise component, i.e., $w = 0$. A first approach to the problem would therefore be to consider w of $\mathcal{O}(\varepsilon)$, assuming that spanwise flows are triggered by weak three-dimensional modulations of two-dimensional waves, thus simplifying the analysis of the three-dimensional flow dynamics [227]. However, we assume $w = \mathcal{O}(1)$ and hence the velocity components in the x and z directions are treated equally.

For simplicity and as done in the derivation of the two-dimensional models in Chap. 6, we shall treat Re as an $\mathcal{O}(1)$ parameter whereas We must be large, corresponding to the strong surface tension limit, more specifically, $We = \mathcal{O}(\varepsilon^{-2})$. These assumptions can be relaxed but the final equations remain the same, as with the derivation of the two-dimensional models (Sect. 6.4). Truncated at first order, which is when second-order viscous terms are suppressed, system (8.1a)–(8.1f) reduces to the three-dimensional first-order boundary layer equations. Further, by assuming self-similar velocity profiles for the velocity components u and w, substituting these profiles into the x and z components of the momentum equation of the three-dimensional second-order boundary layer equations and averaging these equations across y, i.e., extending effectively the procedure in Sect. 6.2.1 for the Kapitza–Shkadov model for two-dimensional flows, leads to the formulation of the Kapitza–Shkadov model for three-dimensional flows [71]. A gradient expansion of the three-dimensional Kapitza–Shkadov model leads to the three-dimensional BE for a vertical plane (Sect. 5.1). Finally, with appropriate orders of magnitude assignments for the different parameters, a weakly nonlinear expansion of the three-dimensional BE leads to the three-dimensional Kawahara equation (Sect. 5.2.2; see also [233–235]) or the three-dimensional KS equation.

Several numerical studies of the three-dimensional dynamics of film flows were based on these reduced equations [44, 52, 71, 74, 127]. These studies focused on the three-dimensional instability of two-dimensional periodic waves or numerical experiments for the fully three-dimensional problem. Recent theoretical and numerical efforts by use of the three-dimensional Kapitza–Shkadov model focused on the stability of well-separated (isolated) two-dimensional pulses to three-dimensional disturbances, the mechanism by which two-dimensional pulses are destabilized, leading to the formation of three-dimensional pulses, construction of the latter and their stability to three-dimensional disturbances [75, 76]. One of the main findings in [75, 76] was that three-dimensional solitary pulses result from the instability of well-separated two-dimensional solitary pulses. However, all the above studies were generally limited to the vertical case, $Ct = 0$—with the exception of a discussion in [75] for small inclination angles by using the Nepomnyashchy model equation [188] (Sect. 5.2).

Three-dimensional numerical computations and stability analyses for two-dimensional periodic waves based on the Kapitza–Shkadov model or the first-order boundary layer equations revealed only a subharmonic instability leading to checkerboard/herringbone patterns [52, 282]. However, Liu et al. [170] clearly observed the widespread presence of the three-dimensional synchronous secondary instability of the saturated slow γ_1 waves. The subharmonic instability leading to checkerboard/herringbone patterns was observed in a relatively narrow range of parameters. Synchronous instability is therefore not satisfactorily captured by first-order equations.

Here, the stability characteristics of γ_1 slow periodic waves are examined in detail by extending some of the models developed in Chap. 6 to three dimensions while the stability of well-separated fast solitary pulses by using the three-dimensional models developed in this monograph (i.e., extending the studies in [75, 76]), is left to a feature study. An important question that is addressed here is the origin of the synchronous in-phase instability reported by Liu et al. [170].

To account for both second-order inertia and viscous effects the weighted-residual approach at second order detailed in Sect. 6.7 is extended appropriately to three-dimensional flows as was done in [238]. Six fields are needed for the velocity components at second order: both the streamwise and spanwise flow rates, $q_\parallel = \int_0^h u\,dy$ and $q_\perp = \int_0^h w\,dy$, respectively, and four corrections, s_1, s_2, r_1 and r_2, corresponding to the polynomial test functions F_1 and F_2 and accounting for the deviations of the velocity profiles from their zeroth-order parabolic shapes, i.e., the polynomial F_0 (for details see Appendix E.2).

The boundary layer equations are then averaged using the Galerkin method by formulating residuals $\langle E_\parallel, F_i \rangle$ and $\langle E_\perp, F_i \rangle$ where $\langle f, g \rangle = \int_0^h fg\,dy$, and $E_\parallel$ and $E_\perp$ refer to the streamwise (8.1a) and spanwise (8.1b) momentum balances, respectively. These residuals yield a system of six evolution equations for h, $q_\parallel$, s_1, s_2, $q_\perp$, r_1 and r_2, completed with the mass balance obtained through integration of (8.1c) across the layer depth, $\partial_t h + \partial_x q_\parallel + \partial_z q_\perp = 0$. This system is referred to as the *full second-order model* for three-dimensional flows and is given explicitly in Appendix E.2 in terms of the Shkadov scaling.

The regularization procedure developed in Sect. 6.9.2 is extended here to three dimensions with the aim of reducing the three-dimensional full second-order model to only three equations for h, $q_\parallel$ and $q_\perp$. We shall demonstrate that the three-dimensional regularized model does capture the synchronous patterns observed by Liu et al. The same is also true for the full second-order model, but it is more complicated and its numerical implementation is much more involved. The simplified model does not capture the synchronous instability but it does capture the herringbone one, which is also captured by the regularized model. Hence, the regularized model serves as a useful prototype for the mathematical and numerical scrutiny of three-dimensional effects on film flows.

First-order expressions of the fields s_1, s_2, r_1 and r_2 are readily obtained from the truncation at $\mathcal{O}(\varepsilon)$ of the residuals corresponding to the weights F_1 and F_2. Substitution of these expressions into the first residuals, $\mathcal{R}_{0,\parallel} = \langle E_\parallel, F_0 \rangle$ and $\mathcal{R}_{0,\perp} = \langle E_\perp, F_0 \rangle$, produces second-order inertia terms, formally denoted as $\mathcal{R}_{0,\parallel}^{(2),\delta}$ and $\mathcal{R}_{0,\perp}^{(2),\delta}$. These terms contain nonlinearities that may lead to nonphysical singularities and hence are removed by adjusting algebraic preconditioners, as was done in the two-dimensional case in Sect. 6.9.2. Residuals $\mathcal{R}_{0,\parallel}$ and $\mathcal{R}_{0,\perp}$ are then sought in the form $\mathcal{G}_{\|}^{-1}\mathcal{F}_{\|}$ and $\mathcal{G}_{\perp}^{-1}\mathcal{F}_{\perp}$ where $\mathcal{F}_{\|}$ and $\mathcal{F}_{\perp}$ correspond to the expressions of the residuals $\mathcal{R}_{0,\|}$ and $\mathcal{R}_{0,\perp}$ when a parabolic velocity profile for the components u and w is assumed; i.e., when the corrections s_i and r_i are neglected. Isolating inertia terms, one thus sets

$$\mathcal{G}_{\|}\left(\varepsilon\mathcal{R}_{0,\|}^{(1),\delta} + \varepsilon^2\mathcal{R}_{0,\|}^{(2),\delta}\right) = \varepsilon\mathcal{R}_{0,\|}^{(1),\delta} \quad\text{and}$$

$$\mathcal{G}_{\perp}\left(\varepsilon\mathcal{R}_{0,\perp}^{(1),\delta} + \varepsilon^2\mathcal{R}_{0,\perp}^{(2),\delta}\right) = \varepsilon\mathcal{R}_{0,\perp}^{(1),\delta}, \tag{8.2}$$

where superscripts refer to first-order and second-order inertia terms. Zeroth-order expressions of the flow rates, $q_{\|} = h^3/3 + \mathcal{O}(\varepsilon)$ and $q_{\perp} = \mathcal{O}(\varepsilon)$, i.e., the Nusselt flat film flow, are next invoked to reduce the degree of nonlinearities of the regularization factors $\mathcal{G}_{\|}$ and $\mathcal{G}_{\perp}$. Consequently, from the inertia terms $\mathcal{R}_{0,\perp}^{(2),\delta}$ induced by deviations of the spanwise velocity field from the parabolic profile appearing asymptotically at order $\mathcal{O}(\varepsilon^3)$, one merely gets, $\mathcal{G}_{\perp} = 1 + \mathcal{O}(\varepsilon^2)$. Similarly, the asymptotic expression of $\mathcal{R}_{0,\|}^{(2),\delta}$ corresponds exactly to the one obtained for a two-dimensional flow. Hence,

$$\mathcal{G}_{\perp} \equiv 1 \quad\text{and}\quad \mathcal{G}_{\|} \equiv \left(1 - \frac{3\varepsilon Re}{70}q_{\|}\partial_x h\right)^{-1}, \tag{8.3}$$

where $\mathcal{G}_{\|}$ is identical to the expression (6.91).

Introducing now the Shkadov scaling, $\{3\varepsilon Re \to \delta,\ \varepsilon Ct \to \zeta,\ \varepsilon^3 We \to 1\}$, the *three-dimensional regularized model* finally reads

$$\partial_t h = -\partial_x q_{\|} - \partial_z q_{\perp}, \tag{8.4a}$$

$$\delta\partial_t q_{\|} = \delta\mathcal{I}_{\|}^{2D} + \mathcal{G}_{\|}\left\{\frac{5}{6}h - \frac{5}{2}\frac{q_{\|}}{h^2} + \delta\mathcal{I}_{\|}^{3D} + \eta[\mathcal{D}_{\|}^{2D} + \mathcal{D}_{\|}^{3D}] + \frac{5}{6}h\partial_x\mathcal{P}\right\}, \tag{8.4b}$$

$$\delta\partial_t q_{\perp} = \delta\mathcal{I}_{\perp}^{2D} - \frac{5}{2}\frac{q_{\perp}}{h^2} + \delta\mathcal{I}_{\perp}^{3D} + \eta\left(\mathcal{D}_{\perp}^{2D} + \mathcal{D}_{\perp}^{3D}\right) + \frac{5}{6}h\partial_z\mathcal{P}, \tag{8.4c}$$

where the terms $\mathcal{I}$ and $\mathcal{D}$ originate from inertia and viscous dispersion, and $\mathcal{P} = \zeta(1-h) + (\partial_{xx} + \partial_{zz})h$ is the pressure distribution. In (8.4b), we have also separated terms already present in the two-dimensional model (superscript $2D$) from those arising from the spanwise dependence (superscript $3D$). Subscripts $\|$ and $\perp$ indicate terms that are symmetric under the exchange $\{q_{\|} \leftrightarrow q_{\perp}, x \leftrightarrow z\}$. The three equations, (8.4a), (8.4b) and (8.4c), correspond to the mass conservation and averaged momentum balances in the directions x and z, respectively. Viscous drag is represented by the terms $\frac{5}{2}q_{\|}/h^2$ in (8.4b) and $\frac{5}{2}q_{\perp}/h^2$ in (8.4c). As with system (8.1a)–(8.1c), the gravity acceleration contributes only to the streamwise momentum balance through the term $\frac{5}{6}h$ in (8.4b). The full three-dimensional second-order regularized model in (8.4a)–(8.4c) is given explicitly in Appendix E.3.

The regularized model (8.4a)–(8.4c) is fully consistent with the BE long wave expansion up to second order given in Sect. 5.1.2, while the *three-dimensional simplified model* (corresponding to the averaging of the momentum balance equations in the cross-stream direction assuming both parabolic velocity profiles and weights) is not, much like its two-dimensional counterpart (see Sect. 6.8.2). The latter can be recovered from the former by replacing the factor $\mathcal{G}_\parallel$ with unity, or, equivalently, by assuming the order of s_i, r_i to be higher than ε, so that their derivatives can be neglected in the full second-order model (E.6a)–(E.6c) given in Appendix E.2.

8.3 Floquet Analysis: Three-Dimensional Stability of γ_1 Waves

We are interested in the stability of two-dimensional periodic waves to transverse perturbations. The aim is to understand the experimental observations by Liu et al. [170]. We note that the two main types of instabilities observed for small inclination angles in [170], namely a synchronous transverse modulation and a herringbone-pattern instability, are connected with the instability of two-dimensional periodic waves and not well-separated (isolated) two-dimensional solitary pulses. As a matter of fact, by imposing a periodic forcing at the inlet, Liu et al. observed two-dimensional periodic waves with the same periodicity in time, at least prior to the onset of secondary instabilities. Integrating the two-dimensional mass balance, $\partial_t h + \partial_x q = 0$, over a period shows that the temporal mean flow rate, $\langle q \rangle_t = \tau^{-1} \int_0^\tau q\, dt$, with τ being the period, is constant at each location on the plate, at least prior to secondary instabilities, and is therefore equal to its value $1/3$ at the inlet (see also Sect. 5.3.1). Nonlinear saturation of the spatial growth of the inlet signal yields waves that are periodic in time and remain periodic in space over a long distance. These waves are nearly stationary in a frame of reference moving at a constant speed c and approach the periodic traveling wave solutions considered in Chaps. 5 and 7.

Experimentally obtained wave profiles indicate that the traveling waves selected by the linear inception are of γ_1 type (see Sect. 7.3.1 and Fig. 7.39), i.e., slow waves with deep troughs and bumped crests. Our effort is accordingly concentrated on the stability analysis of the γ_1 traveling waves. These waves have been computed using AUTO-07P [79] by continuation, starting from infinitesimal sinusoidal waves at linear threshold and increasing the period (see Appendix F.2). The constant flux condition $\langle q \rangle = 1/3$ was enforced by adjusting the flow rate in the moving frame, $q_0 = \int_0^h (u - c)\, dy$ (since traveling waves are stationary in their moving frame, spatial average over a wavelength and temporal average over a period are identical, and it is not necessary to invoke the subscripts used in Sect. 5.3.1 to identify the variable used in the definition of the average when traveling waves are considered).

The stability of traveling wave solutions can be analyzed within the framework of Floquet theory [54, 93]. In the moving frame $\xi = x - ct$, the system of equations reduces to a set of ordinary differential equations. Each field X, e.g., the film thickness h, the flow rate q or the amplitudes s_1, s_2 of the corrections to the parabolic velocity, is expanded in the moving frame as

$$X(\xi, z, t) = X_0(\xi) + \epsilon \tilde{X}(\xi, z, t), \qquad (8.5)$$

where $\epsilon \ll 1$, X_0 is the value of the field for the base two-dimensional traveling wave and $\tilde{X}$ is the perturbation. The equations to be solved for the perturbations can be written as

$$\partial_t \tilde{\mathbf{X}} = \mathbf{L}(\mathbf{X}_0; \partial_\xi, \partial_z)\tilde{\mathbf{X}}, \tag{8.6}$$

where $\tilde{\mathbf{X}}$ and $\mathbf{X}$ are the vectors formed by the sets of perturbations $\tilde{X}$ and solution fields X. The linear matrix-differential operator $\mathbf{L}$ is periodic in ξ with the periodicity of the traveling wave solution $\mathbf{X}_0(\xi + 2\pi/k_x, z, t) = \mathbf{X}_0(\xi, z, t)$, where k_x is its wavenumber.

Floquet theory of ordinary differential equations with a periodic operator suggests the inclusion of the periodicity of the base state X_0 into the normal mode projection of the perturbation $\tilde{X}$, which is then expanded as

$$\tilde{X}(\xi, z, t) \equiv \sum_{\varphi, k_z} \tilde{X}_{\varphi, k_z}(\xi) \exp\big[i(\varphi k_x \xi + k_z z - \omega t)\big], \tag{8.7}$$

where $\tilde{X}_{\varphi, k_z}$ is periodic in ξ with period $2\pi/k_x$, k_x is the wavenumber of the basic two-dimensional stationary wave, k_z is the wavenumber of the perturbation in the transverse direction and ω is the complex angular frequency. The "detuning parameter" φ is the ratio of the streamwise wavenumber of the perturbation to that of the base state, hence $\varphi \in [0, 1]$. $\varphi \in \mathbb{Q}$ signals a subharmonic mode, especially $\varphi = 1/2$, and $\varphi \notin \mathbb{Q}$ an *incommensurate modulated mode*.

The linearized set of equations can then be formally written as

$$\omega \tilde{\mathbf{X}}_{\varphi, k_z} = \mathbf{L}_{\varphi, k_z}(\mathbf{X}_0, \partial_\xi; c, q_0, \delta, \zeta, \eta, \varphi, k_z)\tilde{\mathbf{X}}_{\varphi, k_z}, \tag{8.8}$$

where $\mathbf{L}_{\varphi, k_z}$ is a linear operator. Equation (8.8) constitutes an eigenvalue problem for ω with $\tilde{\mathbf{X}}_{\varphi, k_z}$ the corresponding eigenvector. The maximum imaginary part of ω, denoted by ω_{i}^M, corresponding to the fastest growing temporal perturbation, is of interest from the experimental point of view.

The parameter space $\varphi \times k_z$ can be reduced by invoking two symmetries: (i) the reflection of the waves in the spanwise direction, which allows us to consider only positive k_z; (ii) the nature of the base equations, which makes (8.8) invariant under the transformation, $(\varphi, k_z, \omega, \tilde{\mathbf{X}}_{\varphi, k_z}) \to (-\varphi, -k_z, \omega^\star, \tilde{\mathbf{X}}_{\varphi, k_z}^\star)$, with the star denoting complex conjugation. Thus, the parameter space $\varphi \times k_z$ can be limited to $[0, \frac{1}{2}] \times [0, \infty]$. For convenience, the eigenvalue problem (8.8) can be solved in Fourier space. For this purpose up to 256 real modes are enough to represent the two-dimensional waves and up to 128 complex modes to represent the perturbation (limited to 32 for the full second-order model owing to its complexity). The two-dimensional waves are computed with AUTO-07P, but the eigenvalue problem is solved in Fourier space where both the basic solution and the perturbation are decomposed into a series of Fourier modes. Convergence of the discretized solution in Fourier space to the solution of (8.8) has been confirmed by doubling the number of modes.

Liu et al. [170] considered a falling film of a glycerol–water mixture ($\rho = 1070\ \mathrm{kg\,m^{-3}}$, $\nu = 2.3 \times 10^{-6}\ \mathrm{m^2\,s^{-1}}$ and $\sigma = 67 \times 10^{-3}\ \mathrm{N\,m^{-1}}$), with $\beta = 6.4°$ and $Re = 56$. They measured the wavelength of the two-dimensional base state λ_x

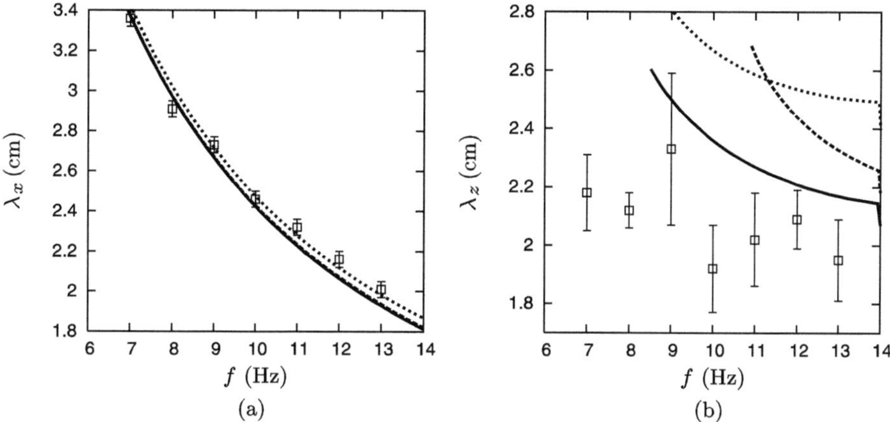

Fig. 8.5 Streamwise wavelengths λ_x of two-dimensional periodic waves (**a**) and spanwise wavelengths λ_z (**b**) of the most amplified three-dimensional perturbations versus forcing frequency f, with $\beta = 6.4°$, $Re = 56$ and $\Gamma = 2002$. *Open squares* correspond to the experimental findings [170]. *Solid*, *dashed* and *dotted lines* correspond to the full second-order model (E.6a)–(E.6c), the regularized model (8.4a)–(8.4c) and the simplified model, respectively. Notice that in panel (**a**), *solid* and *dashed lines* are superimposed

as well as the wavelength of the transverse modulations λ_z as a function of frequency of the periodic forcing. Results of the Floquet analysis using the full second-order, regularized and simplified models are presented in Fig. 8.5 by use of dimensional units and are contrasted with experiments. The agreement with experiments turns out to be better when streamwise and spanwise velocities are assumed to be of the same order, as done here, than when the spanwise velocity field w is assumed to be of $\mathcal{O}(\varepsilon)$. The computed wavelengths λ_x of γ_1 waves are in good agreement with experimental findings. As with the results reported by Liu et al., the computations also indicate relatively small variations of λ_z with frequency. The transverse wavelengths of the fastest growing perturbations for the regularized and the full second-order models are close to each other as f increases, whereas the Floquet analysis of the simplified model indicates larger wavelengths in the transverse direction as f increases.

These observations emphasize the important role played by the second-order inertia terms—induced by the deviations of the velocity profile from its parabolic shape—in the mechanism of the three-dimensional secondary instability. Yet, at low frequencies, the fastest growing perturbation is spanwise-uniform (two-dimensional, λ_z tends to infinity) while the experimental findings suggest a three-dimensional instability with a finite wavelength λ_z in the transverse direction. Another difference between the results of the Floquet analysis and the experimental findings is that the detuning parameter for the most amplified perturbation (not shown) systematically corresponds to a subharmonic secondary instability ($\varphi = \frac{1}{2}$), whereas Liu et al. reported a synchronous instability ($\varphi \approx 0$). These differences may be explained by (i) the relatively weak selection mechanism of the evolving wave pattern by the secondary instability, with the growth rate being weakly sensitive to

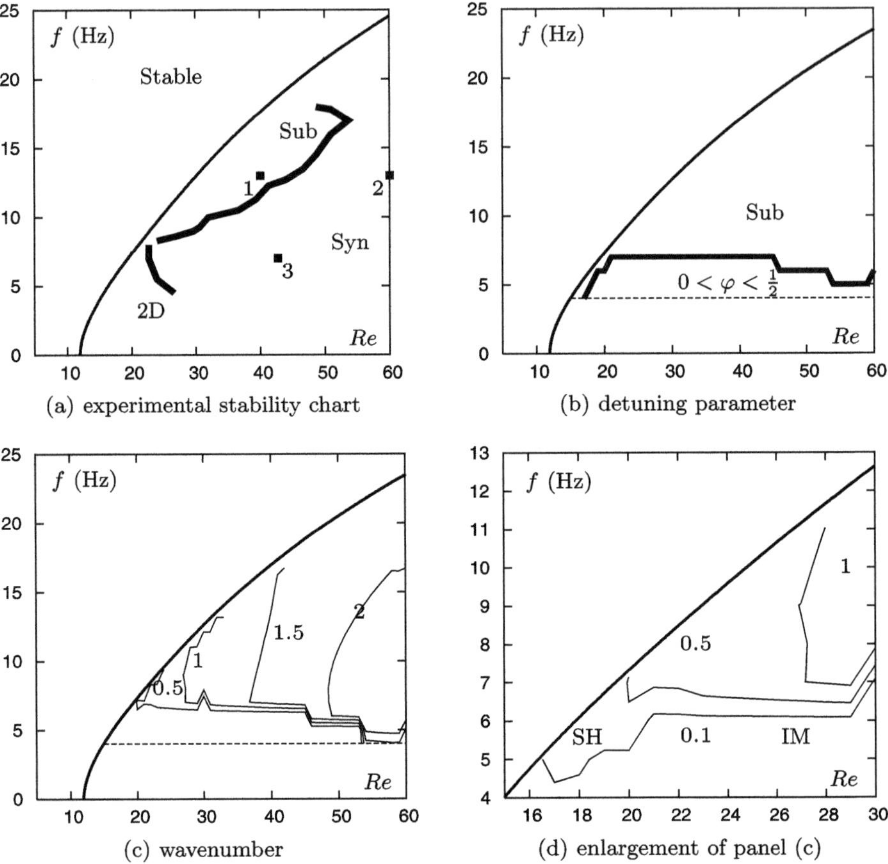

Fig. 8.6 Stability of the γ_1 waves to three-dimensional disturbances as function of the Reynolds number Re and the frequency f for $\beta = 4°$ and $\Gamma = 2340$ [170, Fig. 6]. (**a**) Experimental stability chart. Stability zones are bounded by *thick lines*: "2D" where no three-dimensional instability was observed, "Sub" for three-dimensional subharmonic instability, and "Syn" for three-dimensional synchronous instability. The neutral stability curve is represented by a *thin solid line* (Orr–Sommerfeld analysis). *Squares* refer to parameter sets in Table 8.1. (**b**) Detuning parameter, where the synchronous (Syn) and subharmonic (Sub) instability regions correspond to $\varphi = 0$ and 0.5, respectively. (**c**) Wavenumber k_z of the fastest growing transverse modulation (in cm^{-1}). (**d**) Enlargement of panel (**c**): "SH" subharmonic two-dimensional instability ($\varphi = \frac{1}{2}$), "IM" incommensurate modulated two-dimensional mode ($0 < \varphi < \frac{1}{2}$). *Dashed lines* indicate the limit (4 Hz) of the computations in panels (**b**, **c**). The results presented in panels (**b**)–(**d**) have been obtained using the full second-order model (6.1), (6.78)

variations in φ and k_z; (ii) the simultaneous saturation of the two-dimensional base traveling waves and growth of the three-dimensional instability, an effect that is not taken into account by our Floquet analysis, which presumes that two-dimensional waves saturate before undergoing an instability.

Figure 8.6(a) summarizes the experimental findings by Liu et al. in the $Re \times f$-plane for the glycerol-water mixture of Fig. 8.5 and with $\beta = 4°$. The stability zones are bounded by thick solid lines. In the "2D" region no three-dimensional instability was observed: there are γ_1 waves followed by a two-dimensional instability leading to γ_2 waves. The regions "Sub" and "Syn" correspond to three-dimensional subharmonic and three-dimensional synchronous instabilities, respectively. Corresponding results for the stability of γ_1 waves are presented in Figs. 8.6(b–d) obtained from the full second-order model. The results for the solutions to the regularized and simplified models are very similar to those obtained with the full second-order model and thus are not shown. The detuning parameter (Fig. 8.6(b)) and the spanwise wavenumber (Fig. 8.6(c, d)) of the fastest growing perturbation have been computed with a Reynolds number step of 1 and a frequency step of 1 Hz (the lowest frequency considered is 4 Hz owing to the large number of modes necessary to represent the solution). It appears that, for a given frequency, k_z decreases steadily as Re is lowered and goes to zero in a region close to the neutral stability curve (see Fig. 8.6(d)).

On the other hand, for a fixed Reynolds number lowering the frequency induces a rapid decrease of k_z in a small frequency interval corresponding to the boundary separating two- and three-dimensional secondary instabilities. This agrees well with the results of Liu et al., who reported two-dimensional flows rather close to the threshold of the primary instability.[2] In this small region, the γ_1 waves undergo a subharmonic two-dimensional instability ($\varphi = \frac{1}{2}$). At low frequency and moderate Reynolds number, the instability is also found to be two-dimensional ($k_z = 0$) but corresponds to an incommensurate mode ($\varphi \in]0, \frac{1}{2}[$). This provides an indication that a frontier between two-dimensional and three-dimensional flows actually exists and is not just an experimental artifact due to finite-size effects. At low frequency and moderate Reynolds number, the Floquet stability analysis of γ_1 waves predicts a two-dimensional region wider than reported in experiments. This discrepancy may arise from the limited length of the experimental plane, from the level of ambient noise in the experiments or because γ_2 waves are observed (γ_2 waves are likely to develop in this region of the parameter plane in place of γ_1 waves).

It should be emphasized at this point that γ_1 waves are in fact unstable to both two- and three-dimensional disturbances (and in fact in all cases we found at least one unstable mode for purely two-dimensional disturbances and purely three-dimensional ones). In some cases, the growth rate of two-dimensional disturbances can be larger than the growth rate of three-dimensional ones and the instability will be two-dimensional instead of three-dimensional (this happens in Fig. 8.6 at small frequencies). The finding that γ_1 waves are unstable to both two- and three-dimensional disturbances is consistent with the experimental observations by Liu et al. They never observed γ_1 waves over all the plane. Beyond the inception region

---

[2]The experimental boundary separating two- and three-dimensional secondary instabilities is the bottom solid line in Fig. 8.6(a); it does not correspond exactly to the boundary found numerically (this is the accumulation of the different k_z curves in Fig. 8.6(c) as $k_z \to 0$). This boundary has not been resolved very accurately due to the relatively coarse f step of 1 Hz.

Table 8.1 Dimensionless wavenumber k, phase speed c and averaged thickness $\langle h \rangle$ of the computed γ_1 two-dimensional traveling waves corresponding to the experimental conditions in [170]. The constant mean flow rate condition $\langle q \rangle = 1/3$ was enforced. The control parameters are the Reynolds number Re, the inclination β, the Kapitza number Γ and the forcing frequency f

Set	Re	β (deg)	Γ	f (Hz)	k	c	$\langle h \rangle$
1	40.0	4.0	2340	13	1.565	0.824	0.987
2	60.0	4.0	2340	13	1.494	0.689	0.970
3	42.7	4.0	2340	7	0.953	0.703	0.975
4	48.0	6.4	2002	10	0.980	0.628	0.965

and the formation of a primary γ_1 wavetrain, this train always undergoes a secondary instability, which could lead to γ_2 waves or three-dimensional negative waves (later referred to also as "depressions"), which subsequently reorganize themselves into three-dimensional horseshoe waves. Which types of waves, γ_2 or three-dimensional negative, emerge from γ_1 waves depends on the distance from the instability threshold: Close to the threshold, γ_2 waves are born from γ_1 ones. Somewhat far from the threshold, three-dimensional negative waves are observed, which eventually reorganize themselves into horseshoe waves. Of course we need forcing (as without forcing, γ_1 waves are not even observed) and noise: Without noise the instability of γ_1 waves is not excited, as first noted in Sect. 7.3.1 (see also Fig. 7.43). In fact, as was pointed out there, this does not mean that γ_1 waves are stable; rather, they behave more as "metastable"-like states. In addition to the absence of noise, there is one more case where γ_1 waves could be observed all along the plane: Very close to the instability threshold and with sufficiently strong viscous dispersion, γ_1 waves could be stable. But this corresponds to more like an "academic" possibility. In practice, being very close to the threshold would imply that the wave amplitude is so small that it is practically impossible to detect any waves (as in some of the experiments by Kapitza, in fact).

As already mentioned, the computations predict an overwhelming presence of the subharmonic scenario ($\varphi = \frac{1}{2}$) whereas Liu et al. observed it only close to the neutral stability curve at large frequencies and moderate Reynolds numbers. In fact, our Floquet analysis predicts a region of synchronous three-dimensional instability at large Reynolds numbers, thus in agreement with experimental findings, only when the regularized model is employed (not shown). Figure 8.7 shows the isocontours of the growth rate ω_i of the fastest growing perturbation in the plane $\varphi \times k_z$ for the three models, corresponding to set 2 of Table 8.1. The full second-order and the regularized models agree well with each other for the selection of the fastest growing spanwise wavenumber, whereas the simplified model predicts longer spanwise wavelengths, a difference that is already noticeable in Fig. 8.5(b) (for sufficiently large f, more specifically $f \gtrsim 11$ Hz in Fig. 8.5(b); $f = 13$ Hz in Fig. 8.7, but the values of β and Re for the two figures are similar). Moreover, Figs. 8.7(a), and 8.7(b) show that ω_i varies very little with the detuning parameter φ. Indeed, for the full second-order and regularized models, the growth rates for $\varphi = 0$ and $\varphi = \frac{1}{2}$ are close to each other so that the instability does not distinguish between them. On

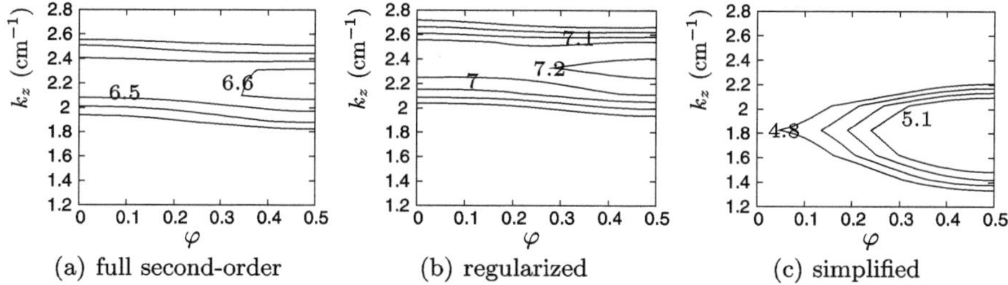

(a) full second-order   (b) regularized   (c) simplified

Fig. 8.7 Isocontours of maximum temporal growth rate in s^{-1} as function of the detuning parameter φ and the transverse wavenumber k_z in cm^{-1}, computed with the different models for set 2 (Table 8.1)

Fig. 8.8 Same caption as for Fig. 8.7 with parameter set 3 (Table 8.1)

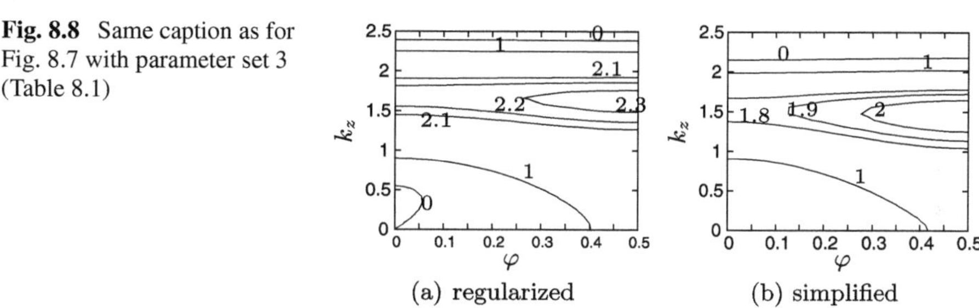

(a) regularized   (b) simplified

the contrary, the simplified model is more selective (see Fig. 8.7(c)) and clearly predicts a subharmonic instability. This result points out again the subtle role played by the second-order inertia terms included in the regularized model in the process of pattern selection. Figure 8.8 presents isocontours of ω_i in the $\varphi \times k_z$-plane for the parameter set 3 of Table 8.1. Results obtained with the full second-order model are not shown since they are too close to those corresponding to the regularized model. It is clear that the detuning parameter does not significantly affect the growth rate and the Floquet stability analysis again does not make a distinction between a synchronous or a subharmonic instability. However, by comparing Fig. 8.8(a) to Fig. 8.8(b), one can see that the growth rate of the fastest growing perturbations of the γ_1 waves is again more sensitive to changes in the detuning parameter φ for the simplified model than for the regularized one.

Finally, let us note that the comparison between results from the Floquet analysis and the experiments is based on three assumptions. First, the γ_1 waves emerge from the primary instability. Second, a broadband white noise is assumed. Third, γ_1 waves are assumed to saturate before the onset of the three-dimensional instability. As indicated by Liu et al. the inlet forcing signal induced by the distributor has time-independent geometric irregularities which preferentially trigger in-phase modulations of the evolving three-dimensional patterns. Therefore, experimental noise contains a larger part of in-phase perturbations than out-of-phase ones, which in turn may trigger the synchronous instability easier than the subharmonic mode, given that they have growth rates close to each other. In fact, Liu et al. were compelled to apply controlled perturbations to subharmonic instabilities. It is precisely because inlet noise may contain significant spanwise perturbations that three-dimensional instabilities may arise before the saturation of γ_1 waves has a chance to develop. Such a sensitivity to changes in inlet conditions can only be checked by numerical simulations of the models in time and space.

8.4 Simulations of Three-Dimensional Flows

The Floquet analysis predicts that the subharmonic scenario is predominant, which contradicts experimental observations. This discrepancy has been attributed to the inability of the secondary instability to discriminate well between a subharmonic instability and a synchronous one, since the maximum growth rate is nearly the same over the whole range $0 \leq \varphi \leq 1/2$ of the detuning parameter. This property makes the three-dimensional instability strongly dependent on the initial conditions, and thus prevents one to relate unequivocally the results of the Floquet analysis to the experimental findings.

We therefore turn to spatio-temporal simulations of the three-dimensional dynamics promoted by an inlet forcing to recover the experimental results by Liu et al. In this section, we discuss time-dependent integrations of the full second-order model (E.6a)–(E.6c), the regularized model (8.4a)–(8.4c) and the simplified model obtained when taking $\mathcal{G}_\parallel = 1$. Periodic boundary conditions in both x and z directions are imposed. This allows us to make use of a pseudo-spectral scheme and thus exploit the good convergence properties of spectral methods. The derivatives are evaluated in Fourier space and the nonlinearities in physical space. The time dependence is accounted for by a fifth-order Runge–Kutta method, which allows controlling the error from the difference with an embedded fourth-order scheme (see [213]). Details of the scheme, such as representation of the variables in Fourier space as well as the treatment of the associated *aliasing phenomenon*, can be found in Appendix F.4. In practice, the time step is adapted to limit the relative error on each variable to 10^{-4}. The explicit character of the algorithm makes it easy to implement for the different models.[3] The computational domain of size $L_x \times L_z$ is discretized with $M \times N$ regularly spaced grid points with coordinates $x_k = k L_x / M$ and $z_j = j L_z / N$. We also define the *energy of deformation* in each direction [127, 213] as the quadratic sum of the Fourier coefficients obtained from the Fourier transform of each streamwise and spanwise profile scanned over the whole computational

---

[3] However, this is not the case when one tries to simulate precisely the spatio-temporal wave dynamics on the whole plane with open-flow conditions as we did in Chap. 7 for two-dimensional flows (Sect. 7.3 and Appendix F.3). Since now nonreflective downstream boundary conditions must be considered, pseudo-spectral schemes using fast Fourier transforms are unavailable. The numerical domain should be sufficiently large to account for the actual physical plane, which means a large number of space steps. To limit the computational time then, implicit schemes should be used.

domain:

$$E_x(t) \equiv \frac{1}{MN} \sum_{j=1}^{N} \left(\sum_{m=1}^{M/2-1} |\hat{A}_m(z_j, t)|^2 \right)^{1/2}, \qquad (8.9a)$$

$$E_z(t) \equiv \frac{1}{MN} \sum_{l=1}^{M} \left(\sum_{n=1}^{N/2-1} |\hat{B}_n(x_l, t)|^2 \right)^{1/2}, \qquad (8.9b)$$

where the spatial Fourier coefficients $\hat{A}_m$ and $\hat{B}_n$ are defined by

$$\hat{A}_m(z, t) = \sum_{l=0}^{M-1} h(x_l, z, t) \exp\left(\frac{2\pi i m l}{M} \right), \qquad (8.9c)$$

$$\hat{B}_n(x, t) = \sum_{j=0}^{N-1} h(x, z_j, t) \exp\left(\frac{2\pi i n j}{N} \right). \qquad (8.9d)$$

Due to the spatial periodicity in the streamwise direction, the simulations correspond to a closed flow, as explained in Sect. 5.3.1. But as mentioned there, the closed-flow condition cannot be achieved experimentally and the open flow condition should be used instead. In fact, the conservation condition in the moving frame

$$\langle h \rangle = \frac{\langle q \rangle - q_0}{c} \qquad (8.10)$$

shows that the averaged thickness $\langle h \rangle$ can be significantly lower than the inlet thickness, depending on the wave characteristics c and q_0. Therefore, in order to improve comparisons of numerical results to experimental data, one can turn to its advantage the closed-flow condition inherent in the numerical scheme by imposing a film thickness tuned to the value obtained from (8.10) for two-dimensional traveling waves at the corresponding forcing frequency using AUTO-07P. Doing so ensures that the right amount of liquid is "inserted" in the computational domain. Since the local flow rate varies as the cube of the local film thickness, this trick can be crucial in recovering experimental results. Thus, the development of two-dimensional waves undergoing three-dimensional instabilities is simulated by enforcing initial conditions in the form

$$h(x, z, 0) = \langle h \rangle + A_x \cos\left(\frac{2\pi n_x x}{L_x} \right) + A_z \cos\left(\frac{2\pi n_z z}{L_z} \right) + A_{\text{noise}}\tilde{r}(x, z), \quad (8.11)$$

where $A_x, A_z, A_{\text{noise}}$ are small amplitudes, $n_x, n_z \in \mathbb{N}$ represent the numbers of harmonic waves in each direction, and $\tilde{r}$ is a random function with values in the interval $[-1, 1]$. The last term of (8.11) accounts for ambient white noise whose amplitude is set to $A_{\text{noise}} = 10^{-3}$. Moreover, to facilitate the comparison with the experimental results, we keep the aspect ratio of the computational domain equal to unity by setting, $L_x = L_z \equiv L$. The value of L must be taken large enough to allow complex flow dynamics. The general form of (8.11) enables us to explore a wide range of experimental results on three-dimensional waves emerging from two-dimensional ones. In the following, we consider three-dimensional modulations of traveling waves of the γ_1 and γ_2 types that are induced by inlet forcing, as well as three-dimensional modulations of natural waves, i.e., waves driven by inlet noise. In practice, the length of the numerical domain is adjusted to fit an integer number of the two-dimensional traveling waves under consideration.

Nonlinearities are well known to generate *aliasing* errors, i.e., distortions of high frequency Fourier modes due to the truncation in Fourier space (sampling) when pseudo-spectral methods are used (see Appendix F.4 where the aliasing phenomenon is discussed in detail). Let us denote by σ_{NL} the order of the nonlinearities involved in the equations to be solved. σ_{NL} is an integer defined as the maximum number of elements involved in a combination of products or divisions. In the case of the Navier–Stokes equations, the advection term $\mathbf{u} \cdot \nabla \mathbf{u}$ is a quadratic nonlinearity. In the case of the simplified model, the advection terms, e.g., $(q_{\parallel}/h)^2 \partial_x h$, give $\sigma_{\mathrm{NL}} = 5$. $\sigma_{\mathrm{NL}} = 5$ also for the full second-order model, whereas for the regularized model, $\sigma_{\mathrm{NL}} = 7$.

As shown in Appendix F.4, applying a low-pass filter such that only the first $2/(\sigma_{\mathrm{NL}} + 1)$ modes are kept, suppresses aliasing errors. This means that in the case of the regularized model, $3/4$ $[2/(7 + 1) = 1/4]$ of the modes should be set to zero at each time step and thus not be used to represent the corresponding solution. Consequently, to maintain the spatial resolution and at the same time to avoid aliasing errors, it is sufficient to increase by 4 the number of mesh points in each direction. This is a numerically costly procedure though unnecessary. In practice, setting only $2/3$ of the modes at each time step appears to be sufficient for all simulations presented in this chapter, except for two cases (sets 7 and 8, Table 8.2) when the wave structure becomes too sharp and needs higher frequencies to be properly resolved, in which case $1/2$ of the modes were set to zero at each time step (this was checked by monitoring the amplitudes of the highest frequency Fourier modes).

8.4.1 Three-Dimensional Modulations of γ_1 Waves

We first consider the transition from γ_1 waves to three-dimensional patterns, which corresponds to the experimental results by Liu et al. [170]. These well-controlled experiments will also serve as a benchmark for a systematic evaluation of the accuracy and usefulness of the different models.

Liu et al. have imposed a spanwise-uniform periodic forcing. To mimic their experiments, we set $A_x = 0.1$ and $A_z = 0$ in (8.11). L is set equal to five times the wavelength $2\pi/k$ of the precursor two-dimensional traveling wave, i.e., $n_x = 5$. The number of grid points for the simulations in this section is $M \times N = 128 \times 64$, hence 64×32 Fourier modes, or effectively 42×21 modes due to the aliasing treatment.

The values of the parameters for the different numerical experiments are indicated in Table 8.1. The flow conditions for an inclination angle $\beta = 4°$ and Kapitza

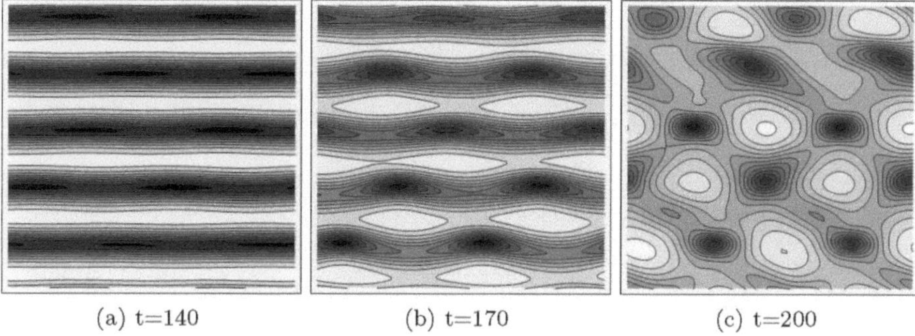

(a) t=140   (b) t=170   (c) t=200

Fig. 8.9 Snapshots of free surface deformations giving rise to a herringbone pattern, computed for the parameter set 1 in Table 8.1 with the regularized model at different times. Isocontours of the thickness are separated by an elevation step of 0.06. The number of grid points is $M \times N = 128 \times 64$ and $L = 2n_x\pi/k$. The amplitudes of the initial periodic forcing are $A_x = 0.1$ and $A_z = 0$, with $n_x = 5$. The *dark and bright zones* correspond to depressions and elevations, respectively

number $\Gamma = 2340$ are first considered (sets 1–3 in Table 8.1 and in Fig. 8.6(a)). Each chosen pair of (frequency, Reynolds number) is indicated by a cross in Fig. 8.6(a). Set 1 (Table 8.1) corresponds to the region of the plane (f, Re) where herringbone patterns were observed experimentally, i.e., the region of subharmonic instability. Simulations of the full second-order, regularized and simplified models agree with both the Floquet analysis and the experimental data. Isocontours of the thickness of the wave patterns are shown at different times in Fig. 8.9 for the regularized model: At its final stage (Fig. 8.9(c)), the film evolves toward a staggered arrangement of round and large humps and thin and deep depressions (three-dimensional negative waves) that agrees well with the experimental observations.

Using the parameter set 2 in Table 8.1, we move next to the region in Fig. 8.6(a) where synchronous secondary instability has been reported in [170], whereas the Floquet analysis predicts a subharmonic instability (compare Fig. 8.6(a) to Fig. 8.6(d)). Time integrations of the different models, given in Fig. 8.10 for the same spanwise energy of deformation E_z, show disagreement: The full second-order model (panel a) shows a *sideband instability*, corresponding to a resonance between the frequency f of the two-dimensional pattern, a low frequency φf (the detuning parameter φ is small), and $(1 - \varphi)f$. The sideband instability observed with the full second-order model then leads to a synchronous pattern, while from the simplified model (panel c) one gets staggered troughs and more deformed crests indicating a subharmonic instability, $\varphi = \frac{1}{2}$. The solution to the regularized model (Fig. 8.10(b)) corresponds to a combination of synchronous and subharmonic modulations, but it is closer to the solution to the full second-order model (and experimental observations) than to the solution of the simplified one. More specifically: (i) crests are hardly deformed whereas troughs tend to form deep isolated depressions; (ii) spanwise and streamwise wavelengths have values close to each other (four spanwise modulations for the full second-order and regularized model, in contrast with three for the simplified one); (iii) nonlinear docking of two neighboring

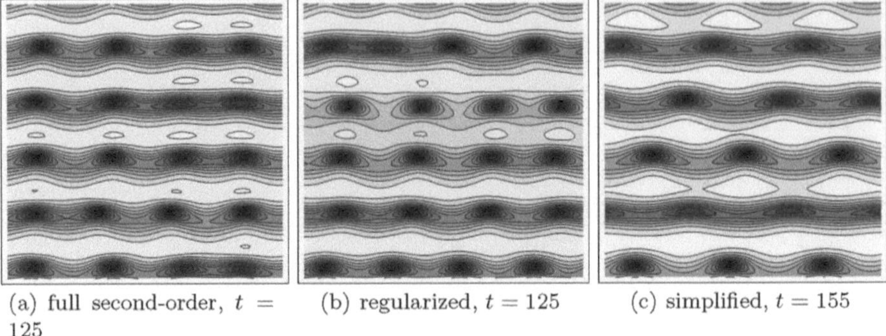

(a) full second-order, $t =$ 125   (b) regularized, $t = 125$ (c) simplified, $t = 155$

Fig. 8.10 Snapshots of free surface deformations computed for parameter set 2 at $E_z \approx 0.05$ for the three models. Isocontours of the thickness are separated by a level difference of 0.08. See also caption of Fig. 8.9

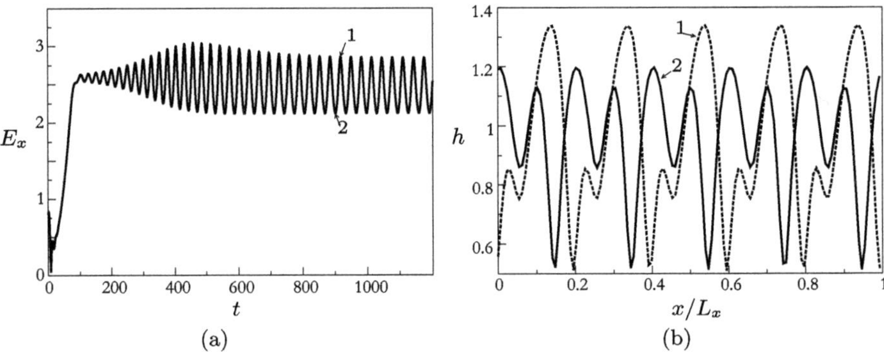

(a)   (b)

Fig. 8.11 (a) Energy of streamwise deformations E_x computed for parameter set 4 in Table 8.1 as function of time; (b) corresponding two-dimensional wave profiles. The full second-order model (E.6a)–(E.6c) has been used for computations and $A_x = 0.1$, $A_z = 0$, $A_{\text{noise}} = 0$, $n_x = 5$, $L_x = 10\pi/k$ for the initial condition

depressions (the two neighboring crests in the top right part of Fig. 8.10(a) and in the bottom left part of Fig. 8.10(b)). Recall that in Figs. 8.7(a) and 8.7(b), the secondary instability does not discriminate between a synchronous ($\varphi = 0$) or a subharmonic ($\varphi = 1/2$) scenario for the parameter set 2. On the other hand, as expected from the linear prediction (Fig. 8.7(c)), the simplified model clearly selects the subharmonic instability, ending in a herringbone pattern (Fig. 8.10(c)). Similar behaviors of the three models (not shown here) also have been found for parameter set 3 (Table 8.1).

Parameter set 4 (Table 8.1) corresponds to a more pronounced inclination angle ($\beta = 6.4°$) and thus to a smaller Kapitza number ($\Gamma = 2002$). The simulations indicate that if the initial excitation is spanwise uniform ($A_z = A_{\text{noise}} = 0$), the two-dimensional steady state corresponds to an oscillatory mode instead of a traveling wave. This is illustrated in Fig. 8.11, where is plotted (a) the time evolution of the streamwise deformation energy E_x and (b) the wave profiles at two different times

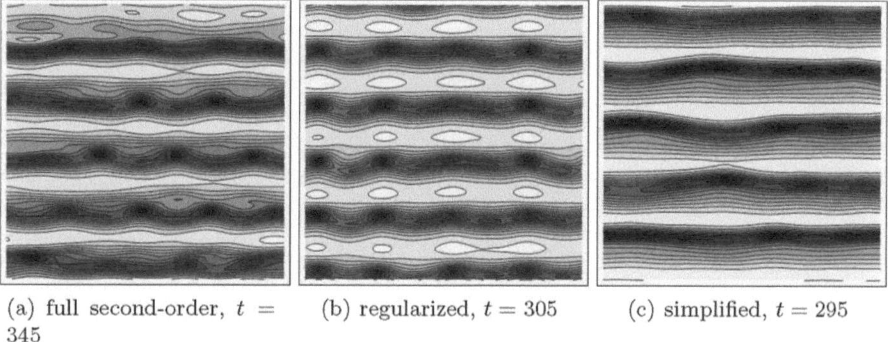

(a) full second-order, $t =$ 345   (b) regularized, $t = 305$ (c) simplified, $t = 295$

Fig. 8.12 Free surface deformations computed for the parameter set 4 at $E_z \approx 0.05$ for the three models. Isocontours of the thickness are separated by an elevation step of 0.08. The amplitude of the initial forcing here is $A_x = 0.2$

corresponding to a maximum (label 1) and a minimum (label 2) of E_x during one oscillation period. Such an oscillatory mode has also been evidenced numerically in the DNS study by Ramaswamy et al. [218] and it is referred to as a *quasi-periodic mode*. Their study also indicates that the quasi-periodic regime is widely present in the case of a vertical wall when the Reynolds number increases. In the language of dynamical systems theory, the flow in the phase space tends to a torus (quasi-periodic regime), instead of evolving toward a limit cycle (traveling wave). This behavior is generated by the destabilization of the existing limit cycle and can be predicted by looking at the maximum growth rate ω_i of Floquet perturbations, which was also found to be positive for the parameter set 4 in Table 8.1.

The wave patterns for the different models are shown in Fig. 8.12. The amplitude of the initial streamwise modulations is increased to $A_x = 0.2$ ($A_z = A_{\text{noise}} = 0$). Both the full second-order and the simplified models yield herringbone patterns whereas the regularized model yields a synchronous pattern, in agreement with experimental data [170]. The observed discrepancy between the results obtained with the full second-order model and the experiment does not invalidate the weighted residuals approach but rather underlines the sensitivity of the three-dimensional dynamics to the boundary conditions at the inlet. In fact, it appears that the onset of the three-dimensional pattern is strongly influenced by the presence of the two-dimensional oscillatory mode and then by the exchange of energy between this mode and the growing spanwise modulations. This exchange depends on the initial conditions and in particular on the amplitude A_x of the initial streamwise modulations. Figure 8.13 shows three-dimensional wave patterns computed with the regularized model for two different values of A_x. Significant qualitative differences can be noted by comparing these patterns to Fig. 8.12(b): At low initial amplitude A_x, the final transverse modulations seem to have longer wavelengths than at larger values of the initial amplitude A_x. Further, crests display out-of-phase modulations for smaller values of A_x and in-phase ones for larger values of A_x. Time evolutions of the energies E_x and E_z are shown in Fig. 8.14. When $A_x = 0.1$, the system clearly approaches the two-dimensional traveling wave solution and remains close to it for a

Fig. 8.13 Free surface
deformations computed for
the parameter set 4 in
Table 8.1 at $E_z \approx 0.05$:
(**a**) $t = 300$, (**b**) $t = 220$

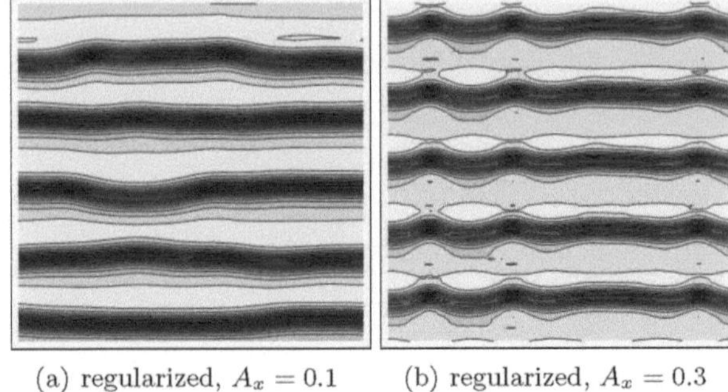

(a) regularized, $A_x = 0.1$ (b) regularized, $A_x = 0.3$

Fig. 8.14 Deformation
energies computed for
parameter set 4 in Table 8.1
using the regularized model
(8.4a)–(8.4c) and various
values of A_x. *Solid and
dashed lines* correspond to
E_x and E_z, respectively.
Figures 8.13(a), 8.12(b) and
8.13(b) correspond to
snapshots taken at the time
instants when E_z crosses the
level 0.05

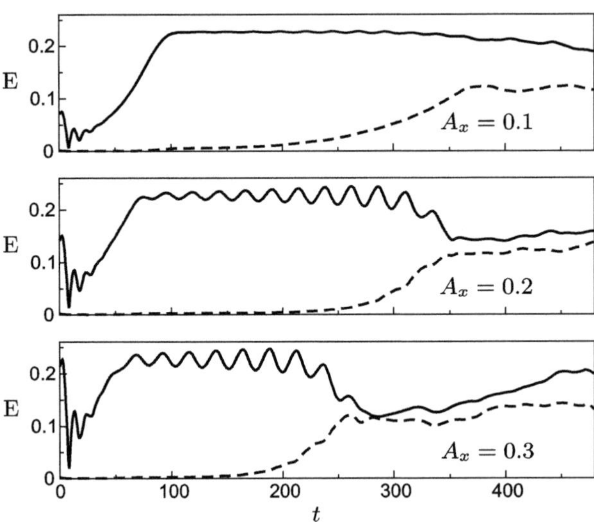

long time. Therefore, the Floquet analysis still applies and the obtained herringbone
pattern corresponds to the predicted subharmonic instability. This is no longer the
case for larger values of A_x where the modulation of the two-dimensional wavetrain
occurs prior to the development of the three-dimensional instability. The observed
synchronous pattern is thus a complex result of two ingredients: The growing two-
dimensional oscillations and the three-dimensional instability.

We have already noticed how sensitive the pattern formation process is to the
initial conditions, a consequence of the poor selection of the synchronous or subharmonic secondary instability. Nevertheless, experiments show a clear selection
of the synchronous instability, most probably triggered by small defects of the inlet distributor. In order to mimic such inlet inhomogeneities in computer simulations, an x-independent noise $\tilde{r}'(z)$ has been added to the initial condition (8.11),
whose amplitude $A^{(z)}_{\mathrm{noise}}$ represents the "inlet roughness." A realistic estimate of
about 1 μm roughness gives an amplitude of $A^{(z)}_{\mathrm{noise}} = 0.01$ for a typical film thickness of 100 μm. Figures 8.15 and 8.16 depict results obtained with the regularized
model together with those obtained experimentally [170, Figs. 7 and 11]. The fig-

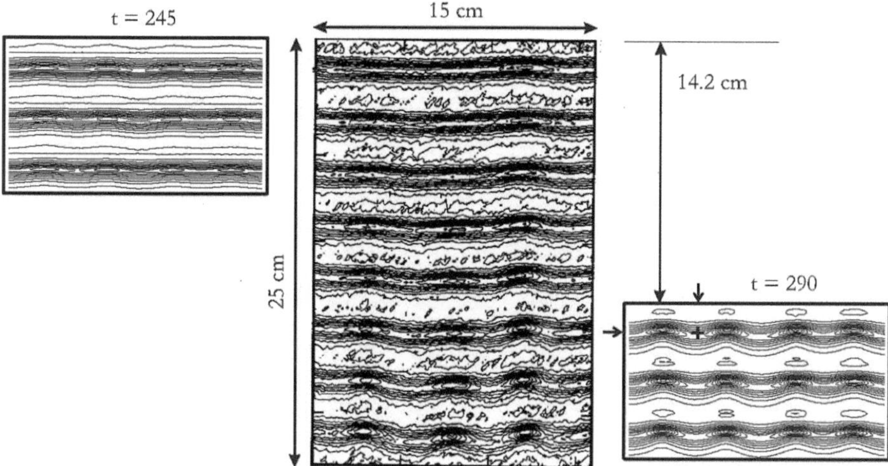

Fig. 8.15 Snapshots of the film free surface obtained using the regularized model (8.4a)–(8.4c) at two different times, along with an experimental snapshot in the center [170, Fig. 7]. Parameter values correspond to set 3 in Table 8.1. $A_x = 0.2$, $n_x = 5$, $A_z = 0$, $L = 2n_x\pi/k$, $A_{\mathrm{noise}} = 10^{-3}$ and an x-independent noise with amplitude $A^{(z)}_{\mathrm{noise}} = 10^{-2}$ is added to mimic the effect of inlet roughness. The size of the computational domain is 89×148 mm. Isocontours of the thickness are separated by an elevation step of 0.06. The location of a saddle point in the right snapshot (see text) is indicated by a *cross and two arrows*

ures reveal the influence of the perturbation, which effectively biases the evolution in favor of the synchronous instability. To facilitate comparison to experimental results, numerical snapshots are separated in the vertical direction by the distance covered by the waves between the two time instants at which the snapshots were taken (roughly 14.2 cm and 5.8 cm in the case of Figs. 8.15 and 8.16, respectively). There is rather good agreement with experiments even though some differences can still be noticed, mostly because of the choice of periodic boundary conditions. The spanwise wavelength selected in the simulation shown in Fig. 8.15 seems to be a little smaller than in the experiment (37 mm instead of about 46 mm), whereas in the case of Fig. 8.16, the simulation and the experiment give essentially the same answer (28 mm, very close to the experimental one of 26 mm).

In fact, experiments and simulations share several common qualitative features. Isocontours of the thickness agree well with each other, and strong modulations of the troughs are observed, whereas the crests remain nearly undeformed, which leads to the formation of isolated depressions (all depressions in Figs. 8.9, 8.10, 8.12, 8.13, 8.15 and 8.16 are three-dimensional negative waves, but the corresponding patterns are different). In particular, as experimentally observed by Liu et al., the numerical simulations here indicate the occurrence of local saddle points on the wave pattern corresponding to minima in the spanwise direction and maxima in the streamwise direction (see the right panel of Fig. 8.15, where one such saddle point is indicated by a cross and two arrows). Liu et al. have measured the difference of height between the minima of the thickness at a trough and the height of the nearby saddle

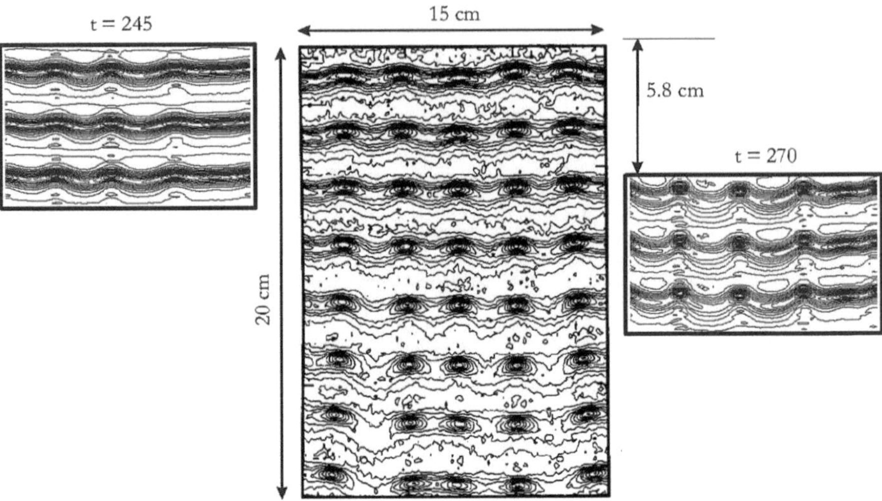

Fig. 8.16 Same caption as for Fig. 8.15 with the parameter set 4 in Table 8.1 [170, Fig. 11]. The size of the computational domain is 71×118 mm. Isocontours of the thickness are separated by an elevation step of 0.08

point. They called it "trough transverse modulation amplitude," denoted $\Delta h_{\min}(x)$. From the measurement of $\Delta h_{\min}(x)$ at different locations for the experimental data corresponding to the parameter set 3, i.e., their Fig. 7 (center of Fig. 8.15), they computed a spatial growth rate of approximately 0.11 cm^{-1}. Following a similar procedure, $\Delta h_{\min}(t)$ is defined as the height difference between the minimum of the thickness in the entire computational domain and the closest saddle point at a given time t. From the measurement of $\Delta h_{\min}(t)$ in the simulation, the temporal growth rate is found to be approximately 2.6 s^{-1} with the help of the speed of the corresponding γ_1 waves, 20.8 cm s^{-1}, which is converted into a spatial growth rate, 0.125 cm^{-1}, hence of the correct order of magnitude.

Let us emphasize that a good agreement between computer simulations and experiments is achieved provided that initial conditions are appropriately tuned. The widespread observation of the synchronous instability in experiments thus seems to result from the presence of spanwise nonuniformities at the inlet, which favors in-phase modulations of the wavefronts. In fact, the synchronous instability of the slow γ_1 branch has not been found in studies focusing on vertically falling films and where the viscous dispersion effects were neglected [42, 282]. Both the small inclination of the plate and viscous dispersion play an important role in the onset of the synchronous scenario.

Moreover, comparison of simulations of the full second-order, regularized and simplified models to the experimental results in [170] shows that the streamwise second-order inertia terms, which result from the departure of the velocity profile from its parabolic flat film shape, also play a crucial role in the onset of the synchronous instability. In the case of the simplified model, which does not take into account the second-order inertia corrections, the secondary instability is much more

Table 8.2 Parameter values of the simulations corresponding to the experiments in [203] for a vertical wall and pure water at 25°C. $\lambda_{z,\text{ndl}}$ is the spanwise wavelength of the needle array and k_z is the corresponding dimensionless wavenumber. The dimensionless wavenumber k, phase speed c and average thickness $\langle h \rangle$ of the corresponding γ_2 waves are also given

Set	Re	β (deg)	Γ	f (Hz)	$\lambda_{z,\text{ndl}}$ (mm)	k	c	$\langle h \rangle$	k_z
5	20.7	90	3375	15.0	10	0.3461	0.900	0.899	0.699
6	20.9	90	3375	19.0	30	0.4720	0.832	0.911	0.233
7	40.8	90	3375	19.1	20	0.3845	0.714	0.912	0.377
8	59.3	90	3375	17.0	20	0.3126	0.630	0.955	0.393

selective in favor of the subharmonic scenario and the synchronous instability was not observed. The full second-order model (seven equations) and the regularized model (three equations) give results in reasonably good agreement with experimental data in all cases. As far as the regularized model is concerned, this agreement is likely due to the incorporation of the second-order inertia corrections using the regularization technique described in Sect. 6.9.2 and extended here to three dimensions. This procedure ensures that the second-order terms remain small compared to the first-order ones for the widest possible ranges of parameter values. For this reason, the regularized model is an accurate and simple alternative to the full model or DNS for a large range of parameter values. It contains the main ingredients of the film flow dynamics. Therefore, the regularized model (8.4a)–(8.4c) will be the only model used for the remainder of the chapter to compare theory with experimental findings.

8.4.2 Three-Dimensional Modulations of γ_2 Waves

We now consider the experimental conditions investigated by Park and Nosoko [203]. These authors observed three-dimensional wave patterns emerging from two-dimensional waves of γ_2-type on a vertical wall. Parameter values corresponding to the different numerical experiments are given in Table 8.2. Park and Nosoko imposed a periodic modulation in the spanwise direction, which biased the selection toward synchronous patterns. They placed an array of regularly spaced needles with period $\lambda_{z,\text{ndl}}$ at the inlet. The initial conditions (8.11) corresponding to the inlet conditions imposed by Park and Nosoko and adapted to the present simulations are taken to be $A_x = 0.2$, $A_z = 0.05$ and $A_{\text{noise}} = 0$.

Figure 8.17 shows snapshots for parameter set 5 in Table 8.2. Initial spanwise modulations of period $\lambda_{z,\text{ndl}} = 10$ mm ($n_z = 6$) are quickly damped, i.e., $E_z \rightarrow 0$, and the pattern evolves to two-dimensional traveling waves, i.e., $E_x \rightarrow$ const, whose profile is given in Fig. 8.17(c). It corresponds to a γ_2 wave with a large hump preceded by capillary ripples, since, when the forcing frequency is small, the γ_1 slow waves are not observed and the linear inception region is immediately followed by the formation of fast γ_2 waves. Such genuine two-dimensional waves were observed

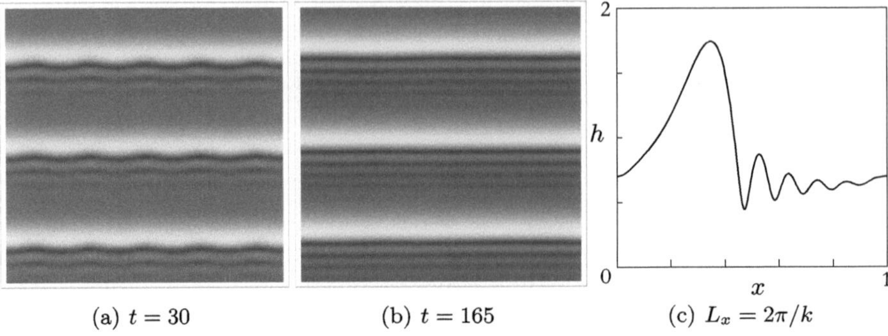

(a) $t = 30$ (b) $t = 165$ (c) $L_x = 2\pi/k$

Fig. 8.17 (**a**), (**b**) Snapshots of the film free surface at two different times computed with the regularized model and for set 5, Table 8.2. Initial conditions are $A_x = 0.2$, $A_z = 0.05$, $A_{\text{noise}} = 10^{-3}$, $n_x = 3$, $n_z = 6$ and $L = 2n_x\pi/k$. The computational domain is 60×60 mm with 128×128 grid points. *Bright/dark zones* correspond to elevations/depressions, respectively. (**c**) Two-dimensional wave profile of (**b**)

by Park and Nosoko in the downstream part of their test section (their Fig. 7a), while in the upstream part they observed large spanwise modulations with a wavelength of about $3\lambda_{z,\text{ndl}}$. These modulations can be recovered (not shown here) by increasing the period $\lambda_{z,\text{ndl}}$ to 30 mm ($n_z = 2$). However, they also decay (with $E_z \to 0$) but at a much smaller rate, indicating that the wavelength $\lambda_z = 3$ cm is close (but still below) the cut-off wavelength for spanwise instability.

Figure 8.18 shows the results for parameter set 6 in Table 8.2. In this case, the initial spanwise modulation is unstable and Figs. 8.18(a) and 8.18(b) give patterns equivalent to those observed experimentally [203, Fig. 7b]. To allow comparison of the evolution in time of the computer simulations to the evolution in space of the experimental waves, we need a way to convert locations in the laboratory frame to dimensionless time in the computations. This is gotten by exploiting the fact that a wave traveling at speed c reaches location x at time x/c. The speeds of the two-dimensional traveling waves corresponding to the experimental conditions have thus been computed using AUTO-07P. The test section in the experiments is 20 cm long, which corresponds approximately to 200 dimensionless time units in the computer simulations. After running the simulation for a much longer time (1500 time units), time oscillations of the spanwise modulations can be observed. Figure 8.19 shows that the energy of spanwise deformations E_z varies with a periodicity of about 300 time units. The region of the experimental domain corresponding to $t \approx 300$ is thus located beyond the test section, which explains why Park and Nosoko did not observe any time oscillations of the spanwise modulations as suggested by our computations. Oscillations of shorter period (about 60 time units) can also be seen, more pronounced for E_x than for E_z in Fig. 8.19. Their amplitudes are small at the beginning so that it is difficult to observe their effects on the three-dimensional wave pattern. However, they grow for $t > 900$ where they begin to influence the pattern evolution in a complex way, as illustrated by the panels i–l of Fig. 8.18. As time proceeds, spanwise modulations of the fronts depart more and more from their initial

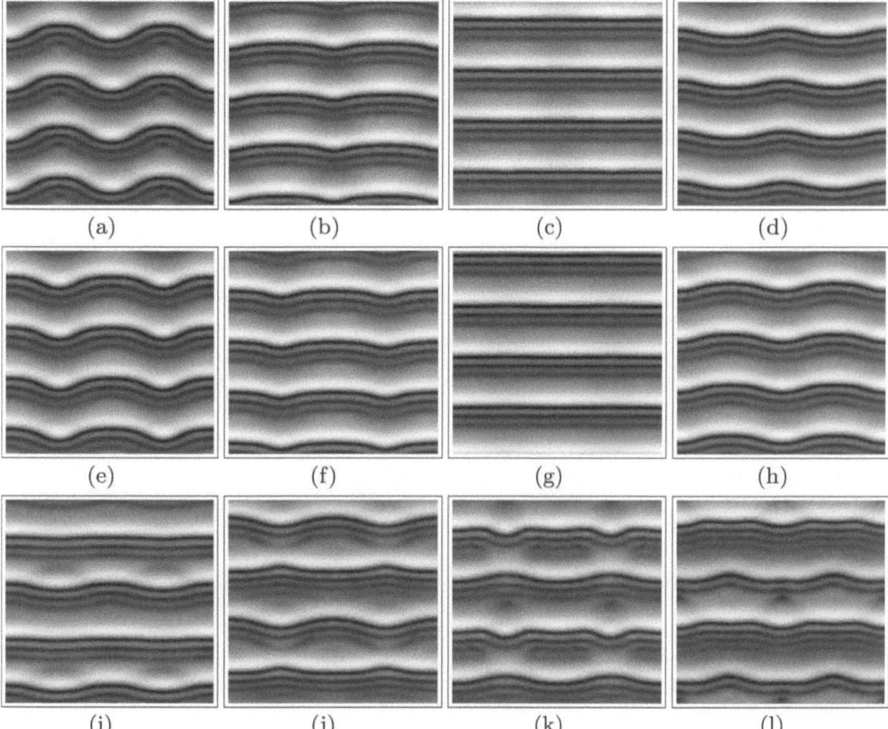

(a)   (b)   (c)   (d)

(e)   (f)   (g)   (h)

(i)   (j)   (k)   (l)

Fig. 8.18 Simulations for the parameter set 6 in Table 8.2. Remaining parameters are given in the caption of Fig. 8.17, except for $n_x = 4$, $n_z = 2$ and $L = 2n_x\pi/k$. Corresponding times are given in Fig. 8.19

harmonic shape. The fronts start to develop rounded tips separated by flat regions. At least two symmetry breakings can be observed. The first one corresponds to a streamwise period doubling of the modulated fronts triggered by a two-dimensional subharmonic instability, since two identical fronts are observable in panel i instead of four in panel h. The second one corresponds to the development of a phase shift of π, observable between the tips of two successive fronts (compare panel l to panel k).

Simulation results for a larger Reynolds number, $Re = 40.8$, are depicted in Fig. 8.20 for parameter set 7 in Table 8.2 and compared to experimental findings [203, Fig. 7c]. Like for $Re = 20.7$, harmonic spanwise modulations of the two-dimensional waves are first observed. However, they rapidly evolve into rugged modulations, made of nearly flat rears and rounded fronts. The pattern then saturated for a while (at least during 30 time units), traveling downstream in a quasi-steady state. These rugged-modulated waves were also observed at smaller Reynolds numbers (not shown), when streamwise and spanwise initial perturbations have comparable wavelengths. In this case, they remain steady for longer times. To facilitate qualitative comparison to the spatial evolution observed in experiments, snapshots of only a third of the numerical domain, corresponding to one streamwise wavelength, are displayed in Fig. 8.20 at increasing times. The interval of time sepa-

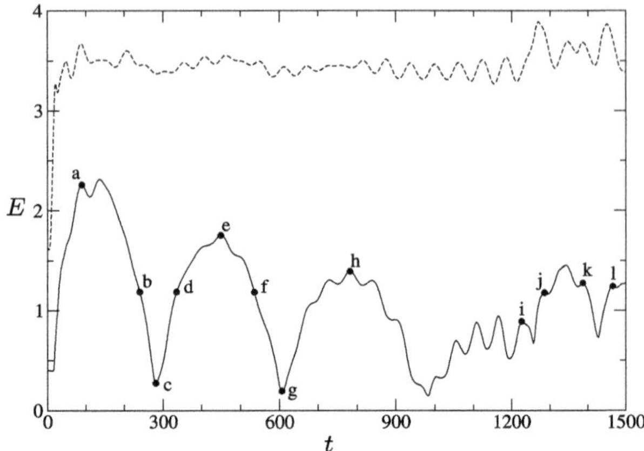

Fig. 8.19 Deformation energies for simulations of set 6 in Table 8.2: *dashed line* for E_x and *solid line* for E_z. Letters refer to the snapshots of Fig. 8.18

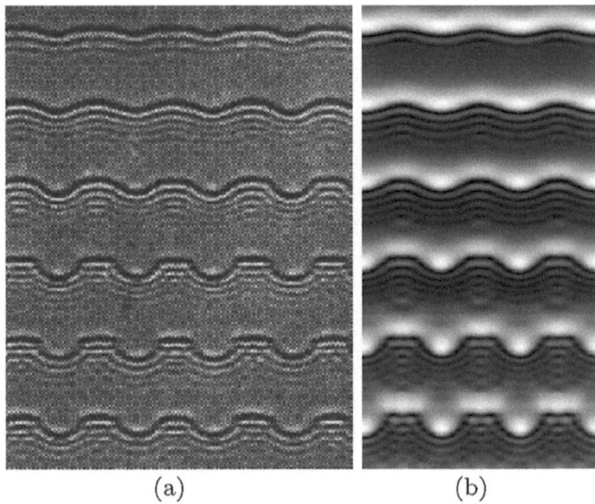

(a)   (b)

Fig. 8.20 (a) Experimental snapshot (real size 85×130 mm) for set 7 in Table 8.2 [203, Fig. 7c]. Reprinted with permission from C.D. Park and T. Nosoko, *AIChE J.*, 49(11):2715–2727, Copyright 2003, John Wiley and Sons; (b) Simulations with $n_x = 3$, $n_z = 3$ and $L = 2n_x\pi/k$. The domain size is 60×60 mm with 256×256 grid points

rating each pair of snapshots roughly corresponds to traveling of the fronts over a distance equal to one wavelength. Despite the use of periodic boundary conditions, the resemblance with the experimental findings [203, Fig. 7c] is convincing. For instance, the checkerboard interference pattern of the capillary ripples preceding the flat zones is recovered.

Above $Re \approx 40$, Park and Nosoko [203] observed a breaking of the modulated fronts leading to horseshoe waves having pointed fronts and long oblique legs. Simulation results for $Re = 59.3$ are presented in Fig. 8.21 for parameter set 8 in Table 8.2, where they are also compared to the experimental findings [203, Fig. 7d]. Due to computational limitations, the computational domain was limited to only one and two wavelengths in the streamwise and spanwise directions, respectively ($n_x = 1$ and $n_z = 2$). As compared to the case with $Re = 40.8$, the rugged modulations develop faster and do not saturate. Instead, the bulges of the wavefront contin-

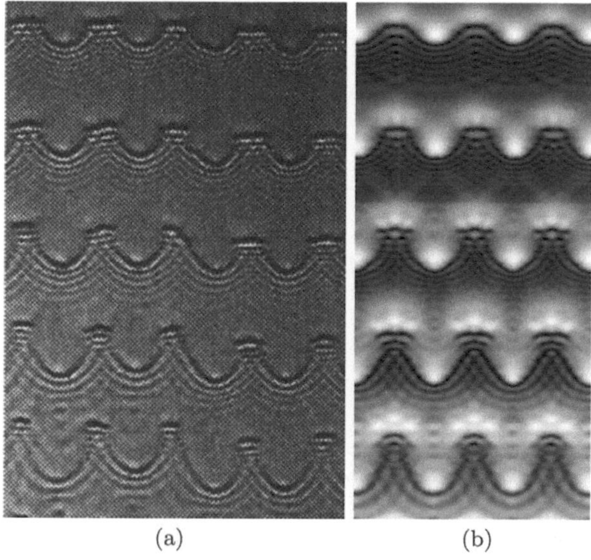

Fig. 8.21 (**a**) Experimental snapshot (real size 90 × 120 mm) for set 8 (*Re* = 59.3) [203, Fig. 7d]. Reprinted with permission from C.D. Park and T. Nosoko, *AIChE J.*, 49(11):2715–2727, Copyright 2003, John Wiley and Sons. (**b**) Snapshots of the simulated free surface. The domain size is 40 × 25 mm with 256 × 256 grid points

(a)   (b)

uously expand into horseshoe shapes, reducing the span of the flat parts at the rear. As time proceeds, the legs of the horseshoes extend and split off into "dimples," in qualitative agreement with experimental observations. The growth of the spanwise perturbations in the simulation is, however, faster than in the experiment.

In contrast with the experiments by Liu et al. [170] focusing on the secondary instabilities of the slow γ_1 waves, one can observe that secondary instabilities of the γ_2 waves lead neither to herringbone patterns—made of bumpy crests and deep troughs—nor to an array of isolated depressions when the instability is synchronous, but rather to modulated wave fronts.

8.4.3 Three-Dimensional Natural (Noise-Driven) Waves

We are now interested in the formation of natural (noise-driven) three-dimensional waves in the absence of periodic forcing. To match the experimental conditions reported in [3], initial conditions (8.11) correspond to white noise of amplitude $A_{\text{noise}} = 10^{-3}$ and $A_x = A_z = 0$. Parameter values for the different numerical experiments are given in Table 8.3. The experimental pictures are shown in Fig. 8.1.

Snapshots of the free surface deformation computed with the regularized model are reported in Fig. 8.22, where the three columns correspond to different Reynolds numbers (sets 9–11 of Table 8.3). Each row in Fig. 8.22 corresponds to a particular transient regime: The first row to mostly two-dimensional waves, the second row to coalescence processes, and the two last rows to the evolution of three-dimensional solitary waves. Since these regimes are time-dependent in computer simulations but space-dependent in experiments, both the dimensionless time t and the approximate location of the numerical domain on the experimental plate are given in Fig. 8.22;

Table 8.3 Parameter values of the simulations corresponding to experiments in [3] for an inclined plate and a 16% water–ethanol solution at 25°C ($\rho = 972$ kg m^{-3}, $\nu = 1.55 \times 10^{-6}$ m^2 s^{-1} and $\sigma = 40.8 \times 10^{-3}$ N m^{-1}). The two-dimensional wave characteristics k, c and $\langle h \rangle$ were computed from the wavelength λ_x, which was estimated by the average streamwise separation of the three-dimensional waves observed in the experimental pictures. See also the caption of Table 8.1

Set	Re	β (deg)	Γ	λ_x (mm)	k	c	$\langle h \rangle$
9	8	75	1106	40	0.15	1.322	0.906
10	16	75	1106	30	0.21	1.062	0.876
11	45	75	1106	25	0.28	0.749	0.904

the distance is again estimated from the phase speed c of the two-dimensional waves (see Table 8.3).

Close to the inlet (first row in Fig. 8.22), the waves are mostly two-dimensional. For $Re = 8$ (panel a) their profile is quasi-harmonic—bright and dark zones occupy equivalent areas—and for $Re = 16$ (panel b) they become of γ_2-type with steep humps of large amplitude. For $Re = 45$ (panel c) the waves have larger crests and thinner and deeper troughs. These waves are of the slow γ_1-type. Connections between two wavefronts in the patterns are observed for the three sets.

Further downstream (second row of Fig. 8.22), three-dimensional secondary instabilities of the primary wavetrain show up. The large amplitude waves travel faster and catch up with the preceding slower ones, they coalesce with them and absorb their mass leaving an increasing flat zone behind them. As time proceeds (third row of Fig. 8.22), fast γ_2 waves are clearly observable. The coalescence process yields solitary waves with preceding capillary ripples and large flat zones in between. Snapshots g, j and h, k of Fig. 8.22 share many similar features with experimental findings. The unsteady experimental pattern is characterized by seemingly interacting quasi-steady three-dimensional solitary waves separated by portions of constant thickness of length 10 to 50 cm. This is precisely the stage of interfacial turbulence or soliton gas we discussed in the introduction of this chapter.

At $Re = 8$ the mean distance between solitary waves tends to saturate for $t > 890$, which indicates that solitary waves have reached their fully developed regime. Alekseenko et al. [3] did not observe such a saturation in their experiments, either because the length of their test section was not long enough to observe saturation, or because this behavior is a consequence of the streamwise periodic boundary condition imposed in the computer simulations. The fact that coarsening of natural waves apparently terminates suggests that the system approaches a fully developed quasi-turbulent three-dimensional wavy regime.

At $Re = 16$ saturation is not observed at all, at least during the 1500 time units of the computer simulation. In that case, the final stage corresponds to interacting oblique fronts rather than three-dimensional horseshoe waves. At $Re = 45$, the three-dimensional waves tend to form localized structures rather than extended wavefronts, as was the case for lower values of Re. This is in agreement with results of [3] and [203], who observed horseshoe-like solitary waves with a sharp pointed front and long tails under similar conditions.

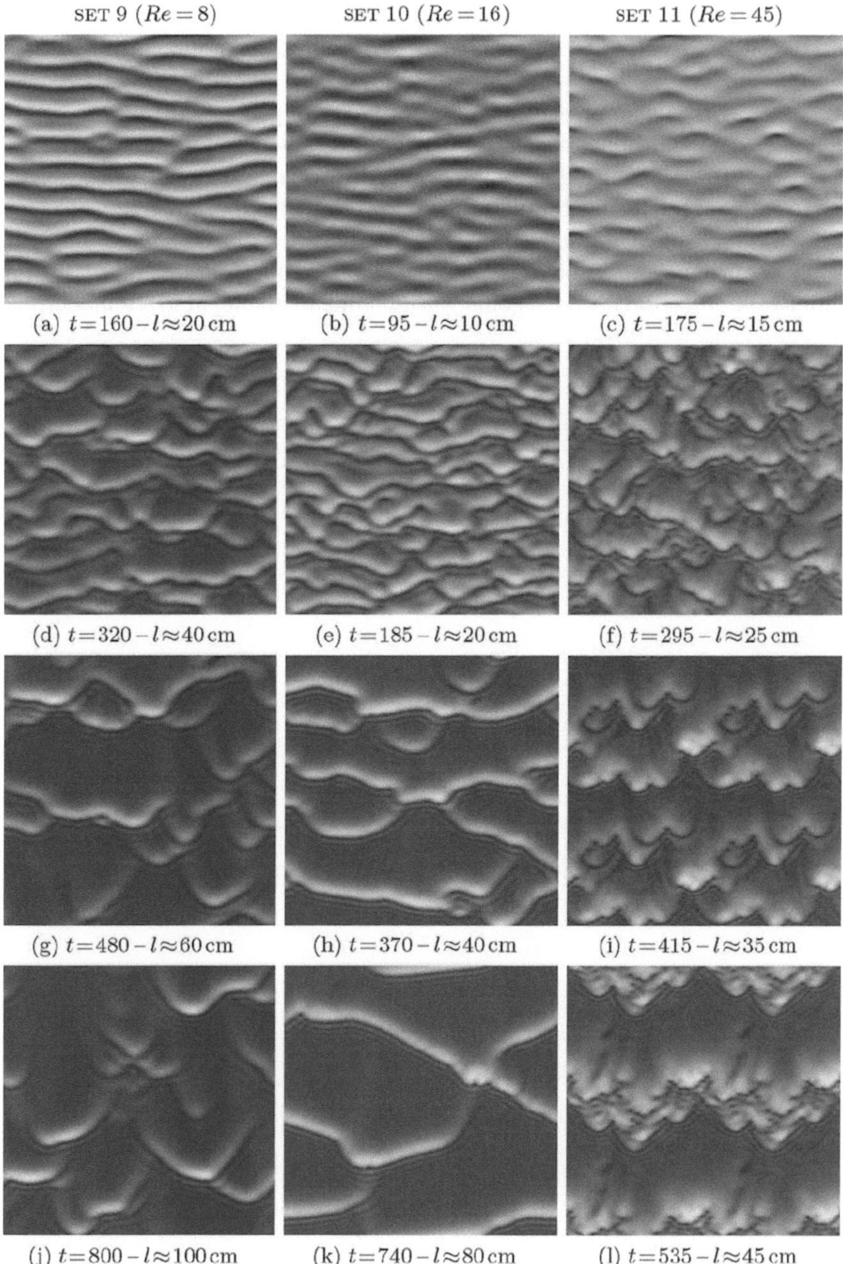

Fig. 8.22 Numerical simulations of natural (noise-driven) three-dimensional wave patterns corresponding to the experiments shown in Fig. 8.1 (see Table 8.3). The computational domain is $100 \times 100\,\mathrm{mm}^2$ with 256×256 grid points for set 9 and 10 and 512×256 for set 11 in Table 8.3 except for panel (**i**) and (**l**) where it corresponds to $50 \times 50\,\mathrm{mm}^2$ and 256×256 grid points: the obtained snapshot is repeated four times. l is the estimated distance from the inlet. The *bright/dark zones* correspond to elevations/depressions, respectively

Yet, several features of the transition from modulated waves to horseshoe-like solitary waves are far from being understood, e.g., the region of the parameter space were oblique solitary waves are present has not been delineated. Some issues related to the mechanism of the transition from two-dimensional waves to fully developed three-dimensional horseshoe-like solitary waves and the linear stability of three-dimensional waves to three-dimensional disturbances, including questions related to the convective instability of such waves, have been examined in [75, 76] by the three-dimensional Kapitza–Shkadov model, whose several shortcomings have already been flagged. Hence, there is a need for a full-scale careful and detailed examination of the three-dimensional wave regime by the regularized model.

Chapter 9
Nonisothermal Case:
Two- and Three-Dimensional Flow

We have now accumulated enough tools to extend the modeling of falling liquid films developed in Chaps. 6 and 7 to the case of a uniformly heated inclined plate, i.e., the specified temperature (ST) case. As with the isothermal case, we shall model the nonisothermal one by a gradient expansion combined with a Galerkin projection using polynomial test functions for both the velocity and temperature fields. The outcome is a set of equations for the evolution of the velocity and temperature amplitudes at first and second order in the gradient expansion parameter. We shall then proceed to a regularization of the second-order model in the same manner with the isothermal case. Solitary wave solutions of the two-dimensional regularized model to this system will be constructed. Through numerical computations in the three-dimensional regime we shall explore the complex interaction between the thermocapillary/S- and the hydrodynamic/H-modes, which under certain conditions can give rise to three-dimensional rivulet structures aligned with the flow. These rivulets can channel two-dimensional solitary waves riding on them.

9.1 Formulation

As with Chap. 6, our starting point is the two-dimensional boundary layer equations (4.11a)–(4.11d) for ST. We shall apply the polynomial expansion approach and the method of weighted residuals outlined in detail in Chap. 6 for isothermal films (Sect. 6.5). The basic idea then is to separate the variables and to expand the velocity and temperature fields on a set of test functions of the reduced coordinate $\bar{y} = y/h$, a naturally rescaled variable as it converts the boundary value problem in the interval $[0, h]$ to the interval $[0, 1]$. To satisfy the Dirichlet boundary conditions

$$u|_0 = v|_0 = 0 \quad \text{and} \quad T|_0 = 1, \tag{9.1}$$

we write the velocity and temperature distributions in the form

$$u(x, y, t) = \sum_{i=0}^{i_{\max}} a_i(x, t) f_i\left(\frac{y}{h(x, t)}\right), \tag{9.2a}$$

S. Kalliadasis et al., *Falling Liquid Films*, Applied Mathematical Sciences 176, DOI 10.1007/978-1-84882-367-9_9, © Springer-Verlag London Limited 2012

$$T(x, y, t) = 1 + \sum_{i=0}^{i_{max}} b_i(x, t) g_i\left(\frac{y}{h(x, t)}\right), \tag{9.2b}$$

where $f_i(0) = g_i(0) = 0$.

Let us next choose

$$f_0 = \bar{y} - \frac{1}{2}\bar{y}^2, \tag{9.3a}$$

$$g_0 = \bar{y}, \tag{9.3b}$$

corresponding to the base state (3.1a), (3.1d) and complete the set of test functions with

$$f_1(\bar{y}) = \bar{y}, \qquad f_i(\bar{y}) = \bar{y}^{i+1}, \quad i \geq 2, \tag{9.3c}$$

$$g_i(\bar{y}) = \bar{y}^{i+1}, \quad i \geq 1, \tag{9.3d}$$

to obtain the polynomial bases for the projection. Note that the basis for the velocity field here is different to that in Sect. 6.6 where at first order we chose a basis to fulfill all boundary conditions from the outset. In the present case the tangential stress boundary condition on the free surface (4.6h) is nonhomogeneous at first order due to the Marangoni effect. Hence, unlike the isothermal case at first order, we cannot identify from the outset a basis of test functions to satisfy the tangential stress boundary condition. We therefore choose the simplest basis possible. The tangential stress condition then becomes a constraint on the amplitudes a_i and can be satisfied through a tau method as introduced in Sect. 6.8 for the second-order isothermal formulation.

Since $(2i_{max} + 3)$ unknowns have been introduced, namely h, a_i and b_i, $2i_{max} + 3$ equations are required to determine them. The first one is the kinematic condition at the interface (4.6g), which can be replaced by integrating the continuity equation (4.6c) along the normal coordinate to give

$$\partial_t h + \partial_x q = 0, \tag{9.4}$$

as we have done several times in this monograph and where $q = \int_0^h u\, dy$ is the flow rate in the streamwise direction. Two additional equations are the boundary conditions (4.6h), (4.6i), and by defining $2i_{max}$ weight functions $w_j(\bar{y})$, the closure equations are obtained by the vanishing residuals from (4.6a)–(4.6b):

$$\mathcal{R}_q(w_j) \equiv \int_0^h w_j(\bar{y})\Big[3\varepsilon Re(\partial_t u + u\partial_x u + v\partial_y u) - \left(\partial_{yy} + 2\varepsilon^2 \partial_{xx}\right)u$$

$$- 1 + \varepsilon Ct\partial_x h - \varepsilon^2 \partial_x(\partial_x u|_h) - \varepsilon^3 We\partial_{xxx}h\Big]dy = 0, \tag{9.5a}$$

$$\mathcal{R}_\theta(w_{i_{max}+j}) \equiv \int_0^h w_{i_{max}+j}\Big[3\varepsilon Pe(\partial_t T + u\partial_x T + v\partial_y T)$$

$$- \left(\varepsilon^2 \partial_{xx} + \partial_{yy}\right)T\Big]dy = 0. \tag{9.5b}$$

with $0 \le j \le i_{\max} - 1$ and where u and T are given by the expansions (9.2a), (9.2b) while v is obtained from the continuity equation, $v = -\int_0^y \partial_x u \, dy$.

At this point, the method we are using is simply one of the numerous weighting residual strategies, which differ from each other only by the specific choice of the weights w_j. As demonstrated in Sect. 6.7, it is not necessary to specify the weighted residuals method we are applying if the number $i_{\max}$ of test functions and residuals is large enough. In fact, requiring the momentum and the energy equations (4.6a) and (4.6b) to be satisfied everywhere—and not simply on average—and inserting the expansions (9.2a), (9.2b) using (9.3a), (9.3b) leads to the cancellation of two polynomials in the reduced normal coordinate $\bar{y}$. By examining then the order of magnitude with respect to ε of each term in (4.6a) and (4.6b), it can be proved that the number of independent conditions on the unknowns a_i and b_i provided by the cancellation of these two polynomials is equal to the number of the residuals (9.5a), (9.5b), provided that $i_{\max}$ is chosen large enough (see Sect. 6.5). Any choice then of the weighting functions would lead to equivalent systems of equations and then to the same reduced model for the dynamics of the flow, as was shown for an isothermal film.

Nevertheless, it is important to keep in mind that we are not simply applying a numerical method. The aim is rather to combine some well-known numerical strategy with a perturbation technique from the flat film base state (3.1a)–(3.1d) corresponding to

$$a_0 = h^2, \qquad b_0 = \frac{-Bh}{1 + Bh}, \qquad a_i = b_i = 0, \quad i \ge 1.$$

9.2 Formulation at First Order

To illustrate the procedure we restrict our attention to the formulation consistent at $\mathcal{O}(\varepsilon)$, with all terms of higher order neglected. The aim is to develop the simplest possible methodology for the projection of the velocity and temperature fields to the amplitudes of the polynomial expansion appearing at first order. For simplicity, we also assume as we did with the derivation of the isothermal models in Chap. 6, that the parameters, Re, Pe and M are of $\mathcal{O}(1)$ and that We is of $\mathcal{O}(\varepsilon^{-2})$, corresponding to the strong surface tension limit. These assumptions can be relaxed and strict orders of magnitude assignments are not required, but the final equations remain the same as they were with the isothermal formulation.

The system of equations (4.6a)–(4.6i) at $\mathcal{O}(\varepsilon)$ reduces to

$$\partial_x u + \partial_y v = 0, \tag{9.6a}$$

$$3\varepsilon Re(\partial_t u + u\partial_x u + v\partial_y u) = \partial_{yy}u + 1 - \varepsilon Ct\partial_x h + \varepsilon^3 We\partial_{xxx}h, \tag{9.6b}$$

$$3\varepsilon Pe(\partial_t T + u\partial_x T + v\partial_y T) = \partial_{yy}T, \tag{9.6c}$$

$$v|_h = \partial_t h + u\partial_x h, \tag{9.6d}$$

$$\partial_y u|_h = -\varepsilon M \partial_x [T|_h], \tag{9.6e}$$

$$\partial_y T|_h = -BT|_h, \tag{9.6f}$$

together with the Dirichlet conditions (9.1). Thus, the $2i_{\max}$ residuals to be evaluated are simplified to

$$\int_0^h w_j(\bar{y})\big[3\varepsilon Re(\partial_t u + u\partial_x u + v\partial_y u) - \partial_{yy}u\big]\,dy$$

$$+ h\big(-1 + \varepsilon Ct\partial_x h - \varepsilon^3 We\partial_{xxx}h\big)\int_0^1 w_j(\bar{y})\,d\bar{y} = 0, \tag{9.7a}$$

$$\int_0^h w_{i_{\max}+j}(\bar{y})\big[3\varepsilon Pe(\partial_t T + u\partial_x T + v\partial_y T) - \partial_{yy}T\big]\,dy = 0. \tag{9.7b}$$

The amplitudes a_i and b_i, $i \geq 1$, result from the slow space and time modulations of the free surface so that they are first-order quantities in ε. Therefore, the space and time derivatives of a_i and b_i, $i \geq 1$, are negligible. One then is led to a linear system of $2i_{\max}+2$ conditions for a_i and b_i—consisting of the $2i_{\max}$ residuals (9.7a), (9.7b) and the two stress balances (9.6e) and (9.6f)—whose coefficients depend at most on a_0, b_0, h and with a right hand side that depends on h, a_0, b_0 and their derivatives

$$\sum_{j'=0}^{2i_{\max}+1} \alpha_{jj'}A_{j'} = \beta_j(h, a_0, b_0, \partial_{x,t}h, \partial_{x,t}a_0, \partial_{x,t}b_0), \quad 0 \leq j \leq 2i_{\max} + 1, \tag{9.8}$$

where $A_j \equiv a_j$ and $A_{i_{\max}+1+j} \equiv b_j$, $0 \leq j \leq i_{\max}$. Solving for the A_j then leads to explicit expressions for the amplitudes a_j, b_j, $j \geq 1$, as functions of a_0, b_0, h and their derivatives, making clear their slaving to the film thickness, the parabolic velocity profile and the linear temperature distribution. As in Sect. 6.6, the inversion of (9.8) also provides two solvability conditions for a_0 and b_0 forming a set of two evolution equations for a_0 and b_0 which, together with the conservation equation (9.4) for h, describe the entire dynamics of the film flow.

As noted earlier, inserting the expansion (9.2a), (9.2b), (9.3a), (9.3b) into (9.6a)–(9.6f) leads to the cancellation of two polynomials in the rescaled normal coordinate $\bar{y}$, say $\mathcal{P}(\bar{y})$ and $\mathcal{Q}(\bar{y})$, corresponding to the momentum and energy equation, respectively. Because the advection terms $\partial_t u + u\partial_x u + v\partial_y u$ and $\partial_t T + u\partial_x T + v\partial_y T$ are first-order quantities, their truncation at $\mathcal{O}(\varepsilon)$ involves only the parabolic and linear profiles corresponding to a_0 and b_0. Consequently, it can be shown that the advection terms are polynomials in $\bar{y}$ of only degree four and three, respectively. Therefore, the monomials of highest degree appearing in $\mathcal{P}(\bar{y})$ and $\mathcal{Q}(\bar{y})$ originate from the terms $\partial_{yy}u$ and $\partial_{yy}T$ such that $\mathcal{P}(\bar{y})$ and $\mathcal{Q}(\bar{y})$ are of degree $i_{\max} - 1$. Canceling those two polynomials gives $2i_{\max}$ independent relationships, i.e., the same as the number of residuals (9.7a), (9.7b). Completing these $2i_{\max}$ relationships with the two boundary conditions (9.6e) and (9.6f) gives a system of equations equivalent to (9.8) leading to the same evolution equations for a_0 and b_0 (provided $i_{\max}$ is large enough).

The solution to system (9.8) will be explicitly given later in this section and in the next (the solution to (9.8) is written in (9.12), (9.16) and (9.18a)–(9.18h) formulated using the definition of the flow rate $q = \int_0^h u \, dy$). The algebra required to solve (9.8) is cumbersome but doable. However, turning to the formulation of the model at second order—as will be done in the next section—significantly increases the algebraic manipulations required to invert the system of equations and a shortcut is desirable. Since each different weighting residual technique only differs by its specific choice for the weighting functions w_j, it is relevant to look for the best choice of w_j that would simplify the algebraic manipulations leading to the two solvability conditions that complete the set of evolution equations for the $\mathcal{O}(1)$ unknowns h, a_0 and b_0. Another reason for the search for the most efficient weighted residual technique is to obtain a suitable set of test functions at second order on which we project the velocity and temperature fields (with again the will to reduce algebraic manipulations). Indeed, in Chap. 6 the construction of the set of test functions F_j given in (6.74) was performed by imposing an orthogonality condition, which enabled us to drastically reduce the algebra when the Galerkin method was next applied.

Let us first consider the residuals (9.7a). Because $\partial_t u + u \partial_x u + v \partial_y u$ are first-order terms, the unknowns a_i, $i \geq 1$, may enter into their evaluation only through the integral $\int_0^h w_j \partial_{yy} u$. Two integrations by parts give

$$\int_0^h w_j\left(\frac{y}{h}\right)\partial_{yy}u \, dy = \left[w_j\left(\frac{y}{h}\right)\partial_y u\right]_0^h - \frac{1}{h}\left[w_j'\left(\frac{y}{h}\right)u\right]_0^h + \frac{1}{h^2}\int_0^h w_j''\left(\frac{y}{h}\right)u \, dy.$$

$$(9.9)$$

These two integrations by parts enable us to bypass the constraint on the amplitudes a_j brought by the stress balance at the free surface (9.6e). As $\partial_y u|_h$ in (9.6e) is proportional to $\partial_x[T|_h]$, it can only involve h, a_0 and b_0 at first order. Making also use of the no-slip condition on the solid plate $u|_0 = 0$, only three terms are left to consider, namely $w_j(0)\partial_y u|_0$, $w_j'(1)u|_h$ and $\int_0^h w_j''(y/h)u \, dy$. With the introduction of the flow rate $q \equiv \int_0^h u \, dy$, this suggests we choose w_0 so as to have $w_0(0) = 0$, $w_0'(1) = 0$ and w_0'' a constant. Interestingly, f_0 in (9.3a) has precisely these features[1] and we can readily take $w_0 \propto f_0$.

As discussed in Sect. 6.6, it is appropriate to relate the amplitude of the parabolic profile a_0 to the flow rate q, which is a quantity appearing explicitly in the integral form of the kinematic condition (9.4). To introduce q explicitly into our expansion,

---

[1] Recall that the parabolic profile f_0 corresponds to the zeroth-order formulation of the problem for the velocity:

$$\partial_{yy}u = -1, \qquad u|_0 = 0, \qquad \partial_y u|_h = 0. \tag{9.10}$$

Therefore, considering the two integrations by parts performed in (9.9), the similitude between the weight function w_0 and the test function f_0 is related to the linear operator $L \equiv \partial_{yy}$ being self-adjoint in the space of functions satisfying the boundary conditions (9.10).

let us integrate (9.2a) between 0 and h to obtain the expression

$$a_0 = 3\frac{q}{h} - \frac{3}{2}a_1 - \sum_{i=2}^{i_{max}} \frac{3}{i+2}a_i.$$ (9.11)

Therefore, evaluating the residual (9.7a) corresponding to $j = 0$ with $w_0 \propto f_0$, and using the boundary condition (9.6e) leads to

$$3\varepsilon Re\left(\frac{2}{5}\partial_t q - \frac{23}{40}\frac{q}{h}\partial_t h - \frac{18}{35}\frac{q^2}{h^2}\partial_x h + \frac{111}{280}\frac{q}{h}\partial_x q\right) + \frac{q}{h^2}$$

$$+ \frac{1}{3}h\left(-1 + \varepsilon Ct\partial_x h - \varepsilon^3 We\partial_{xxx}h\right) + \frac{1}{2}\varepsilon M\,\partial_x(T|_h) = 0,$$ (9.12)

where the unknowns a_i do not appear, a consequence of the weighting strategy. (Although (9.12) involves projection onto more than one test function, we could have obtained it by projecting u onto the single test function f_0, i.e., by substituting $u = (3q/h)f_0$ in the expression (9.9) obtained from integrations by parts.) We therefore obtain straightforwardly the solvability condition for a_0 we were looking for, written here in terms of the flow rate q. Choosing the weight functions to be the test functions themselves is the essence of the Galerkin method, which in turn is equivalent to a variational method (whenever a variational formulation is available [92]). As in Chap. 6, the Galerkin method is then identified as the most efficient one providing the momentum averaged equation with the minimum of algebra.

Turning to the weighted residuals for the energy equation (9.7b) and with the same arguments used for the treatment of the momentum equation, the unknowns b_i, $i \geq 1$, may play a role only through the integral, $\int_0^h w_j\partial_{yy}T$. Two integrations by parts give

$$\int_0^h w_j\left(\frac{y}{h}\right)\partial_{yy}T\,dy = \left[w_j\left(\frac{y}{h}\right)\partial_y T\right]_0^h - \frac{1}{h}\left[w_j'\left(\frac{y}{h}\right)T\right]_0^h$$

$$+ \frac{1}{h^2}\int_0^h w_j''\left(\frac{y}{h}\right)T\,dy.$$ (9.13)

Again, the two integrations by part enable us to bypass the condition on the coefficients b_j brought by the heat flux balance at the free surface (9.6f). Making use of (9.6f) and the constant temperature distribution at the solid wall $T|_0 = 1$, we get

$$\int_0^h w_j\left(\frac{y}{h}\right)\partial_{yy}T\,dy = -Bw_j(1)T|_h - w_j(0)\partial_y T|_0 + \frac{1}{h}\left[w_j'(0) - w_j'(1)T|_h\right]$$

$$+ \frac{1}{h^2}\int_0^h w_j''\left(\frac{y}{h}\right)T\,dy.$$ (9.14)

Following exactly the same approach as before would lead to the choice for the first weight function, $w_{i_{max}}(0) = 0$, $w_{i_{max}}'(1) = 0$. Setting $w_{i_{max}}''$ to a constant would introduce the average temperature across the flow, $(1/h)\int_0^h T\,dy$. This choice would obviously be problematic since the term involving $T|_h$ would remain in (9.14) and in (9.12) through the Marangoni effect. On the other hand, it is rather the heat flux at the surface $\partial_y T|_h$ or the temperature at the surface $T|_h$ that have physical meaning. Because it already appears in (9.12), we prefer to put the emphasis on $\theta \equiv T|_h$ by choosing $w_{i_{\max}}(0) = 0$, $w''_{i_{\max}} = 0$ so that $w_{i_{\max}} \propto \bar{y} = g_0$. This choice has the advantage of dissociating the coupling term $(1/2)M\partial_x[T|_h]$ in (9.12) from the definition of any other amplitudes needed to describe the temperature distribution. Again we find that the weight function is identical to the test function itself and that the Galerkin method is the most effective one. It is therefore appropriate to replace the physically meaningless unknown b_0 by θ through the substitution

$$b_0 = \theta - 1 - \sum_{i=1}^{i_{\max}} b_i. \tag{9.15}$$

From the residual (9.7b) corresponding to $w_{i_{\max}} = g_0 = \bar{y}$, we then get the second solvability condition rewritten in terms of θ,

$$3\varepsilon Pe\left(\frac{1-\theta}{3}\partial_t h + \frac{1}{3}h\partial_t\theta + \frac{11}{40}(1-\theta)\partial_x q + \frac{9}{20}q\partial_x\theta\right) + \frac{\theta-1}{h} + B\theta = 0. \tag{9.16}$$

Using now the equivalence $\partial_t h = -\partial_x q$ from (9.4), a model consistent at $\mathcal{O}(\varepsilon)$ can be formulated in terms of three coupled evolution equations for h, q and θ written here in terms of the Shkadov scaling:

$$\partial_t h = -\partial_x q, \tag{9.17a}$$

$$\delta\partial_t q = \frac{5}{6}h - \frac{5}{2}\frac{q}{h^2} + \delta\left(\frac{9}{7}\frac{q^2}{h^2}\partial_x h - \frac{17}{7}\frac{q}{h}\partial_x q\right) - \frac{5}{4}M\partial_x\theta$$

$$\qquad - \frac{5}{6}\zeta h\partial_x h + \frac{5}{6}h\partial_{xxx}h, \tag{9.17b}$$

$$Pr\delta\partial_t\theta = 3\frac{[1-(1+Bh)\theta]}{h^2} + Pr\delta\left[\frac{7}{40}\frac{(1-\theta)}{h}\partial_x q - \frac{27}{20}\frac{q}{h}\partial_x\theta\right]. \tag{9.17c}$$

Equations (9.17a)–(9.17c) will be hereinafter called the *first-order model* for the heated falling film problem.

Finally, we can easily see how to relax the order of magnitude assignments for the different parameters. For example let us relax the order of magnitude assignment $Pe = \mathcal{O}(1)$, or $Pr = \mathcal{O}(1)$, since $Re = \mathcal{O}(1)$. εPr must be at most of $\mathcal{O}(1)$, the maximum order in the right hand side of the energy equation (9.6c). At the same time the convective terms in the energy equation must dominate over the neglected terms in the first-order boundary layer equations (9.6a)–(9.6f), or $Pr \gg \varepsilon$. The amplitudes of the test functions then for the projection of the temperature field (9.2b) become of $\mathcal{O}(\varepsilon Pr)$ (see (9.18a)–(9.18h)), with the exception of b_0, which is of $\mathcal{O}(1)$.

9.3 Formulation at Second Order

The aim now is to derive a model consistent at second order, i.e., to account for all ε^2 terms of the boundary layer equations (4.6a)–(4.6i) such as the ε^2-order viscous and thermal diffusion terms of the momentum and energy equations (4.6a), (4.6b). For this purpose, we explicitly need the solution of (9.8) for the amplitudes of the projections:

$$a_1 = 3\varepsilon Re\left[-\frac{6}{5}h\partial_x\left(\frac{q^2}{h}\right) - h\partial_t q\right] - \varepsilon Mh\partial_x\theta, \tag{9.18a}$$

$$a_2 = 3\varepsilon Re\left[q\partial_x q + \frac{1}{2}h\partial_t q\right], \tag{9.18b}$$

$$a_3 = 3\varepsilon Re\left[-\frac{3}{4}\frac{q^2}{h}\partial_x h - \frac{1}{8}h\partial_t q\right], \tag{9.18c}$$

$$a_4 = 3\varepsilon Re\left[-\frac{3}{40}h^6\partial_x\left(\frac{q^2}{h^6}\right)\right], \tag{9.18d}$$

$$a_5 = 3\varepsilon Re\left[\frac{1}{80}h^6\partial_x\left(\frac{q^2}{h^6}\right)\right], \tag{9.18e}$$

$$b_2 = \frac{1}{6}3\varepsilon Peh\left[(\theta-1)\partial_x q + h\partial_t\theta\right], \tag{9.18f}$$

$$b_3 = \frac{1}{8}3\varepsilon Peh\left[-(\theta-1)\partial_x q + 2q\partial_x\theta\right], \tag{9.18g}$$

$$b_4 = \frac{1}{40}3\varepsilon Peh\left[(\theta-1)\partial_x q - 3q\partial_x\theta\right], \tag{9.18h}$$

$$b_1 = 0, \qquad a_i = b_j = 0, \quad i \geq 6, j \geq 5,$$

completed by the two solvability conditions (9.12), (9.16).[2] Note that the amplitudes a_i of the monomials of degree greater or equal to seven are identically vanishing at first order. This can be shown by examination of the degree of the polynomial in $\bar{y}$ corresponding to the left hand side of (9.6b). Because f_0 is of degree two, the above polynomial is of degree four, so that the right hand side of (9.6b) is also a polynomial of degree four. Hence, the amplitude a_n corresponding to $f_n = \bar{y}^{n+1}$ is equal to zero if $n \geq 6$, with the operator ∂_{yy} decreasing the degree in $\bar{y}$ of the left hand side of (9.6b) by two. The same argument can be applied to (9.6c) where the convection terms $\partial_t T + u\partial_x T + v\partial_y T$ on the left hand side constitute a polynomial in $\bar{y}$ of degree three only.

Consequently, the derivatives of the fields a_i with $i \geq 6$ and b_j with $j \geq 5$ are of order higher than ε^2 and can be dropped at this stage of the approximation.

---

[2]Compared to (6.44), system (9.18a)–(9.18h) contains one more nonzero amplitude a_i (five instead of four). In fact, this is a consequence of the need for a supplementary amplitude to fulfill the boundary condition (9.6e) using the tau method.

Their dynamics are thus slaved to the dynamics of the other variables. From the expressions (9.18a)–(9.18h) we can now obtain $a_4 = -6a_5$, $a_2 = -4a_3 + 40a_5$ and $a_1 = 8a_3 - 96a_5 - \varepsilon M h \partial_x \theta$ so that eliminating these amplitudes in (9.11) yields $a_0 = 3q/h - \frac{48}{5}a_3 + \frac{816}{7}a_5 + \frac{3}{2}\varepsilon M h \partial_x \theta$. The velocity field at first order can then be written as

$$u = 3\frac{q}{h} f_0(\bar{y}) + \varepsilon M h \partial_x \theta \, \tilde{f}_1(\bar{y}) + a_3 \tilde{f}_3(\bar{y}) + a_5 \tilde{f}_5(\bar{y}), \tag{9.19}$$

where $\tilde{f}_1 = -\frac{3}{4}\bar{y}^2 + \frac{1}{2}\bar{y}$, $\tilde{f}_3 = \bar{y}^4 - 4\bar{y}^3 + \frac{24}{5}\bar{y}^2 - \frac{8}{5}\bar{y}$ and $\tilde{f}_5 = \bar{y}^6 - 6\bar{y}^5 + 40\bar{y}^3 - \frac{408}{7}\bar{y}^2 + \frac{144}{7}\bar{y}$. Therefore, u is a combination of four independent fields, q/h, a_3, a_5 and $h\partial_x \theta$, rather than six, as might be expected at first. Similarly, T can be written at first order as a combination of four independent fields, namely, θ, b_2, b_3 and b_4. As a consequence, a consistent model for the dynamics of the flow at second order would require nine unknowns corresponding to the introduction of eight independent fields to correctly represent the velocity and temperature distributions as well as the film thickness h.

Since the degree of the polynomials $\tilde{f}_1$, $\tilde{f}_3$, $\tilde{f}_5$ is smaller or equal to six, the second-order diffusive term $\partial_{yy}u$ and the quadratic nonlinearities of the Navier–Stokes equations imply that the description of the velocity field at $\mathcal{O}(\varepsilon^2)$ involves polynomials of degree up to 10. Therefore, the set of test functions for the velocity field needs to be completed by six other functions in order to obtain a basis for the set of polynomials of degree up to 10 satisfying the Dirichlet condition at the solid wall. Turning now to the modeling of the energy equation at second order, a basis for the set of polynomials of degree up to nine verifying the Dirichlet condition is required to fully describe the temperature field at that order. This means that six amplitudes for the velocity field and five for the temperature field need to be eliminated—through a slaving procedure—to obtain a set of eight evolution equations for the eight unknowns, plus the conservation equation (9.4). Needless to say, such a task would require some cumbersome algebraic manipulations and hence a shortcut is desirable.

Such a shortcut is possible by following the same approach with the isothermal case, i.e., by constructing a new set of polynomial test functions F_i satisfying the orthogonality condition $\int_0^1 F_i F_j \, d\bar{y} \propto \delta_{ij}$ with the help of a Gram–Schmidt orthogonalization procedure so that $F_0 \equiv f_0$, F_1, F_2 and F_3 are linear combinations of f_0, $\tilde{f}_1$, $\tilde{f}_3$ and $\tilde{f}_5$. The result is

$$F_0 = \bar{y} - \frac{1}{2}\bar{y}^2, \tag{9.20a}$$

$$F_1 = \bar{y} - \frac{17}{6}\bar{y}^2 + \frac{7}{3}\bar{y}^3 - \frac{7}{12}\bar{y}^4, \tag{9.20b}$$

$$F_2 = \bar{y} - \frac{13}{2}\bar{y}^2 + \frac{57}{4}\bar{y}^3 - \frac{111}{8}\bar{y}^4 + \frac{99}{16}\bar{y}^5 - \frac{33}{32}\bar{y}^6, \tag{9.20c}$$

$$F_3 = \bar{y} - \frac{531}{62}\bar{y}^2 + \frac{2871}{124}\bar{y}^3 - \frac{6369}{248}\bar{y}^4 + \frac{29601}{2480}\bar{y}^5 - \frac{9867}{4960}\bar{y}^6. \tag{9.20d}$$

The functions F_1 and F_2 have been chosen so that they correspond exactly to the polynomials introduced in the isothermal case (Sect. 6.8). The introduction of the polynomial F_3 is made necessary by the presence of the Marangoni effect, which modifies the stress condition (9.6e) at the interface.

Similarly, a set of orthogonal test functions for the temperature field is constructed from linear combinations of g_0, g_2, g_3 and g_4, with $G_0 \equiv g_0$ [3]:

$$G_0 = \bar{y}, \tag{9.21a}$$

$$G_1 = \bar{y} - \frac{5}{3}\bar{y}^3, \tag{9.21b}$$

$$G_2 = \bar{y} - 7\bar{y}^3 + \frac{32}{5}\bar{y}^4, \tag{9.21c}$$

$$G_3 = \bar{y} - \frac{56}{3}\bar{y}^3 + \frac{192}{5}\bar{y}^4 - 21\bar{y}^5. \tag{9.21d}$$

Therefore, the velocity and temperature fields at $\mathcal{O}(\varepsilon)$ are as follows:

$$u = \frac{3}{h}(q - s_1 - s_2 - s_3)F_0 + 45\frac{s_1}{h}F_1 + 210\frac{s_2}{h}F_2 + 434\frac{s_3}{h}F_3, \tag{9.22a}$$

$$T = 1 + (\theta - 1 - t_1 - t_2 - t_3)G_0 - \frac{3}{2}t_1 G_1 + \frac{5}{2}t_2 G_2 - \frac{15}{4}t_3 G_3. \tag{9.22b}$$

In line with our previous derivation of a second-order model for the isothermal case (Sect. 6.8), the first-order fields s_i, $1 \leq i \leq 3$, have been introduced so that u preserves the definition of the flow rate $q = \int_0^h u\,dy$, as it should. These fields correspond to corrections to the amplitude of the parabolic velocity profile and at the same time their role in the velocity profile is similar to that of q so that the final evolution equations for q and s_i will have similar functional forms. In the same spirit, the introduction of the fields t_i, $1 \leq i \leq 3$, preserves the definition of the temperature at the surface, $\theta = T|_{y=h}$.

To complete our set, now, of test functions so that a basis for the set of polynomials of degree up to 10 satisfying the no-slip condition can be obtained, we write:

$$u = \frac{3}{h}(q - s_1 - s_2 - s_3)F_0(\bar{y}) + 45\frac{s_1}{h}F_1(\bar{y}) + 210\frac{s_2}{h}F_2(\bar{y})$$

$$+ \frac{434}{h}\left(s_3 - \sum_{i=4}^{9} s_i\right)F_3(\bar{y}) + \sum_{i=4}^{9}\frac{1}{\int_0^1 F_i(\bar{y})d\bar{y}}\frac{s_i}{h}F_i(\bar{y}). \tag{9.23}$$

As it will be shown below, the explicit formulations of the polynomials F_i, $4 \leq i \leq 9$, will not be required so that in practice the Gram–Schmidt orthogonalization procedure is limited to the determination of F_1, F_2 and F_3.

---

[3]Note that G_0 and $-\frac{3}{2}G_1$ are Legendre polynomials. Such polynomials form an orthogonal basis with respect to the scalar product $\langle f|g\rangle = \int_{-1}^{1} fg\,d\bar{y}$ instead of $\langle f|g\rangle = \int_0^1 fg\,d\bar{y}$.

We are now ready to apply a Galerkin projection. Let us consider closely the first four residuals for the momentum equation. Being of $\mathcal{O}(\varepsilon^2)$ or higher, the corrective fields s_i, $4 \leq i \leq 9$, can contribute to the formulation only through the evaluation of the zeroth-order viscous term $\int_0^h F_i(y/h)\partial_{yy}u\,dy$, which after integrating twice by parts becomes $\int_0^h F_i''(y/h)u\,dy$. Notice that $F_0'' = -1$, $F_1'' = 14F_0 - \frac{17}{3}$, $F_2'' = \frac{1485}{28}F_1 + \frac{909}{28}F_0 - 13$ and $F_3'' = \frac{88803}{868}F_1 + \frac{31779}{868}F_0 - \frac{531}{31}$, are linear combinations of 1, F_0 and F_1. Consequently, by making use of the orthogonality of the polynomials F_i, the first four residuals of the momentum equation (9.5a) written as $\mathcal{R}_q(F_i)$, $0 \leq i \leq 3$, do not involve the second-order fields s_i, $i \geq 4$. Thus, after some algebraic manipulations, they lead to a set of evolution equations for q, s_1, s_2, s_3 only, which are given in Appendix E.4 in terms of the Shkadov scaling, namely system (E.8d)–(E.8e).

The same argument applies to the temperature field so that the set of test functions G_i must be completed at second order with five polynomials of degree up to nine. Nevertheless, since G_i'', $0 \leq i \leq 3$, are not linear combinations of G_i, $0 \leq i \leq 3$, the first four residuals do not form a closed set of equations for θ, t_1, t_2 and t_3. Yet, a basis for the set of polynomials of degree up to five satisfying the Dirichlet condition at the solid wall can be obtained by introducing only one polynomial orthogonal to the first four G_i. This polynomial, G_4, is given explicitly by

$$
G_4(\bar{y}) = \bar{y} - \frac{128}{15}\bar{y}^2 + 24\bar{y}^3 - \frac{192}{7}\bar{y}^4 + 11\bar{y}^5. \tag{9.24}
$$

The temperature field can now be written to second order as

$$
T = 1 + (\theta - 1 - t_1 - t_2 - t_3 - t_4)G_0(\bar{y}) - \frac{3}{2}t_1 G_1(\bar{y}) + \frac{5}{2}t_2 G_2(\bar{y})
$$

$$
- \frac{15}{4}\left(t_3 - \sum_{i=5}^{8} t_i\right)G_3(\bar{y}) + \frac{105}{4}t_4 G_4(\bar{y}) + \sum_{i=5}^{8} t_i \frac{G_i(\bar{y})}{G_i(1)}. \tag{9.25}
$$

This formulation ensures that the evaluation of $\int_0^h G_i''(\bar{y})T\,dy$, $0 \leq i \leq 4$, does not require the definitions of G_i, $i \geq 5$. By applying next a Galerkin projection to the energy equation, the first five residuals (9.5b) written as $\mathcal{R}_\theta(G_i)$, $0 \leq i \leq 4$, constitute a closed set. Since the amplitude t_4 is of $\mathcal{O}(\varepsilon^2)$, its space and time derivatives can be neglected so that an explicit formulation as function of h, θ, t_1, t_2 and t_3 can be obtained, thus expressing the slaving of the former to the latter. After some lengthy algebraic manipulations, one gets a set of evolution equations for θ, t_1, t_2, t_3, which is given in Appendix E.4 in terms of the Shkadov scaling (system (E.8f)–(E.8i)).

The system of nine coupled equations (E.8a)–(E.8i) that we have obtained in this section will be referred to hereinafter as the *full second-order model* for the heated falling film problem. This model is unique, unlike the reduced models that will be obtained in the next sections.

We close this section with a comment on (9.19). So far we have followed a systematic procedure by using the Galerkin method and integrations by parts, while

below (9.12) we offered a shortcut for the averaged momentum equation of the
first-order model by projecting u onto the test function f_0 only. The same argument
can also be applied to the temperature field: Projecting onto the single test function
g_0, $T = 1 + (\theta - 1)g_0$, and with the above projection for the velocity, we obtain
the averaged energy equation of the first-order model. An alternative shortcut also
exists [138, 139, 276]. One could guess a priori a velocity profile that satisfies all
boundary conditions: It would consist of (9.19) truncated after the first two terms
corresponding precisely to a tau method, which "homogenizes" the first-order tan-
gential stress balance $u = 3\frac{q}{h} f_0(\bar{y}) + \varepsilon M h \partial_x \theta \, \tilde{f}_1(\bar{y})$. Together with the above simple
expression for the temperature, $T = 1 + (\theta - 1)g_0$, the momentum and energy equa-
tions of the first-order model (9.17a)–(9.17c) are obtained straightforwardly: Substitute the above profiles into the momentum and energy equations (4.6a) and (4.6b)
of the boundary layer approximation followed by their averaging with weight functions $f_0(\bar{y})$ and $\bar{y}$, respectively. This is the essence of the Galerkin method. Note
that since the expression for u satisfies all boundary conditions, we do not need to
incorporate them into the momentum equation through integrations by parts as before. The expression for T, however, does not satisfy the leading-order free-surface
boundary condition in (4.6i). But this condition is incorporated into the averaged energy equation during integrations by parts as was done in (9.13). Note also that when
we substitute the above two-term expansion for u into the momentum and energy
equations (4.6a) and (4.6b), respectively, the contribution of the Marangoni effect
to the convective terms of these equations is of $\mathcal{O}(\varepsilon^2)$ and hence negligible. The
Marangoni effect only contributes to the cross-stream viscous and thermal diffusion
terms.

9.4 Reduced Models

Admittedly, the full second-order model, derived from a systematic weighted residuals approach, is cumbersome to use because of its complexity and large dimensionality. It is hence necessary to obtain simpler models and in terms of a reduced
number of independent variables, but still retaining the dynamic features of the full
second-order model.

9.4.1 Gradient Expansion

A significant reduction can be achieved by performing a gradient expansion of the
full second-order model with $(\partial_t, \partial_x, \partial_{xx}) \rightarrow (\varepsilon \partial_t, \varepsilon \partial_x, \varepsilon^2 \partial_{xx})$ and

$$q = q^{(0)} + \varepsilon q^{(1)} + \varepsilon^2 q^{(2)} + \cdots,$$

$$\theta = \theta^{(0)} + \varepsilon \theta^{(1)} + \varepsilon^2 \theta^{(2)} + \cdots,$$

$$s_i = \varepsilon s_i^{(1)} + \varepsilon^2 s_i^{(2)} + \cdots,$$

$$t_i = \varepsilon t_i^{(1)} + \varepsilon^2 t_i^{(2)} + \cdots.$$

At $\mathcal{O}(\varepsilon^0)$ we recover the Nusselt flat film solution:

$$q^{(0)} = \frac{h^3}{3} \quad \text{and} \quad \theta^{(0)} = \frac{1}{1 + Bih}. \tag{9.26}$$

Inserting $q = q^{(0)}$ into the kinematic equation $\partial_t h + \partial_x q = 0$ yields a single evolution equation for the film thickness (a Burgers equation with zero viscosity—see (5.6)). As the heat transfer and the mechanical equilibrium of the flat film are two decoupled problems in this limit, this equation does not involve the Marangoni effect that appears at first order through the term $M\partial_x\theta$—i.e., the first-order contribution to the tangential stress (4.6h) at the interface. At $\mathcal{O}(\varepsilon)$ we obtain the first-order correction to the flow rate and the surface temperature, respectively,

$$q^{(1)} = \left[Re\frac{2}{5}h^6 - \frac{Ct}{3}h^3 + \frac{BMh^2}{2(1 + Bh)^2} \right]\partial_x h + \frac{1}{3}\varepsilon^2 Weh^3\partial_{xxx}h, \tag{9.27}$$

$$\theta^{(1)} = BiPeh^4\partial_x h\frac{7Bih - 15}{120(1 + Bih)^3}, \tag{9.28}$$

where the expression for $\theta^{(0)}$ has been used. Since the surface temperature θ is coupled through its gradient $\partial_x\theta$ in the momentum balance equations, the expression for $q^{(2)}$ can be obtained without a need to solve for the gradient expansion up to second order of the temperature fields θ and t_i:

$$q^{(2)} = (\partial_x h)^2\left\{ \frac{7}{3}h^3 - \frac{8}{5}CtReh^6 + \frac{127}{35}Re^2h^9 + \frac{3ReBMh^5}{(1 + Bh)^4}\left[\frac{11}{20} + \frac{1}{4}Pr \right.\right.$$

$$\left.\left. + Bh\left(\frac{5}{8} - \frac{1}{12}Pr \right) + B^2h^2\left(\frac{3}{40} - \frac{7}{120}Pr \right) \right]\right\}$$

$$+ \partial_{xx}h\left\{ h^4 - \frac{10}{21}CtReh^7 + \frac{12}{21}Re^2h^{10} + \frac{3ReBMh^6}{(1 + Bh)^3}\left[\frac{19}{80} + \frac{1}{16}Pr \right.\right.$$

$$\left.\left. + Bh\left(\frac{19}{80} - \frac{7}{240}Pr \right) \right]\right\} + 3\varepsilon^2 WeReh^5\left\{ \frac{8}{5}(\partial_x h)^2\partial_{xx}h + \frac{4}{5}h(\partial_{xx}h)^2 \right.$$

$$\left. + \frac{4}{3}h\partial_x h\,\partial_{xxx}h + \frac{10}{63}h^2\partial_{xxxx}h \right\}. \tag{9.29}$$

Finally, the second-order evolution equation for the free surface reads

$$\partial_t h + \partial_x\left(q^{(0)} + \varepsilon q^{(1)} + \varepsilon^2 q^{(2)} \right) = 0, \tag{9.30}$$

which fully agrees with the second-order Benney equation (BE) (5.13). Hence, the full second-order model fully resolves the behavior close to the instability threshold.

However, as shown in Chap. 5, the BE suffers from a very serious drawback as it exhibits an unphysical finite-time blow up behavior at some Reynolds number, $Re - Re_c > 1$. Hence, the BE is necessarily restricted close to criticality, i.e., for

$Re - Re_\mathrm{c}$ up to an $\mathcal{O}(1)$ value. It is then essential that any reduced model obtained from the full second-order model not suffer from the drawback of the BE. For this purpose the reduced model should have a higher degree of complexity than the BE in (9.30) and should consist of at least two or more equations. But at the same time, it should be a lower degree of complexity than the full system having nine unknowns in (E.8a)–(E.8i). Further, it should fully resolve conditions near criticality and hence not only correct all critical quantities but also give the BE with an appropriate gradient expansion, much like the full second-order model does. Finally, the reduced model should also include the second-order viscous and thermal effects.

9.4.2 Reduction of the Full Second-Order Model

To render these arguments explicit, let us start the reduction procedure by considering the projections of the velocity and temperature fields given by (9.23) and (9.25). The corrective fields s_i and t_i correspond to polynomials of increasing degree and, hence, they exhibit increasingly abrupt variations. Therefore, viscosity and thermal diffusivity will tend to damp them. This can be shown, as was done in Sect. 6.9 for the isothermal case, by linearizing the full second-order model (E.8a)–(E.8i) around the Nusselt flat film solution, assuming no spatial dependence of the perturbations, i.e., by setting their wavenumber equal to zero. With this hypothesis, $dh/dt = 0$, and the film thickness is constant. Further, both systems for (q, s_1, s_2, s_3) and (θ, t_1, t_2, t_3) are decoupled and by writing $q = 1/3 + \varepsilon \tilde{q}$, $s_i = \varepsilon \tilde{s}_i$, $\theta = (1 + \varepsilon \tilde{\theta})/(1 + B)$, $t_i = \varepsilon \tilde{t}_i$ with $\varepsilon \ll 1$, one obtains two linear systems in the form

$$\delta \frac{dV}{dt} = \mathbf{A} V, \qquad Pr\delta \frac{dW}{dt} = \mathbf{B} W, \tag{9.31}$$

where $V = (\tilde{q}, \tilde{s}_1, \tilde{s}_2, \tilde{s}_3)^t$, $W = (\tilde{\theta}, \tilde{t}_1, \tilde{t}_2, \tilde{t}_3)^t$ and $\mathbf{A}$ and $\mathbf{B}$ are two square matrices of dimension 4×4. The eigenvalues of $\mathbf{A}$ and $\mathbf{B}$ are $-190.8, -87.7, -22.3, -2.47$ and $-267.3, -63.0, -22.2, -2.47$, respectively. Therefore, there is a large gap between the least stable (largest) eigenvalues and the other eigenvalues. The spectra are hence well separated and the perturbations associated with the eigenvalues far from zero are quickly damped. The evolution of the flow in the limit of long waves is therefore dominated by the eigenvectors corresponding to the eigenvalues closest to zero. These are $(\tilde{q}, \tilde{s}_1, \tilde{s}_2, \tilde{s}_3)^t = (-1.00, 1.33 \times 10^{-2}, -1.38 \times 10^{-4}, 2.22 \times 10^{-7})^t$ and $(\tilde{\theta}, \tilde{t}_1, \tilde{t}_2, \tilde{t}_3)^t = (0.976, -0.219, 8.08 \times 10^{-3}, 7.52 \times 10^{-4})^t$. In both eigenvectors, the coefficients corresponding to the corrections $\tilde{s}_i$ and $\tilde{t}_i$ are negligible except for $\tilde{t}_1$ which is, however, four times smaller than the coefficient corresponding to $\tilde{\theta}$. Then, even if nine amplitudes, h, q, θ, s_i and t_i, $1 \leq i \leq 3$, are needed to describe the dynamics of the flow at second order, only q, h and θ will play a significant role and the other ones will be virtually slaved to q, h and θ, at least for some range of Reynolds and Péclet numbers. Therefore, it seems that at least in principle it is possible to obtain a reduced model in terms of h, q and θ only, reproducing reliably the dynamics of the film up to moderate Reynolds and Péclet numbers.

Let us now consider the two residuals corresponding to the parabolic velocity profile $\mathcal{R}_q(F_0)$ and to the linear temperature distribution $\mathcal{R}_\theta(G_0)$. In these two residuals, the fields s_i and t_i appear through inertia terms involving their space and time derivatives or through products with derivatives of h or q, which are terms of $\mathcal{O}(\varepsilon^2)$. Therefore, the fields s_i and t_i can be eliminated at second order provided explicit expressions of them as functions of h, q and θ and their derivatives are available at first order. Such relations can easily be obtained if we drop all second-order terms from residuals $\mathcal{R}_q(F_i)$ and $\mathcal{R}_\theta(G_i)$ ($i = 1, 2, 3$) and then solve for s_i and t_i. We get

$$s_1 = 3\varepsilon Re\left(\frac{1}{210}h^2\partial_t q - \frac{19}{1925}q^2\partial_x h + \frac{74}{5775}hq\partial_x q\right) + \frac{1}{40}\varepsilon M h^2\partial_x\theta, \quad (9.32a)$$

$$s_2 = 3\varepsilon Re\left(\frac{2}{5775}q^2\partial_x h - \frac{2}{17325}hq\partial_x q\right) - \frac{299}{53760}\varepsilon M h^2\partial_x\theta, \quad (9.32b)$$

$$s_3 = \frac{5}{3584}\varepsilon M h^2\partial_x\theta, \quad (9.32c)$$

$$t_1 = 3\varepsilon Pe\left[\frac{1}{15}h^2\partial_t\theta + \frac{133}{5760}h(\theta-1)\partial_x q + \frac{73}{960}hq\partial_x\theta\right], \quad (9.32d)$$

$$t_2 = 3\varepsilon Pe\left[-\frac{111}{22400}h(\theta-1)\partial_x q + \frac{79}{11200}hq\partial_x\theta\right], \quad (9.32e)$$

$$t_3 = 3\varepsilon Pe\left[\frac{1}{3150}h(\theta-1)\partial_x q - \frac{1}{1050}hq\partial_x\theta\right]. \quad (9.32f)$$

Substituting these expressions into the first residuals of the momentum and energy equations $\mathcal{R}_q(F_0)$ and $\mathcal{R}_\theta(G_0)$, corresponding to a parabolic and a linear weight, respectively, and making use of the kinematic equivalence $\partial_t h = -\partial_x q$ yields

$$3\varepsilon Re\partial_t q = \frac{5}{6}h - \frac{5}{2}\frac{q}{h^2} + 3\varepsilon Re\left(\frac{9}{7}\frac{q^2}{h^2}\partial_x h - \frac{17}{7}\frac{q}{h}\partial_x q\right) - \frac{5}{6}\varepsilon Cth\partial_x h - \frac{5}{4}\varepsilon M\partial_x\theta$$

$$+ \varepsilon^2\left[4\frac{q}{h^2}(\partial_x h)^2 - \frac{9}{2h}\partial_x q\partial_x h - 6\frac{q}{h}\partial_{xx}h + \frac{9}{2}\partial_{xx}q\right] + 9\varepsilon^2 Re^2\mathcal{K}[h, q]$$

$$+ 3\varepsilon^2 M Re\mathcal{K}^M[h, q, \theta] + \frac{5}{6}\varepsilon^3 Weh\partial_{xxx}h, \quad (9.33a)$$

$$3\varepsilon Pe\partial_t\theta = 3\frac{(1-\theta-Bh\theta)}{h^2} + 3\varepsilon Pe\left[\frac{7}{40}\frac{(1-\theta)}{h}\partial_x q - \frac{27}{20}\frac{q}{h}\partial_x\theta\right]$$

$$+ \varepsilon^2\left[\left(1-\theta-\frac{3}{2}Bh\theta\right)\left(\frac{\partial_x h}{h}\right)^2 + \frac{\partial_x h\partial_x\theta}{h} + (1-\theta)\frac{\partial_{xx}h}{h} + \partial_{xx}\theta\right]$$

$$+ 3\varepsilon^2 Pe\{3Re\mathcal{K}_{\theta q}[h, q, \theta] + 3Pe\mathcal{K}_\theta[h, q, \theta] + M\mathcal{K}_\theta^M[h, q, \theta]\}, \quad (9.33b)$$

where $\mathcal{K}$, $\mathcal{K}^M$, $\mathcal{K}_{\theta q}$, $\mathcal{K}_\theta$ and $\mathcal{K}_\theta^M$ contain the second-order inertia terms introduced by the corrections to the flat film solution, namely s_i and t_i given in Appendix E.5. $\mathcal{K}$ contains terms of the momentum equation produced by the advection of the first-order corrections of the velocity profile as in isothermal conditions (see (6.83)). $\mathcal{K}^M$ denotes the terms of the momentum equation associated with the Marangoni flow produced by the temperature gradient at the free surface. Similarly, $\mathcal{K}_\theta$ contains terms originating from the averaged energy equation through the advection of the first-order corrections of the temperature profile (t_1, t_2, t_3). The terms contained in $\mathcal{K}_{\theta q}$ and $\mathcal{K}_\theta^M$ originate from the advection of the linear flat-film temperature distribution by the first-order corrections of the velocity profile induced by the deformation of the free surface and the Marangoni flow, respectively.

Although the explicit expressions of $\mathcal{K}$, $\mathcal{K}_{\theta q}$, $\mathcal{K}_\theta$, $\mathcal{K}^M$ and $\mathcal{K}_\theta^M$ are complicated and involve time derivatives (see Appendix E.5), they can be drastically simplified by using the relations provided by the zeroth-order flat film solution:

$$q = \frac{h^3}{3} + \mathcal{O}(\varepsilon) \quad \text{and} \quad \theta = \frac{1}{1 + Bih} + \mathcal{O}(\varepsilon). \tag{9.34}$$

In fact, as shown for the isothermal case in Sect. 6.9, the second-order terms appearing in (9.33a), (9.33b) do not have a unique formulation since a large number of asymptotically equivalent expressions is possible via the expressions in (9.34) (note, however, that the full second-order model (E.8a)–(E.8i) is unique). Hence, we do not end up with a single three-field model fully compatible with the second-order BE (9.30) but with a whole family of such models.

Nevertheless, even if all of them are asymptotically equivalent, they might not necessarily behave in the same manner. In isothermal conditions, for instance, the branch of solitary wave solutions to (9.33a), with $\mathcal{K}^M = 0$ and $\mathcal{K}$ given by (E.9a), shows a turning point (Sect. 6.9.1). As earlier noted, this unphysical behavior is related to the high-order nonlinearities present in (E.9a). The second-order corrections in nonisothermal conditions, as given in Appendix E.5, also possess such high-order nonlinearities, suggesting that the corresponding branch in nonisothermal conditions will have the same kind of unphysical catastrophic behavior. However, like in the isothermal case, we can apply a regularization procedure that reduces the order of nonlinearities and prevents singularities, so that a model that also accounts accurately for the drag-inertia regime can be obtained, i.e., a model valid in the widest possible range of both Reynolds and Péclet numbers. The aim then is to elucidate the interaction between the H- and S-modes in the widest possible range of these parameters.

9.4.3 Padé-Like Regularization

We shall extend the Padé-like regularization procedure developed in Sect. 6.9.2 to the nonisothermal case. Let us first consider the residual $\mathcal{R}_{q,0} = \mathcal{R}_q(F_0)$ from (9.5a) obtained by averaging the momentum equation with weight F_0, which can be written as $\mathcal{R}_{q,0}^{(0)} + \mathcal{R}_{q,0}^{(1)} + \mathcal{R}_{q,0}^{(2),\eta} + \mathcal{R}_{q,0}^{(2),\delta}$, where the numbers in the superscripts refer to the different orders in the gradient expansion. In addition, we separate the second-order terms into those having a viscous origin (superscript η) from those accounting for the convective acceleration induced by the departures of the velocity profile from the parabolic shape, including here the Marangoni effect (superscript δ). So $\mathcal{R}_{q,0}$ is sought in the form $\mathcal{G}^{-1}\mathcal{F}$, where $\mathcal{G}$ is now simply a function of h, q, θ and their derivatives, and $\mathcal{F}$ is reduced to $\mathcal{R}_{q,0}^{(0)} + \mathcal{R}_{q,0}^{(1)} + \mathcal{R}_{q,0}^{(2),\eta}$, i.e., the residual obtained by assuming a parabolic velocity profile (i.e., (9.33a), (9.33b) with all $\mathcal{K}$ terms set to zero). Formally, the regularization factor has the same form as in the isothermal case (see (6.88)):

$$\mathcal{G}_q = \left(1 + \frac{\mathcal{R}_{q,0}^{(2),\delta}}{\mathcal{R}_{q,0}^{(1),\delta}}\right)^{-1}, \tag{9.35}$$

with

$$\mathcal{R}_{q,0}^{(1),\delta} = 3\varepsilon Re\left(\frac{2}{5}\partial_t q - \frac{18}{35}\frac{q^2}{h^2}\partial_x h + \frac{34}{35}\frac{q}{h}\partial_x q\right), \tag{9.36}$$

$$\mathcal{R}_{q,0}^{(2),\delta} = -\frac{6}{5}\varepsilon^2 Re\left(3Re\mathcal{K} + M\mathcal{K}^M\right), \tag{9.37}$$

where $\mathcal{K}$ and $\mathcal{K}^M$ are given explicitly in Appendix E.5. An asymptotically equivalent expression for $\mathcal{G}_q$ can be found using the zeroth-order expressions (9.34) and the kinematic equivalence $\partial_t \equiv -h^2\partial_x + \mathcal{O}(\varepsilon)$ that applies to the three variables h, q and θ. Therefore, $\partial_{xt}\theta$ in $\mathcal{K}^M$ is asymptotically equivalent to $\partial_x(-h^2\partial_x\theta) + \mathcal{O}(\varepsilon^2)$. We then obtain

$$\mathcal{R}_{q,0}^{(1),\delta} = -\frac{2}{5}\varepsilon Re h^4 \partial_x h + \mathcal{O}(\varepsilon^2),$$

$$\mathcal{R}_{q,0}^{(2),\delta} = \frac{\varepsilon^2 Re^2}{175}h^7(\partial_x h)^2 - 3\varepsilon^2 M Re\left(\frac{1}{84}h^3\partial_x h\partial_x\theta + \frac{1}{6080}h^4\partial_{xx}\theta\right) + \mathcal{O}(\varepsilon^3),$$

which, when substituted in (9.35), yields

$$\mathcal{G}_q = \left[1 - \frac{\varepsilon Re}{70}h^3\partial_x h + \varepsilon M\left(\frac{5}{56}\frac{\partial_x\theta}{h} + \frac{1}{224}\frac{\partial_{xx}\theta}{\partial_x h}\right)\right]^{-1} + \mathcal{O}(\varepsilon^2). \tag{9.38}$$

Though h cannot be zero (assuming that no dry spots are possible), $\partial_x h$ can vanish. The last term in (9.38) will thus lead to singularity and should be avoided in its present form. It turns out that it is not possible to obtain an expression asymptotically equivalent to the right hand side of (E.9b) in the form of a second-order correction of first-order inertia terms, i.e., like

$$\mathcal{K}^M \propto \frac{\partial_x\theta}{h}\left(\partial_t q - \frac{9}{7}\frac{q^2}{h^2}\partial_x h + \frac{17}{7}\frac{q}{h}\partial_x q\right), \tag{9.39}$$

due to the presence of the last two terms in (E.9b), $\frac{1}{48}h^2\partial_{xt}\theta$ and $\frac{15}{224}hq\,\partial_{xx}\theta$. Hence, the closest form to (9.39) asymptotically equivalent to (E.9b) is

$$\mathcal{K}^M = \frac{5}{56}\frac{\partial_x\theta}{h}\left(\partial_t q - \frac{9}{7}\frac{q^2}{h^2}\partial_x h + \frac{17}{7}\frac{q}{h}\partial_x q\right) + \frac{1}{224}qh\partial_{xx}\theta. \tag{9.40}$$

Expression (9.40) will be used hereinafter.

To achieve a maximum reduction of the order of nonlinearities, $\mathcal{G}_q$ is finally rewritten in terms of the local slope $\partial_x h$ and the local Reynolds number $3Req$ (defined first in Sect. 5.4):

$$\mathcal{G}_q = \left(1 - \frac{3}{70}\varepsilon Req\,\partial_x h + \varepsilon M\frac{5}{56}\frac{\partial_x\theta}{h}\right)^{-1}. \tag{9.41}$$

As far as the averaged energy balance (9.33b) is concerned, its regularization appears to be very cumbersome because of the coupling between the different physical effects (momentum and energy advections, free-surface deformations and Marangoni effect). In fact, for the second-order inertia and thermocapillary terms represented by $\mathcal{K}_\theta$, $\mathcal{K}_{\theta q}$ and $\mathcal{K}_\theta^M$ (induced by the deviations of the velocity and temperature profiles from the Nusselt flat film solution), it has not been possible to obtain asymptotically equivalent formulations analogous to (9.40), if the temperature field is assumed to be slaved to the free surface temperature θ only. This failure suggests that we should try to describe the temperature field by allowing at least the first correction t_1 to θ, to have its own dynamics. Notice that we already reached the same conclusion from the result of the linear eigenvalue problem associated with (9.31). However, for the sake of simplicity, we will not extend here the model to include the additional field t_1. Instead, $\mathcal{K}_{\theta q}$, $\mathcal{K}_\theta$ and $\mathcal{K}_\theta^M$ will be simply neglected. The argument supporting this option is twofold: (i) there is still consistency with the gradient expansion at second order since the interfacial temperature is coupled to the local flow rate through its gradient only which is of $\mathcal{O}(\varepsilon)$; (ii) other limitations intrinsic to the energy equation will prevent in any case the extension of the range of validity of any of its regularized forms (see Sect. 9.6).

In conclusion, we only apply the regularization procedure to the momentum equation (9.33a) to ensure, despite high-order nonlinearities, the smallness of the second-order corrections $\mathcal{K}$ and $\mathcal{K}^M$ relative to the corresponding first-order terms, thus avoiding unphysical blow-up behaviors. The model thus has the following form:

$$\partial_t h = -\partial_x q, \tag{9.42a}$$

$$3\varepsilon Re\partial_t q = 3\varepsilon Re\left(\frac{9}{7}\frac{q^2}{h^2}\partial_x h - \frac{17}{7}\frac{q}{h}\partial_x q\right) + \left(1 - \frac{3\varepsilon Re}{70}q\,\partial_x h + \varepsilon M\frac{5}{56}\frac{\partial_x\theta}{h}\right)^{-1}$$

$$\times\left[\frac{5}{6}h - \frac{5}{2}\frac{q}{h^2} + \varepsilon^2\left(4\frac{q}{h^2}(\partial_x h)^2 - \frac{9}{2h}\partial_x q\,\partial_x h - 6\frac{q}{h}\partial_{xx}h + \frac{9}{2}\partial_{xx}q\right)\right.$$

$$-\frac{5}{6}\varepsilon\, Cth\,\partial_x h + \frac{5}{6}\varepsilon^3 Weh\,\partial_{xxx} h$$

$$-\varepsilon M\left(\frac{5}{4}\partial_x\theta - \frac{\delta}{224}hq\,\partial_{xx}\theta\right)\Bigg],\tag{9.42b}$$

$$3\varepsilon Pe\,\partial_t\theta = 3\frac{(1-\theta-Bh\theta)}{h^2} + 3\varepsilon Pe\left[\frac{7}{40}\frac{(1-\theta)}{h}\partial_x q - \frac{27}{20}\frac{q}{h}\partial_x\theta\right]$$

$$+\varepsilon^2\left[\left(1-\theta-\frac{3}{2}Bh\theta\right)\left(\frac{\partial_x h}{h}\right)^2 + \frac{\partial_x h\,\partial_x\theta}{h}\right.$$

$$\left.+(1-\theta)\frac{\partial_{xx}h}{h} + \partial_{xx}\theta\right],\tag{9.42c}$$

and will be referred to hereafter as the *regularized model* for the heated falling film problem. As expected, a gradient expansion of (9.42a)–(9.42c) recovers exactly the expressions in (9.27) and (9.29) of $q^{(1)}$ and $q^{(2)}$ and hence leads to the BE (9.30). Using the Shkadov scaling, (9.42a)–(9.42c) becomes

$$\partial_t h = -\partial_x q,\tag{9.43a}$$

$$\delta\partial_t q = \delta\left(\frac{9}{7}\frac{q^2}{h^2}\partial_x h - \frac{17}{7}\frac{q}{h}\partial_x q\right) + \left(1 - \frac{\delta}{70}q\partial_x h + M\frac{5}{56}\frac{\partial_x\theta}{h}\right)^{-1}$$

$$\times\left[\frac{5}{6}h - \frac{5}{2}\frac{q}{h^2} + \eta\left(4\frac{q}{h^2}(\partial_x h)^2 - \frac{9}{2h}\partial_x q\,\partial_x h - 6\frac{q}{h}\partial_{xx}h + \frac{9}{2}\partial_{xx}q\right)\right.$$

$$\left.-\frac{5}{6}\zeta h\partial_x h + \frac{5}{6}h\partial_{xxx}h - M\left(\frac{5}{4}\partial_x\theta - \frac{\delta}{224}hq\,\partial_{xx}\theta\right)\right],\tag{9.43b}$$

$$Pr\delta\partial_t\theta = 3\frac{(1-\theta-Bh\theta)}{h^2} + Pr\delta\left[\frac{7}{40}\frac{(1-\theta)}{h}\partial_x q - \frac{27}{20}\frac{q}{h}\partial_x\theta\right]$$

$$+\eta\left[\left(1-\theta-\frac{3}{2}Bh\theta\right)\left(\frac{\partial_x h}{h}\right)^2 + \frac{\partial_x h\,\partial_x\theta}{h}\right.$$

$$\left.+(1-\theta)\frac{\partial_{xx}h}{h} + \partial_{xx}\theta\right].\tag{9.43c}$$

Noteworthy is that the momentum equation (9.43b) with $\mathcal{M}=0$ reduces to its isothermal version (6.92).

In the next two sections we shall demonstrate that this model satisfies a number of criteria, both linear and nonlinear: (i) good agreement of its linear stability characteristics with Orr–Sommerfeld; (ii) its nonlinear solutions, in particular single-hump solitary pulses, exist for the widest possible range of parameters.

The regularized model for the heat flux (HF) case is developed in Appendix E.6.

9.5 Linear Stability

We now examine the linear stability of the Nusselt flat film solution by using the regularized model (9.42a)–(9.42c) and we compare the results to those obtained from the Orr–Sommerfeld eigenvalue problem of the full Navier–Stokes and Fourier equations presented in Chap. 3. It is also instructive here to include the linear stability analysis obtained from: (i) the second-order boundary layer equations (4.5a), (4.5b) together with the continuity equation (4.2a), the wall and free surface boundary conditions (4.2f)–(4.2i), (4.2k)–(4.2m); after all, the models developed here are based on the boundary layer approximation; (ii) the full second-order model (E.8a)–(E.8i) to assess the depth-averaging approach leading to equations independent on the cross-stream coordinate y; (iii) the first-order model (9.17a)–(9.17c) to assess the influence of the second-order effects included in the regularized model—i.e., viscous and thermal effects as well as second-order corrections due to inertia; (iv) the first-order BE.

For a temporal stability analysis, the dispersion relation corresponding to (9.42a)–(9.42c) is obtained by first expressing the perturbations to the flat film solution in the form of normal modes with real wavenumber k and complex angular frequency ω:

$$\begin{pmatrix} h \\ q \\ \theta \end{pmatrix} = \begin{pmatrix} 1 \\ 1/3 \\ 1/(1+B) \end{pmatrix} + \varsigma \begin{pmatrix} 1 \\ A_q \\ A_\theta \end{pmatrix} \exp\left[i(kx - \omega t)\right]. \tag{9.44}$$

Expressions (9.44) are next substituted into (9.42a)–(9.42c), which are subsequently linearized for $\varsigma \ll 1$. For the resulting system of linear algebraic equations to have nontrivial solutions, it is necessary and sufficient that its principal determinant be equal to zero. Likewise, substituting (9.44), $s_i = \varsigma A_{s_i} \exp[i(kx - \omega t)]$ and $t_i = \varsigma A_{t_i} \exp[i(kx - \omega t)]$ in (E.8a)–(E.8i) leads to the dispersion relation for the full second-order model.

A small wavenumber expansion of the dispersion relations shows that all models developed in this chapter lead to the following neutral stability results obtained from the Navier–Stokes and Fourier equations (or equivalently from the boundary layer equations) and the BE in Chaps. 3 and 5, respectively:

$$c \equiv \frac{\omega_{\mathrm{r}}}{k} = 1, \qquad k_{\mathrm{c}} = \left[\frac{2}{5}\delta - \zeta + \frac{3B\mathcal{M}}{2(1+B)^2}\right]^{1/2}, \tag{9.45}$$

where c is the linear phase speed and k_{c} is the cut-off wavenumber. The critical condition for the instability can be obtained by setting $k_{\mathrm{c}} = 0$ in (9.45), which yields

$$\frac{2}{5}\delta + \frac{3B\mathcal{M}}{2(1+B)^2} = \zeta. \tag{9.46}$$

To simplify comparisons with the linear stability analysis presented in Chap. 3 obtained from full Navier–Stokes and Fourier equations, it is convenient one rewrite the expressions in (9.45) and (9.46) using the Nusselt scaling. As already done several times in this monograph, the conversion from the Shkadov scaling to the Nusselt one simply consists by destretching the x-coordinate through the transformation $(k, \omega) \to \kappa(k, \omega)$ and using the definition of the reduced variables, $\delta = 3Re/\kappa$, $\zeta = Ct/\kappa$, $\mathcal{M} = M/\kappa$, where $\kappa^3 = We$. Equations (9.45) and (9.46) are then

$$c = 1, \qquad k_c = \frac{1}{We^{1/2}} \left[\frac{6}{5}Re - Ct + \frac{3M^\star}{2(1+B)} \right]^{1/2}, \tag{9.47}$$

$$\frac{6}{5}Re + \frac{3M^\star}{2(1+B)} = Ct, \tag{9.48}$$

where the film Marangoni number $M^\star = M \Delta T_s$ is based on the temperature difference across the uniform fluid layer, and $\Delta T_s = B/(1+B)$ (see Sect. 2.5). Notice that the change of scales does not affect the speed—it is equal to unity so that the linear waves propagate with a velocity three times the averaged velocity—and does not introduce any modifications to the coefficients of the expressions for the cut-off wavenumber and critical condition except for the coefficient 3 along with the Reynolds number.

The expressions for the cut-off wavenumber and criticality condition in (9.47) and (9.48) coincide with (3.35) and (3.31) obtained from Orr–Sommerfeld and (5.18) obtained from the BE. Note that for k_c in (5.18a) we need the additional transformation $k_c \to k_c/\varepsilon$ to remove ε present in (5.18a) due to the presence of ε in the BE; for the average models on the other hand, ε is scaled away prior to the introduction of the Shkadov scaling.

From the definitions of the different parameters in Chap. 2, one has $We \propto Re^{-2/3}$, $B \propto Re^{1/3}$ and $M^\star \propto Re^{-1/3} \propto 1/h_N$. Therefore, in the region of large Re, i.e., large film thicknesses or equivalently large flow rates, the interfacial forces due to both the Marangoni effect and surface tension are not important compared to the dominant inertia forces, and the H-mode prevails. Conversely, in the limit of vanishing Reynolds numbers, i.e., vanishing film thicknesses or equivalently vanishing flow rates, the inertia effects are negligible and the Marangoni effect becomes dominant, i.e., the S-mode prevails: for small film thicknesses, the destabilizing inertia forces are vanishing, but the destabilizing interfacial forces due to the Marangoni effect are still present—the interfacial forces due to capillarity are always stabilizing. Also since $We \propto Re^{-2/3}$ and $M^\star/We \propto Re^{1/3}$, from (9.47) the cut-off wavenumber tends to zero like $\sqrt{M^\star/We} \propto Re^{1/6}$. The fidelity of this scaling law will be confirmed with computations of the different models next.

9.5.1 Neutral Stability Curves

Our aim now is to decipher the neutral stability characteristics of a falling film heated uniformly from below. We assume that the gas–liquid–solid system (physical

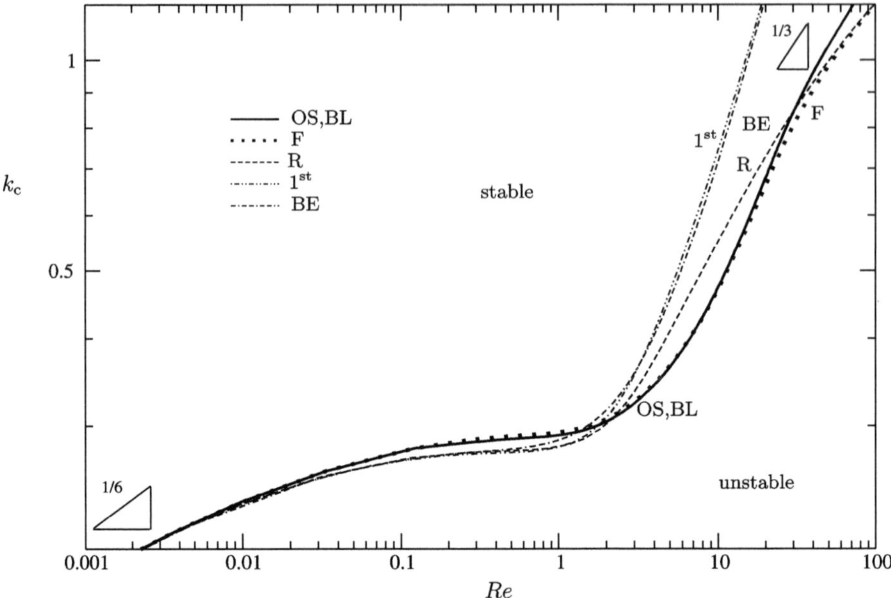

Fig. 9.1 Neutral stability curves in a log–log plot for a heated falling film with $\Gamma = 250$, $Ct = 0$, $Pr = 7$, $Ma = 50$ and $Bi = 1$ obtained from various models: OS, Orr–Sommerfeld; BL, second-order boundary layer equations; F, full second-order model (E.8a)–(E.8i); R, regularized model (9.42a)–(9.42c); 1st, first-order model (9.17a)–(9.17c); BE, first-order BE and (9.47). For the region of Reynolds numbers in the figure, OS and BL are indistinguishable

properties of the gas–liquid system and wall heating conditions i.e., wall temperature) and inclination angle β are fixed, as in a real experiment, and the only control parameter is the inlet flow rate (see Appendix D.1). Fixing the gas–liquid–solid system and β means fixing the viscous-gravity set of parameters, Ct, Γ, Ma, Bi and Pr (see again Appendix D.1). Therefore, we only vary the Nusselt film thickness h_N or, equivalently, the Reynolds number $Re \propto h_N^3$.

Figure 9.1 depicts the neutral stability curves in the wavenumber–Reynolds number plane for $Pr = 7$, $\Gamma = 250$, $Ct = 0$, $Ma = 50$ and $Bi = 1$ computed with the various models we have developed for the heated falling film as well as the Orr–Sommerfeld eigenvalue problem. The neutral curves are given in a log–log plot, which helps us separate them in the region $Re \to 0$. The parameter values are chosen so that the differences between the various systems of equations can be clearly identified; hence, the choice $Bi = 1$, an unrealistically large Biot number, which amplifies the Marangoni effect. (In the derivation of the averaged models we assumed $Bi = \mathcal{O}(1)$, but as we emphasized in Sect. 9.2, this order of magnitude can be relaxed; for example, for the first-order model the most general order of magnitude assignment for this parameter would be $Bi \gg \varepsilon^2$ and Bi at most of $\mathcal{O}(\varepsilon)$.) The wavenumber k is obtained from the dimensional wavenumber $\bar{k}$ nondimensionalized with $1/\bar{h}_N$.

As noted earlier, as Re tends to zero, the wavenumber k given in (9.47) tends to zero like $\sqrt{M^\star/We} \propto Re^{1/6}$, as also shown on the bottom left corner of Fig. 9.1. As Re increases, the first-order models (1st, BE) deviate significantly from the other models, showing the importance of the second-order viscous and thermal effects, as first noted in Sect. 4.3. Notice that the curve for the first-order BE is precisely the one given by (9.47). On the contrary, the full second-order model (F) compares very well with the boundary layer approximation (BL)—or equivalently the Orr–Sommerfeld solution (OS), which is indistinguishable from the BL solution for the region of Reynolds numbers in the figure—but it slightly underpredicts the neutral wavenumber in the region of large Reynolds numbers. This discrepancy is most likely due to the limited radius of convergence of the perturbation scheme used to obtain the full second-order model, as is the case with any approximate method. Notice also that for large Re, the neutral stability curves obtained with any of the second-order models behave like $k \propto Re^{1/3}$, as shown on the upper right corner of Fig. 9.1. This means that the dimensional wavenumber $\bar{k}$ becomes independent of the film thickness $\bar{h}_N$ for large Re. On the other hand, at low Reynolds numbers all models yield results in agreement with the asymptotic result (9.47). In this region the dynamics of the flow is slaved to its kinematics, i.e., both flow rate and interfacial temperature are adiabatically slaved to the film thickness and they depend on time only through the dependence of the film thickness on time. This is precisely the region where the BE long wave expansion applies.

The dramatic change of slope in Fig. 9.1 with increasing Re is connected with the transition from the drag-gravity regime at low Re to the drag-inertia regime at moderate Re (Sect. 4.9.1). The two instability modes are closely connected with these regimes: the thermocapillary S-mode is predominant for small Re and the hydrodynamic H-mode is predominant for large Re. For the conditions in Fig. 9.1 the transition between the drag-gravity and the drag-inertia regimes occurs at $Re \approx 3$ (or equivalently $\delta \approx 2.3$, i.e., an $\mathcal{O}(1)$ value of δ as expected) and thus corresponds to an intermediate regime where both the S- and H-modes are of same order of magnitude. Figure 9.1 also indicates that at the transition the full second-order model (F) is in good agreement with the boundary layer approximation (BL), but the regularized model (R) predicts a smaller neutral wavenumber (still, the agreement with more exact models is qualitatively satisfactory). This clearly shows the limitation of the regularized model to correctly take into account the convective heat transport effects at the transition where the S-mode and the H-mode compete with each other. Moreover, as we shall demonstrate in Sect. 9.6 in the nonlinear regime and for large-amplitude waves, the regularized model fails to describe correctly the temperature field as the Péclet number increases, in which case the convective heat transport effects become increasingly important.

9.5.2 Growth Rate Curves

Of particular interest are also the growth rates in the range $0 < k < k_c$ of unstable wavenumbers defined by the neutral stability curves. Figure 9.2 shows typical

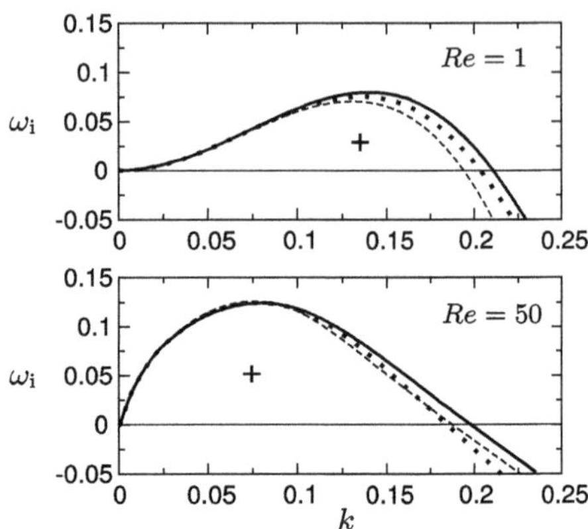

Fig. 9.2 Dispersion relations for the growth rate as a function of wavenumber, for two different Reynolds numbers obtained from the Orr–Sommerfeld eigenvalue problem (*solid lines*), the full second-order model (E.8a)–(E.8i) (*dotted lines*) and the regularized model (9.42a)–(9.42c) (*dashed lines*). Parameter values are given in the caption of Fig. 9.1

growth rate curves ω_i as functions of k for two different Reynolds numbers, $Re = 1$ and $Re = 50$. The curves feature a band of unstable wavenumbers in $0 < k < k_c$, which contains the maximum growing wavenumber with the largest growth rate. The figure also shows that the growth rates predicted by the regularized model (dashed line) and the full second-order model (dotted line) are fairly close to the growth rate obtained by the Orr–Sommerfeld eigenvalue problem (solid line), even for the relatively large value $Re = 50$, where the H-mode is dominant. This demonstrates the significance of having a model consistent at second order, as it has already been demonstrated for isothermal conditions (see, e.g., Sect. 7.1). In fact, the second-order viscous and thermal effects are crucial for a good agreement with Orr–Sommerfeld. Yet, the regularized model is of lower complexity than the full second-order model—three variables instead of nine—which makes it an attractive prototype for mathematical and numerical scrutiny. It will hence be used in the rest of this section to investigate the influence of Bi, Ma, Pr, Γ and Ct on the neutral curves. It will also form the basis for the computations of nonlinear solutions in Sect. 9.6.

9.5.3 Influence of Bi, Ma, Pr, Γ on the Neutral Stability Curves

We now turn our attention to the influence of the parameters Bi, Ma, Pr, Γ on the neutral curves. These parameters depend on the properties of the gas–liquid–solid system and their influence is reported in Fig. 9.3 for the regularized model (9.42a)–(9.42c). The results are for a vertical wall ($Ct = 0$) and are presented in logarithmic plots, as in Fig. 9.1. Figure 9.3a shows the influence of heat transfer through the Biot number. For $Bi = 1$ the influence of the Marangoni effect is large at small and moderate Reynolds numbers. In fact, if Bi tends to zero or infinity, the free surface temperature becomes uniform (hence independent of thickness variations so that any perturbations on h do not affect the free-surface temperature—see also Sect. 2.5) and the Marangoni effect is simply not an issue. For the other plots of Fig. 9.3,

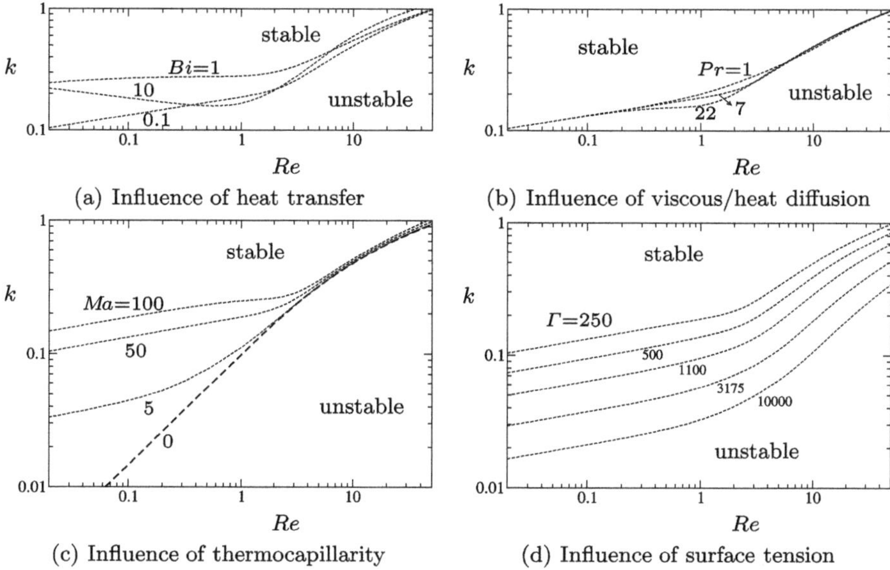

Fig. 9.3 Influence of Bi, Ma, Pr, Γ on the neutral stability curves for a film falling down a vertical wall, obtained with the regularized model (9.42a)–(9.42c): (**a**) various Bi with $Pr = 7$, $Ma = 50$ and $\Gamma = 250$; (**b**) various Pr with $Bi = 0.1$, $Ma = 50$ and $\Gamma = 250$; (**c**) various Ma with $Bi = 0.1$, $Pr = 7$ and $\Gamma = 250$; (**d**) various Γ with $Bi = 0.1$, $Ma = 50$ and $Pr = 7$. Neutral curve corresponding to isothermal conditions ($Ma = 0$) is plotted in (**c**) (*thick dashed line*)

we choose a smaller Biot number, $Bi = 0.1$, which corresponds to more realistic values as are encountered in experiments (see, e.g., Appendix D.4). Figure 9.3(b) depicts the influence of viscous and heat diffusion through the Prandtl number on the neutral stability curves. Clearly, the Prandtl number has little influence for large Re, i.e., when the H-mode predominates, whereas the curves are strongly affected by the Prandtl number for $Re = \mathcal{O}(1)$, i.e., when the S-mode is of the same order of magnitude with the H-mode. After all, the origin of the S-mode is the gradient of temperature at the interface. This gradient may be weakened by the transport of heat from the troughs to the crests of a free surface deformation due to the motion of the fluid, a process that is intensified with large Prandtl numbers (see also Sect. 9.6.3).

Figure 9.3(c) depicts the influence of the thermocapillary effect through the Marangoni number. As expected, for $Ma = 0$, we recover the classical H-mode, with the corresponding curve starting from the origin. For $Pr = 7$, increasing the Marangoni number increases the range of unstable wavenumbers especially at low Reynolds numbers where the Marangoni effect is predominant (S-mode). If Re is sufficiently large, the hydrodynamic H-mode predominates and the thermocapillary effect as measured by Ma should not modify significantly the cut-off wavenumber, and the neutral curves should all converge to the curve for $Ma = 0$. Therefore, the small disparity of the curves as compared with the curve for isothermal conditions (thick dashed line) is a consequence of the increased inaccuracy of the regularized model as the Reynolds number increases.

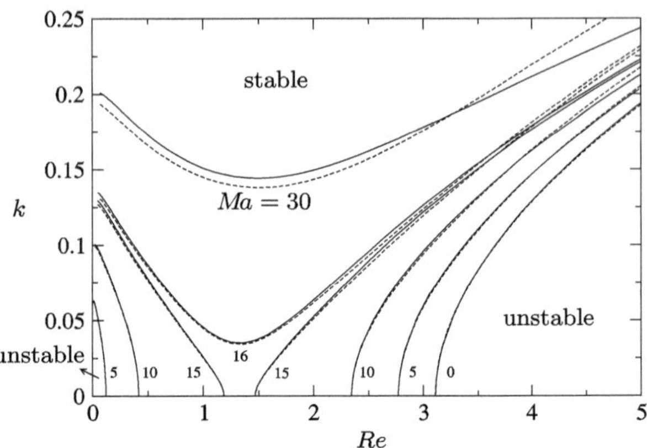

Fig. 9.4 Influence of the Marangoni number on the neutral stability for an inclined plate forming an angle $\beta = 15°$ with the horizontal direction with $\Gamma = 250$, $Pr = 7$ and $Bi = 1$, and computed with the Orr–Sommerfeld eigenvalue problem (*solid lines*) and the regularized model (9.42a)–(9.42c) (*dashed lines*)

Finally, Fig. 9.3(d) shows the influence of the Kapitza number Γ on the neutral stability curves. Decreasing the value of Γ increases the range of unstable wavenumbers. It is precisely because we wish to show the effect of viscous dispersion and thus emphasize the differences between the different models, that Γ ($= 250$) has been deliberately chosen relatively small as compared to usual values ($\Gamma = 3175$ for water at 18°C).

9.5.4 Influence of Inclination

For inclined walls the critical condition (9.48) shows that there is a critical value of the Reynolds number, denoted Re_c, above which the Nusselt flat film solution is unstable, first given in (3.31). Recall that if the critical condition is rewritten using the viscosity-gravity scaling introduced in Chap. 2, with $Re = h_N^3/3$, $M = Ma/h_N^2$ and $B = Bih_N$, that is,

$$\frac{2}{5}h_N^3 + \frac{3BiMa}{2h_N(1 + Bih_N)^2} = Ct, \tag{9.49}$$

it can have two real and positive roots in terms of h_N for certain parameter values. The film is then stable for thicknesses between the two roots and unstable otherwise, with the small thickness root corresponding to the thermocapillary S-mode and the large thickness root corresponding to the hydrodynamic H-mode. This is also illustrated in Fig. 9.4 for a plate inclined at an angle $\beta = 15°$ with respect to the horizontal direction. The parameters are identical to those in Fig. 3.2. For $Ma < 15.11$ two distinct unstable regions are observed. Noteworthy is the excellent agreement of the curves corresponding to the regularized model (9.42a)–(9.42c) with Orr–Sommerfeld in the vicinity of the two thresholds resulting precisely from a correct representation of the instability threshold. The agreement persists even far from the threshold, a direct consequence of taking into account the second-order terms in the formulation.

9.6 Solitary Waves

In the previous section we showed that the regularized model compares well in the linear regime with the Orr–Sommerfeld eigenvalue problem for a wide range of parameters. Hence, the regularized model satisfies the first of the two criteria listed at the end of Sect. 9.4.3.

In this section, we shall seek traveling wave solutions of the regularized model. We restrict our attention to single-hump solitary waves. After all, by now we know that for isothermal films the long-time evolution is characterized by a train of soliton-like coherent structures each of which resembling the infinite-domain solitary pulses. Therefore, by analogy with the isothermal case, we anticipate that for the nonisothermal problem, the long-time evolution is also dominated by solitary waves.

Before we examine traveling waves, it is worth noting that all parameters of the regularized model written in terms of the Shkadov scaling (9.43a)–(9.43c) vanish as the Reynolds number tends to zero except for $\mathcal{M}$, which diverges: With $Re \sim h_{\mathrm{N}}^{1/3}$, $Re \to 0$ or $h_{\mathrm{N}} \to 0$ implies $We = \Gamma/h_{\mathrm{N}}^2 \to \infty$ and hence $\delta, \zeta, \eta \to 0$. Therefore, for low flow rates the terms multiplied by these parameters can be neglected such that (9.43b) and (9.43c) become

$$q = \frac{h^3}{3}(1 + \partial_{xxx}h) - \frac{\mathcal{M}}{2}h^2\partial_x\theta, \qquad \theta = \frac{1}{1 + Bh} = 1 - Bh + \mathcal{O}(B^2), \quad (9.50)$$

where we have retained the leading-order term in θ involving B. The mass conservation equation (9.43a) then gives

$$\partial_t h + \partial_x\left[\frac{h^3}{3}(1 + \partial_{xxx}h) + \frac{\mathcal{M}B}{2}h^2\partial_x h\right] = 0, \qquad (9.51)$$

where $\mathcal{M}B \propto Re^{-1/9}$, which diverges as $Re \to 0$. By rescaling x and t in (9.51) as

$$\partial_t h = \phi\partial_\tau, \qquad \partial_x = \phi\partial_\xi,$$

the coefficients of the Marangoni and surface tension terms are equal when $\phi = (\mathcal{M}B/2)^{1/2}$ and the equation becomes:

$$\partial_\tau + \partial_\xi\left[\frac{h^3}{3}\left(1 + \phi^3\partial_{\xi\xi\xi}h\right) + \phi^3 h^2\partial_\xi h\right] = 0$$

(suggesting a universal behavior for h in the region of $h_{\mathrm{N}} \to 0$). Hence, in the region of small Reynolds numbers where the S-mode dominates over the H-mode, the stabilizing surface tension terms are still present (after all both Marangoni and capillary forces are surface forces), as they should be for the formation of nonlinear structures.

It can be shown that the homoclinic solutions to (9.51) blow up when $\mathcal{M}B$ tends to infinity, i.e., the Reynolds number tends to zero [139]. However, this does not correspond to a true singularity formation: in this region of small flow rates and hence small film thicknesses, the film is expected to form isolated drops separated by very thin layers of fluid for which forces of nonhydrodynamic origin such as van der

Waals forces not included here may become important. Such forces, if stabilizing, are expected to arrest the singularity formation observed for the homoclinic orbits in the region of low Reynolds numbers. Inversely, if Re tends to infinity, both $\mathcal{M}$ and $\mathcal{M}B$ tend to zero and the velocity and temperature fields are decoupled in this limit. Therefore, at large Reynolds numbers, the shape of the waves should be unaffected by the Marangoni effect. In this region, the H-mode dominates over the S-mode. These two limits will enable us to elucidate the influence of Reynolds number on the shape of the waves.

In what follows, we discuss in detail the properties of the solitary wave solutions of the system (9.43a)–(9.43c) as well as the influence of the different physical effects and different parameters, primarily Re, Pr and Ma, on these waves. In all cases the wall is taken to be vertical.

9.6.1 Bifurcation Diagrams

Consider traveling wave solutions propagating at constant speed c and hence stationary in the moving frame of coordinate, $\xi = x - ct$. In this frame, the set of equations (9.43a)–(9.43c) can be written in a dynamical system form as

$$\frac{d\mathbf{U}}{d\xi} = \mathbf{F}(\mathbf{U};\ \delta, \zeta, \eta, B, \mathcal{M}, q_0), \tag{9.52}$$

where $\mathbf{U} = (h, h', h'', \theta, \theta')^t$. The constant q_0 is the mass flux under the wave in the moving frame and is obtained after one integration of the mass conservation equation: $-ch' + q' = 0$ or $q_0 = q - ch$. For solitary pulses, the Nusselt flat film solution, $h = 1$, should be approached far from the solitary humps which gives, $q_0 = 1/3 - c$. Since the speed of the waves is larger than the maximum velocity in the liquid, q_0 is a negative constant.

As in previous chapters, single-hump solitary wave solutions, whose phase-plane analogues are referred to as *principal homoclinic orbits* [103], are computed using the continuation software AUTO-07P with the HOMCONT option for tracing homoclinic orbits [79]. Their numerical construction also involves periodic boundary conditions in an extended finite domain. We focus on positive-hump waves. In fact, much like the isothermal falling film (see Sects. 7.3.1 and 8.3) negative-hump waves are unstable in time-dependent computations.

In Fig. 9.5 we present the maximum amplitude and speed of the single-hump solitary wave family of the regularized model as function of Re for two different values of Prandtl and Marangoni numbers. For comparison we also show in the same figure the wave family corresponding to isothermal flows ($Ma = 0$). In all computations in this section we take the values $\Gamma = 250$ and $Bi = 0.1$ for the Kapitza and Biot numbers, respectively. The Kapitza number is chosen much smaller than its value for common liquids in order to clearly demonstrate the role of the second-order viscous dispersion and inertia effects. Notice that for a given point (Ma, Re) in Fig. 9.5, the reduced parameters $(\delta, \eta, \mathcal{M}, B)$ can be obtained easily with the help of Appendix D.3.

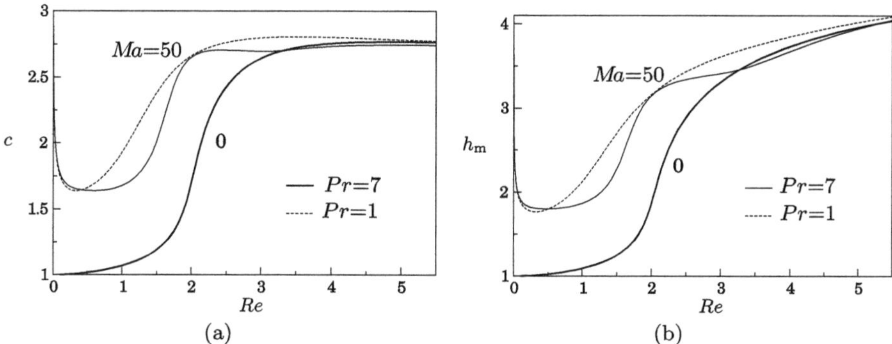

Fig. 9.5 (a) Speed and (b) maximum height of single-hump solitary wave solutions of the regularized model (9.43a)–(9.43c) with $Ct = 0$, $Bi = 0.1$ and $\Gamma = 250$. The transition from the drag–gravity to the drag-inertia regime occurs for $Re \approx 1.5$–2 corresponding to $\delta \approx 1$–1.4, in agreement with Chap. 4

The single-hump solitary wave solution branch obtained from (9.43a)–(9.43c) seems to exist for all Reynolds numbers, i.e., it does not present any turning points with branch multiplicity connected to finite-time blow up behavior as occurs with the BE. Different reduced second-order formulations (Sect. 9.4) were also tested (not shown) and their solitary-wave solution branches do exhibit turning points. Hence, the regularized model in (9.43a)–(9.43c) satisfies also the second of the two criteria listed at end of Sect. 9.4.3, and thus it is a well-behaved low-dimensional model.

Increasing the Marangoni number leads to larger amplitudes and speeds, showing that the S-mode reinforces the H-mode. This effect is more pronounced at low Reynolds numbers ($\mathcal{M}$ being proportional to $Re^{-4/9}$). This is also consistent with the linear stability analysis presented in the previous section which suggests that the Marangoni effect destabilizes the film for all Reynolds numbers, but its influence is enhanced in the region of small Reynolds numbers. On the other hand, in the region of large Re the different curves merge with the isothermal one. In this region the destabilizing interfacial Marangoni forces are weaker compared to the dominant inertia forces.

The effect of the Prandtl number is more subtle. At low Reynolds numbers, $Re \lesssim 0.5$, larger values of Pr seem to slightly favor instability, whereas for any larger Re, we have the opposite effect. To elucidate the influence of the Prandtl number, we shall compute the streamlines and isotherms in the moving frame by calculating the velocity and temperature fields from their respective polynomial expansions and by utilizing the first-order approximation of the corrections s_i and t_i in (9.32a)–(9.32f). The second-order corrections for both fields can also be computed from the residuals associated with the corresponding test functions followed by an inversion of the resulting linear system. Nevertheless, due to the complexity of this procedure, we assume here that the velocity and temperature fields are described sufficiently accurately by their representation at first order, at least for the purposes of a qualitative discussion. In all computations the Marangoni number has been fixed at $Ma = 50$.

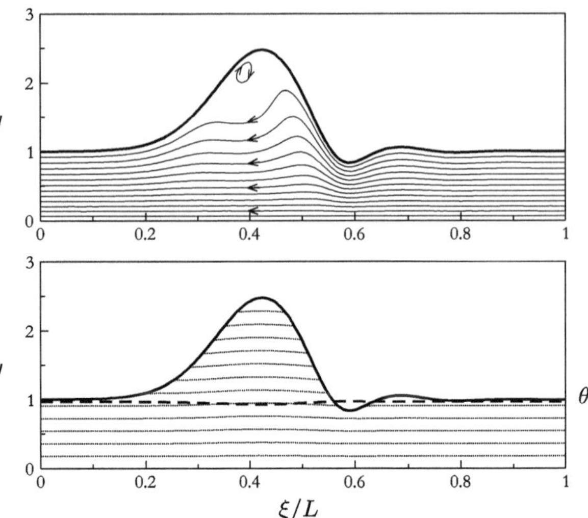

Fig. 9.6 Streamlines (*above*) and isotherms (*below*) of a solitary wave in its moving frame, ξ, normalized with the length of the computational domain L. The wave is computed for the point $(Re, Ma) = (0.01, 50)$ in Fig. 9.5 and for $Pr = 7$. The reduced parameter values are $\delta = 0.0022$, $\eta = 0.0053$, $\mathcal{M} = 37.7$ and $B = 0.031$. θ represents the interfacial temperature (*dashed line*). There are 12 isotherms separating 13 equally spaced intervals, ranging from $T = 1$ on the wall to $T_{\min} = 0.929$ at the crest of the wave

9.6.2 Drag-Gravity Regime

Figure 9.6 shows the streamlines and isotherms for $Re = 0.01$ of a solitary wave solution in the moving coordinate, $\xi = x - ct$. Once again, though solitary waves have, strictly speaking, an infinite wavelength numerically, and as we did before in this monograph, they are calculated in an extended periodic domain of wavelength L. For all solutions given in the rest of this section, $L = 250$. Figure 9.6(b) also depicts the free-surface temperature distribution, which is also a solitary pulse, but of rather small amplitude for the conditions in the figure. Since the parameters $Pr\delta$, η and B are small, the film flow evolution is well approximated by the evolution equation for the free surface (9.51). We also have $\partial_{yy}T \approx 0$ so that the temperature field is nearly linear, $T \approx 1 - By$. Therefore, the isotherms are practically aligned with the wall. Notice also from Fig. 9.6 that the interfacial temperature θ is almost uniform since $B \ll 1$. For the small Reynolds number used, inertia effects play little if any role at all and the large-amplitude solitary hump for the free surface (with a large phase speed $c = 2.35$) is due to the Marangoni effect. This large-amplitude wave then exhibits a recirculation zone (turning clockwise), which in turn transports the "trapped" fluid mass downstream (and in that respect solitary pulses carry mass). This behavior induced by the Marangoni effect is very similar to that triggered by inertia for larger Reynolds numbers as we will see.

The streamlines and isotherms computed for $Re = 1$ are shown in Fig. 9.7. For $Pr = 1$, the isotherms are nearly aligned (with both B and $Pr\delta$ being relatively small). Conversely, at $Pr = 7$, the isotherms are deflected upward by the motion of the fluid in the crest. Therefore, the minimum of temperature (which is achieved at the crest of the solitary wave) is not as low as for $Pr = 1$ and consequently the Marangoni effect is reduced, which in turn reduces the amplitude and the phase speed of the wave. Hence, with this mechanism the transport of heat by convection has a stabilizing effect in the drag-gravity regime.

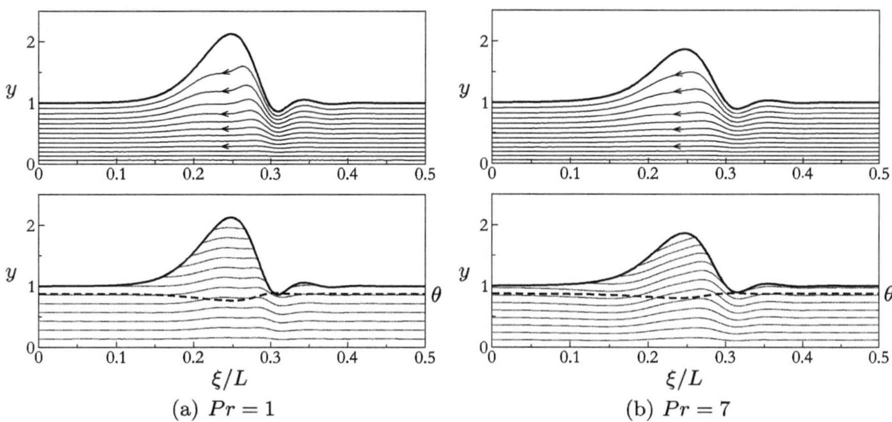

Fig. 9.7 Streamlines (*above*) and isotherms (*below*) for the point $(Re, Ma) = (1, 50)$ in Fig. 9.5 and for two different values of *Pr*. The reduced parameter values are $\delta = 0.61$, $\eta = 0.041$, $\mathcal{M} = 4.87$ and $B = 0.14$. (**a**) $T_{\min} = 0.765$; (**b**) $T_{\min} = 0.8$. In all cases, a total of 12 isotherms separating 13 equally spaced intervals between $T = 1$ and $T = T_{\min}$ are shown

9.6.3 Drag-Inertia Regime

For larger Reynolds numbers, corresponding to the drag-inertia regime, inertia becomes increasingly dominant, and the speed and amplitude of solitary waves increase substantially, as shown in Fig. 9.5. As a consequence, a recirculation zone can now be present inside the main solitary hump, much like in Sect. 9.6.2, but there the large amplitude and speed of the solitary wave and hence recirculation zone in the main hump were due to the action of the Marangoni effect.

Streamlines computed for $Re = 2$ and $Re = 3$ shown in Figs. 9.8 and 9.9, respectively, do exhibit such recirculation zones, turning clockwise. These zones suggest the existence of two stagnation points at the free surface, one at the rear and one at the front of the hump. Again, these recirculation zones transport the "trapped" fluid mass downstream. On the other hand, for $Re = 2$ and $Ma = 0$, there is no recirculation zone, so the recirculation zone for $Ma = 0$ should be born somewhere between $Re = 2$ and $Re = 3$.

As noted earlier, the presence of a recirculation zone seems to be related to the amplitude of the waves. This then would suggest that since we have a recirculation zone for $Re = 3$ and $Ma = 0$ we should have one also for $Re = 2$ and $Ma = 50$, since, as we see from Fig. 9.5 the two cases have roughly the same amplitude and speed. It turns out that is indeed the case.

Hence, in the presence of the Marangoni effect recirculation zones appear for smaller *Re*. This early appearance is also connected with the abrupt increase of amplitude and speed of the solitary waves corresponding to the transition from the drag-gravity to the drag-inertia regimes occurring for smaller values of the Reynolds number if the Marangoni effect is present, as shown in Fig. 9.5.

The role of the Marangoni effect coupled to that of heat transport is pretty intricate. Consider the case $Re = 3$ depicted in Fig. 9.9. A recirculation zone is already

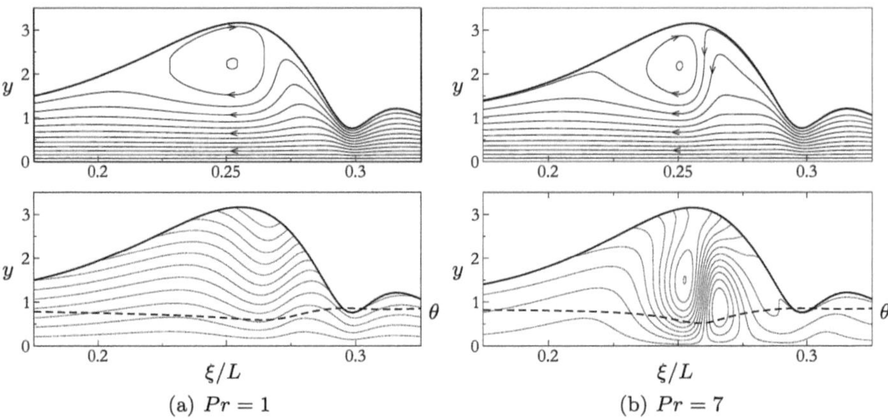

(a) $Pr = 1$ (b) $Pr = 7$

Fig. 9.8 Streamlines (*above*) and isotherms (*below*) for the point $(Re, Ma) = (2, 50)$ in Fig. 9.5. The reduced parameter values are $\delta = 1.42$, $\eta = 0.056$, $\mathcal{M} = 3.58$ and $B = 0.18$. (**a**) $T_{\min} = 0.591$; (**b**) $T_{\min} = 0.429$ and $T_{\max} = 1.26$

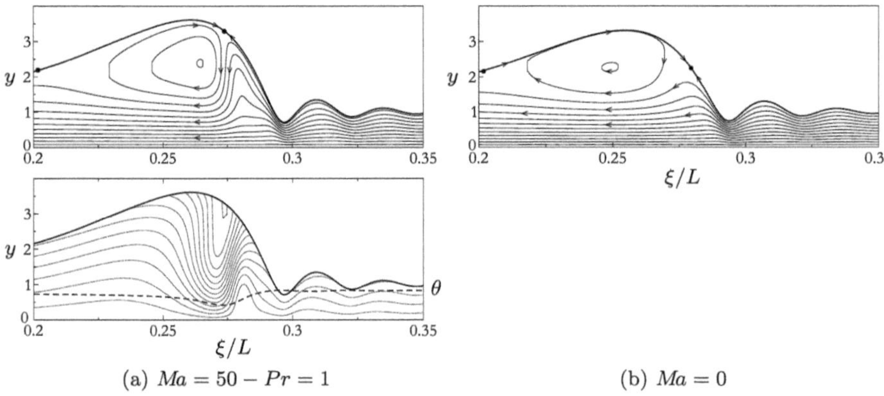

(a) $Ma = 50 - Pr = 1$ (b) $Ma = 0$

Fig. 9.9 (**a**) Streamlines (*above*) and isotherms (*below*) for the point $Re = 3$. The reduced parameter values are $\delta = 2.33$, $\eta = 0.067$, $\mathcal{M} = 2.99$ and $B = 0.21$ with $T_{\min} = 0.414$. *Black dots* represent stagnation points; (**b**) same parameter values for δ, η as in (**a**) but for the isothermal case

present at $Ma = 0$ as shown in Fig. 9.9(b). When the wall is heated, Fig. 9.9(a), heat is transported upward at the front, which then seems to deflect upward the streamlines there; this in turn is accompanied by an upward displacement of the front stagnation point. At the same time, the temperature minimum $T_{\min}$ occurs at the front stagnation point (this, however, is not always the case; see, e.g., Fig. 9.8(a)) and because this point is now higher it enhances the thermocapillary flow, which in turn increases the amplitude and speed of the wave. But as a consequence of the increase of the amplitude of the wave, the recirculation zone is enhanced, which helps even more the transport of heat at the front (with Pr the coupling parameter).

Notice also that the influence of the Marangoni effect from Fig. 9.9(b) to Fig. 9.9(a) is noticeable due to the relatively small value $Re = 3$, which is still at the transition between the drag-gravity and drag-inertia regimes: increasing Re diminishes the influence of the Marangoni effect and eventually the H-mode dominates the S-mode (see Fig. 9.5).

Figure 9.8 indicates that, given a recirculation zone, increasing Pr (for a fixed Re) enhances heat transport due to enhanced mixing of the temperature field, which in turn contributes to homogenizing the temperature field (this point is further discussed in Sect. 9.6.4). This then reduces the temperature gradients at the free surface, and hence the Marangoni effect, which in turn reduces the amplitude of the wave (actually a small effect for the value $Re = 2$ in Fig. 9.8), and much like with the drag-gravity regime (see end of Sect. 9.6.2), the transport of heat by convection has a stabilizing effect in the drag-inertia regime also. This observation is also consistent with the linear stability analysis in Sect. 9.5.3. Hence, unlike the Marangoni number, which increases the temperature gradient on the free surface, the Prandtl number decreases these gradients. But both shift the stagnation point toward the crest—see Figs. 9.8 and 9.9—a process that involves an intricate combination of both Marangoni effect and heat transport, as noted in our discussion above of Fig. 9.9; however, the Marangoni effect is a surface one, unlike the Prandtl number effect, which is a bulk effect.

We also note that comparison of Figs. 9.8(a) and 9.8(b) indicates that increasing the Prandtl number from $Pr = 1$ to $Pr = 7$ at $Re = 2$ enhances the cooling process of the crest and reduces the temperature minimum from $T_{\min} \equiv \theta_{\min} = 0.591$—which appears on the surface and very close to the front stagnation point, again as a result of enhancing the mixing of the temperature field—to 0.429. Similarly, comparing Figs. 9.8(a) and 9.9(a), we notice that $T_{\min} \equiv \theta_{\min}$ drops from to 0.591 to 0.414 when Re increases from 2 to 3 at $Pr = 1$ (in both cases mixing is enhanced due to increasing the Péclet number).

Finally, much as in the isothermal case, where viscous dispersion effects determine the amplitude and frequency of the radiation oscillations in front of the free surface pulses (Sect. 4.3), for the heated film case the second-order viscous and thermal effects are important for the amplitude and frequency of the radiation oscillations in front of the free surface and interfacial temperature pulses.

9.6.4 Limitations Related to the Surface Temperature Equation

For $R = 2$ and $Pr = 7$, the temperature maximum is $T_{\max} = 1.216$ and is no longer located at the wall (not shown). At larger values of the Reynolds number, negative values of the dimensionless temperature appear in the fluid. Turning back to dimensional quantities, this would lead to a temperature in the fluid that can be locally higher than the temperature of the wall or lower than the temperature of the air. This is physically unacceptable, as the temperature everywhere in the fluid should be bounded between the wall and air temperatures.

To understand the appearance of this unphysical behavior when a recirculation zone is present, i.e., for large-amplitude waves, let us consider the influence of the heat transport convective effects in the high-Péclet number limit, $Pe = RePr \gg 1$. More specifically, since the heat transport convective effects are multiplied by $\varepsilon RePr$ (e.g., energy equation (9.6c)) we are interested in the case $\varepsilon RePr$ of $\mathcal{O}(1)$ or larger (recall, however, from Sect. 9.2 that strictly speaking $\varepsilon RePr$ is at most of $\mathcal{O}(1)$). For such values of $\varepsilon RePr$ we start seeing the formation of a thermal boundary layer at the front stagnation point and also part of the interface associated with the recirculation zone [279]. If $\varepsilon RePr$ is small, we do not have to worry about boundary layers, even though we might have a recirculation zone. This is precisely the case in Fig. 9.6. Conversely, if $\varepsilon RePr$ is of $\mathcal{O}(1)$ or larger but we do not have a recirculation zone, then again we have no thermal boundary layers.

Therefore, the presence of a recirculation zone and $\varepsilon RePr$ of $\mathcal{O}(1)$ or larger, is a sufficient and necessary condition for the initiation of the formation of thermal boundary layers at the front stagnation point of a solitary pulse. In Sect. 9.6.3, we observed that the presence of a recirculation zone is connected with the amplitude of the solitary wave: In general, large-amplitude pulses have recirculation zones and small ones do not. Therefore, for a fixed Re increasing Ma increases the amplitude of the pulses (see Figs. 9.5 and 9.9) and promotes the creation of recirculation zones and hence the formation of thermal boundary layers (again provided we have $\varepsilon RePr$ of $\mathcal{O}(1)$ or larger).

The boundary layer becomes fully developed when $\varepsilon RePr \gg 1$. In this case, transport of heat via molecular diffusion can be neglected except in the (fully developed) boundary layer of thickness $(\varepsilon RePr)^{-1/2}$. Hence, cross-stream convection associated with the recirculation zone dominates diffusion; The temperature field in the recirculation zone is simply transported by the flow, and the streamlines are identical to the temperature contours (see, e.g., [252, 278]). This means that the temperature along each streamline is constant due to the strong advection mixing. The temperature field becomes a passive scalar and is simply transported by the flow. Hence, within the recirculation zone the isotherms are closed curves, as shown in Fig. 9.8(b) (the isotherms are not identical to the streamlines in the figure as the Prandtl number is not sufficiently large; besides, for the conditions in the figure our model cannot capture accurately the temperature field). Consequently, the hypothesis $\partial_y T \gg \partial_x T$ necessary for the derivation of the models would be violated in these regions.[4]

At the same time we have neglected in the averaged heat balance (9.43c) the transport of heat due to the Marangoni flow, $\mathcal{K}_\theta^M$ (see (9.33b)). Though these terms are formally of second order, they could be quite significant due to the enhancement of the Marangoni flow by the hydrodynamics. This might also contribute to

---

[4]Notice, however, that the presence of recirculation zones does not invalidate the assumption $u \gg v$ necessary for any boundary layer approach. In fact, the computed streamlines correspond to the envelopes of the velocity field in the moving frame, $(u - c, v)$. Recirculation zones indicate regions where $v \gg U$, e.g., points where the flow returns with $U = 0$ and the stagnation points where $U = v = 0$. However, in the laboratory frame, these regions have $u = c \gg v$ so that the conditions $u \gg v$ in the laboratory frame and $v \gg U$ in the moving frame can both hold at the same time.

the appearance of negative temperatures. There are different possibilities to cure this strong limitation of the three-equations regularized model (9.43a)–(9.43c). One such possibility is to consider more unknowns, such as t_1, for the description of the heat transfer process in the flow. Another possibility is to relax the assumption $\partial_y T \gg \partial_x T$ and use instead the original energy equation without any approximations. In this case the original energy equation should be solved numerically to obtain the temperature distribution within the film. And another possibility explored in [279] consists of the introduction of appropriately modified weight functions for the energy equation prior to averaging.

Finally, notice that since the appearance of unphysical temperatures at large Prandtl numbers is connected with the formation of recirculation zones in solitary waves, the regularized model (9.43a)–(9.43c) should give results in reasonable agreement with experiments for waves of smaller amplitude for which no recirculation zones are present.

9.7 Three-Dimensional Regularized Model

The methodology outlined in Chap. 8 for three-dimensional isothermal films can be readily used to extend the regularized model for ST in (9.43a)–(9.43c) for two-dimensional flows to three-dimensional ones. The outcome is a four-field model for the film thickness h, the streamwise and spanwise flow rates q and p, respectively, and the interfacial temperature θ, which in terms of the Shkadov scaling read

$$\partial_t h = -\partial_x q - \partial_z p, \tag{9.53a}$$

$$\begin{aligned}
\delta \partial_t q = {}& \delta\left(\frac{9}{7}\frac{q^2}{h^2}\partial_x h - \frac{17}{7}\frac{q}{h}\partial_x q\right) + \left[\delta\left(-\frac{8}{7}\frac{q\partial_z p}{h} - \frac{9}{7}\frac{p\partial_z q}{h} + \frac{9}{7}\frac{qp\partial_z h}{h^2}\right) + \frac{5}{6}h\right.\\
& - \frac{5}{2}\frac{q}{h^2} + \eta\left(4\frac{q(\partial_x h)^2}{h^2} - \frac{9}{2}\frac{\partial_x q\partial_x h}{h} - \frac{13}{16}\frac{\partial_z p\partial_x h}{h} - \frac{43}{16}\frac{\partial_x p\partial_z h}{h}\right.\\
& - \frac{73}{16}\frac{p\partial_{xz}h}{h} - 6\frac{q\partial_{xx}h}{h} - \frac{\partial_z q\partial_z h}{h} + \frac{3}{4}\frac{q(\partial_z h)^2}{h^2} + \frac{13}{4}\frac{p\partial_x h\partial_z h}{h^2}\\
& - \frac{23}{16}\frac{q\partial_{zz}h}{h} + \frac{9}{2}\partial_{xx}q + \partial_{zz}q + \frac{7}{2}\partial_{xz}p\Big) - \mathcal{M}\left(\frac{5}{4}\partial_x\theta - \frac{\delta}{224}hq\,\partial_{xx}\theta\right)\\
& \left.+ \frac{5}{6}\zeta h\partial_x h + \frac{5}{6}h(\partial_{xxx}h + \partial_{xzz}h)\right]\left(1 - \frac{\delta}{70}q\partial_x h + \mathcal{M}\frac{5}{56}\frac{\partial_x\theta}{h}\right)^{-1},
\end{aligned}$$

$$\tag{9.53b}$$

$$\delta \partial_t p = \delta\left(\frac{9}{7}\frac{p^2}{h^2}\partial_z h - \frac{17}{7}\frac{p}{h}\partial_z p - \frac{8}{7}\frac{p\partial_x q}{h} - \frac{9}{7}\frac{q\partial_x p}{h} + \frac{9}{7}\frac{qp\partial_x h}{h^2}\right) - \frac{5}{2}\frac{p}{h^2}$$

$$+ \eta \left(4 \frac{p(\partial_z h)^2}{h^2} - \frac{9}{2} \frac{\partial_z p \partial_z h}{h} - \frac{13}{16} \frac{\partial_x q \partial_z h}{h} - \frac{43}{16} \frac{\partial_z q \partial_x h}{h} - \frac{73}{16} \frac{q \partial_{xz} h}{h} \right.$$

$$- 6 \frac{p \partial_{zz} h}{h} - \frac{\partial_x p \partial_x h}{h} + \frac{3}{4} \frac{p(\partial_x h)^2}{h^2} + \frac{13}{4} \frac{q \partial_x h \partial_z h}{h^2} - \frac{23}{16} \frac{p \partial_{xx} h}{h}$$

$$\left. + \frac{9}{2} \partial_{zz} p + \partial_{xx} p + \frac{7}{2} \partial_{xz} q \right) - \frac{5}{4} \mathcal{M} \partial_z \theta + \frac{5}{6} \zeta h \partial_x h$$

$$+ \frac{5}{6} h (\partial_{xxz} h + \partial_{zzz} h), \tag{9.53c}$$

$$Pr \delta \partial_t \theta = 3 \frac{1 - \theta - Bh\theta}{h^2} + Pr\delta \left[\frac{7}{40} (1 - \theta) \frac{\partial_x q + \partial_z p}{h} - \frac{27}{20} \frac{q \partial_x \theta + p \partial_z \theta}{h} \right]$$

$$+ \eta \left[\left(1 - \theta - \frac{3}{2} Bh\theta \right) \left(\frac{(\partial_x h)^2}{h^2} + \frac{(\partial_z h)^2}{h^2} \right) + \frac{\partial_x h \partial_x \theta}{h} + \frac{\partial_z h \partial_z \theta}{h} \right.$$

$$\left. + (1 - \theta) \frac{\partial_{xx} h + \partial_{zz} h}{h} + \partial_{xx} \theta + \partial_{zz} \theta \right]. \tag{9.53d}$$

Equation (9.53a) is the integral version of the continuity equation in three dimensions expressing mass conservation, (9.53b) and (9.53c) are the averaged momentum equations in the x and z directions, respectively, and (9.53d) is the averaged energy equation.

9.7.1 Small-Size Domain

We first investigate the three-dimensional dynamics of a uniformly heated falling film for low Reynolds numbers, i.e., when a one-field model, such as the BE (5.11), can still be valid (recall that the BE is valid up to $\delta \sim 1$). The aim is to scrutinize the formation of rivulet structures due to transverse thermocapillary effects [129, 219]. Results of simulations with the three-dimensional regularized model (9.53a)–(9.53d) are shown in Fig. 9.10 (corresponding to the case of Fig. 9 in the DNS study by Ramaswamy et al. [219]). The computational method is similar to the pseudo-spectral scheme in a periodic domain discussed in Sect. 8.4 for three-dimensional isothermal flows.

This simulation uses as an initial condition a simple harmonic perturbation of the form

$$h(x, z, 0) = 1 + 0.1 \cos(k_x x) + 0.1 \cos(k_z z), \tag{9.54}$$

as shown in Fig. 9.10(a). The wavenumbers $k_x = k_z = 0.335$ correspond precisely to the wavenumbers chosen in Fig. 9 in the DNS study by Ramaswamy et al. [219]. The wavenumbers corresponding to the maximum linear growth rate in each direction are $k_{x_{\max}} = 0.56$ and $k_{z_{\max}} = 0.53$, respectively. The wavenumber k_z is chosen so that

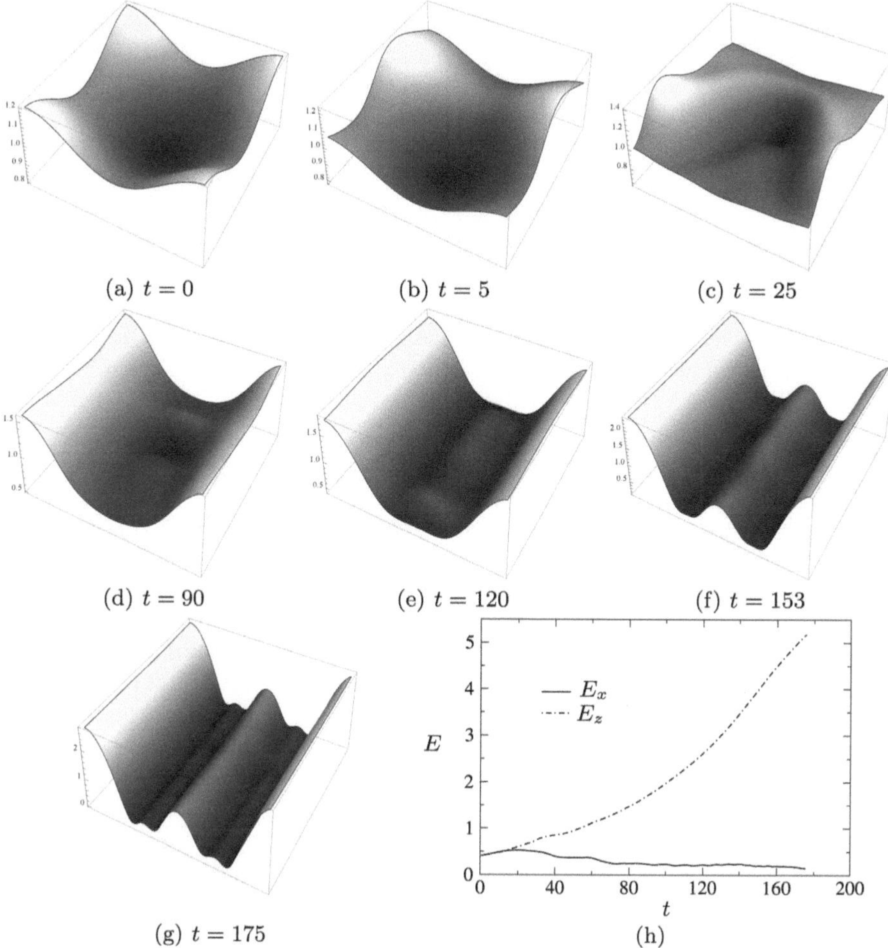

Fig. 9.10 (a)–(g) Inception and development of a rivulet aligned with the flow computed with (9.53a)–(9.53d) for $\delta = 0.15$, $\eta = 0.022$, $\zeta = 0$, $\mathcal{M} = 1.49$, $B = 1$ and $Pr = 7$, corresponding to $Re = 1/3$, $\Gamma = 300$, $Ma = 10$ and $Bi = 1$. The domain size is $2\pi/k_x \times 2\pi/k_z$ where $k_x = k_z = 0.335$. The computational mesh consists of 32×32 points. The flow direction in each box is *from top to bottom* as indicated by the *arrow*; (h) streamwise (E_x) and spanwise (E_z) energies of deformations versus time

it is at most $k_{z_c}/2$ where k_{z_c} is the cut-off wavenumber[5] or $k_z < k_{z_c}/2$, i.e., $2k_z < k_{z_c}$ which then gives at least two unstable modes in the linear regime, thus allowing for

---

[5]In the long-wave expansion of the Orr–Sommerfeld eigenvalue problem in Chap. 3, the cut-off wavenumbers in both directions are

$$k_{x_c} = \left[\frac{2}{5}\delta - \zeta + \frac{3}{2} \frac{\mathcal{M}B}{(1+B)^2} \right]^{1/2}, \qquad k_{z_c} = \left[-\zeta + \frac{3}{2} \frac{\mathcal{M}B}{(1+B)^2} \right]^{1/2}, \qquad (9.55)$$

interesting secondary flow development in the spanwise direction. Ramaswamy et al. chose a square computational box of size $2\pi/k_x \times 2\pi/k_z$, where $k_x = k_z$ which then fixes $k_z = 0.335$. The initial perturbation corresponds to a trough in the center of the domain (a). Then, thermocapillarity sets in, displacing the fluid from this hotter trough toward the surrounding colder crests. However, the growth rate of the hydrodynamic mode is dominant at the beginning and surface waves develop (b). As the local phase speed is proportional to the square of the local film thickness, the crests travel faster than the troughs, leading to steepening of the wave as it grows (c). Due to the absence of mean flow in the spanwise direction, the liquid is more easily displaced laterally due to thermocapillarity. Hence, as time progresses the thinning of the liquid layer persists in the trough and forms a valley surrounded by rivulets aligned with the flow (d) (see also Introduction and Fig. 1.11). This process is similar to the evolution of a heated thin film on a horizontal plate [156, 197]. Likewise, the inclined film exhibits the formation of a secondary rivulet between the main ones (e). As found in [29] for horizontal layers, a "cascade of structures" takes place in thinner zones (g), prior to film rupture.

It should be emphasized here that the computation with the BE (5.11) compares well with the DNS study of the full Navier–Stokes and energy equations as shown by Ramaswamy et al. [219] up to $t = 120$ and then diverges at $t = 146$ when rupture occurs. Beyond this time, the BE fails to properly account for the dynamics of the flow. On the contrary, the last stage before rupture obtained by Ramaswamy et al. with DNS at $t = 153$ is in excellent agreement with Fig. 9.10f. Most impressively, our computation with the regularized model (9.53a)–(9.53d) can continue beyond this time (up to $t = 175$) and reveals finer structures in the thin film region and just prior to rupture[6] (see Fig. 9.10g). It is possible that the DNS performed in [219] was not capable of resolving the evolution of the film past $t = 153$ due to the choice of the number of mesh points in the direction normal to the wall. Quite likely, with a re-fined grid resolution, the authors would have been able to compute the evolution for larger times. However, this would have been at the expense of computational time, which demonstrates the significant advantage—especially for moderate Reynolds numbers and large system sizes as considered in the next section—of working with a model of reduced dimensionality and in terms of interfacial and averaged variables (for instance, the time necessary for computing the case of Fig. 9.10, with an accuracy of 10^{-4} for each variable, is about one hour on a standard desktop computer).

(see (3.21a) and (3.35)) which fully agree with those obtained from the regularized model. The wavenumber then corresponding to the maximum linear growth rate is

$$k_{i_{\max}} = \frac{k_{i_c}}{\sqrt{2}} \quad \text{with } i = x, z.$$

[6]The computations are terminated when the film thickness reaches a "minimal thickness" of about $h \sim 10^{-3}$ for which long-range van der Waals intermolecular forces cannot be neglected (~ 100 nm). Consequently, as far as the computations with the regularized model are concerned, we describe *rupture* to be when the film reaches this minimal thickness. The dynamics subsequent to rupture can only be resolved by including a disjoining pressure term in (9.53b), (9.53c) as is typically done in thin film studies (see, e.g., [24, 237]).

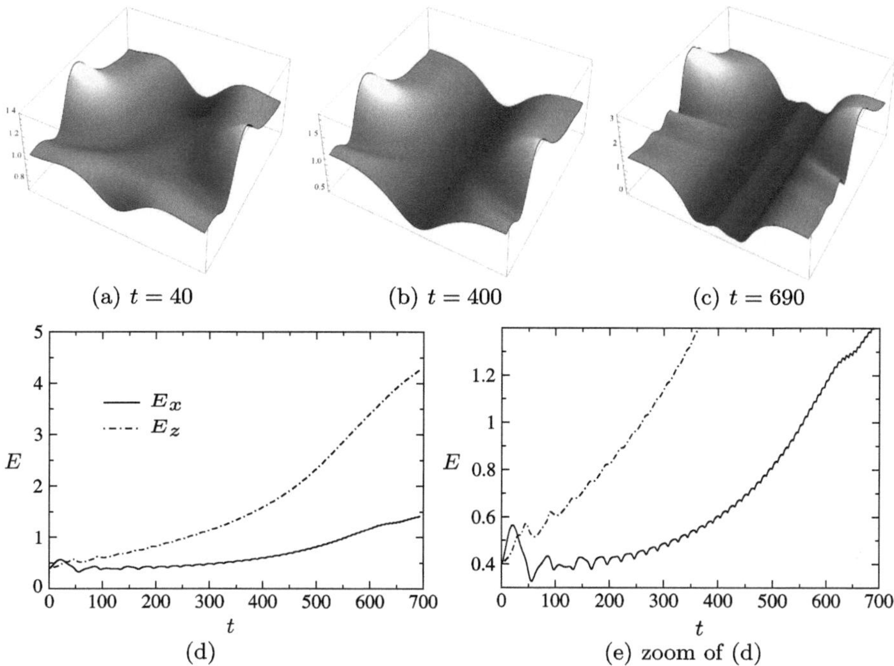

(a) $t = 40$ (b) $t = 400$ (c) $t = 690$

(d) (e) zoom of (d)

Fig. 9.11   Same as for Fig. 9.10 but for $Re = 2$

Figure 9.10h depicts the energy of streamwise and spanwise deformations, E_x and E_z, given by (8.9). While E_z increases until the film ruptures, E_x increases first and then decreases continuously, showing that the presence of rivulets damps the evolution of hydrodynamic waves as time progresses and can eventually suppress them altogether: for small Reynolds numbers the system is thus dominated by the thermocapillary Marangoni effect.

Let us now increase the Reynolds number to $Re = 2$ and keep the other physical parameter values, Γ, Ma, Bi and Pr, the same. The reduced Reynolds number becomes $\delta = 1.33$, which now lies outside the range of validity of the BE in (5.11), restricted up to $\delta \simeq 1$. The streamwise and spanwise wavenumbers are chosen as before, i.e., $k_x = k_z = 0.335$, so that the computational domain is the same with that in Fig. 9.10. But now since Re is different to that in Fig. 9.10, the maximum growing wavenumbers have changed to $k_{x\,\mathrm{max}} \approx 0.62$ and $k_{z\,\mathrm{max}} \approx 0.34$, so that $k_x = k_z = 0.335 \approx k_{x\,\mathrm{max}}/2 \approx k_{z\,\mathrm{max}}$. Hence, the wavenumber k_x in the streamwise direction is sufficiently far from the cut-off one k_{x_c} allowing for a sufficiently long domain, which is necessary for the development of a solitary wave.

Figure 9.11 shows that the hydrodynamic H-mode quickly generates a large-amplitude deformation (a, b) leading to a solitary-like wave with preceding capillary ripples (c). However, the thermocapillary S-mode causes film rupture before this wave fully develops. This is depicted in Fig. 9.11(d), where both components of the energy of the streamwise and spanwise deformations, E_x and E_z, increase continuously in time but $E_x < E_z$ and the system is still dominated by the Marangoni effect. A remarkable interaction between the two instability modes is observed: As the thermocapillary flow feeds the core of the rivulet, the rivulet grows and thus the mean film thickness at the crest increases and so does the local flow rate. Hence, the wave profile at the crest of the rivulet does not saturate but rather follows the change of the local Reynolds number by increasing its amplitude and its phase speed. This process terminates at $t > 620$ when the film is sufficiently close to the wall for the viscous stresses to slow down the lateral thermocapillary flow. The hydrodynamic wave and the transverse rivulet are found to coexist over a long time before the film ruptures.

9.7.2 Large-Size Domain

We now present large-size computations for a water film at 20°C ($\Gamma = 3375$, $Pr = 7$) with a temperature difference between the vertical wall and the ambient air of 5°C ($Ma = 50$) and a high transfer coefficient of $1000\,\mathrm{W\,m^{-2}\,K^{-1}}$ ($Bi = 0.1$). The time-dependent simulations are started with white noise of maximum amplitude $1/1000$ relative to the flat film thickness $\bar{h}_\mathrm{N}$, which is itself varied between 85 and 146 μm (i.e., for $2 < Re < 10$). Aliasing is treated by applying a low-pass filter whose optimum cut-off frequency has been found from a trial–error process which ensures the convergence of the numerical solution. Effectively, we keep only the first 2/3 of the Fourier modes in each direction (i.e., the 42 first modes) before each iteration (see Appendix F.4 for details).

Figure 9.12 shows as before the formation of rivulets due to the Marangoni effect: After the development of a parallel wave train (a), drop-like accumulation breaks the two-dimensional wave structure into a fully developed three-dimensional pattern (b, c), prior to rivulet-like patterns aligned with the flow (d, e). As shown before in Fig. 9.11, the liquid then accumulates into rivulets, which increases the local Reynolds number and fosters two-dimensional solitary-like waves of larger amplitude and phase speed than in isothermal conditions (f).

The rivulet formation shown in Fig. 9.12 continues until rupture of the film, whose snapshot is shown in Fig. 9.13(a). Similar rivulet patterns, but with larger wavelength, are also found for $Re = 5$ (b). If the Reynolds number is further increased to $Re = 10$, the flow rate is sufficiently large to prevent the formation of rivulets (c) and the film behaves like in isothermal conditions (see Sect. 8.4.3).

We can draw here a correlation between the different wave patterns shown in Fig. 9.13 and the branches of homoclinic solutions that are plotted in Fig. 9.14 for $Ma = 50$ as well as for isothermal conditions, $Ma = 0$: For low Reynolds numbers and small phase speeds, i.e., in the drag-gravity regime, inertia effects are small relative to thermocapillary effects, while the opposite is true for large Reynolds numbers and large phase speeds, i.e., in the drag-inertia regime, where inertia is dominant. The resulting patterns plotted in Fig. 9.13 are consequently radically different. In the drag-gravity regime, quasi-regularly spaced rivulets arise and grow up until rupture (a). Meanwhile, the rivulets confine the flow in such a way that waves riding

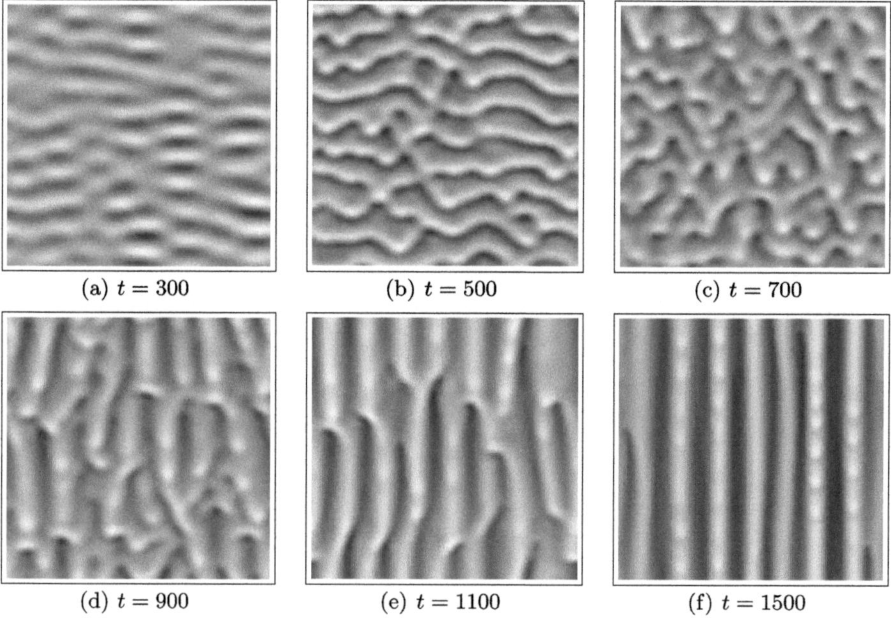

Fig. 9.12 Water film free surface at different times computed with the three-dimensional regularized model (9.53a)–(9.53d). Parameter values are $Re = 2$ for $Ma = 50$, $Bi = 0.1$, $Pr = 7$, $Ct = 0$ and $\Gamma = 3375$. The domain size is $2\pi/k_x \times 2\pi/k_z$ where $k_x = k_z = 0.05$. The spatial mesh is fixed to 128×128 points. *Bright/dark zones* correspond to elevations/depressions, respectively

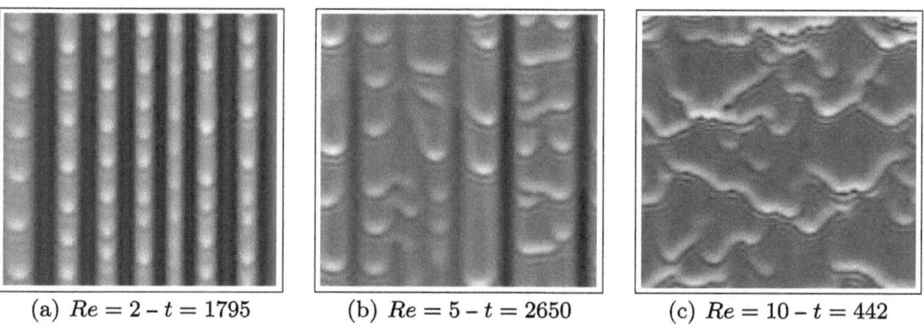

Fig. 9.13 Wave patterns for $Ma = 50$, $Bi = 0.1$, $Pr = 7$ and $\Gamma = 3375$ for different Re. Times given for (**a**) and (**b**) are close to rupture

them behave like two-dimensional solitary waves, but of higher flow rate because of the local increase of the Reynolds number. On the contrary, no qualitative influence of the Marangoni effect has been observed in the drag-inertia regime, at least during the time of the computer simulations, showing that inertia fully dominates the dynamics of the film (c). The transition between these two regimes for $4 < Re < 6$ shows a complex cooperative behavior between both hydrodynamic H- and thermo-

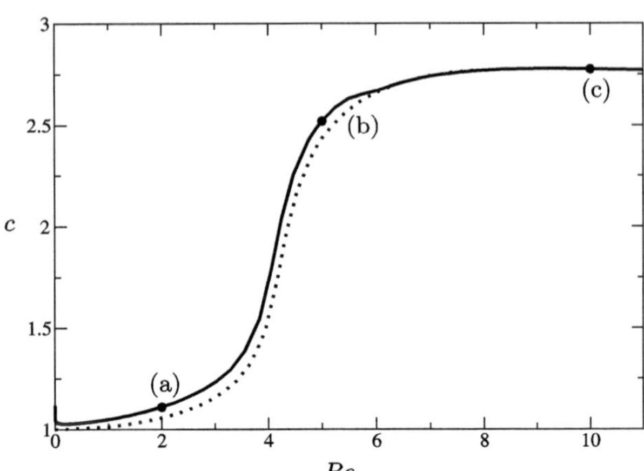

Fig. 9.14 Phase speed c of homoclinic solutions (i.e., solitary waves) versus Re for $Ma = 50$ (*solid line*) and $Ma = 0$ (*dotted line*). The letters refer to Fig. 9.13

capillary S-modes, as illustrated for $Re = 5$ in Fig. 9.13(b) (additional details can be found in the study by Scheid et al. [242]).

Chapter 10
Open Questions and Suggestions
for Further Research

The fascinating dynamics of a water film flowing down a vertical plane, certainly the most common example of falling films, easily observed on windows or in streets during rainfalls, is not yet fully understood, despite the considerable attention and intense scrutiny it has received for several decades now since Kapitza's pioneering work more than 60 years ago. The long-time evolution is dominated by a three-dimensional wavy process, a weakly turbulent flow, characterized by the presence of three-dimensional coherent structures which interact continuously with each other as quasi-particles. The precise details of this regime still remain elusive. Open questions include the number per area, or "density," of the horseshoe-like solitary waves that structure and organize the flow, or the role of viscous second-order effects on solitary pulse interaction, in particular coalescence or repulsion processes, especially as most previous studies have ignored viscous effects [44]. A noted exception is the recent effort in [212], which developed a coherent structure theory for the simplified second-order model and scrutinized the effect of viscous dispersion on solitary pulse interaction. But this study was restricted to two dimensions. The models developed here, can reproduce not only qualitatively but also quantitatively both the two- and three-dimensional wave dynamics. One may then expect some decisive achievements on the understanding of the three-dimensional turbulent-like regime in the years to come by using these models.

As far as the influence of additional effects and complexities are concerned, for pedagogical reasons we have limited ourselves to the consideration of heated falling films on planar substrates, which more or less corresponds to our own contributions to the field. We have endeavored to treat and to critically revise and reexamine the Marangoni effect as comprehensively as possible. But clearly, several problems related to heat transport and encountered in most chemical engineering processes have not been addressed. For instance, the consequences of the coupling between the hydrodynamic H-mode and the thermocapillary S-mode are far from being fully understood, in particular when three-dimensional flows are considered. Indeed, Chap. 9 reports a preliminary study, the "tip of the iceberg" that underlines the richness of the problem. In fact, in most problems involving heat transfer, phase changes through either evaporation or condensation or both, also have to be taken into account [260].

S. Kalliadasis et al., *Falling Liquid Films*, Applied Mathematical Sciences 176,
DOI 10.1007/978-1-84882-367-9_10, © Springer-Verlag London Limited 2012

As far as mass transport is concerned there is a large class of problems that involve transport of a species from a gas to a falling film. An important question in these problems is the development of systematic ways for obtaining effective heat/mass transport coefficients.

Another important class of falling film flows is that of reactive falling films. Long wave theories for the problem of a falling film in the presence of an exothermic chemical reaction were developed in [274, 277], while the weighted residuals methodology of Chap. 6 has been extended to the same problem in [275]. Chemical reactions, exo- or endothermic may exhibit intricate dynamics that could further complicate the already quite complex flow features due to the underlying hydrodynamics. As a result the studies in [274, 275, 277] were limited to a paradigmatic model problem with first-order kinetics, with the coupling of heat to flow through the thermocapillary Marangoni effect. Yet, another case of interest is the coupling of the free surface deformation to chemical reactions by the solutocapillary Marangoni effect. A step in this direction is the recent studies in [207, 251] on a nonreactive falling film with surfactants. Of particular interest would also be the evolution of a falling film in the presence of oscillatory or even chaotic chemical kinetics such as the Belousov–Zhabotinskii [245] and the CDIMA reactions [16] or excitable media [181]. In fact, the recent study in [208, 209] on a horizontal reactive film with insoluble reactive surfactants participating in a reaction described by the FitzHugh–Nagumo prototype reveals a quite intricate evolution.

The generation of Marangoni stresses through gradients of concentrations can also be used to increase the spreading rates of droplets deposited on hydrophobic substrates, which is desirable in the manufacturing of chemical sprays or in microprinting/patterning (e.g., pattern replication through deposition of liquids in a regular manner determined by the template). The resulting fingering instability at the contact line of spreading droplets is an example of a solutocapillary instability of thin film flows [36]. This system belongs to a wide variety of problems related to both solutocapillary and thermocapillary Marangoni effects and where inertia is not important. Such problems have been treated before within the framework of the lubrication approximation. But the methodologies outlined in this monograph should not only be applicable when inertia is important, but also when inertia is small but the convective terms of the concentration or energy equation become increasingly important, i.e., when the corresponding mass or heat Péclet numbers are no longer small (e.g., due to slow diffusion of a solute). On the other hand, the methodologies developed here should be applicable in the treatment of contact line problems without any soluto/thermocapillary Marangoni effects but in the presence of inertia (previous studies utilized the BE long wave expansion, e.g., [174]). We note here that the incompatibility between a moving contact line and the no-slip boundary condition at the wall, known as the "contact-line paradox," remains as of yet an open question [68, 78, 88, 89, 122, 183]. In practice, a popular but ad hoc approach is to replace the no-slip boundary condition with a Navier slip model. As far as the control of the fingering instability of the contact line is concerned, one approach is to impose a surface anisotropy [143, 152].

Actually, a great variety of systems involve thin film flows in one fashion or another, several of which are detailed in [61]. Some of them are related to the substrate

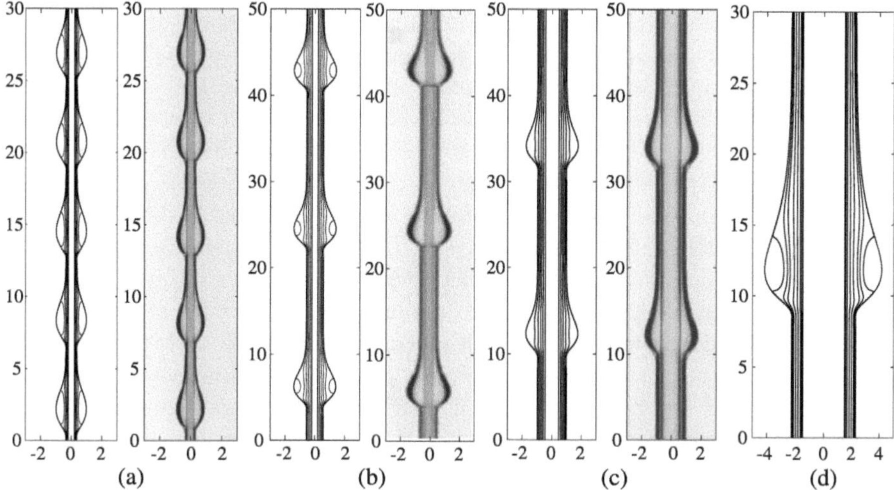

Fig. 10.1 Viscous beads on the surface of a film flowing down a vertical fiber of radius R: streamlines in the moving frame obtained from an average model indicating the appearance of recirculation zones inside the beads and experimental snapshots. The flow is excited at the inlet with a forcing frequency f; (**a**) silicon oil v100, $R = 0.2$ mm, $\bar{h}_N = 0.52$ mm and $f = 4$ Hz; (**b**) silicon oil v50, $R = 0.2$ mm, $\bar{h}_N = 0.64$ mm and $f = 8$ Hz; (**c**) v50, $R = 0.475$ mm, $\bar{h}_N = 0.76$ mm, and $f = 7$ Hz; (**d**) solitary pulse with $h_{max} = 2.66$ mm for v50 and $R = 1.5$ mm. Reprinted with permission from Duprat, Ruyer-Quil and Giorgiutti-Dauphine, *Phys. Fluids* 21(4):042109, Copyright 2009, American Institute of Physics

geometry, for instance when the wall is corrugated, a configuration that is commonly used to enhance heat or mass transport in the design of two-phase heat exchangers, absorption columns using flooding and distillation trays and other chemical engineering processes and devices. The coupling between the instability mechanism and the wall corrugation is presently the subject of intense research (see, e.g., [11] and [63] and references therein). A first shot to the problem of flow over a corrugated substrate using the Kapitza–Shkadov approach was done in [236], while a successful application of the weighted residuals method including the development of a regularized model for this problem was done in [112].

Another geometry of interest is the flow of a film outside a cylinder [60, 136, 149, 217, 280]. When the radius of the cylinder R is comparable to the film thickness $\bar{h}_N$, the hydrodynamic instability mode is coupled to a capillary Rayleigh–Plateau instability induced by the curvature of the cylinder; see Fig. 10.1. The weighted residual methodology presented here has been successfully extended to this problem [231]. The corresponding average model for this flow has been validated experimentally in [85]. A film coating a vertical fiber is an excellent prototype for the study of absolute and convective instabilities [86] and formation of bound states [284, 286]. The latter studies demonstrated that a coherent structures theory of solitary pulse solutions of the Kawahara equation is in qualitative agreement with experiments. An interesting future research direction would be the extension of the theory to the weighted residuals model developed in [231].

In some applications, adsorption of the fluid to the substrate is desirable [204, 205]. Indeed, for ink jet printing and in paper coating processes, the spreading of droplets and thin films on unsaturated porous substrates are of significance (e.g., [267]). In civil engineering, certain coating is sometimes used to treat porous concrete surfaces [7]. Let us also note the cases of moving solid walls, either vibrating [165, 166, 184] or rotating [64–66].

A related topic is the problem of thin films on soft deformable substrates that is encountered, for example, in lung airways [110]. It should be emphasized that most biological fluids present a non-Newtonian rheology and mucus flows in bronchi do not constitute exception. An amusing example of non-Newtonian biological films is the adhesive locomotion of snails on mucus [40]. But, the biological thin film problem that has attracted the most interest is probably tear film dynamics because of its importance for the contact lens industry and ophthalmology (the "dry-eye syndrome"). Experiments have confirmed that in that case too, tear films are made of a non-Newtonian shear-thinning fluid [272].

Non-Newtonian fluids, generally shear-thinning also, are also frequently encountered in coating processes [6, 148, 297]. The difficulties for the practitioners of the "art of coating" include the control of the dynamic wetting line of liquid curtains on the web, with possible air entrainment and development of laminar boundary layers, instabilities of multilayered film flows as well as chemically induced surface activity, contamination of the free surface and temperature variations due to rapid evaporation, all of which could generate surface tension-driven flows. Dewetting and fingering instabilities associated with a contact line are also frequently encountered phenomena in coating processes. A graduate-level introduction to the problem of the fingering instability of a thin film flowing on an incline can be found in [151] (see also [201]).

Surface patterning is a rapidly developing area of thin film applications, e.g., in the design of microstructures at microscales for ultra miniaturized optoelectronic devices. The basic principle is to generate self-organized patterns by applying a spatial varying force field, either by using electric field across a dielectric thin film [247, 285, 293] or by employing chemically patterned substrates. In the latter case, the idea is to replicate the substrate pattern on a soft material obtained after spinodal dewetting of a polymer solution (see, e.g., [142]).

The strategy followed in the case of isothermal and heated films here should be applicable to the flows of non-Newtonian fluids, dewetting and fingering phenomena as well as surface patterning phenomena listed above if inertia is significant and/or heat/mass transport convective effects are important. And more general, it should be applicable to other configurations or geometries, which can be connected to the "thin film" case.

One example that falls in this category is flows of liquids in a Hele–Shaw cell, made of two rigid walls separated by a small gap. Hele–Shaw cells constitute the simplest geometry of porous media. They are, therefore, model systems for the study of phenomena occurring at the scale of the pores, e.g., diffusion of pollutants and chemical reactions [77, 302]. Another reason for the enduring interest for the Hele-Shaw cells is the existing analogy with quasi-static solidification. The "Saffman–

Taylor instability" of a driven fluid–fluid interface is an analogue to the "Mullins–Sekerka instability" of a solidification front [153, 157]. An example of an inertia-driven instability in a Hele–Shaw cell, to which the methodologies detailed here can be applied, is the Kelvin–Helmholtz instability of a gas–liquid interface [105, 225]. An interesting extension would be the inclusion of chemical reactions which affect the fluid flow through solutal and/or thermal contributions to the density.

Another large class of problems is the long wave instability of two-phase flows, and more general multiphase flows, of particular interest in industrial applications. A classical example is a vapor-generating channel in which a vapor and a liquid flow simultaneously. The relative motion of the liquid and vapor result in an additional shear that affects the dynamics of the interface. Another example is the core-annular flow of two liquids inside a tube, a problem that is typically found in oil extraction. The so-called "bamboo waves" are typically observed when the viscosity of the outer liquid is smaller than the viscosity of the inner liquid. In horizontal pipes the generation of waves at the interface separating core and outer flows is a necessary condition for the levitation from the wall of the core fluid, either lighter of heavier than the lubricating outer one. Experiments suggest that inertia plays a key role in the levitation process [130]. We believe that the strategy outlined here can be fruitfully transposed to the core-annular flow problem.

Geophysics also offers several problems to which the methodologies developed here can be straightforwardly applied. Amongst these we have already mentioned torrential flows in rivers and the dynamics of roll waves frequently observed in dam spillways (see also [43]). Roll waves can also be observed in overland flows, especially in rill flows, with a potential increase of the soil erosion [172], and in mud flows [14]. Notice that rill bottoms may have a typical staircase geometry—as observed in the Loess Plateau area of China—with a possible effect on the instability. Rivers may carry large amounts of clay, in particular in estuaries. The resulting mud flows are frequently pulsating and resemble roll waves in turbulent flows of clear water [171]. Roll waves in mud flows can also be observed soon after torrential rains in mountain streams. Induced debris flows can move stones, boulders and even trees, with destructive power on their path. A related geophysical flow, one which can be modeled as non-Newtonian gravitational flow, is lava flow from a volcanic eruption. In this case heat transfer and the dependence of the rheological properties with temperature must be accounted for. Research on these problems should be rather exciting.

Appendix A
Historical Notes

A.1 Piotr Leonidovitch Kapitza (1894–1984)

Kapitza together with his son Sergey Petrovitch Kapitza (born in 1928) conducted the first well-controlled experiment on a falling film [141]. Kapitza's theoretical contribution [140] to the subject contains some innovative and pioneering ideas, such as the averaging of the governing equations across the film thickness. The resulting averaged equations were later on developed with great success.

However, Kapitza is better known for his experiments on intense magnetic fields, helium liquefaction and the discovery of helium superfluidity (see Fig. A.1), for which he earned the Nobel Prize for Physics in 1978. His pioneering work on falling liquid films is actually minor in comparison to his contributions to low temperature physics, high-power electronics and plasma physics, not to mention his contributions to air liquefaction and fractionation technologies (his design of a high efficiency compressed gas turbo engine is still used in the large scale production of oxygen). Yet, how Kapitza got interested in the problem of falling films is an interesting story by itself [28].

Kapitza was born in 1894 at Kronstad, an island fortress near St. Petersburg. He graduated from Petrograd/St. Petersburg Polytechnic Institute in 1919. In 1916

Fig. A.1 P.L. Kapitza (*right*) and his assistant S.I. Filimonov conducting an experiment on helium superfluidity at the Institute for Physical Problems, Moscow, 1940. Reprinted with permission from Kapitza Memorial Museum, Kapitza Institute for Physical Problems, Moscow

S. Kalliadasis et al., *Falling Liquid Films*, Applied Mathematical Sciences 176, DOI 10.1007/978-1-84882-367-9, © Springer-Verlag London Limited 2012

he married Nadezhda Kirillovna Chernosvitova. In the years 1919–1920 he lost his wife and two children. Their death was due to the terrible conditions following the revolution and the civil war, augmented by the so-called Spanish influenza. This dramatic period was overcome thanks to the encouragement of eminent colleagues and friends like A.F. Ioffe, A.N. Krylov and N.N. Semenov, who firmly believed in Kapitza's brilliance for science. As member of a scientific committee, Kapitza embarked on a trip to England in 1921, eventually ending up working on his PhD with the eminent E. Rutherford at the Cavendish Laboratory of Cambridge University. He completed his doctorate in 1923 and continued working with Rutherford for several years after that, and although the original plan was for him to stay only over the winter of 1921, he remained in Cambridge for 13 years. In 1927 he married Anna Alekseyevna Krylova, daughter of the earlier mentioned Krylov. The couple had two sons, Sergey (born in 1928) who became a physicist, and Andrei (born in 1931), who became a geoscientist. In 1929 Kapitza was elected to the Fellowship of the Royal Society.

During his time in England, Kapitza frequently returned to Russia to give seminars. From 1926 to 1934 he visited Russia nearly every summer and was always granted a return visa to come back to England, a very unusual practice at that time. In autumn 1934, on one of his trips back to Russia, his unusual status came to an end and his passport was seized at Stalin's order. The reasons of Kapitza's retention in Russia are unclear. According to Rutherford, "Kapitza in one of his expansive moods in Russia told the Soviet engineers that he himself would be able to alter the whole face of electrical engineering in his lifetime." The need of talented researchers in physical sciences to support the Soviet economy at that time supports Rutherford's testimony [28, p. 46]. Kapitza had to wait 32 years before permission was granted him to visit England again. In Moscow he was ordered to open a new laboratory and to found the Institute for Physical Problems, of which he was the first director and which since then has borne his name. Kapitza's equipment at the Royal Society Mond laboratory was purchased by the Soviet government with the help of Rutherford and then shipped to Russia. It is at that time that Kapitza conducted his work on low temperature physics and discovered the superfluidity of helium II (1937). In 1978 he was awarded the Nobel Prize in Physics "for his basic inventions and discoveries in the area of low-temperature physics" (unusual indeed, as these results had been obtained four decades earlier).

After the war, Kapitza refused to work under Beria (also head of the state police, the N.K.V.D., soon renamed M.V.D., after the war) on the Russian nuclear weapon project. Due to Beria's hostility, he was fired from the post of director at his Institute and he had no other choice but to retire to his dacha in Nikolina Gora (Nicholas Hill) near Moscow to the end of the Stalin era (1946–1953). It was not the first time that Kapitza had to face the all powerful N.K.V.D. Already in 1938, in the midst of one of the worst purges preceding Word War II, he wrote directly to Stalin and to Beria and obtained the release of his friend Landau, who was then accused of spying on behalf of Germany.

According to Landau's own testimony:

"Kapitza went to the Kremlin and announced that he would have to leave his Institute if I wasn't released. It is hardly necessary to say that such an action in those years required no little courage, great humanity and crystal-clear honesty" [28, p. 67].[1]

Not without a good sense of humor, Kapitza rebaptized his small dacha "Izba for Physical Problems." It was there that Kapitza conducted amongst other pieces of work his pioneering experiments on falling film instabilities with the help of his son Sergey and, at times, of his other son, Andrei.

Here is what the father Kapitza wrote about their falling film house experiments in one of his letters dated 2 December 1949 and addressed to one of his friends and colleagues, Vladimir Engelhardt, molecular biologist and member of the Russian Academy of Sciences:

"I then thought that in the position I found myself, the only possibility of continuing scientific research work was to take up biology. I thought there was no branch of physics where it would be possible to look for really new and significant phenomena. But I was wrong. Almost immediately after I was deprived of my Institute and its facilities for low temperatures and high magnetic fields, I came across an interesting question in hydrodynamics—the flow of a thin layer of a viscous liquid. Ever since the time of Poiseuille this was considered as a classical case of laminar flow, but I realized that there are a number of indications that this is not so. [...] It is rare to discover a new form of wave motion and I decided to look into it. With the modest means at my disposal in my dacha, and of course with the help of my son [Sergey], I succeeded in observing and studying this type of wave flow and in confirming my theory" [28, p. 388].

Would Kapitza have ever considered the problem of falling liquid films if he had not been compelled to stop working on low temperature physics? Probably not. In another letter addressed directly to Stalin, Kapitza complained about his work conditions at Nikolina Gora. Referring to his experiment on falling films, he wrote:

"But the work goes slowly since I have to do everything myself, even making the necessary apparatus with my own hands, helped only by my family" [28, p. 386].

It is not an overstatement to say that Kapitza was an outstanding experimentalist. A favorite hobby was to dismantle watches and to repair them, manufacturing himself the spare parts.[2] Indeed, reading Kapitza's own description of his falling film experimental set-up, one wonders how he achieved such accuracy while assigned to residence in his country home. The fluid, water or alcohol, was injected at the top of a vertical glass tube of 2.5 cm in diameter and 20 to 25 cm in length. To ensure the axisymmetry of the observed waves, a great accuracy was necessary in manufacturing the surface of the tube and of the supply unit. In fact, an accuracy to within one micron was necessary in the design of the conical shape mandrel placed

---

[1] All quotations in Appendix A.1, are reprinted from "Kapitza in Cambridge and Moscow: life and letters of a Russian scientist," J.W. Boag, P.E. Rubinin and D. Shoenberg, North-Holland, pp. 67, 386 and 388, Copyright (1990).

[2] An anecdote underlines his interest on watchmaking: On a trip to Strasbourg in 1926, Kapitza asked the clockmaker in charge of the maintenance of the famous cathedral clock to show him the details of the clock mechanism.

at the top of the tube to maintain a regular distribution of the fluid. Kapitza designed a clever stroboscopic device to illuminate the free surface of the film and to detect the waves. A task that was not easy considering that the film thickness was no more than a few tenths of a millimeter. The vibrations of the motor rotating the stroboscope were transferred to the tube at a small but sufficient level to synchronize the wave dynamics of the film and the stroboscope so that standing permanent images of the traveling waves on the film surface were produced. Up to now, the quality of the experimental results obtained by Kapitza and his son has been rarely reached in other experiments and their data are still used as benchmarks for the numerical studies devoted to falling film flows.

A.2 Carlo Giuseppe Matteo Marangoni (1840–1925)

Marangoni's influence on interfacial phenomena in liquids through two seminal contributions [179, 180] is so great that his name is nearly always associated with this area of research. He was born in Pavia, Italy, and graduated from the University of Pavia ("Laurea in Mathematiche pure e applicate"—somewhat equivalent to a Masters thesis) in 1865, under the supervision of Professor Giovanni Cantoni, with a dissertation entitled "Sull' espansione delle goccie liquide" (On the spreading of liquid drops) [178]. He then moved to Florence where he eventually became a high school physics teacher at the high school "Liceo Classico Dante," where he taught for four decades (1869–1916) (see Fig. A.2). He died in Florence in 1925.

He dedicated much of his time to the development of the Department of Physics, especially the Laboratory of Physics. He was recognized by his colleagues and students as a consummate teacher but also as a skilled investigator and scientist,[3] patient and ingenious, always stimulating interest and curiosity.

Marangoni left many writings attesting to the high scientific quality of his works. The most significant ones are those on capillarity, drops, on certain optical illusions and various educational experiments. Though James Thompson [271] initially understood that the formation of "tears of wine" at the wall of a glass was promoted by the difference of surface tensions between water and alcohol, it was Marangoni who gave the first rigorous explanation of this phenomenon. He formulated a rather complete theory for flows driven by surface tension gradients due to variations in temperature or composition, an effect that now bears his name. Despite the explanation of the "tears of wine" he was a fervent teetotaler.

Marangoni also contributed to meteorology and invented an apparatus to observe clouds. Anecdotally, the formation of hail was for him an assiduous topic, for which he tried to give an explanation for many years. Though he did not manage to give a satisfactory theory, his numerous observations and ideas were shown to be useful for subsequent generations.

---

[3]During the time, a great part of scientific research was actually conducted in high school laboratories, which were often equipped better than university ones.

Fig. A.2 Carlo Marangoni at the Liceo Classico Dante in Florence, Italy [173]. This picture comes from a larger one showing a group of high school teachers and students celebrating final year high school graduation in 1909. The original photograph belongs to a private collection (Dr. Valleri). A photographic reproduction of the original photograph is deposited in the archive of Liceo Classico Dante. Photo courtesy of Prof. G. Loglio

Noteworthy is that he had the singular merit of being one of the very first (since 1882) to claim that the "future and the economic wealth of Italy was in its mountain water, and forests that protect water and fuel." Being an apostle of reforestation, he was also involved in agricultural economics.

Appendix B
On the Surface Tension Constitutive Relation and Newton's Law of Cooling

B.1 Surface Tension Relation

The linear approximation for the surface tension in (2.1) is the basic equation used to model the Marangoni effect and can be viewed as an equation of state for the interface. The rate $(d\sigma/dT_s)_{T_\infty}$ is the agent for this effect. Notice that although generally the surface tension decreases when the temperature increases (this is the case of "normal thermocapillarity," $\gamma > 0$), there are systems like some water–alcohol solutions and liquid crystals that display the opposite behavior or even exhibit a minimum for $\sigma = \sigma(T_s)$ ("anomalous thermocapillarity" [189], $\gamma < 0$ or both $\gamma > 0$ and $\gamma < 0$). For most of the monograph, when we examine the influence of the Marangoni effect, i.e., in Chaps. 3, 5 and 9 where we consider the case of a film heated uniformly from below, we restrict our attention to the normal thermocapillarity case, i.e., $\gamma > 0$.

B.2 Newton's Law of Cooling

Around 1700 Newton considered the convective cooling of a warm body by a cool gas and suggested that cooling would be such that the temperature T of the body changes according to $dT/dt \propto T - T_\infty$, where T_∞ is the temperature of the incoming fluid, but he never wrote the expression (2.2). However, from the first law of thermodynamics for a closed system and in the absence of work, $Q = dU/dt \propto dT/dt$, where Q is the heat transfer rate between the system and its surroundings and U its internal energy. Hence, $Q \propto T - T_\infty$, which can be rewritten in terms of the heat flux $q = Q/A$ with A the body's area as $q = \alpha(T - T_\infty)$.

Newton's law of cooling is not based on fundamental principles, as, e.g., Newton's law of viscosity, but it is phenomenological in that it is the relation that defines the heat transfer coefficient α. It is introduced for simplicity and mathematical convenience as it allows us to bypass the substantially more involved conjugated heat transfer problems in the air–liquid, liquid–solid and solid–air interfaces. (Although

S. Kalliadasis et al., *Falling Liquid Films*, Applied Mathematical Sciences 176,
DOI 10.1007/978-1-84882-367-9, © Springer-Verlag London Limited 2012

substantially more involved, by assuming continuity of temperatures and normal components of heat fluxes in any boundary, in principle one could solve for the motion and temperature distribution in all phases.)

In general, the factors influencing the heat transfer coefficient α will depend on the particular mode of heat transfer [162]:

(i) "Forced convection," e.g., flow over a flat plate. In this case α depends on the physical properties of the phase to which heat is being transported. (The properties are in general a function of temperature, but for gases and relatively small temperature differences, as is the case here, this dependence is weak.) α also depends on the physical state of this phase, i.e., a fast moving fluid will in general have better heat transfer characteristics than a slowly moving one.

In fact, α increases with the average velocity of the phase to which heat is transported, e.g., it is directly proportional to the average velocity raised to the 1/2 power for flow over a flat plate. In our case the gas is dragged by the liquid, but due to the relatively slow motion of the liquid (in this monograph the Reynolds numbers are small to moderate) most likely forced convection in the gas has a minimal effect on heat transfer.

(ii) "Natural convection" caused by density changes, which is quite likely the case here for the gas in contact with both the liquid and the solid; however, in the formulation of the basic equations we adopt a one-fluid approach in which the motion of the gas does not influence the motion of the liquid (from H5 in Chap. 2).

α depends on the physical properties of the phase to which heat is being transported (again, for gases the properties have a weak dependence on temperature) and the temperature difference $T - T_0$ (e.g., it is directly proportional to $(T - T_0)^{1/4}$ for natural convection from a vertical or horizontal heated plane); hence it depends indirectly on the heat resistance of the phase under consideration or equivalently its thermal conductivity—large thermal conductivity of the phase under consideration will in general lead to a larger T and hence larger α.

These observations imply that the heat transfer characteristics of the liquid–gas interface will in general be different from those of the solid–gas one, leading to different heat transfer coefficients for the two interfaces. More precisely, since in general the temperature, say T_{liq}, of the liquid at the liquid–gas interface will be smaller than the temperature of the solid at the solid–gas interface and since the heat transfer coefficient grows with $T - T_0$, the liquid–gas coefficient will be smaller than the solid–gas one.

Further, since by nature a falling film evolves in both time and space, α would also depend on both time and position. For simplicity we shall assume that wavy regimes more or less homogenize the heat transfer process at the liquid–gas interface. In addition, changing the liquid flow rate will also influence the temperature difference in the liquid and hence the liquid–gas heat transfer coefficient. However, provided that the temperature difference in the liquid is small relative to the temperature difference $T_{\mathrm{liq}} - T_\infty$ to begin with, the effect of liquid flow rate on the liquid–gas coefficient will be small. As an example, consider a situation where the temperature difference in the liquid is 2°C, $T_{\mathrm{liq}} = 30°C$ and $T_\infty = 0°C$. Assume now that the flow rate is quadrupled, which from our discussion above suggests that for convection heat transfer in the liquid, the heat transfer coefficient in the liquid is doubled leading to a 50% reduction of the temperature difference in the liquid. That is, it results in a difference of 1°C, which in turn leads to a new $T_{liq} = 31°C$ (assuming that the liquid temperature at the solid–liquid interface is not affected). The heat transfer coefficient at the liquid–gas interface then changes only by a factor of $(31/30)^{1/4} \approx 1.008$, and hence it remains practically unaltered. A small temperature difference in the liquid can be achieved either with a high liquid conductivity and/or by increasing the flow rate, which for simplicity we shall assume is the case here. The liquid–gas heat transfer coefficient then becomes practically independent of the liquid and depends only on the physical properties of the gas.

Appendix C
Definitions and Derivations

C.1 Heat Flux Boundary Condition (HF)

The HF thermal boundary condition is obtained by solving the steady state energy equation (2.5) for the wall temperature $T_{\rm w}$ augmented with a heat generation term on its right hand side, $q_{\rm w}/h_{\rm w}$, representing the heat flux generated by the heater per unit wall thickness:

$$\lambda_{\rm w}\partial_{yy}T_{\rm w} + \frac{q_{\rm w}}{h_{\rm w}} = 0, \tag{C.1}$$

where $\lambda_{\rm w}$ is the wall thermal conductivity. This is a Poisson equation in a region bounded by two surfaces: the top surface where we impose continuity of temperatures:

$$y = 0 : T_{\rm w} = T, \tag{C.2a}$$

with T the liquid temperature on the wall, and the bottom surface where Newton's law of cooling in (2.2) applies:

$$y = -h_{\rm w} : \lambda_{\rm w}\partial_y T_{\rm w} = \alpha_{\rm w}(T_{\rm w} - T_\infty), \tag{C.2b}$$

where $\alpha_{\rm w}$ is the heat transfer coefficient between the wall and the gas. The geometry is sketched in Fig. 2.2. At $y = -h_{\rm w}$, $\mathbf{n} = -\mathbf{j}$, with $\mathbf{j}$ the unit vector having the direction of the positive y-axis. Notice that the heat conduction term $\lambda_{\rm w}\partial_{xx}T_{\rm w}$ has been neglected in (C.1): Once we obtain the solution for $T_{\rm w}$ the wall thickness will be shrunk to zero as the wall is just a heat source while the liquid and the gas are just heat sinks, and hence streamwise heat conduction in the wall is negligible. This would not be the case, however, for nonuniform heating, $q_{\rm w} = q_{\rm w}(x)$, e.g., localized heating/point source, which would induce conduction in the streamwise direction so that the term $\lambda_{\rm w}\partial_{xx}T_{\rm w}$ in (C.1) should be retained.

The solution of (C.1) is readily obtained to be

$$T_{\rm w} = -\frac{q_{\rm w}}{2\lambda_{\rm w}h_{\rm w}}y^2 + {\sf A}y + {\sf B}, \tag{C.3a}$$

S. Kalliadasis et al., *Falling Liquid Films*, Applied Mathematical Sciences 176, DOI 10.1007/978-1-84882-367-9, © Springer-Verlag London Limited 2012

where the two integration constants A and B are determined from the boundary conditions (C.2a), (C.2b):

$$A = \frac{-q_{\mathrm{w}}(1 + \frac{\alpha_{\mathrm{w}} h_{\mathrm{w}}}{2\lambda_{\mathrm{w}}}) + \alpha_{\mathrm{w}}(T - T_{\infty})}{\lambda_{\mathrm{w}} + \alpha_{\mathrm{w}} h_{\mathrm{w}}}, \tag{C.3b}$$

$$B = T. \tag{C.3c}$$

Continuity of fluxes at the substrate demands

$$y = 0 : \lambda \partial_y T = \lambda_{\mathrm{w}} \partial_y T_{\mathrm{w}}. \tag{C.4a}$$

By using (C.3a), (C.3b), the right hand side of (C.4a) becomes

$$y = 0 : \lambda_{\mathrm{w}} \partial_y T_{\mathrm{w}} = \lambda_{\mathrm{w}} A. \tag{C.4b}$$

We then shrink the wall thickness to zero: As already noted, the wall is just a boundary that serves as a heat source and its thickness is of no consequence. The resulting solution will be then used in the development of a boundary condition for the liquid. Taking the limit of (C.3b) as $h_{\mathrm{w}} \to 0$ gives

$$A = \frac{-q_{\mathrm{w}} + \alpha_{\mathrm{w}}(T - T_{\infty})}{\lambda_{\mathrm{w}}},$$

so that (C.4a) becomes

$$y = 0 : \lambda \partial_y T = -q_{\mathrm{w}} + \alpha_{\mathrm{w}}(T - T_{\infty}). \tag{C.5}$$

Of course at the liquid–solid interface we also have Newton's law of cooling in (2.2) with $\mathbf{n} = \mathbf{j}$,

$$y = 0 : -\lambda_{\mathrm{w}} \partial_y T_{\mathrm{w}} = \alpha^{\mathrm{w}}(T_{\mathrm{w}} - T_{\mathrm{liq}}), \tag{C.6}$$

where α^{w} is the heat transfer coefficient between the wall and the liquid and T_{liq} is the liquid temperature right outside the thermal resistance layer in the immediate vicinity of the wall (see also our discussion on Newton's law of cooling (2.2) in Sect. 2.1); so it happens that the temperature gradient in the liquid is linear throughout for the Nusselt flat film solution in (2.15a)–(2.15f) and strictly speaking we do not have a thermal resistance layer in the immediate vicinity of the wall. At this stage, however, we do not know a priori what the temperature distribution in the liquid will be; as a matter of fact we are utilizing the concept of a thermal resistance layer precisely so we may obtain the thermal boundary condition for the HF case. Besides, we shall eventually demonstrate that for the boundary condition we are after, the heat transfer characteristics at the liquid–solid interface are good to the point that there is no resistance to heat transfer in the immediate vicinity of the wall and the layer there does not exist.

The constants A and B in (C.3a) are now replaced by

$$A = \frac{-q_{\mathrm{w}}(1 + \frac{h_{\mathrm{w}} \alpha_{\mathrm{w}}}{2\lambda_{\mathrm{w}}}) + \alpha_{\mathrm{w}}(T_{\mathrm{liq}} - T_{\infty})}{\lambda_{\mathrm{w}}(1 + \alpha_{\mathrm{w}} h_{\mathrm{w}} + \frac{\alpha_{\mathrm{w}}}{\alpha^{\mathrm{w}}})}, \tag{C.7a}$$

$$B = -\frac{\lambda_w}{\alpha^w}A + T_{\text{liq}}. \tag{C.7b}$$

The true boundary condition felt by the liquid is (C.4a), which with (C.6) becomes

$$y = 0 : \lambda \partial_y T = \alpha^w (T_{\text{liq}} - T_w). \tag{C.8}$$

Via (C.3a) and (C.7b) the right hand side of this equation can be written as $\alpha^w (T_{\text{liq}} - T_w) = \lambda_w A$ at $y = 0$. As before we take the limit of (C.7a) as $h_w \to 0$ which yields

$$\lambda_w A = \frac{-q_w + \alpha_w (T_{\text{liq}} - T_\infty)}{1 + \frac{\alpha_w}{\alpha^w}}.$$

Further, it is quite natural to assume that α^w is large so that the above expression is further simplified to

$$\lambda_w A = -q_w + \alpha_w (T_{\text{liq}} - T_\infty).$$

The boundary condition in (C.8) then becomes

$$y = 0 : \lambda \partial_y T = -q_w + \alpha_w (T_{\text{liq}} - T_\infty), \tag{C.9}$$

which is the same with (C.5) but with T_{liq} instead of T on the right hand side. However, from (C.8), $T_{\text{liq}} - T_w|_{y=0} = (1/\alpha^w)\lambda \partial_y T|_{y=0}$, which for large α^w gives $T_{\text{liq}} \to T_w|_{y=0}$ (α^w is large and $T_{\text{liq}} - T_w|_{y=0}$ small, but the product $\partial_y T|_{y=0}$ is finite). But $T_w|_{y=0} = T|_{y=0}$—continuity of temperatures at the substrate always holds. Hence, the temperature variation in the thermal resistance layer in the immediate vicinity of the wall is negligible and $T_{\text{liq}} \to T|_{y=0}$ so that (C.9) and (C.5) are identical.

The condition of large α^w implies that the liquid thermal resistance layer is very thin. Indeed, using a linear profile to approximate the liquid temperature in the layer, we can estimate the layer's thickness, say δ_{TRL}: from $\delta_{\text{TRL}} \sim (T_{\text{liq}} - T|_{y=0})/(\partial_y T|_{y=0}) \to 0$ as $T_{\text{liq}} \to T|_{y=0}$. Physically, the heat transfer process in the immediate vicinity of the wall is good to the point that the thermal resistance layer does not exist.

To summarize, it is by approximating T_{liq} with $T|_{y=0}$ in the second derivation, that we obtain (C.5). But this requires one additional condition, that of large α^w, compared to the first derivation, where only $h_w \to 0$ is assumed. This additional condition then effectively gets rid of α^w and the boundary condition at $y = 0$ in (C.6) that involves α^w. And that is precisely why the final forms of the wall boundary condition obtained from the two derivations are the same.

C.2 Surface Gradient Operator

The *surface gradient operator* is defined as

$$\nabla_s = (\mathbf{I} - \mathbf{n} \otimes \mathbf{n}) \cdot \nabla, \tag{C.10}$$

where $\mathbf{I}$ is the identity matrix and $\mathbf{n} \otimes \mathbf{n}$ is the dyadic product of the normal vector $\mathbf{n}$ with itself, i.e.,

$$\mathbf{n} \otimes \mathbf{n} = \frac{1}{n^2} \begin{pmatrix} (\partial_x h)^2 & -\partial_x h & \partial_x h \partial_z h \\ -\partial_x h & 1 & -\partial_z h \\ \partial_x h \partial_z h & -\partial_z h & (\partial_z h)^2 \end{pmatrix}.$$

The tensor $\mathbf{I} - \mathbf{n} \otimes \mathbf{n}$ singles out the tangential projection of a vector, e.g., $(\mathbf{I} - \mathbf{n} \otimes \mathbf{n}) \cdot \mathbf{v} = \mathbf{v}_s$: $(\mathbf{I} - \mathbf{n} \otimes \mathbf{n}) \cdot \mathbf{v} = \mathbf{I} \cdot \mathbf{v} - (\mathbf{n} \otimes \mathbf{n}) \cdot \mathbf{v} = \mathbf{v} - (\mathbf{v} \cdot \mathbf{n})\mathbf{n} = \mathbf{v} - v_n \mathbf{n} = \mathbf{v} - \mathbf{v}_n = \mathbf{v}_s + \mathbf{v}_n - \mathbf{v}_n = \mathbf{v}_s$, where $\mathbf{v}_n$ is the velocity component normal to the surface and we have used the identity $(\mathbf{n} \otimes \mathbf{n}) \cdot \mathbf{v} \equiv (\mathbf{v} \cdot \mathbf{n})\mathbf{n}$.

C.3 On the Choice of the Unit Vectors Tangential to the Surface

When we gave the governing equations and boundary conditions in Chap. 2, we deliberately did not choose the set of orthogonal tangential vectors $(1/\tau_1)(1, 0, \partial_x h)$ and $(1/\tau_2)(0, 1, \partial_z h)$, since the only requirement for τ_1 and τ_2 is that they be linearly independent but not necessarily normal to each other. For example, $(1, 0, \partial_x h)$ and $(0, 1, \partial_z h)$ (with appropriate normalization coefficients) are two such vectors which also could have been chosen. To see this, let us write the tangential stress balance in the form $\mathbf{f} \cdot \boldsymbol{\tau}_i = 0$, where $\mathbf{f} = \mathbf{P} \cdot \mathbf{n} - \nabla_s$. Consider now the two linearly independent tangential vectors $\mathbf{s}_i$; $i = 1, 2$. We then write $\boldsymbol{\tau}_1 = a_1 \mathbf{s}_1 + a_2 \mathbf{s}_2$ and $\boldsymbol{\tau}_2 = b_1 \mathbf{s}_1 + b_2 \mathbf{s}_2$, so that $\mathbf{f} \cdot \boldsymbol{\tau}_1 = a_1(\mathbf{f} \cdot \mathbf{s}_1) + a_2(\mathbf{f} \cdot \mathbf{s}_2) = 0$ and $\mathbf{f} \cdot \boldsymbol{\tau}_2 = b_1(\mathbf{f} \cdot \mathbf{s}_1) + b_2(\mathbf{f} \cdot \mathbf{s}_2) = 0$. But $\left| \begin{smallmatrix} a_1 & a_2 \\ b_1 & b_2 \end{smallmatrix} \right| \neq 0$ due to the linear independence of the vectors $\mathbf{s}_i$. Hence, the only solution to this system is the trivial solution, $\mathbf{f} \cdot \mathbf{s}_1 = \mathbf{f} \cdot \mathbf{s}_2 = 0$. The general form of the tangential stress balance is then preserved. But the two new tangential boundary conditions will have in general a different functional form than the previous ones. Nevertheless, each new condition is simply a linear combination of the old equations, and so we can recover the old from the new. Indeed, with the decompositions $\mathbf{s}_1 = c_1 \boldsymbol{\tau}_1 + c_2 \boldsymbol{\tau}_2$ and $\mathbf{s}_2 = d_1 \boldsymbol{\tau}_1 + d_2 \boldsymbol{\tau}_2$, we have $\mathbf{f} \cdot \mathbf{s}_1 = c_1(\mathbf{f} \cdot \boldsymbol{\tau}_1) + c_2(\mathbf{f} \cdot \boldsymbol{\tau}_2) = 0$ and $\mathbf{f} \cdot \mathbf{s}_2 = d_1(\mathbf{f} \cdot \boldsymbol{\tau}_1) + d_2(\mathbf{f} \cdot \boldsymbol{\tau}_2) = 0$, which, since $\left| \begin{smallmatrix} c_1 & c_2 \\ d_1 & d_2 \end{smallmatrix} \right| \neq 0$ due to the linear independence now of $\boldsymbol{\tau}_1$ and $\boldsymbol{\tau}_2$, respectively give $\mathbf{f} \cdot \boldsymbol{\tau}_1 = 0$ and $\mathbf{f} \cdot \boldsymbol{\tau}_2 = 0$. In other words, $\mathbf{f} \cdot \mathbf{s}_i = 0$ if and only if $\mathbf{f} \cdot \boldsymbol{\tau}_i = 0$, so that in all cases we end up with the same set of equations for the tangential stress balance.

C.4 On the Evaluation of the Right Hand Side of the Tangential Stress Balance (2.13)

Here we comment on the evaluation of $\boldsymbol{\tau}_i \cdot \nabla_s \sigma$ in (2.13). Using the definition of the surface gradient operator (C.10) we obtain

$$\boldsymbol{\tau}_1 \cdot \nabla_s = \frac{1}{\tau_1}(\partial_x + \partial_x h \partial_y)$$

and

$$\boldsymbol{\tau}_2 \cdot \mathbf{V}_s = \frac{1}{\tau_2}(\partial_z + \partial_z h \partial_y).$$

If we now operate on σ we have $\boldsymbol{\tau}_1 \cdot \mathbf{V}_s \sigma = (1/\tau_1)(\partial_x \sigma + \partial_x h \partial_y \sigma)$ and $\boldsymbol{\tau}_2 \cdot \mathbf{V}_s \sigma = (1/\tau_2)(\partial_z \sigma + \partial_z h \partial_y \sigma)$ at $y = h(x, z, t)$. From H6 in Chap. 2, $\sigma = \sigma(T_s)$, with $T_s = T(x, h(x, z, t), z, t) \equiv T_s(x, z, t)$. Hence, σ is only a function of x, z and t. As a result

$$\boldsymbol{\tau}_1 \cdot \mathbf{V}_s \sigma = \frac{1}{\tau_1}\partial_x \sigma$$

and

$$\boldsymbol{\tau}_2 \cdot \mathbf{V}_s \sigma = \frac{1}{\tau_2}\partial_z \sigma.$$

By using the equation of state in (2.1) we obtain,

$$\partial_x \sigma = \frac{d\sigma}{dT_s}\partial_x T_s = -\gamma \partial_x T_s,$$

and

$$\partial_z \sigma = \frac{d\sigma}{dT_s}\partial_z T_s = -\gamma \partial_z T_s.$$

The gradients of the interfacial temperature $\partial_x T_s$ and $\partial_z T_s$ can be further related to the temperature field T at $y = h(x, z, t)$. Using the chain rule,

$$\partial_x T_s = (\partial_x T + \partial_x h \partial_y T)|_{y=h}$$

and

$$\partial_z T_s = (\partial_z T + \partial_z h \partial_y T)|_{y=h},$$

where it is understood that we first take the derivatives on the right hand side with respect to x, y and z and then we substitute $y = h$. The final form of $\boldsymbol{\tau}_i \cdot \mathbf{V}_s \sigma$ in (2.13) is then

$$\boldsymbol{\tau}_1 \cdot \mathbf{V}_s \sigma = -\frac{\gamma}{\tau_1}(\partial_x T + \partial_x h \partial_y T)|_{y=h}$$

and

$$\boldsymbol{\tau}_2 \cdot \mathbf{V}_s \sigma = -\frac{\gamma}{\tau_2}(\partial_z T + \partial_z h \partial_y T)|_{y=h}.$$

The same expressions can also be obtained by noting that $\boldsymbol{\tau}_i \cdot \nabla \sigma$ is the "directional derivative" $\partial_{t_i} \sigma$ on the direction t_i defined by $\boldsymbol{\tau}_i$. We then have

$$\boldsymbol{\tau}_1 \cdot \mathbf{V}_s \sigma = \partial_{t_1} \sigma = -\gamma \partial_{t_1} T_s = -\gamma \boldsymbol{\tau}_1 \cdot \nabla T_s = -\frac{\gamma}{\tau_1}\partial_x T_s = -\frac{\gamma}{\tau_1}(\partial_x T + \partial_x h \partial_y T)|_{y=h},$$

and

$$\boldsymbol{\tau}_2 \cdot \nabla_{\!s}\sigma = \partial_{t_2}\sigma = -\gamma\,\partial_{t_2}T_s = -\gamma\,\boldsymbol{\tau}_2 \cdot \nabla T_s = -\frac{\gamma}{\tau_2}\partial_z T_s = -\frac{\gamma}{\tau_2}(\partial_z T + \partial_z h\,\partial_y T)|_{y=h}.$$

C.5 Short Library of Weakly Nonlinear Model Equations: Bottom-up Dispersion Relation Approach

Several studies have examined the film flow dynamics within the framework of weakly nonlinear analyses [19, 20, 41, 42, 44, 53, 164, 185, 199]. The basic idea underlying these studies is that, sufficiently close to onset, the flow dynamics are determined by the properties of the linear stability of the base flow and that deviations from it remain small. Fluctuations are then decomposed into elementary instability waves of the "normal mode" form, $a(x, t)\exp(\lambda t + ikx)$, with k their (real) wavenumber and $\lambda = \lambda_r + i\lambda_i$ a complex eigenvalue whose real part λ_r is the (temporal) growth rate (subscripts "r" and "i" are used to denote real and imaginary parts, respectively) and imaginary part λ_i is the negative value of the (real angular) frequency (precise definitions of these terms are given in Chap. 3). The structure of the weakly nonlinear equations, or *amplitude equations*, giving the evolution of the envelope $a(x, t)$ is then determined by the dispersion relation, which expresses λ as a function of the wavenumber k.

It is then possible to invert the procedure by starting from the possible structure of the dispersion relation at the onset of the instability. Hence, writing from the outset generic amplitude equations without deriving them from the fully nonlinear system is a "bottom-up dispersion relation approach," a heuristic approach that guides qualitatively the understanding of the competing generic linear and nonlinear effects for small amplitude disturbances. In Chap. 5 the pertinent amplitude equations for the falling film problem are derived systematically with weakly nonlinear expansions from the BE.

Considering the hydrodynamic surface wave instability and the Marangoni instability described earlier, the transition from the uniform laminar flow to a wavy one can be understood as a transition from a steady to an oscillatory flow. Typically, "oscillatory instabilities" occur when "negative feedback" exists and hence a perturbation switches on some compensating mechanisms that try to diminish it. However, these mechanisms do not always suppress the perturbation, but sometimes they lead to "overshooting oscillations" ("overstability"). It seems essential that the destabilizing and stabilizing factors have different time scales, so that their counteraction is characterized by a certain effective "time delay." The asynchronous changes of fields of different physical variables leads to the appearance of oscillations instead of a monotonic growth or decay of the perturbation, in which case the instability is referred to as "monotonic" or "stationary." We give the precise definitions of the two basic types of instabilities, oscillatory and stationary, in Chap. 3.

A way to classify instabilities, oscillatory and otherwise, is obtained via the minimum of the neutral stability curve, that defines the critical value of the control parameter (see Chap. 3 for the definitions of these terms). If the "critical wavenumber," i.e., the wavenumber of the fastest growing perturbation at the instability onset, k_0, is nonzero, the oscillatory instability is, generally, short wavelength; otherwise (if $k_0 = 0$), it is a long wavelength instability. For the long wavelength instability, it is necessary to further distinguish between two cases.

In the first case, the eigenvalue $\lambda(k, \Sigma)$ that determines the (temporal) growth rate λ_r and the (real angular) frequency of the oscillations $-\lambda_i$ as a function of a given control parameter Σ (used as a generic quantity whose definition now is not needed), can be expanded into a Taylor series near the critical point ($k_0 = 0$, Σ_c) as

$$\lambda_r = (\partial_\Sigma \lambda_r)_0 (\Sigma - \Sigma_c) + \frac{1}{2}(\partial_{kk}\lambda_r)_0 k^2 + \cdots , \tag{C.11a}$$

$$\lambda_i = (\lambda_i)_0 + (\partial_k \lambda_i)_0 k + (\partial_\Sigma \lambda_i)_0 (\Sigma - \Sigma_c) + \frac{1}{2}(\partial_{kk}\lambda_i)_0 k^2 + \cdots , \tag{C.11b}$$

where the subscript 0 denotes that the corresponding quantity is evaluated at the critical point. In this case, the spatially homogeneous perturbation with $k = 0$ oscillates and grows with the largest growth rate, $(\partial_\Sigma \lambda_r)_0 (\Sigma - \Sigma_c)$. This type of oscillatory instability occurs frequently in reaction–diffusion systems.

However, there is a wide class of problems where growth of a spatially homogeneous disturbance is forbidden by a "conservation law." An example is the falling film problem. It is a system that involves liquid flow from an inlet and the conservation law is imposed by the inlet flow rate. A homogeneous change of the base state film thickness is only possible through a change in the flow rate, but it is not allowed if the flow rate remains fixed. The mode associated with a homogeneous change of the base state is called *Goldstone mode* in condensed matter physics (e.g., [58]; we return to this mode on several occasions in this monograph) and is characterized by

$$\lambda(0, \Sigma) = 0 \tag{C.12}$$

for any Σ, which can produce a long wavelength instability for small but nonzero k. In the latter case, it is not $\lambda_r(0, \Sigma)$ but $\partial_{kk}\lambda_r(0, \Sigma)$ that changes sign at the threshold of the instability, $\Sigma = \Sigma_c$:

$$\lambda_r = \frac{1}{2}(\partial_{kk\Sigma}\lambda_r)_0 k^2 (\Sigma - \Sigma_c) + \frac{1}{24}(\partial_{kkkk}\lambda_r)_0 k^4 + \cdots , \tag{C.13a}$$

$$\lambda_i = (\partial_k \lambda_i)_0 k + (\partial_{k\Sigma}\lambda_i)_0 k(\Sigma - \Sigma_c) + \frac{1}{6}(\partial_{kkk}\lambda_i)_0 k^3 + \cdots . \tag{C.13b}$$

Past the instability threshold, the growth rate is proportional to k^2 at small k (this effect appears akin to the role of an effective "negative viscosity").

Equations (C.13a), (C.13b) correspond precisely to the H-mode of instability for a falling liquid film. The dispersion relation for the growth rate λ_r features a band of unstable modes extending from zero up to a "cut-off wavenumber" k_c, above which the system is stable, $k_c = [-12(\partial_{kk\Sigma}\lambda_r)_0 (\Sigma - \Sigma_c)/(\partial_{kkkk}\lambda_r)_0]^{1/2}$. The unstable band $0 \le k \le k_c$ contains the "maximum growing wavenumber" k_{max} with the largest positive growth rate, $\lambda_r(k_{max})$. We notice that even though at the critical wavenumber $k = 0$, λ_i vanishes, the instability is oscillatory. As a matter of fact, $k \to 0$ is a "degenerate" limit where from (C.12) the exponential part of the disturbances reduces to $\exp 0 = 1$ corresponding to a uniform shift of the base state (by changing the flow rate), the Goldstone mode defined earlier (it is neither damped nor amplified). Infinitely long waves are a mathematical artifact and finite-size effects (finite length of the channel) will remove the degeneracy (because of the discrete spectrum of modes imposed by the finite size, while the smallest wavenumber scales as $k \sim 1/L$ with L the channel's length)[1] so that a true "Hopf bifurcation" with $\lambda_i \neq 0$ occurs. In the linear regime the disturbance will grow at rate $\lambda_r(k_{max})$ and at the same time it is periodic in space with wavenumber k_{max} and oscillates in time with frequency $-\lambda_i(k_{max})$. The combination of periodicity in space and oscillatory behavior in time leads to a traveling wave.

The behavior of the eigenvalue in the vicinity of the point (k_0, Σ_c) is crucial for the evolution of the weakly nonlinear waves generated by the oscillatory instability. In the case of a short wavelength instability where the fastest growing disturbance has a finite wavenumber, $k_0 \neq 0$, the *complex Ginzburg–Landau equation*

$$\partial_t a = \gamma_0 a + \gamma_2 \partial_{xx} a - \delta_0 |a|^2 a \tag{C.14}$$

is a universal equation for a complex envelope function $a(x, t)$ describing the modulation of the waves; here γ_0, γ_2 and δ_0 are complex constant coefficients. The space-independent case (formally $\gamma_2 = 0$) is called the "Landau equation" in several areas of physics such as phase transitions. It was introduced by Stuart (e.g., [263]) to study flow instabilities with transition between steady fluid motions and is also frequently referred to as the "Landau–Stuart equation."

The Ginzburg–Landau equation (C.14) can also be used to describe the transition when an unstable wave motion of given ("fundamental") wavenumber interacts with its first stable harmonics (in fact, the Ginzburg–Landau equation is only valid when the fundamental wavenumber is weakly unstable—i.e., just below k_c—while the overtone with a wavenumber twice that of the fundamental is stable [53, 274]). This situation occurs for monochromatic waves excited on a falling film by forcing at the inlet with a frequency close to the cut-off frequency, which then yields a wavenumber k close to the cut-off wavenumber k_c (details are given in Chap. 7) determined by the balance of the H-instability mechanism and the damping effect of surface tension. Using (C.14), Lin [164] showed that such monochromatic waves are sideband stable.[2] This, however, was in contradiction with the experimental results by Liu and Gollub [168]. Later on, Oron and Gottlieb [199] corrected Lin's

---

[1] In a finite-size container, e.g., the annular container used to investigate the propagation of waves due to solutocapillary Marangoni effect in [189, 300], the $\lambda = k = 0$ mode is removed. Clearly, due to the conservation law, i.e., conservation of fluid volume, a homogeneous change of the layer's thickness is not allowed.

[2] In general, the term "sideband instability" of a wave refers to a resonance between three frequencies, the frequency of the wave, say f, and two frequencies, $f + \delta f$ and $f - \delta f$ with δf small. Sideband instabilities in three dimensions are examined in Chap. 8.

analysis and demonstrated that excited monochromatic waves at $k \lesssim k_c$ can be side-band unstable.

It is then evident that for the falling liquid film, and more general long wave instabilities with $k_0 = 0$ and with the dependence $\lambda(k, \Sigma)$ determined by (C.13a), (C.13b), the Ginzburg–Landau equation (or its simplified version, the Landau–Stuart equation) has a limited applicability, since as already emphasized above it is only valid when the fundamental wavenumber is weakly unstable. For example, the maximum growing wavenumber $k_{\max} \in [0, k_c]$, which, in general, is not weakly unstable (unless we are very close to criticality), cannot be captured by (C.14). As a matter of fact, due to the multiplicity of the dominant Fourier modes more than one amplitude equation must be considered [42]. Alternatively, the Kuramoto–Sivashinsky equation introduced later (which is not necessarily valid only for conditions very close to criticality—Chap. 4) can accommodate a large band of Fourier modes.

Let us consider now the case of a negative viscosity (dispersion relations (C.13a), (C.13b)) caused by a conservation law. The structure of an amplitude equation compatible with a conservation law is

$$\partial_t a = \mathcal{L}_x a + \partial_x \mathcal{F}(a), \qquad (C.15a)$$

where the linear operator $\mathcal{L}_x$ involves derivatives with respect to x only and has the structure

$$\mathcal{L}_x = \gamma_1 \partial_x + \gamma_2(\Sigma - \Sigma_c)\partial_{xx} + \gamma_3 \partial_{xxx} + \gamma_4 \partial_{xxxx} + \cdots, \qquad (C.15b)$$

where

$$\gamma_1 = (\partial_k \lambda_i)_0 + (\partial_{k\Sigma} \lambda_i)_0(\Sigma - \Sigma_c), \qquad \gamma_2 = -\frac{1}{2}(\partial_{kk\Sigma} \lambda_r)_0, \qquad (C.15c)$$

$$\gamma_3 = -\frac{1}{6}(\partial_{kkk} \lambda_i)_0, \qquad \gamma_4 = \frac{1}{24}(\partial_{kkkk} \lambda_r)_0, \ldots, \qquad (C.15d)$$

and

$$\mathcal{F}(a) = \delta_1 a^2 + \delta_2 \partial_x(a^2) + \cdots. \qquad (C.15e)$$

It should be noted that if the conservation law is an approximate one, the amplitude equation can contain additional small terms not differentiated with respect to x. The term containing γ_1 can be removed by a *Galilean transformation* to a suitable moving reference frame. It should also be noted that the nonlinearity in the amplitude equation (C.15a) is actually a guess, while the linear operator in (C.15a) gives precisely the dispersion relation (C.13a), (C.13b) (which can be either exact, i.e., obtained from the fully nonlinear system, or a guess).

First, let us consider the generic case where all the coefficients γ_n, $n = 1, 2, \ldots$ in (C.15a)–(C.15e) are of $\mathcal{O}(1)$. A weakly nonlinear prototype can be obtained from (C.15a)–(C.15e) by utilizing multiple scale-type arguments. Taking $\Sigma - \Sigma_c = \mathcal{O}(\epsilon^2)$ where $\epsilon \ll 1$, the expression for the cut-off wavenumber, $k_c =$

$[-12(\partial_{kk}\Sigma\lambda_{\mathrm{r}})_0(\Sigma - \Sigma_{\mathrm{c}})/(\partial_{kkkk}\lambda_{\mathrm{r}})_0]^{1/2}$, shows that $k_{\mathrm{c}} \sim \epsilon$ so that $x \sim \epsilon^{-1}$, a long scale, or equivalently $\partial_x = \mathcal{O}(\epsilon)$ (the order of magnitude of the spatial gradient is dictated by the order of the cut-off wavelength, which is the wavelength of the instability pattern emerging at the onset). Assuming that the largest nonlinear term $\delta_1 \partial_x (a^2)$ is balanced by the dispersion term $\gamma_3 \partial_{xxx} a$ (the term "dispersion" is used to denote that the wave velocity depends on wavelength or "color" as in optics—this is clarified in Sects. 5.1.4, 5.2.1), we find that $a = \mathcal{O}(\epsilon^2)$, and the amplitude equation is

$$\partial_t a = \gamma_3 \partial_{xxx} a + \delta_1 \partial_x (a^2) + \gamma_2 (\Sigma - \Sigma_{\mathrm{c}}) \partial_{xx} a + \gamma_4 \partial_{xxxx} a + \delta_2 \partial_{xx} (a^2). \quad (\text{C.16})$$

The first two terms on the right hand side of this equation are of $\mathcal{O}(\epsilon^5)$; the last three terms, describing the instability at long wavelengths, the stabilization at short wavelengths, and a nonlinear correction to the negative viscosity coefficient, are of $\mathcal{O}(\epsilon^6)$ (such that very close to criticality, instability at long wavelengths balances stability at short ones). Further, the time derivative balances the dispersive and nonlinear terms on the long time scale $t \sim \epsilon^{-3}$. Thus, we find that the generic amplitude equation in (C.16) is a *driven-dissipative Boussinesq–Korteweg–de Vries equation* (BKdV) which is effectively a perturbed "BKdV equation". The equation with $\gamma_4 = \delta_2 = 0$ is the "BKdV–Burgers equation." The terminology used here deserves a remark. We use *BKdV equation*, instead of the standard terminology "Korteweg–de Vries equation," because the former equation was first obtained by Boussinesq—actually written in a footnote: p. 360, Eq. (283 bis) in [30]!

In fact, it is not necessary to introduce the small parameter ϵ. We need only $\Sigma - \Sigma_{\mathrm{c}} \ll 1$, a condition required to obtain the dispersion relation (C.13a), (C.13b). Now $k_{\mathrm{c}} \sim (\Sigma - \Sigma_{\mathrm{c}})^{1/2} \ll 1$ and the appropriate long scale should be $x \sim k_{\mathrm{c}}^{-1} (\Sigma - \Sigma_{\mathrm{c}})^{-1/2} \gg 1$. Balancing the instability with the stability terms, $a(\Sigma - \Sigma_{\mathrm{c}})/x^2 \sim a/x^4$, gives $x \sim (\Sigma - \Sigma_{\mathrm{c}})^{-1/2}$, consistent with the above assignment (to be expected, as the balance between the instability and stability terms is precisely what determines k_{c}). Balancing the nonlinearity with the dispersion term, $a\partial_x a \sim \partial_{xxx} a$ or $a \sim \Sigma - \Sigma_{\mathrm{c}}$. The order of magnitude of the nonlinearity and dispersion terms then is $a\partial_x a \sim (\Sigma - \Sigma_{\mathrm{c}})^{5/2}$, while that of instability and stability, $a(\Sigma - \Sigma_{\mathrm{c}})/x^2 \sim (\Sigma - \Sigma_{\mathrm{c}})^3 \ll (\Sigma - \Sigma_{\mathrm{c}})^{5/2}$, and the relevant weakly nonlinear prototype is the BKdV equation. In other words, the BKdV equation is always the relevant weakly nonlinear prototype when $\Sigma - \Sigma_{\mathrm{c}} \ll 1$ and all coefficients γ_i of the dispersion relation are of $\mathcal{O}(1)$.

If the dispersion coefficient is small, $\gamma_3 = \mathcal{O}(\epsilon)$ (for instance, this feature is characteristic of the modulational instability of periodic waves with small wavenumbers), it is balanced by the nonlinearity if $a = \mathcal{O}(\epsilon^3)$, and one obtains the *Kawahara equation* (e.g., [144]):

$$\partial_t a = \gamma_2 (\Sigma - \Sigma_{\mathrm{c}}) \partial_{xx} a + \gamma_3 \partial_{xxx} a + \gamma_4 \partial_{xxxx} a + \delta_1 \partial_x (a^2), \quad (\text{C.17})$$

with all terms on the right hand side of $\mathcal{O}(\epsilon^7)$. The time derivative now balances all terms on the right hand side on the long time scale $t \sim \epsilon^{-4}$.

If the dispersion coefficient vanishes ($\gamma_3 = 0$), the Kawahara equation is reduced to the *Kuramoto–Sivashinsky equation* (KS)

$$\partial_t a = \gamma_2(\Sigma - \Sigma_c)\partial_{xx}a + \gamma_4\partial_{xxxx}a + \delta_1\partial_x(a^2), \tag{C.18}$$

first derived by Homsy [118, 119] and typical for instabilities in a wide variety of nonlinear processes. Note that the Kawahara equation in (C.17) is also frequently referred to as the *generalized KS equation*.

The KS equation can also be obtained in the case where the instability region is no longer narrow as above because of large γ_4: This allows increase in the order of magnitude of the stability and hence instability (the two should always balance) as well as nonlinearity terms compared to dispersion. Consider, e.g., the case $\Sigma - \Sigma_c = \mathcal{O}(1)$, $\gamma_4 = \mathcal{O}(\epsilon^{-2})$ (of course the dispersion relation in (C.15a)–(C.15e) was obtained with an expansion for $\Sigma - \Sigma_c \ll 1$, but here we consider (C.15a)–(C.15e) as a model exact dispersion relation). The cut-off wavenumber then is $k_c \sim \sqrt{(\Sigma - \Sigma_c)/\gamma_4} \sim \epsilon$ so that the corresponding long scale is $x \sim 1/k_c \sim 1/\epsilon$. At this long scale the two terms representing instability at long wavelengths and stability at short ones balance as expected: $a(\Sigma - \Sigma_c)/x^2 \sim a\gamma_4/x^4$ or $x \sim 1/\epsilon$. To balance the nonlinearity with these two terms, $a\partial_x a \sim a/x^2$ or $a \sim \epsilon$. The order of instability, stability and nonlinearity then is $aa_x \sim \epsilon^3$ while that of dispersion is $\partial_{xxx}a \sim \epsilon^4$.

On the other hand, having large γ_4, small $\Sigma - \Sigma_c$ and $\gamma_3 = \mathcal{O}(1)$ gives the Kawahara equation. Consider, e.g., the case, $\Sigma - \Sigma_c = \mathcal{O}(\epsilon^2)$ and $\gamma_4 = \mathcal{O}(\epsilon^2)$. Then $k_c \sim \epsilon^2$ and $x \sim 1/k_c \sim \epsilon^{-2}$. Balance instability with stability, $(\Sigma - \Sigma_c)/x^2 \sim 1/x^4$ or $x \sim \epsilon^{-2}$, consistent with the above order (as expected). Balance the nonlinearity with the instability and stability terms, $a\partial_x a \sim a(\Sigma - \Sigma_c)/x^2$ or $a \sim \epsilon^4$. The order of instability, stability and nonlinearity terms then is $a\partial_x a \sim \epsilon^{10}$ while that of dispersion is $\partial_{xxx}a \sim \epsilon^{10}$, so that all terms are of the same order.

In Chap. 5 we demonstrate that both KS and Kawahara equations can be obtained from the governing equations of a falling liquid film via a weakly nonlinear expansion and multiple scale-type arguments similar to the ones adopted here. We also demonstrate in Chap. 5 that the second-order viscous effects are responsible for the dispersion term in the Kawahara equation.

The KS equation, which has both locally stable regular wavy solutions and spatio-temporal chaotic regimes, provides a paradigmatic example of the transition between regular and chaotic patterns, as well as a reference model for the application of the "dynamical systems" approach to spatio-temporal chaos. It suffices at this point to say that the path allowing the use of the theory of dynamical systems is the following. By a suitable Galilean boost we can look at the expected waves in their moving frame of reference. This change of reference frame converts a partial differential equation into an ordinary differential one, which underlines the corresponding dynamical system. A solitary wave then corresponds to a *homoclinic orbit* or trajectory of the dynamical system, while the presence of dissipation exhibits repelling and attractive directions on the orbit (these points are clarified in Chap. 7). When periodicity in time is present, as when a given steady wave solution becomes unstable through a (new) *Hopf bifurcation*, the new wave solution corresponds to a *limit cycle* of the transformed problem. Other possibilities exist including chaotic solutions, and they are studied in this monograph.

It should be noted that the KS equation is formally equally valid for both oscillatory and stationary instabilities. The only "sign" of the "wavy" origin of this equation is the lack of the invariance of the nonlinear term to the parity (reflection) transformation $x \to -x$: The presence of $\partial_x(a^2)$ breaks this symmetry and is connected with the fact that the system has a preferred direction (this is a "mean flow" term in the context of fluid mechanics). However, the transformation $a = \partial_x h$ provides another form of the KS equation,

$$\partial_t h = \gamma_2(\Sigma - \Sigma_c)\partial_{xx}h + \gamma_4\partial_{xxxx}h + \delta_1(\partial_x h)^2, \tag{C.19}$$

which is invariant to the parity transformation.

Finally, it is noteworthy that the analysis of the long wavelength instability just given was based on the assumption of the analyticity of the function $\lambda(k, \Sigma)$ at the point $k = 0$. If the analyticity condition does not hold, the amplitude equation can contain some nonlocal integral terms.

C.6 Negative Polarity in the BKdV Equation

Let us consider the BKdV equation with a dispersion coefficient δ_K:

$$\partial_u + u\partial_x u + \delta_K\partial_{xxx}u = 0. \tag{C.20a}$$

This equation is invariant under the transformation $u \to -u$, $x \to -x$ and $\delta_K \to -\delta_K$ (t is always > 0 and we do not change it). Alternatively, consider (C.20a) in the moving frame $x \to x - ct$,

$$-c\partial_x u + u\partial_x u + \delta_K\partial_{xxx}u = 0, \tag{C.20b}$$

which is invariant under the transformation $u \to -u$, $x \to -x$, $c \to -c$ and $\delta_K \to -\delta_K$. This symmetry shows that (C.20a) admits negative-hump waves traveling backward, but can we have such waves for $\delta_K > 0$ much like with the Kawahara equation as we point out in Sect. 5.3.2.

It is the sign of δ_K that determines the direction of propagation: $\delta_K > 0$ means that positive-hump solitary pulses travel forward while negative-hump ones travel backward. To see this consider the solution of (C.20a), $H = 3c\,\mathrm{sech}^2[(1/2)\sqrt{c/\delta_K}(x - ct)]$, corresponding to a solitary wave traveling with speed c. It is then clear that both c and δ_K must have the same sign for solitary waves to exist. Hence, as we note in Sect. 5.3.2, negative waves of the Kawahara equation with $\delta_K > 0$ no longer exist as the Kawahara equation approaches the perturbed BKdV one.

As far as the above mentioned symmetry is concerned, it is lost with the transformation,

$$u = \delta_K^{1/3}\bar{u}, \qquad x = \delta_K^{1/3}\bar{x}, \tag{C.21a}$$

which converts (C.20a) into the BKdV equation

$$\partial_t \overline{u} + \overline{u}\partial_x \overline{u} + \partial_{xxx}\overline{u} = 0 \qquad\qquad \text{(C.21b)}$$

with a dispersion coefficient of unity. This equation is not invariant under the transformation $\overline{u} \to -\overline{u}$ and $\overline{x} \to -\overline{x}$ and hence it does not admit negative-hump pulses traveling backward. In fact, integrating numerically (C.21b) with an initial condition a negative-hump wave obtained by simply inverting the positive-hump wave of the equation, shows that the negative-hump wave collapses and degenerates into a wave train that disperses and at the same time travels backward [82]. Accordingly, the equation $\partial_t \overline{u} + \overline{u}\partial_x \overline{u} - \partial_{xxx}\overline{u} = 0$ admits only negative-hump waves propagating backward. In essence, the transformation in (C.21a) collapses the two families of solitary wave solutions, positive-hump ones with $\delta_K > 0$ and negative-hump ones with $\delta_K < 0$ of (C.20a), into the single family of positive-hump solutions with $\delta_K = 1$ of (C.21b).

C.7 Padé Approximants

To remedy the singularity of the BE, Ooshida [196] developed a resummation technique inspired from the *Padé approximants technique* (see, e.g., [18, 115]). The technique relies on the basic idea that the divergence of a power series representation of a function $Q(x)$, namely, $Q(x) = \sum_{k=0,1,2,\ldots} Q_k x^k$ on $[0, a]$, is due to the hidden presence of poles. The divergence reflects the inability of the polynomial representation to approximate the function adequately near a singularity. This leads us to express $Q(x)$ as the ratio of two polynomials $F_N(x)$ and $G_M(x)$ of degrees N and M, respectively, where the zeros of G_M are supposed to capture the causes of the divergence. Let us define the rational function

$$R_{N,M} = \frac{F_N(x)}{G_M(x)} \quad \text{for } 0 \leq x \leq a.$$

We wish to make the maximum error between this function and $Q(x)$ as small as possible. The Padé approximants technique then requires that $Q(x)$ and its derivatives be continuous at a point, say, $x = 0$. The polynomials employed are of the form

$$F_N(x) = f_0 + f_1 x + \cdots + f_N x^N \quad \text{and} \quad G_M(x) = 1 + g_1 x + \cdots + g_M x^M.$$

They are constructed so that $Q(x)$ and $R_{N,M}$ and their derivatives up to degree $N + M$ agree at $x = 0$. For a fixed value of $N + M$ (the degrees of the polynomials F and G are open to choice) the error is smallest when $N = M$ or $N = M + 1$. The easiest way to obtain these coefficients is by writing, $Z(x) = Q(x)G_M(x) - F_N(x)$ and ensuring that the coefficients of x^k in $Z(x)$ vanish for $k = 0, \ldots, N + M$.

C.8 Center Manifold Projection for a Scalar Equation

Here we illustrate how to implement the center manifold approach for a scalar non-linear partial differential equation. Assume that the evolution of a physical variable $u(x, t)$ is described by

$$\partial_t u = \mathcal{L}u + N(u, \varepsilon), \tag{C.22}$$

where ε is a control parameter, $\mathcal{L} \equiv \mathcal{L}_x$ is a linear differential operator that describes the "flow" close to the origin $(u, \varepsilon) = (0, 0)$ and N is a nonlinear functional. We define the eigenvalue problem, $\mathcal{L}\Phi_k = \lambda_k \Phi_k; k = 0, 1, 2, \ldots$ associated with $\mathcal{L}$, where Φ_k and λ_k are the eigenfunctions and eigenvalues of $\mathcal{L}$, respectively, and where appropriate boundary conditions depending on the physical problem have been imposed. We also define the adjoint eigenvalue problem, $\mathcal{L}^* \hat{\Phi}_k = \bar{\lambda}_k \hat{\Phi}_k; k = 0, 1, 2, \ldots$, where the overbar is used to denote complex conjugation. The adjoint operator $\mathcal{L}^*$ can be defined with respect to an appropriately chosen inner product, e.g., in an infinite domain the $L^2(-\infty, +\infty)$ inner product, $\langle f | g \rangle = \int_{-\infty}^{+\infty} f \bar{g} \, dx$, for any two functions f and g decaying sufficiently fast at infinity and such that $\langle \mathcal{L}f | g \rangle = \langle f | \mathcal{L}^* g \rangle$.

Assume now for simplicity that the linear operator $\mathcal{L}$ has a single zero eigenvalue, $\lambda_0 = 0$, while all other eigenvalues have negative real parts (all eigenvalues assumed simple). We then decompose u into eigenmodes as $u = a\Phi_0 + \hat{u}$ where $\hat{u}$ is the "complement" with respect to the null eigenfunction Φ_0 and is given by

$$\hat{u} = \sum_{i \geq 1} u_i \Phi_i. \tag{C.23a}$$

By introducing the "projection operator", $\mathcal{P}$ to denote projection onto the null space,

$$\mathcal{P}f = \langle f | \hat{\Phi}_0 \rangle \Phi_0,$$

and the "complementary projection operator", $\mathcal{F} - \mathcal{P}$ to denote projection onto the complement of Φ_0,

$$(\mathcal{F} - \mathcal{P})f = \sum_{i=1}^{\infty} \langle f | \hat{\Phi}_i \rangle \Phi_i$$

for any function f in the domain of $\mathcal{L}$, u can be written as

$$u = \mathcal{P}u + (\mathcal{F} - \mathcal{P})u. \tag{C.23b}$$

The above expansion is a real-valued function even when complex eigenvalues and eigenfunctions are present: For real operators complex eigenvalues and eigenfunctions appear in conjugate pairs and hence if $\langle u | \hat{\Phi}_k \rangle$ and $\langle u | \hat{\Phi}_{k+1} \rangle$ are the coefficients of Φ_k and Φ_{k+1} in the projection for u, respectively, with $\lambda_{k+1} = \bar{\lambda}_k$ and $\Phi_{k+1} = \bar{\Phi}_k$, we must have $\langle u | \hat{\Phi}_{k+1} \rangle = \int_{-\infty}^{\infty} u \bar{\hat{\Phi}}_{k+1} \, dx = \langle \hat{\Phi}_k | u \rangle = \overline{\langle u | \hat{\Phi}_k \rangle}$.

Substituting the projection for u into (C.22) and taking inner products of both sides with the adjoint eigenfunctions $\hat{\Phi}_0$, $\hat{\Phi}_i$ with $i \neq 0$ and using the orthogonality condition $\langle \Phi_i | \hat{\Phi}_j \rangle = \delta_{ij}$ leads to

$$\partial_t a = F(\hat{u}, a, \varepsilon) \tag{C.24a}$$

$$\hat{u} = \mathcal{L}\hat{u} + H(a, \hat{u}, \varepsilon), \tag{C.24b}$$

where the functions F and H are given from

$$F(\hat{u}, a, \varepsilon) = \langle N(a\Phi_0 + \hat{u}, \varepsilon) | \hat{\Phi}_0 \rangle \tag{C.24c}$$

and

$$H(a, \hat{u}, \varepsilon) = (\mathcal{F} - \mathcal{P})N(a\Phi_0 + \hat{u}, \varepsilon). \tag{C.24d}$$

Assume now that we wish to study small solutions of (C.22) for $\varepsilon \ll 1$. The center manifold approach requires that the eigenvalues of $\mathcal{L}$ have zero or negative real parts and in addition the eigenvalues with zero real parts must be well isolated from those with negative ones. However, for the system in (C.24a)–(C.24d) it can very well happen that the associated linearized problem has eigenvalues that are positive (and small) for $\varepsilon \neq 0$ (see [38] for examples of application of the center manifold theorem to systems of nonlinear ordinary differential equations and to singular perturbation problems). Nevertheless, we can write the above system in the following "extended manner":

$$\partial_t a = F(\hat{u}, a, \varepsilon) \tag{C.25a}$$

$$\hat{u} = \mathcal{L}\hat{u} + H(a, \hat{u}, \varepsilon) \tag{C.25b}$$

$$\partial_t \varepsilon = 0, \tag{C.25c}$$

thus ensuring that all eigenvalues are either zero or negative. The addition of (C.25c) introduces a second zero eigenvalue.

According to the center manifold approach then, the system (C.25a)–(C.25c) has a two-dimensional center manifold given by the "adiabatic coupling" $\hat{u} = \hat{u}(a, \varepsilon) = \mathcal{O}(2)$ to leading order with respect to a, ε. To obtain the flow onto the center manifold we note that $\partial_t a = \mathcal{O}(2)$ so that $\partial_t \hat{u} = \mathcal{O}(3)$ and the leading-order center manifold projection, the so-called "tangent space approximation," is simply $\partial_t \hat{u} = 0$ or $\mathcal{L}\hat{u} = -H_2$ where H_2 contains the $\mathcal{O}(2)$ terms of H (terms of $\mathcal{O}(3)$ and higher are neglected). Note that $\mathcal{L}$ is a singular operator and hence the right hand side of (C.25b) must satisfy the Fredhölm alternative solvability condition (see Sect. 5.1), but it does so automatically since the null eigenfunction has already been removed in the projection that leads to (C.24a)–(C.24d) (see, e.g., [135] for an example on how this is done in practice). By inverting $\mathcal{L}$, i.e., taking inner products with the adjoint eigenfunctions, one then obtains explicitly $\hat{u}$:

$$\hat{u} = -\mathcal{L}^{-1}H_2 = -\sum_{i=1}^{\infty} \frac{1}{\lambda_i} \langle H_2 | \hat{\Phi}_i \rangle \Phi_i.$$

Substituting this result into (C.25a) eliminates $\hat{u}$ from the problem and leads to a partial differential equation for a:

$$\partial_t a = F\left(-\mathcal{L}^{-1} H_2, a, \varepsilon\right) \equiv G(a, \varepsilon).$$

Appendix D
Scalings, Dimensionless Groups and Physical Parameters

D.1 The Viscous-Gravity Scaling

In the falling liquid film the gravitational acceleration g causes flow, and the kinematic viscosity ν (friction) resists the flow. The balance between the two, which gives rise to the Nusselt flat film solution in (2.15a)–(2.15f), can be rendered explicit with the viscous-gravity length and time scales introduced in Sect. 2.2. These scales can be obtained from simple physical considerations without prior knowledge of the specific details of the system. In fact, straightforward dimensional analysis dictates that

$$g \sin \beta \sim \frac{l_\nu}{t_\nu^2} \quad \text{and} \quad \nu \sim \frac{l_\nu^2}{t_\nu},$$

from which l_ν and t_ν, can be readily obtained:

$$l_\nu = \left(\frac{\nu^2}{g \sin \beta} \right)^{1/3} \quad \text{and} \quad t_\nu = \left(\frac{\nu}{(g \sin \beta)^2} \right)^{1/3}.$$

Their ratio gives the characteristic velocity for viscous-gravitational drainage:

$$u_\nu \sim \frac{l_\nu}{t_\nu} \sim \frac{g \sin \beta l_\nu^2}{\nu} \sim (\nu g \sin \beta)^{1/3}.$$

With this definition of u_ν, viscous diffusion in the y direction and gravity in (2.4) balance automatically, i.e., $\mu \partial_{yy} u \sim \rho g \sin \beta$. The pressure scale is selected from balancing the pressure gradient with viscous forces in (2.4), i.e., $\partial_x p \sim \mu \partial_{yy} u$ or

$$p_\nu \sim \frac{\mu u_\nu}{l_\nu} \sim \rho u_\nu^2 \sim \rho (\nu g \sin \beta)^{2/3}.$$

With $\sin \beta \sim 1$ and for water at 25°C, $\nu \approx 10^{-2}$ cm^2 s^{-1}, which yields $l_\nu \approx 4.7 \times 10^{-3}$ cm. The corresponding Nusselt flat film thickness is then always small, e.g., with $Re = 50$, from (2.37) $\bar{h}_{\mathrm{N}} \approx 0.25$ mm.

S. Kalliadasis et al., *Falling Liquid Films*, Applied Mathematical Sciences 176,
DOI 10.1007/978-1-84882-367-9, © Springer-Verlag London Limited 2012

The above scaling will be referred to as the *viscous-gravity scaling* as it expresses the importance of viscous and gravitational forces in our system. This scaling is relevant for inclined plates for which $\sin\beta$ is of order unity and film thickness $\bar{h}_N$ of the order of the length scale l_ν. It allows us to assess the incompressibility assumption in H1, Sect. 2.1. For a film with thickness $\bar{h}_N \sim l_\nu$, the Reynolds number $Re = g\sin\beta\bar{h}_N^3/(3\nu^2)$ is of $\mathcal{O}(1)$ $(l_\nu^3 \sim \nu^2/(g\sin\beta))$ so that the "Grashof number" $Gr = \gamma_T\Delta T/Re$ is small since the thermal expansion coefficient γ_T is usually small for typical fluids (the "Boussinesq approximation"). The Grashof number decreases further as Re increases, i.e., the film thickness $\bar{h}_N$ becomes large compared to l_ν. Hence, buoyancy can be neglected.

We can also assess the neglect of heat production by viscous dissipation in H8, Sect. 2.1. In fact, this is not just a complementary assumption but a direct consequence of a small $\bar{h}_N$ and large constant pressure heat capacity. By comparing the strength of the dissipation function $\Phi_v{}^1$ to the heat transport by the flow leads to the ratio $\nu^2/(c_p\Delta T\bar{h}_N^2) = BqRe^2$ where $Bq = g\bar{h}_N/(c_p\Delta T)$ is the "Boussinesq number." With $\bar{h}_N$ very small and the constant pressure heat capacity of the liquid being generally high, the Boussinesq number is in general very small so that the group $BqRe^2$ has in general a negligible value, even with $Re > 1$.

By introducing the nondimensionalization,

$$(x, y, z) \to l_\nu(x, y, z), \qquad h \to l_\nu h, \tag{D.1a}$$

$$t \to t_\nu t, \qquad (u, v, w) \to u_\nu(u, v, w),$$

$$p \to p_\infty + p_\nu p, \qquad T \to T_\infty + T\Delta T, \tag{D.1b}$$

where $\Delta T = T_w - T_\infty$ for ST and $\Delta T = q_w l_\nu/\lambda$ for HF (see Sect. 2.2), the dimensionless versions of the equations of motion and energy (2.3)–(2.5) and wall and free surface boundary conditions (2.6)–(2.9) and (2.12)–(2.14) contain the parameters in (2.29)–(2.34). The Reynolds number based on u_ν and l_ν, $u_\nu l_\nu/\nu$ does not appear in the equations as it is equal to unity. That is because with the viscous-gravity scaling inertia balances automatically all other terms in (2.4): $\rho\partial_t u \sim \mu\partial_{yy}u$ and $\rho\partial_t u \sim \rho u\partial_x u$.

However, the Reynolds number in (2.35) based on film thickness $\bar{h}_N$ and average velocity $\bar{u}_N$ is effectively hidden in the inlet boundary condition that defines the Nusselt flat film solution (2.15a)–(2.15f). Therefore, the dimensionless equations of motion and energy and wall and free surface boundary conditions obtained with the viscous-gravity scaling are governed by the dimensionless Nusselt flat film

---

[1]The "viscous dissipation function" is defined as [26]

$$\Phi_v = \mu\big\{2\big[(\partial_x u)^2 + (\partial_y v)^2 + (\partial_z w)^2\big]$$

$$+ (\partial_x v + \partial_y u)^2 + (\partial_y w + \partial_z v)^2 + (\partial_z u + \partial_x w)^2\big\}$$

such that the product $(\nu/c_p)\Phi_v$ represents the viscous heating that would have been added to the energy equation (2.5) if significant. Curiously enough, the minimization of ϕ_v leads to the semiparabolic Nusselt flat film solution, as pointed out in the Introduction.

thickness $h_N = \bar{h}_N / l_\nu$ or, equivalently, the Reynolds number, which appears implicitly through h_N, $Re = h_N^3/3$ (see (2.37)), Ct and the four dimensionless groups, Γ, Ma, Pr and Bi for ST and the five dimensionless groups, Γ, Ma, Pr, Bi and Bi_w for HF. Hence, a complete investigation over the entire parameter space would be very cumbersome. However, for a fixed liquid and inclination angle β, the Prandtl and Kapitza numbers are fixed, thus reducing the number of relevant parameters by three. On the other hand, for given properties of the gas–liquid–solid system (physical properties of the gas–liquid system and wall heating conditions (wall temperature/heat flux)) and β, the only free parameter is the Reynolds number (through the inlet condition), which from Sect. 2.2 is the flow control parameter.

In other words, by using the above nondimensionalization, which is based on viscosity and gravity, we have ended up with only one parameter, Re, that depends on h_N, with the remaining parameters, Ct, Γ, Ma, Pr, Bi and Bi_w are all independent of h_N and fixed for a given gas–liquid–solid system and given β. Hence, the viscous-gravity scaling is experimentally quite relevant: In experiments the film thickness is modified by changing the flow rate and therefore for comparisons with experiments it is useful to have only one parameter that depends on h_N.

However, the Nusselt flat film solution can also be taken as the boundary condition $h \to h_N$ far from a local surface deformation like a solitary hump, which also corresponds to the inlet boundary condition as discussed above. Hence, for numerical purposes the formulation of a model in which the film thickness has been scaled out of the boundary conditions and the Nusselt flat film solution is fixed, thus allowing useful numerical comparisons to be made, seems desirable. Therefore, for convenience another scaling is employed based on h_N through the following transformation of the dimensionless variables in (D.1a)–(D.1b):

$$(x, y, z) \to h_N(x, y, z), \qquad h \to h_N h, \qquad t \to \frac{t}{h_N}, \tag{D.2a}$$

$$(u, v, w) \to h_N^2(u, v, w), \qquad p \to h_N p, \tag{D.2b}$$

$$\text{ST:} \quad T \to T, \tag{D.2c}$$

or

$$\text{HF:} \quad T \to h_N T, \tag{D.2d}$$

which converts the boundary condition $h \to h_N$ far from a solitary hump to $h \to 1$. The combination of (D.1a)–(D.1b) and (D.2a)–(D.2d) is precisely the scaling given in (2.16a)–(2.16f). This scaling is based on the Nusselt flat film solution (2.15a)–(2.15f) and is defined as the *Nusselt scaling*. Notice that the numerical factor of 3 appearing along with the Reynolds and Péclet numbers in the dimensionless momentum and energy equations (2.18)–(2.21) is due to the definition of the Reynolds number (2.35) based on the flow rate.

The Nusselt scaling explicitly scales out $\bar{h}_N$ from the full equations of motion and energy and wall and free surface boundary conditions, but as a consequence all governing dimensionless groups, Pe, We, M, B, B_w, and of course Re, depend on the flow rate (through the dimensionless Nusselt flat film thickness h_N).

As an example, assume that we wish to construct numerically a local surface deformation like a solitary hump and examine the influence of the flow rate and the temperature difference between wall and ambient gas phase or wall heat flux (corresponding to the ST or HF cases, respectively) only on this deformation. We then need to fix β and the physical properties of the liquid–gas system, i.e., fix the viscous-gravity parameters Pr, Γ and Bi or Bi_w for ST or HF, respectively. The only free parameters then are Re or, equivalently, h_N and Ma. In the numerical scheme for the construction of the deformation, the Nusselt parameters Pe, We and B are then varied with h_N according to their definitions in (2.38), (2.39) and (2.41) while M is varied with both h_N and Ma according to (2.40a) for ST and (2.40b) for HF. The results of different characteristics of the deformation, e.g., amplitude, speed are then reported as a function of Re or h_N, different Ma and fixed β, Pr, Γ and Bi or Bi_w for ST or HF, respectively, reflecting precisely how the different parameters are input in the numerical scheme.

Hence, although the Nusselt scaling is the most widely used scaling in the literature, in this monograph it is used as in experiments (where it is much easier to fix the gas–liquid–solid system and β), instead of, e.g., varying independently Re and We, as is often the case in the literature.

D.2 On the Orders of Magnitude for the Different Groups in the Boundary Layer Equations

The orders of magnitude assignments for the different dimensionless groups in the derivation of the boundary layer equations in Sect. 4.1 are made for simplicity and in order to fix ideas. These assignments can be relaxed.

For example, let us relax the order of magnitude assignment on Re while the remaining groups have the assignments used in the derivation of the boundary layer equations in Sect. 4.1. For the second-order boundary layer equations then, the y component of the momentum equation (4.2c) shows that in order to neglect the $\mathcal{O}(\varepsilon^2 Re)$ cross-stream inertia terms on the left hand side of this equation compared to the smallest term on the right hand side, i.e., $\varepsilon \partial_{yy} v$, we must have $\varepsilon Re \ll 1$. Hence Re can only increase at a rate slower than $1/\varepsilon$. This also ensures automatically that $Re \ll We$, the condition for cross-stream inertia to be negligible compared to surface tension in the free surface pressure distribution across the film (4.4). At the same time, in order for the streamwise inertia terms in (4.2b) to be kept compared to the neglected $\mathcal{O}(\varepsilon^3)$ terms on the right hand side of this equation we must have $Re \gg \varepsilon^2$.

For the first-order boundary layer equations, εRe can be at most of $\mathcal{O}(1)$, the maximum order on the right hand side of the streamwise momentum balance (4.2b), which automatically satisfies $\varepsilon^2 Re \ll 1$ for cross-stream inertia to be negligible in the y component of the momentum equation (4.2c) and it satisfies $Re \ll We$ for cross-stream inertia to be negligible compared to surface tension in the pressure distribution across the film (4.4). At the same time, for the streamwise inertia terms in (4.2b) to dominate over the neglected $\mathcal{O}(\varepsilon^2)$ terms of the right hand side in the same equation, $Re \gg \varepsilon$.

As an example, let us take the upper bound on Re for the first-order boundary layer equations, $Re \sim 1/\varepsilon$. Assume also that we are not too close to criticality, more specifically, $Re - Re_c = \mathcal{O}(1)$. As pointed out in Sect. 4.1, in this case $k \sim \varepsilon$. For a wavelength of the waves ~ 1 mm, i.e., of the order of the capillary length (see Introduction) and $\bar{h}_N \sim 0.1$ mm, $k \sim 0.1$ so that $\varepsilon \sim 0.1$, which gives $Re \sim 10$, a moderate value.

Let us now relax the order of magnitude assignments for both Re and We in the second-order boundary layer equations while the remaining parameters remain of $\mathcal{O}(1)$. The y component of the momentum equation (4.2c) then shows that in order to neglect cross-stream inertia, $\varepsilon^2 Re \ll \varepsilon \partial_{yy} v$, i.e., $\varepsilon Re \ll 1$. The neglected terms in the pressure distribution (4.4) then are of $\mathcal{O}(\varepsilon^2, \varepsilon^2 Re)$, while for surface tension we require $\varepsilon^2 We$ at most of $\mathcal{O}(1)$. At the same time, for surface tension to dominate over these neglected terms we need $\varepsilon^2 We \gg \varepsilon^2$, or $We \gg 1$, and $\varepsilon^2 We \gg \varepsilon^2 Re$, or simply $Re \ll We$. The pressure distribution (4.4) is then substituted into the x component of the momentum equation (4.2b). Following the differentiation of this distribution once, the contribution of its neglected terms in (4.2b) is of $\mathcal{O}(\varepsilon^3, \varepsilon^3 Re)$. To keep the inertia terms on the left hand side of (4.2b), we need $\varepsilon Re \gg \varepsilon^3$ or $Re \gg \varepsilon^2$ and $\varepsilon Re \gg \varepsilon^3 Re$, which is automatically satisfied. Also, the viscous terms on the right hand side of (4.2b) must be kept compared to the $\mathcal{O}(\varepsilon^3 Re)$ neglected terms in $\varepsilon \partial_x p$, i.e., $\varepsilon^3 Re \ll \varepsilon^2$ or $\varepsilon Re \ll 1$, a condition we already have.

To summarize:

(i) The conditions on Re are $Re \gg \varepsilon^2$, $Re \ll We$ and $Re \ll \varepsilon^{-1}$ or $Re \ll \min\{We, \varepsilon^{-1}\}$.

(ii) The conditions on We are $\varepsilon^2 We$ at most of $\mathcal{O}(1)$, $We \gg 1$.

Assume now that for the first-order boundary layer equations, B, B_w remain of $\mathcal{O}(1)$, but we relax the orders of magnitude assignments for Re, We and M. The more general orders of magnitude assignments for these groups then are εRe at most of $\mathcal{O}(1)$, $Re \gg \varepsilon$, εM at most of $\mathcal{O}(1)$ and $M \gg \varepsilon$, and $\varepsilon^2 We$ at most of $\mathcal{O}(1)$ and $\varepsilon We \gg 1$ (i.e., $We \gg Re$, a condition that has already been utilized in the derivation of the above orders of magnitude assignments). However, the final (first- or second-order) boundary layer equations remain the same, and in fact the neglected terms are of $\mathcal{O}(\epsilon^2)$ for the first-order boundary layer equations and of $\mathcal{O}(\epsilon^3)$ for the second-order ones, as they should be: e.g., for the first-order equations the neglected terms are of $\mathcal{O}(\varepsilon^2, \varepsilon^3 Re, \varepsilon^3 M) \equiv \mathcal{O}(\varepsilon^2)$ since $\varepsilon Re, \varepsilon M$ are at most of $\mathcal{O}(1)$ [207].

In all cases, We must be large, i.e., $We = \mathcal{O}(\varepsilon^{-2})$, $\mathcal{O}(\varepsilon^{-1})$ and $We = \mathcal{O}(\varepsilon^{-2})$, $\mathcal{O}(\varepsilon^{-3/2})$ are possible orders of magnitude assignments for the second- and first-order boundary layer equations, respectively. Large We is an essential requirement for the validity of the boundary layer approximation. This point is discussed in detail in Sect. 4.4.

D.3 Dimensionless Groups and Their Relationships for the ST Case

Reynolds number:

$$Re = \frac{\delta}{3\eta^{1/2}}$$

Inclination number:

$$Ct = \frac{\zeta}{\eta^{1/2}}$$

Kapitza number:

$$\Gamma = \frac{\delta^{2/3}}{\eta^{11/6}} = We(3Re)^{2/3}$$

Marangoni number:

$$Ma = \frac{\mathcal{M}\delta^{2/3}}{\eta^{5/6}} = M(3Re)^{2/3}$$

Biot number:

$$Bi = \frac{B}{(3Re)^{1/3}}$$

Weber number:

$$We = \frac{1}{\eta^{3/2}} = \frac{\Gamma}{(3Re)^{2/3}}$$

Film Marangoni number:

$$M = \frac{\mathcal{M}}{\eta^{1/2}} = \frac{Ma}{(3Re)^{2/3}}$$

Film Biot number:

$$B = Bi\,(3Re)^{1/3}$$

Reduced Reynolds number:

$$\delta = \frac{3Re}{We^{1/3}} = \frac{(3Re)^{11/9}}{\Gamma^{1/3}}$$

Reduced inclination number:

$$\zeta = \frac{Ct}{We^{1/3}} = \frac{Ct(3Re)^{2/9}}{\Gamma^{1/3}}$$

Viscous dispersion number:

$$\eta = \frac{1}{We^{2/3}} = \frac{(3Re)^{4/9}}{\Gamma^{2/3}}$$

Reduced film Marangoni number:

$$\mathcal{M} = \frac{M}{We^{1/3}} = \frac{Ma}{\Gamma^{1/3}(3Re)^{4/9}}$$

D.4 Physical Parameters

Table D.1 shows typical properties and parameter values for different liquids used in experiments [3, 131, 132, 239]. The Marangoni number is calculated with the

Table D.1 Physical properties of different liquids [296] and corresponding dimensionless parameters, with $\Delta T = 1$ K used for Ma and $\alpha = 100$ W m^{-2} K^{-1} used for Bi

Liquid	l_ν (μm)	t_ν (ms)	Γ	Ma	Bi
Water at 20°C	47	2.2	3375	8.9	0.008
Water at 15°C	50	2.3	2950	7.7	0.009
FC–72 at 20°C	26	1.6	1100	9.7	0.045
MD–3F at 30°C	31	1.8	703	5.8	0.047
25% ethyl alcohol at 20°C	87	3.0	500	1.5	0.02

temperature difference $\Delta T = 1$ K and the Biot number with the heat transfer coefficient $\alpha = 100$ W m^{-2} K^{-1}. These are reference values and in practice, typical values would be 10 times larger for Ma, i.e., $\Delta T = 10$ K, and 5 times larger for Bi, i.e., $\alpha = 500$ W m^{-2} K^{-1}.

Appendix E
Model Details

E.1 Dynamical System Corresponding to the Full Second-Order Model

In the moving frame $\xi = x - ct$, the four-equation system (6.78) is transformed into a set of four ordinary differential equations. One corresponds to the mass balance $q' = ch'$, which after one integration gives $q = ch + q_0$. We can then eliminate q from the other three equations. Solving the system of equations for h''', s_1' and s_2' leads to an autonomous five-dimensional *dynamical system* in the phase space spanned by $\mathbf{U} = (U_1, U_2, U_3, U_4, U_5)$ where $U_1 = h$, $U_2 = h'$, $U_3 = h''$, $U_4 = s_1$ and $U_5 = s_2$:

$$U_1' = U_2, \qquad U_2' = U_3, \tag{E.1a}$$

$$
\begin{aligned}
U_3' = {} & 3\frac{q_0}{U_1^3} + 3\frac{c}{U_1^2} - 1 + \frac{1}{U_1^3(c^2 U_1^2 + \frac{152}{11} c q_0 U_1 - \frac{444}{11} q_0^2)} \\
& \times \left\{ \left[\delta\left(\frac{2160}{121} q_0^4 - \frac{720}{11} c q_0^3 U_1 - \frac{3060}{121} c^2 q_0^2 U_1^2 + \frac{540}{121} c^3 q_0 U_1^3 + \frac{580}{363} c^4 U_1^4 \right) \right.\right. \\
& + B U_1^3 \left(c^2 U_1^2 + \frac{152}{11} c q_0 U_1 - \frac{444}{11} q_0^2 \right) \\
& + \delta\left(\frac{10080}{121} q_0^3 + \frac{7056}{121} c q_0^2 U_1 - \frac{24584}{121} c^2 q_0 U_1^2 - \frac{504}{11} c^3 U_1^3 \right) U_4 \\
& \left. + \delta\left(\frac{25920}{121} q_0^3 + \frac{74376}{121} c q_0^2 U_1 + \frac{27336}{121} c^2 q_0 U_1^2 - \frac{324}{11} c^3 U_1^3 \right) U_5 \right] U_2 \\
& + \eta\left(\frac{909}{4} q_0^3 - \frac{1971}{11} c q_0^2 U_1 - \frac{42589}{352} c^2 q_0 U_1^2 - \frac{4525}{176} c^3 U_1^3 \right) U_2^2 \\
& \left. + \eta\left(-\frac{15957}{88} q_0^3 + \frac{7221}{88} c q_0^2 U_1 + \frac{6727}{704} c^2 q_0 U_1^2 - \frac{9071}{2112} c^3 U_1^3 \right) U_1 U_3 \right\},
\end{aligned}
$$

$$\tag{E.1b}$$

S. Kalliadasis et al., *Falling Liquid Films*, Applied Mathematical Sciences 176, DOI 10.1007/978-1-84882-367-9, © Springer-Verlag London Limited 2012

$$
U_4' = \frac{1}{\delta U_1 \left(c^2 U_1^2 + \frac{152}{11} c q_0 U_1 - \frac{444}{11} q_0^2 \right)}
$$

$$
\times \left\{ \left(\frac{13689}{11} q_0 + \frac{949}{11} c U_1 \right) U_4 + \left(\frac{91143}{22} q_0 + \frac{37479}{11} c U_1 \right) U_5 \right.
$$

$$
+ \left[\delta \left(\frac{45942}{4235} q_0^3 + \frac{24986}{4235} c q_0^2 U_1 - \frac{533}{8470} c^2 q_0 U_1^2 - \frac{16003}{25410} c^3 U_1^3 \right) \right.
$$

$$
+ \delta \left(-\frac{50772}{605} q_0^2 + \frac{5850}{121} c q_0 U_1 + \frac{1069}{55} c^2 U_1^2 \right) U_4
$$

$$
+ \delta \left(-\frac{166644}{4235} q_0^2 - \frac{96825}{847} c q_0 U_1 + \frac{5058}{385} c^2 U_1^2 \right) U_5 \bigg] U_2
$$

$$
+ \eta \left(-\frac{905931}{24640} q_0^2 + \frac{436553}{12320} c q_0 U_1 + \frac{35659}{3080} c^2 U_1^2 \right) U_2^2
$$

$$
\left. + \eta \left(-\frac{1647087}{49280} q_0^2 U_1 + \frac{61217}{9856} c q_0 U_1^2 + \frac{50167}{18480} c^2 U_1^3 \right) U_3 \right\}, \tag{E.1c}
$$

$$
U_5' = \frac{1}{\delta U_1 \left(c^2 U_1^2 + \frac{152}{11} c q_0 U_1 - \frac{444}{11} q_0^2 \right)}
$$

$$
\times \left\{ \left(\frac{4186}{11} q_0 + \frac{728}{33} c U_1 \right) U_4 + \left(\frac{4559}{11} q_0 - \frac{1729}{11} c U_1 \right) U_5 \right.
$$

$$
+ \left[\delta \left(\frac{1404}{605} q_0^3 + \frac{312}{605} c q_0^2 U_1 - \frac{13}{605} c^2 q_0 U_1^2 + \frac{416}{5445} c^3 U_1^3 \right) \right.
$$

$$
+ \delta \left(\frac{6552}{605} q_0^2 + \frac{6188}{363} c q_0 U_1 - \frac{182}{165} c^2 U_1^2 \right) U_4
$$

$$
+ \delta \left(-\frac{51528}{605} q_0^2 + \frac{1146}{121} c q_0 U_1 - \frac{94}{55} c^2 U_1^2 \right) U_5 \bigg] U_2
$$

$$
+ \eta \left(\frac{75179}{1760} q_0^2 + \frac{3913}{5280} c q_0 U_1 - \frac{3497}{2640} c^2 U_1^2 \right) U_2^2
$$

$$
\left. + \eta \left(\frac{16783}{3520} q_0^2 U_1 - \frac{611}{528} c q_0 U_1^2 - \frac{11947}{31680} c^2 U_1^3 \right) U_3 \right\}. \tag{E.1d}
$$

This dynamical system becomes singular with $\mathbf{U}'$ not defined at points where the denominator of (E.1b)–(E.1d) vanishes:

$$
\delta U_1 \left(c^2 U_1^2 + \frac{152}{11} c q_0 U_1 - \frac{444}{11} q_0^2 \right) = 0. \tag{E.2}
$$

$U_1 = 0$ corresponds to the onset of dry patches on the inclined plate. However, physically the formation of dry patches requires forces of nonhydrodynamic origin, such as long-range attractive intermolecular interactions that are not considered in this monograph. The touch down with $U_1 = 0$ of a trajectory in the phase plane therefore is a nonphysical solution. The other two roots of (E.2) are $U_{\text{sing}\pm} = \frac{2}{11} q_0(-38 \pm \sqrt{2665})/c$. At least one of these roots is positive. The presence of the singular planes $U_1 = U_{\text{sing}\pm}$ in the phase space is a sign of the complexity of the full second-order model and does not result from any actual physical limitations. It is rather a direct consequence of the projection of the velocity field on a small set of polynomials.

On the other hand, the three-dimensional dynamical system (7.42) does not have any denominators and the above difficulty is avoided. This is due to the simplicity of the corresponding models. Notice, however, that the choice of only one test function does not necessarily sidestep the onset of singular planes in the phase space. The formulation adopted by Lee and Mei [161] by retaining cross-stream inertial terms while using the assumption of a self-similar parabolic velocity profile also led to the presence of singular planes in a three-dimensional phase plane.

By setting q_0 to $1/3 - c$ as in (7.39), the two fixed points of the flow (E.1a)–(E.1d) verify $U_2 = U_3 = U_4 = U_5 = 0$ and (7.44). To simplify notations, the two fixed points are denoted with the same symbols used for the fixed points $\mathbf{U}_{\text{I}}$ and $\mathbf{U}_{\text{II}}$ of the three-dimensional system (7.42). Heteroclinic orbits must connect the two fixed points without encountering one of the singular planes. For $c > 1/3$, $q_0 < 0$ and only $U_{\text{sing}-} = \frac{2}{11}(38 + \sqrt{2665})(1 - 1/(3c))$ is positive. $U_{\text{sing}-} = 1$ admits a root

$$c_- = \frac{2}{3} \frac{38 + \sqrt{2665}}{65 + 2\sqrt{2665}} \approx 0.355127.$$

Similarly, $U_{\text{sing}-} = h_{\text{II}}$ leads to $c = 1/3$ or to the second-order equation,

$$\frac{16436}{1089} + \frac{304\sqrt{2665}}{1089} - \left(\frac{17272}{363} + \frac{326\sqrt{2665}}{363} \right) c + c^2 = 0,$$

which admits two roots, $c_- \approx 0.315117$ and $c_+ \approx 93.6279$. The first value is lower than the limiting value $1/3$, below which the second fixed point $\mathbf{U}_{\text{II}}$ vanishes. The second one is much larger than the usual speed of observed waves. The limiting speeds c_- and c_+ correspond to extreme positions of the fixed point $\mathbf{U}_{\text{II}}$, either very close to the origin, $\mathbf{U}_{\text{II}} \approx 0.0615879$, or very far from it, $\mathbf{U}_{\text{II}} \approx 16.2372$. Therefore, the condition

$$c_- < c < c_+ \tag{E.3}$$

is not restrictive and does not limit the exploration of the pertinent solutions (limit cycles, homoclinic and heteroclinic orbits) to the dynamical system.

The position of the fixed points influences the orbits in the phase space. Homoclinic orbits connecting $\mathbf{U}_{\text{I}}$ to itself spiral around $\mathbf{U}_{\text{II}}$ such that the maximum amplitudes of $U_2, \ldots, U_5$ increase with the distance separating the two fixed points. The two values $1 - c_-$ and $c_+ - 1$ can be viewed as upper limits on the distance between the two fixed points, which in turn correspond to upper limits on the wave amplitude.

E.2 Three-Dimensional Full Second-Order Model

Following the weighted residuals methodology detailed in Chap. 6, the velocity field is projected onto the polynomials F_0, F_1 and F_2 defined in (6.74) and repeated below for convenience:

$$F_0 = \bar{y} - \frac{1}{2}\bar{y}^2 \tag{E.4a}$$

$$F_1 = \bar{y} - \frac{17}{6}\bar{y}^2 + \frac{7}{3}\bar{y}^3 - \frac{7}{12}\bar{y}^4 \tag{E.4b}$$

$$F_2 = \bar{y} - \frac{13}{2}\bar{y}^2 + \frac{57}{4}\bar{y}^3 - \frac{111}{8}\bar{y}^4 + \frac{99}{16}\bar{y}^5 - \frac{33}{32}\bar{y}^6. \tag{E.4c}$$

The streamwise and spanwise velocity distributions thus read

$$u = \frac{3}{h}(q_\| - s_1 - s_2)g_0(\bar{y}) + 45\frac{s_1}{h}g_1(\bar{y}) + 210\frac{s_2}{h}g_2(\bar{y}) \tag{E.5a}$$

$$w = \frac{3}{h}(p - r_1 - r_2)g_0(\bar{y}) + 45\frac{r_1}{h}g_1(\bar{y}) + 210\frac{r_2}{h}g_2(\bar{y}), \tag{E.5b}$$

where $\bar{y} = y/h$ and the streamwise and spanwise flow rates $q_\| = \int_0^h u\,dy$ and $q_\perp = \int_0^h w\,dy$, respectively, appear with two corrections each, namely s_1, s_2 and r_1, r_2.

Applying the Galerkin method, which consists of integrating the boundary layer equations (4.5a), (4.5b) across the film, substituting the projections (E.5a)–(E.5b) into the integrated equations, taking the test functions (E.4a)–(E.4c) as weight functions and using the boundary conditions (4.2f), (4.2k), (4.2l) yields the full second-order model for three-dimensional flows. Let us define two fictitious parameters, $\epsilon_x \equiv 1$ and $\epsilon_z \equiv 0$. They will be used as "tracers" to identify in-plane gravity terms that are equal to zero for the momentum equations in the transverse z direction. The evolution equations for $q_\|$, s_1 and s_2 read

$$\delta\partial_t q_\| = \epsilon_x \frac{27}{28}h - \frac{81}{28}\frac{q_\|}{h^2} - 33\frac{s_1}{h^2} - \frac{3069}{28}\frac{s_2}{h^2} + \delta\left(-\frac{12}{5}\frac{q_\| s_1 \partial_x h}{h^2} - \frac{126}{65}\frac{q_\| s_2 \partial_x h}{h^2}\right.$$

$$+ \frac{12}{5}\frac{s_1 \partial_x q_\|}{h} + \frac{171}{65}\frac{s_2 \partial_x q_\|}{h} + \frac{12}{5}\frac{q_\| \partial_x s_1}{h} + \frac{1017}{455}\frac{q_\| \partial_x s_2}{h} + \frac{6}{5}\frac{q_\|^2 \partial_x h}{h^2}$$

$$- \frac{12}{5}\frac{q_\| \partial_x q_\|}{h} - \frac{6}{5}\frac{q_\| \partial_z q_\perp}{h} - \frac{6}{5}\frac{q_\perp \partial_z q_\|}{h} + \frac{6}{5}\frac{q_\| q_\perp \partial_z h}{h^2} - \frac{6}{5}\frac{q_\| r_1 \partial_z h}{h^2}$$

$$- \frac{63}{65}\frac{q_\| r_2 \partial_z h}{h^2} - \frac{6}{5}\frac{q_\perp s_1 \partial_z h}{h^2} - \frac{63}{65}\frac{q_\perp s_2 \partial_z h}{h^2} + \frac{6}{5}\frac{s_1 \partial_z q_\perp}{h} + \frac{108}{65}\frac{s_2 \partial_z q_\perp}{h}$$

$$+ \frac{6}{5}\frac{r_1 \partial_z q_\|}{h} + \frac{63}{65}\frac{r_2 \partial_z q_\|}{h} + \frac{6}{5}\frac{q_\| \partial_z r_1}{h} + \frac{576}{455}\frac{q_\| \partial_z r_2}{h} + \frac{6}{5}\frac{q_\perp \partial_z s_1}{h}$$

$$\left.+ \frac{63}{65}\frac{q_\perp \partial_z s_2}{h}\right) + \eta\left(\frac{5025}{896}\frac{q_\|(\partial_x h)^2}{h^2} - \frac{5055}{896}\frac{\partial_x q_\| \partial_x h}{h} - \frac{10851}{1792}\frac{q_\| \partial_{xx} h}{h}\right.$$

$$+ \frac{2027}{448}\partial_{xx}q_\parallel + \partial_{zz}q_\parallel - \frac{2463}{1792}\frac{\partial_z q_\parallel \partial_z h}{h} + \frac{2433}{1792}\frac{q_\parallel(\partial_z h)^2}{h^2} - \frac{5361}{3584}\frac{q_\parallel \partial_{zz}h}{h}$$

$$+ \frac{7617}{1792}\frac{q_\perp \partial_x h \partial_z h}{h^2} - \frac{4749}{3584}\frac{\partial_z q_\perp \partial_x h}{h} - \frac{10545}{3584}\frac{\partial_x q_\perp \partial_z h}{h} - \frac{16341}{3584}\frac{q_\perp \partial_{xz}h}{h}$$

$$+ \frac{1579}{448}\partial_{xz}q_\perp \Bigg) - \frac{27}{28}\zeta h\,\partial_x h + \frac{27}{28}h(\partial_{xxx} + \partial_{xzz})h \tag{E.6a}$$

$$\delta\partial_t s_1 = \epsilon_x \frac{1}{10}h - \frac{3}{10}\frac{q_\parallel}{h^2} - \frac{126}{5}\frac{s_1}{h^2} - \frac{126}{5}\frac{s_2}{h^2} + \delta\Bigg(\frac{1}{35}\frac{q_\parallel \partial_x q_\parallel}{h} - \frac{3}{35}\frac{q_\parallel^2 \partial_x h}{h^2}$$

$$+ \frac{108}{55}\frac{q_\parallel s_1 \partial_x h}{h^2} - \frac{5022}{5005}\frac{q_\parallel s_2 \partial_x h}{h^2} - \frac{103}{55}\frac{s_1 \partial_x q_\parallel}{h} + \frac{9657}{5005}\frac{s_2 \partial_x q_\parallel}{h}$$

$$- \frac{39}{55}\frac{q_\parallel \partial_x s_1}{h} + \frac{10557}{10010}\frac{q_\parallel \partial_x s_2}{h} - \frac{2}{35}\frac{q_\parallel \partial_z q_\perp}{h} + \frac{3}{35}\frac{q_\perp \partial_z q_\parallel}{h}$$

$$- \frac{3}{35}\frac{q_\parallel q_\perp \partial_z h}{h^2} + \frac{54}{55}\frac{q_\parallel r_1 \partial_z h}{h^2} + \frac{54}{55}\frac{q_\perp s_1 \partial_z h}{h^2} - \frac{54}{55}\frac{r_1 \partial_z q_\parallel}{h}$$

$$- \frac{54}{55}\frac{q_\perp \partial_z s_1}{h} - \frac{2511}{5005}\frac{q_\perp s_2 \partial_z h}{h^2} - \frac{2511}{5005}\frac{q_\parallel r_2 \partial_z h}{h^2} + \frac{2511}{5005}\frac{r_2 \partial_z q_\parallel}{h}$$

$$+ \frac{2511}{5005}\frac{q_\perp \partial_z s_2}{h} - \frac{49}{55}\frac{s_1 \partial_z q_\perp}{h} + \frac{7146}{5005}\frac{s_2 \partial_z q_\perp}{h} + \frac{3}{11}\frac{q_\parallel \partial_z r_1}{h}$$

$$+ \frac{1107}{2002}\frac{q_\parallel \partial_z r_2}{h}\Bigg) + \eta\Bigg(\frac{93}{40}\frac{q_\parallel(\partial_x h)^2}{h^2} - \frac{69}{40}\frac{\partial_x h \partial_x q_\parallel}{h} + \frac{21}{80}\frac{q_\parallel \partial_{xx}h}{h}$$

$$- \frac{9}{40}\partial_{xx}q_\parallel - \frac{57}{80}\frac{\partial_z q_\parallel \partial_z h}{h} + \frac{81}{80}\frac{q_\parallel(\partial_z h)^2}{h^2} - \frac{3}{40}\frac{q_\parallel \partial_{zz}h}{h} + \frac{27}{80}\frac{q_\perp \partial_{xz}h}{h}$$

$$+ \frac{21}{16}\frac{q_\perp \partial_x h \partial_z h}{h^2} - \frac{63}{80}\frac{\partial_z q_\perp \partial_x h}{h} - \frac{9}{40}\frac{\partial_z h \partial_x q_\perp}{h} - \frac{9}{40}\partial_{xz}q_\perp\Bigg)$$

$$- \frac{1}{10}\zeta h \partial_x h + \frac{1}{10}h(\partial_{xxx} + \partial_{xzz})h \tag{E.6b}$$

$$\delta\partial_t s_2 = \epsilon_x \frac{13}{420}h - \frac{13}{140}\frac{q_\parallel}{h^2} - \frac{39}{5}\frac{s_1}{h^2} - \frac{11817}{140}\frac{s_2}{h^2} + \delta\Bigg(-\frac{4}{11}\frac{q_\parallel s_1 \partial_x h}{h^2} + \frac{18}{11}\frac{q_\parallel s_2 \partial_x h}{h^2}$$

$$- \frac{2}{33}\frac{s_1 \partial_x q_\parallel}{h} - \frac{19}{11}\frac{s_2 \partial_x q_\parallel}{h} + \frac{6}{55}\frac{q_\parallel \partial_x s_1}{h} - \frac{288}{385}\frac{q_\parallel \partial_x s_2}{h} - \frac{2}{11}\frac{q_\parallel r_1 \partial_z h}{h^2}$$

$$- \frac{2}{11}\frac{q_\perp s_1 \partial_z h}{h^2} + \frac{2}{11}\frac{r_1 \partial_z q_\parallel}{h} + \frac{2}{11}\frac{q_\perp \partial_z s_1}{h} + \frac{9}{11}\frac{q_\parallel r_2 \partial_z h}{h^2} + \frac{9}{11}\frac{q_\perp s_2 \partial_z h}{h^2}$$

$$- \frac{9}{11}\frac{r_2 \partial_z q_\parallel}{h} - \frac{9}{11}\frac{q_\perp \partial_z s_2}{h} - \frac{8}{33}\frac{s_1 \partial_z q_\perp}{h} - \frac{10}{11}\frac{s_2 \partial_z q_\perp}{h} - \frac{4}{55}\frac{q_\parallel \partial_z r_1}{h}$$

$$+ \frac{27}{385} \frac{q_{\parallel} \partial_z r_2}{h} \bigg) + \eta \bigg(-\frac{3211}{4480} \frac{q_{\parallel}(\partial_x h)^2}{h^2} + \frac{2613}{4480} \frac{\partial_x h \partial_x q_{\parallel}}{h} - \frac{2847}{8960} \frac{q_{\parallel}\partial_{xx}h}{h}$$

$$+ \frac{559}{2240}\partial_{xx}q_{\parallel} + \frac{3029}{8960} \frac{\partial_z q_{\parallel}\partial_z h}{h} - \frac{3627}{8960} \frac{q_{\parallel}(\partial_z h)^2}{h^2} + \frac{299}{17920} \frac{q_{\parallel}\partial_{zz}h}{h}$$

$$- \frac{559}{1792} \frac{q_{\perp}\partial_x h \partial_z h}{h^2} + \frac{4927}{17920} \frac{\partial_z q_{\perp}\partial_x h}{h} - \frac{533}{17920} \frac{\partial_x q_{\perp}\partial_z h}{h} - \frac{5993}{17920} \frac{q_{\perp}\partial_{xz}h}{h}$$

$$+ \frac{559}{2240}\partial_{xz}q_{\perp} \bigg) - \frac{13}{420}\zeta h \partial_x h + \frac{13}{420}h(\partial_{xxx} + \partial_{xzz})h. \tag{E.6c}$$

The equations for $q_{\perp}$, r_1 and r_2 are obtained from (E.6a)–(E.6c) through the exchanges $\{x \leftrightarrow z,\; q_{\parallel} \leftrightarrow q_{\perp},\; s_{1,2} \leftrightarrow r_{1,2},\; \epsilon_x \leftrightarrow \epsilon_z\}$ (hence the introduction of the "tracers" $\epsilon_{x,z}$ reduces to a minimum the set of equations to be written). The set of equations is completed by the mass conservation $\partial_t h = -\partial_x q_{\parallel} - \partial_z q_{\perp}$.

E.3 Three-Dimensional Regularized Second-Order Model

$$\partial_t h = -\partial_x q_{\parallel} - \partial_z q_{\perp} \tag{E.7a}$$

$$\delta \partial_t q_{\parallel} = \delta \left[\frac{9}{7}\frac{q_{\parallel}^2}{h^2}\partial_x h - \frac{17}{7}\frac{q_{\parallel}}{h}\partial_x q_{\parallel} \right] + \left[\frac{5}{6}h - \frac{5}{2}\frac{q_{\parallel}}{h^2} \right.$$

$$+ \delta\left(-\frac{8}{7}\frac{q_{\parallel}\partial_z q_{\perp}}{h} - \frac{9}{7}\frac{q_{\perp}\partial_z q_{\parallel}}{h} + \frac{9}{7}\frac{q_{\parallel}q_{\perp}\partial_z h}{h^2} \right)$$

$$+ \eta\left(4\frac{q_{\parallel}(\partial_x h)^2}{h^2} - \frac{9}{2}\frac{\partial_x q_{\parallel}\partial_x h}{h} - 6\frac{q_{\parallel}\partial_{xx}h}{h} + \frac{9}{2}\partial_{xx}q_{\parallel} \right.$$

$$+ \frac{13}{4}\frac{q_{\perp}\partial_x h \partial_z h}{h^2} - \frac{\partial_z q_{\parallel}\partial_z h}{h} - \frac{43}{16}\frac{\partial_x q_{\perp}\partial_z h}{h} - \frac{13}{16}\frac{\partial_z q_{\perp}\partial_x h}{h}$$

$$+ \frac{3}{4}\frac{q_{\parallel}(\partial_z h)^2}{h^2} - \frac{23}{16}\frac{q_{\parallel}\partial_{zz}h}{h} - \frac{73}{16}\frac{q_{\perp}\partial_{xz}h}{h} + \partial_{zz}q_{\parallel} + \frac{7}{2}\partial_{xz}q_{\perp} \bigg)$$

$$\left. - \frac{5}{6}\zeta h\, \partial_x h + \frac{5}{6}h(\partial_{xxx} + \partial_{xzz})h \right]\left(1 - \frac{\delta}{70}q_{\parallel}\partial_x h \right)^{-1} \tag{E.7b}$$

$$\delta \partial_t q_{\perp} = \delta\left(\frac{9}{7}\frac{q_{\perp}^2}{h^2}\partial_z h - \frac{17}{7}\frac{q_{\perp}}{h}\partial_z q_{\perp} \right) - \frac{5}{2}\frac{q_{\perp}}{h^2}$$

$$+ \delta\left(-\frac{8}{7}\frac{q_{\perp}\partial_x q_{\parallel}}{h} - \frac{9}{7}\frac{q_{\parallel}\partial_x q_{\perp}}{h} + \frac{9}{7}\frac{q_{\parallel}q_{\perp}\partial_x h}{h^2} \right)$$

$$+ \eta\left(4\frac{q_{\perp}(\partial_z h)^2}{h^2} - \frac{9}{2}\frac{\partial_z q_{\perp}\partial_z h}{h} - 6\frac{q_{\perp}\partial_{zz}h}{h} + \frac{9}{2}\partial_{zz}q_{\perp} \right.$$

$$+ \frac{13}{4} \frac{q_\| \partial_x h \partial_z h}{h^2} - \frac{\partial_x q_\perp \partial_x h}{h} - \frac{43}{16} \frac{\partial_z q_\| \partial_x h}{h} - \frac{13}{16} \frac{\partial_x q_\| \partial_z h}{h}$$

$$+ \frac{3}{4} \frac{q_\perp (\partial_x h)^2}{h^2} - \frac{23}{16} \frac{q_\perp \partial_{xx} h}{h} - \frac{73}{16} \frac{q_\| \partial_{xz} h}{h} + \partial_{xx} q_\perp + \frac{7}{2} \partial_{xz} q_\| \Big)$$

$$- \frac{5}{6} \zeta h \partial_z h + \frac{5}{6} h (\partial_{xxz} + \partial_{zzz}) h. \tag{E.7c}$$

E.4 Full Second-Order Model for the ST Case

$$\partial_t h = -\partial_x q \tag{E.8a}$$

$$\delta \partial_t q = \frac{30}{31} h - \frac{90}{31} \frac{q}{h^2} - \frac{1050}{31} \frac{s_1}{h^2} - \frac{3690}{31} \frac{s_2}{h^2} - \frac{9066}{31} \frac{s_3}{h^2}$$

$$+ \delta \Big(\frac{6}{5} \frac{q^2 \partial_x h}{h^2} - \frac{12}{5} \frac{q \partial_x q}{h} - \frac{12}{5} \frac{q s_1 \partial_x h}{h^2} - \frac{4248}{2015} \frac{q s_2 \partial_x h}{h^2}$$

$$- \frac{5296}{3875} \frac{q s_3 \partial_x h}{h^2} + \frac{12}{5} \frac{s_1 \partial_x q}{h} + \frac{1026}{403} \frac{s_2 \partial_x q}{h} + \frac{11722}{3875} \frac{s_3 \partial_x q}{h}$$

$$+ \frac{12}{5} \frac{q \partial_x s_1}{h} + \frac{4626}{2015} \frac{q \partial_x s_2}{h} + \frac{1538}{775} \frac{q \partial_x s_3}{h} \Big) - \frac{213}{248} \mathcal{M} \partial_x \theta$$

$$+ \eta \Big(\frac{1569}{248} \frac{q (\partial_x h)^2}{h^2} - \frac{1569}{248} \frac{\partial_x q \partial_x h}{h} - \frac{2847}{496} \frac{q \partial_{xx} h}{h} + \frac{1069}{248} \partial_{xx} q \Big)$$

$$- \frac{30}{31} \zeta h \partial_x h + \frac{30}{31} h \partial_{xxx} h \tag{E.8b}$$

$$\delta \partial_t s_1 = \frac{1}{10} h - \frac{3}{10} \frac{q}{h^2} - \frac{126}{5} \frac{s_1}{h^2} - \frac{126}{5} \frac{s_2}{h^2} - \frac{126}{5} \frac{s_3}{h^2}$$

$$+ \delta \Big(-\frac{3}{35} \frac{q^2 \partial_x h}{h^2} + \frac{1}{35} \frac{q \partial_x q}{h} + \frac{108}{55} \frac{q s_1 \partial_x h}{h^2} - \frac{5022}{5005} \frac{q s_2 \partial_x h}{h^2}$$

$$+ \frac{6}{35} \frac{q s_3 \partial_x h}{h^2} - \frac{103}{55} \frac{s_1 \partial_x q}{h} + \frac{9657}{5005} \frac{s_2 \partial_x q}{h} - \frac{1}{35} \frac{s_3 \partial_x q}{h}$$

$$- \frac{39}{55} \frac{q \partial_x s_1}{h} + \frac{10557}{10010} \frac{q \partial_x s_2}{h} + \frac{19}{70} \frac{q \partial_x s_3}{h} \Big) + \frac{3}{8} \mathcal{M} \partial_x \theta$$

$$+ \eta \Big(\frac{93}{40} \frac{q (\partial_x h)^2}{h^2} - \frac{69}{40} \frac{\partial_x h \partial_x q}{h} + \frac{21}{80} \frac{q \partial_{xx} h}{h} - \frac{9}{40} \partial_{xx} q \Big)$$

$$- \frac{1}{10} \zeta h \partial_x h + \frac{1}{10} h \partial_{xxx} h \tag{E.8c}$$

$$
\begin{aligned}
\delta\partial_t s_2 = {}& \frac{13}{420}h - \frac{13}{140}\frac{q}{h^2} - \frac{39}{5}\frac{s_1}{h^2} - \frac{11817}{140}\frac{s_2}{h^2} - \frac{11817}{140}\frac{s_3}{h^2} \\
& + \delta\left(-\frac{4}{11}\frac{qs_1\partial_x h}{h^2} + \frac{18}{11}\frac{qs_2\partial_x h}{h^2} - \frac{38}{25}\frac{qs_3\partial_x h}{h^2} - \frac{2}{33}\frac{s_1\partial_x q}{h}\right. \\
& \quad - \frac{19}{11}\frac{s_2\partial_x q}{h} + \frac{76}{25}\frac{s_3\partial_x q}{h} + \frac{6}{55}\frac{q\partial_x s_1}{h} - \frac{288}{385}\frac{q\partial_x s_2}{h} \\
& \quad \left. + \frac{73}{70}\frac{q\partial_x s_3}{h}\right) - \frac{13}{64}\mathcal{M}\partial_x\theta \\
& + \eta\left(-\frac{3211}{4480}\frac{q(\partial_x h)^2}{h^2} + \frac{2613}{4480}\frac{\partial_x h\,\partial_x q}{h} - \frac{2847}{8960}\frac{q\partial_{xx}h}{h} + \frac{559}{2240}\partial_{xx}q\right) \\
& - \frac{13}{420}\zeta h\partial_x h + \frac{13}{420}h\partial_{xxx}h
\end{aligned}
\tag{E.8d}
$$

$$
\begin{aligned}
\delta\partial_t s_3 = {}& \frac{8}{868}h - \frac{9}{868}\frac{q}{h^2} - \frac{27}{31}\frac{s_1}{h^2} - \frac{8181}{868}\frac{s_2}{h^2} - \frac{158709}{868}\frac{s_3}{h^2} \\
& + \delta\left(-\frac{342}{2015}\frac{qs_2\partial_x h}{h^2} + \frac{9894}{3875}\frac{qs_3\partial_x h}{h^2} - \frac{171}{2015}\frac{s_2\partial_x q}{h} - \frac{9358}{3875}\frac{s_3\partial_x q}{h}\right. \\
& \quad \left. + \frac{171}{2821}\frac{q\partial_x s_2}{h} - \frac{13653}{10850}\frac{q\partial_x s_3}{h}\right) + \frac{435}{1984}\mathcal{M}\partial_x\theta \\
& + \eta\left(+\frac{19953}{27776}\frac{q(\partial_x h)^2}{h^2} - \frac{19023}{27776}\frac{\partial_x h\,\partial_x q}{h} + \frac{17517}{55552}\frac{q\partial_{xx}h}{h}\right. \\
& \quad \left. - \frac{2973}{13888}\partial_{xx}q\right) - \frac{3}{868}\zeta h\partial_x h + \frac{33}{868}h\partial_{xxx}h
\end{aligned}
\tag{E.8e}
$$

$$
\begin{aligned}
Pr\delta\partial_t\theta = {}& \frac{30009}{1273}\left(\frac{1-(1+Bh)\theta}{h^2}\right) - \frac{130950}{1273}\frac{t_1}{h^2} - \frac{1384362}{6365}\frac{t_2}{h^2} - \frac{1863792}{6365}\frac{t_3}{h^2} \\
& + Pr\delta\left(-\frac{3117}{203680}\frac{(1-\theta)\partial_x q}{h} - \frac{155877}{101840}\frac{q\partial_x\theta}{h} - \frac{3117}{81472}\frac{t_1\partial_x q}{h}\right. \\
& \quad + \frac{801101}{4480960}\frac{t_2\partial_x q}{h} - \frac{1768473}{1323920}\frac{t_3\partial_x q}{h} + \frac{3117}{40736}\frac{q\partial_x t_1}{h} - \frac{364701}{2240480}\frac{q\partial_x t_2}{h} \\
& \quad + \frac{7840671}{14563120}\frac{q\partial_x t_3}{h} - \frac{21819}{101840}\frac{(1-\theta)\partial_x s_1}{h} - \frac{10066647}{21182720}\frac{(1-\theta)\partial_x s_2}{h} \\
& \quad - \frac{949089}{8147200}\frac{(1-\theta)\partial_x s_3}{h} + \frac{245511}{50920}\frac{s_1\partial_x\theta}{h} - \frac{44102157}{10591360}\frac{s_2\partial_x\theta}{h} \\
& \quad \left. + \frac{295506139}{4073600}\frac{s_3\partial_x\theta}{h}\right) + \eta\left(\frac{27463}{1273}\frac{(1-\theta)(\partial_x h)^2}{h^2} - \frac{30009}{2546}B\frac{\theta(\partial_x h)^2}{h}\right. \\
& \quad \left. + \frac{27463}{1273}\frac{\partial_x h\,\partial_x\theta}{h} + \frac{(1-\theta)\partial_{xx}h}{h} + \partial_{xx}\theta\right)
\end{aligned}
\tag{E.8f}
$$

$$Pr\delta\partial_t t_1 = \frac{8176}{1273}\left(\frac{1-(1+Bh)\theta}{h^2}\right) - \frac{40880}{1273}\frac{t_1}{h^2} - \frac{34874}{1273}\frac{t_2}{h^2} - \frac{28784}{1273}\frac{t_3}{h^2}$$

$$+ Pr\delta\left(-\frac{97063}{407360}\frac{(1-\theta)\partial_x q}{h} - \frac{56327}{203680}\frac{q\partial_x\theta}{h} - \frac{166435}{488832}\frac{t_1\partial_x q}{h}\right.$$

$$-\frac{20340413}{26885760}\frac{t_2\partial_x q}{h} - \frac{28729997}{174757440}\frac{t_3\partial_x q}{h} - \frac{225649}{244416}\frac{q\partial_x t_1}{h}$$

$$+\frac{222059}{4480960}\frac{q\partial_x t_2}{h} + \frac{16530119}{43689360}\frac{q\partial_x t_3}{h} + \frac{1849169}{1222080}\frac{(1-\theta)\partial_x s_1}{h}$$

$$+\frac{70164667}{42365440}\frac{(1-\theta)\partial_x s_2}{h} - \frac{11651371}{9776640}\frac{(1-\theta)\partial_x s_3}{h} + \frac{294637}{76380}\frac{s_1\partial_x\theta}{h}$$

$$-\frac{122474147}{21182720}\frac{s_2\partial_x\theta}{h} + \frac{88699283}{4888320}\frac{s_3\partial_x\theta}{h}\right) + \eta\left(\frac{8176}{1273}\frac{\partial_x h\partial_x\theta}{h}\right.$$

$$+\frac{8176}{1273}\frac{(1-\theta)(\partial_x h)^2}{h^2} - \frac{4088}{1273}B\frac{\theta(\partial_x h)^2}{h}\right) \tag{E.8g}$$

$$Pr\delta\partial_t t_2 = \frac{44838}{6365}\left(\frac{1-(1+Bh)\theta}{h^2}\right) - \frac{44838}{1273}\frac{t_1}{h^2} - \frac{3231144}{31825}\frac{t_2}{h^2} - \frac{2306304}{31825}\frac{t_3}{h^2}$$

$$+ Pr\delta\left(\frac{158337}{2036800}\frac{(1-\theta)\partial_x q}{h} + \frac{158337}{1018400}\frac{q\partial_x\theta}{h} + \frac{2674149}{5703040}\frac{t_1\partial_x q}{h}\right.$$

$$+\frac{13678523}{313667200}\frac{t_2\partial_x q}{h} - \frac{446443197}{509709200}\frac{t_3\partial_x q}{h} - \frac{1222929}{2851520}\frac{q\partial_x t_1}{h}$$

$$-\frac{204549663}{156833600}\frac{q\partial_x t_2}{h} - \frac{336704427}{1019418400}\frac{q\partial_x t_3}{h} + \frac{97173}{2036800}\frac{(1-\theta)\partial_x s_1}{h}$$

$$-\frac{490225113}{211827200}\frac{(1-\theta)\partial_x s_2}{h} - \frac{64692369}{81472000}\frac{(1-\theta)\partial_x s_3}{h} - \frac{352833}{254600}\frac{s_1\partial_x\theta}{h}$$

$$-\frac{302607783}{105913600}\frac{s_2\partial_x\theta}{h} + \frac{1304283921}{40736000}\frac{s_3\partial_x\theta}{h}\right) + \eta\left(\frac{44838}{6365}\frac{\partial_x h\partial_x\theta}{h}\right.$$

$$+\frac{44838}{6365}\frac{(1-\theta)(\partial_x h)^2}{h^2} - \frac{22419}{6365}B\frac{\theta(\partial_x h)^2}{h}\right) \tag{E.8h}$$

$$Pr\delta\partial_t t_3 = \frac{45232}{6365}\left(\frac{1-(1+Bh)\theta}{h^2}\right) - \frac{45232}{1273}\frac{t_1}{h^2} - \frac{2818816}{31825}\frac{t_2}{h^2} - \frac{6293056}{31825}\frac{t_3}{h^2}$$

$$+ Pr\delta\left(-\frac{7579}{254600}\frac{(1-\theta)\partial_x q}{h} - \frac{7579}{127300}\frac{q\partial_x\theta}{h} - \frac{260999}{2138640}\frac{t_1\partial_x q}{h}\right.$$

$$+\frac{6804757}{10693200}\frac{t_2\partial_x q}{h} - \frac{62613409}{139011600}\frac{t_3\partial_x q}{h} + \frac{210079}{1069320}\frac{q\partial_x t_1}{h}$$

$$-\frac{555439}{1782200}\frac{q\partial_x t_2}{h} - \frac{58821361}{69505800}\frac{q\partial_x t_3}{h} - \frac{133699}{381900}\frac{(1-\theta)\partial_x s_1}{h}$$

$$
\begin{aligned}
&+ \frac{8441321}{26478400}\frac{(1-\theta)\partial_x s_2}{h} + \frac{121848419}{30552000}\frac{(1-\theta)\partial_x s_3}{h} - \frac{95509}{190950}\frac{s_1\partial_x\theta}{h} \\
&+ \frac{55645211}{13239200}\frac{s_2\partial_x\theta}{h} + \frac{284237129}{15276000}\frac{s_3\partial_x\theta}{h}\Bigg) + \eta\Bigg(\frac{45232}{6365}\frac{\partial_x h\partial_x\theta}{h} \\
&+ \frac{45232}{6365}\frac{(1-\theta)(\partial_x h)^2}{h^2} - \frac{22616}{6365}B\frac{\theta(\partial_x h)^2}{h}\Bigg).
\end{aligned}
\tag{E.8i}
$$

E.5 Second-Order Inertia Corrections to the Regularized Model (9.33a), (9.33b) for the ST Case

$$
\begin{aligned}
\mathcal{K} = &\ \frac{1}{210}h^2\partial_{tt}q + \frac{17}{630}hq\,\partial_{xt}q - \frac{1}{105}q\,\partial_x h\partial_t q + \frac{1}{42}h\partial_x q\,\partial_t q - \frac{26}{231}\frac{q^2\partial_x h\partial_x q}{h} \\
&+ \frac{653}{8085}q(\partial_x q)^2 + \frac{386}{8085}q^2\partial_{xx}q + \frac{104}{2695}\frac{q^3(\partial_x h)^2}{h^2} - \frac{78}{2695}\frac{q^3\partial_{xx}h}{h}
\end{aligned}
\tag{E.9a}
$$

$$
\mathcal{K}^M = \frac{5}{112}q\,\partial_x h\partial_x\theta + \frac{19}{336}h\partial_x q\,\partial_x\theta + \frac{1}{48}h^2\partial_{xt}\theta + \frac{15}{224}hq\,\partial_{xx}\theta
\tag{E.9b}
$$

$$
\begin{aligned}
\mathcal{K}_{\theta q} = &\ -\frac{19}{1400}\big[(1-\theta)\partial_x h - h\partial_x\theta\big]\partial_t q - \frac{19}{2800}(1-\theta)\partial_{xt}q \\
&+ \frac{47}{4800}(1-\theta)\frac{\partial_x h\partial_x q}{h} - \frac{613}{33600}(1-\theta)\big[(\partial_x q)^2 + q\,\partial_{xx}q\big] \\
&- \frac{157}{1600}\frac{q^2\partial_x h\partial_x\theta}{h} + \frac{613}{16800}q\,\partial_x q\,\partial_x\theta + \frac{157}{11200}\frac{(1-\theta)q^2\partial_{xx}h}{h}
\end{aligned}
\tag{E.9c}
$$

$$
\begin{aligned}
\mathcal{K}_\theta = &\ \frac{1}{15}h^2\partial_{tt}\theta + \frac{23}{140}(qh\partial_{xt}\theta + q\,\partial_x h\partial_t\theta) + \frac{23}{280}h\partial_t q\,\partial_x\theta - \frac{33}{280}h\partial_x q\,\partial_t\theta \\
&- \frac{31}{1680}(1-\theta)h\partial_{xt}q - \frac{491}{22400}(1-\theta)\Bigg(\frac{\partial_x h\partial_x q}{h} + q\,\partial_{xx}q\Bigg) \\
&+ \frac{1391}{67200}(1-\theta)(\partial_x q)^2 + \frac{573}{5600}\Bigg(\frac{q^2\partial_x h\partial_x\theta}{h} + q^2\partial_{xx}\theta\Bigg) \\
&+ \frac{113}{2800}q\,\partial_x q\,\partial_x\theta
\end{aligned}
\tag{E.9d}
$$

$$
\mathcal{K}_\theta^M = \frac{3}{40}h(\partial_x\theta)^2 - \frac{3}{40}(1-\theta)\partial_x h\partial_x\theta - \frac{3}{80}(1-\theta)h\partial_{xx}\theta.
\tag{E.9e}
$$

E.6 Weighted Residuals Modeling for the HF Case

Up to now, the weighted residual approach has been applied for the problem of a uniformly heated film corresponding to the ST condition. Here we develop the weighted residuals modeling for the HF condition which takes into account heat losses from the wall to the gas phase in contact with the wall. The condition is given in (2.23b) and is rewritten here for clarity:

$$\partial_y T|_0 = -1 + B_{\rm w} T|_0. \tag{E.10}$$

As emphasized in Sect. 9.4, at zeroth order the heat transfer and mechanical equilibrium of the film are decoupled from each other. The coupling between the two appears at first order through the presence of interfacial deformations. Moreover, the zeroth-order formulation of the surface temperature (5.4b)

$$\theta^{(0)} = \frac{1}{B + B_{\rm w}(1 + Bh)} \tag{E.11}$$

yields the same formulation of the second-order terms $\mathcal{K}^M$ (recall that these terms are induced in the momentum equation by the Marangoni flow produced by the gradient of temperature at the film surface) as obtained in (9.40) for ST. Therefore, the momentum equation (9.43b) of the reduced model will remain unaltered with HF and we hence focus only on the energy equation in what follows. In addition, the Galerkin averaging procedure for the energy equation is overall similar to the one presented in Sect. 9.2, but with some modifications.

Much like with the ST case, we wish to have the film surface temperature θ in the formulation of the averaged model. Let us then rewrite the linear zeroth-order temperature profile across the film (5.3f) in terms of the surface temperature θ:

$$T^{(0)} = \theta + \mathbb{F}(h - y) \quad \text{where } \mathbb{F} = \frac{1 - B_{\rm w}\theta}{1 + B_{\rm w}h}. \tag{E.12}$$

The *effective heat flux* $\mathbb{F}(x, t)$ at the wall decreases with the intensity $B_{\rm w}$ of the heat losses from the liquid to the wall and with the increase of the film surface temperature θ. To satisfy the boundary condition (E.10), we write the temperature field as

$$T(x, y, t) = -\mathbb{F}(x, t)\, y + \sum_{i=0}^{i_{\max}} b_i(x, t) g_i\left(\frac{y}{h(x, t)}\right), \tag{E.13}$$

where $g_0 = 1$ corresponds to the base state, and the set of test functions is completed with $g_i(\bar{y}) = \bar{y}^{i+1}$, $i \le 1$, to obtain the polynomial basis for the projection.

E.6.1 Formulation at First-Order

Turning to the weighted residuals for the energy equation (9.7b) and with the same arguments as in Sect. 9.2, the unknowns b_i, $i \ge 1$, may only play a role through the integral, $\int_0^h w_j \partial_{yy} T$. With two integrations by parts and making use of the boundary condition at the surface (9.6f) and the heat flux condition at the wall (E.10), we obtain

$$\int_0^h w_j\left(\frac{y}{h}\right)\partial_{yy}T\,dy = -Bw_j(1)T|_h + w_j(0)(1 - B_{\mathrm{w}}T|_0)$$

$$+ \frac{1}{h}\left(w_j{}'(0)T|_0 - w_j{}'(1)T|_h\right)$$

$$+ \frac{1}{h^2}\int_0^h w_j{}''\left(\frac{y}{h}\right)T\,dy. \tag{E.14}$$

In order to put the emphasis on $\theta \equiv T|_h$, we choose for the first weight function $w'_{i_{\max}}(0) = 0$, $w''_{i_{\max}} = 0$, so that $w_{i_{\max}} \propto 1 = g_0$. It is next appropriate to replace the physically meaningless unknown b_0 by θ through the substitution

$$b_0 = \mathbb{F}h + \theta - \sum_{i=1}^{i_{\max}} b_i. \tag{E.15}$$

The evaluation of the first residual (9.7b) corresponding to $w_{i_{\max}} = g_0 = 1$ then yields

$$3\varepsilon Pe\left[\mathbb{F}\partial_t h + \frac{1}{2}h\partial_t\mathbb{F} + \partial_t\theta + \frac{3}{8}\left(\frac{\mathbb{F}q\partial_x h}{h} + \mathbb{F}\partial_x q + q\partial_x\mathbb{F}\right) + \frac{q\partial_x\theta}{h}\right]$$

$$- \frac{(\mathbb{F} - B\theta)}{h} = 0. \tag{E.16}$$

Substituting now $\mathbb{F} = (1 - B_{\mathrm{w}}\theta)/(1 + B_{\mathrm{w}}h)$ and using the kinematic equivalence, $\partial_t h = -\partial_x q$, leads to the following equation for θ:

$$3\varepsilon Pe\partial_t\theta = \left\{\frac{(1 - B_{\mathrm{w}}\theta)}{h(1 + B_{\mathrm{w}}h)} - \frac{B\theta}{h} + 3\varepsilon Pe\left[\frac{(5 + B_{\mathrm{w}}h)(1 - B_{\mathrm{w}}\theta)}{8(1 + B_{\mathrm{w}}h)^2}\partial_x q\right.\right.$$

$$\left.\left. - \frac{3}{8}\frac{(1 - B_{\mathrm{w}}\theta)}{(1 + B_{\mathrm{w}}h)^2}\frac{q}{h}\partial_x h - \left(1 - \frac{3}{8}\frac{B_{\mathrm{w}}h}{(1 + B_{\mathrm{w}}h)}\right)\frac{q}{h}\partial_x\theta\right]\right\}$$

$$\times \left(1 - \frac{B_{\mathrm{w}}h}{2 + 2B_{\mathrm{w}}h}\right)^{-1}, \tag{E.17}$$

where the unknowns b_i do not appear. This equation is consistent at $\mathcal{O}(\varepsilon)$ and can be substituted for (9.17c) into (9.17a)–(9.17c) to get the first-order model for the heat flux condition; the model consists of three coupled evolution equations for h, q and θ.

E.6.2 Formulation at Second-Order

As in Sect. 9.3, we extend here the first-order formulation to take into account the second-order thermal effects. For this purpose, we need the explicit expressions for the amplitudes b_j of the projection at first order. This is done by eliminating the coefficients of the polynomial obtained by substituting the temperature field (E.13) and the velocity field (9.2a) into the second-order energy equation (4.6b). Since the amplitudes a_i are known from (9.18a)–(9.18e) and b_0 from (E.15), and the b_i, $i \geq 1$, are at least of $\mathcal{O}(\varepsilon)$, the coefficients of the above polynomial provide the required expressions of the b_i as functions of h, θ, q and their derivatives:

$$b_1 = \frac{3}{2}\varepsilon Peh^2(h\partial_t \mathbb{F} + \partial_t \theta - \mathbb{F}\partial_x q) \tag{E.18a}$$

$$b_2 = -\frac{1}{2}\varepsilon Peh\left[h^2\partial_t \mathbb{F} - 3q(\mathbb{F}\partial_x h + h\partial_x \mathbb{F} + \partial_x \theta)\right] \tag{E.18b}$$

$$b_3 = -\frac{3}{8}\varepsilon Peh\left[-h\mathbb{F}\partial_x q + q(3\mathbb{F}\partial_x h + 3h\partial_x \mathbb{F} + \partial_x \theta)\right] \tag{E.18c}$$

$$b_4 = \frac{3}{40}\varepsilon Peh\left[-h\mathbb{F}\partial_x q + 3q(\mathbb{F}\partial_x h + h\partial_x \mathbb{F})\right] \tag{E.18d}$$

$$b_i = 0, \quad i \geq 5. \tag{E.18e}$$

In contrast to the ST case, here the amplitude b_1 is nonzero. Therefore, the temperature T at first order is a combination of five independent fields, namely, θ, b_1, b_2, b_3 and b_4. As a consequence, a consistent formulation of a model for the dynamics of the flow at second order would require 14 unknowns, instead of 13 for the ST case. However, rather than solving 14 equations, let us use the same approach as for the ST case and construct a set of orthogonal test functions for the temperature field from linear combinations of g_0, g_1, g_2, g_3 and g_4 such that $G_0 \equiv g_0$:

$$G_0 = 1 \tag{E.19a}$$

$$G_1 = 1 - 3\bar{y}^2 \tag{E.19b}$$

$$G_2 = 1 - 15\bar{y}^2 + 16\bar{y}^3 \tag{E.19c}$$

$$G_3 = 1 - 45\bar{y}^2 + 112\bar{y}^3 - 70\bar{y}^4 \tag{E.19d}$$

$$G_4 = 1 - 105\bar{y}^2 + 448\bar{y}^3 - 630\bar{y}^4 + 288\bar{y}^5. \tag{E.19e}$$

Therefore, the temperature field can be accurately described at $\mathcal{O}(\varepsilon)$ from

$$T = -\mathbb{F}y + (\mathbb{F}h + \theta - t_1 - t_2 - t_3 - t_4)G_0 + \frac{1}{2}\sum_{i=1}^{4}(-1)^i t_i G_i. \tag{E.20}$$

The set of test functions G_i must be completed at second order with 10 polynomials of degree up to 14. Nevertheless, since G_i'', $0 \leq i \leq 4$, are not linear combinations of G_i, $0 \leq i \leq 4$, the five first residuals do not form a closed set of equations for θ, t_1, t_2, t_3 and t_4. Yet, a basis for the set of polynomials of degree up to five satisfying the HF condition can be obtained by introducing only one polynomial orthogonal to the first four G_i. This polynomial, G_5, is

$$G_5(\bar{y}) = 1 - \frac{70}{3}\bar{y} + 140\bar{y}^2 - 336\bar{y}^3 + 350\bar{y}^4 - 132\bar{y}^5. \qquad \text{(E.21)}$$

The temperature field can now be written explicitly at second order as

$$T = -\mathbb{F}y + (\mathbb{F}h + \theta - t_1 - t_2 - t_3 - t_4)G_0(\bar{y}) - \frac{1}{2}t_1 G_1(\bar{y})$$

$$+ \frac{1}{2}t_2 G_2(\bar{y}) - \frac{1}{2}t_3 G_3(\bar{y}) + \frac{1}{2}\left(t_4 - \sum_{i=6}^{9} t_i\right)G_4(\bar{y})$$

$$- 3t_5 G_5(\bar{y}) + \sum_{i=6}^{9} t_i \frac{G_i(\bar{y})}{G_i(1)}. \qquad \text{(E.22)}$$

The choice of this formulation ensures that the evaluation of $\int_0^h G_j''(\bar{y})T\,dy$, $0 \le j \le 5$, does not require the definitions of G_j, $j \ge 6$. By applying next the Galerkin method to the energy equation we find that the first six residuals $\mathcal{R}_\theta(G_i)$, $0 \le i \le 5$, constitute a closed set. Since the amplitude t_5 is of $\mathcal{O}(\varepsilon^2)$, its space and time derivatives can be neglected at this order, so that an explicit formulation in terms of h, θ, t_1, t_2, t_3 and t_4 can be obtained, expressing the slaving of the former to the latter. We can then derive a set of five evolution equations for θ, t_1, t_2, t_3, t_4 that couple with the five other evolution equations (E.8a)–(E.8e) to provide the full second-order model, a system of 10 equations with 10 unknowns.

However, we will take a shortcut here based on considerations already developed in the ST case (see Sect. 9.4.3). In fact, the aim is once again to obtain a three-unknown regularized model for h, q and θ that remains asymptotically correct up to $\mathcal{O}(\varepsilon^2)$ with the long-wave theory. Yet, as temperature is coupled through its gradient in the momentum equation, the second-order terms in the energy equation do not enter the second-order gradient expansion. Further, recall that it is not possible to take into account the second-order corrections appearing in the averaged energy equation—which are induced by the deviations of the temperature and velocity profiles from the Nusselt flat film solution—if the temperature field is assumed to be slaved to the free surface temperature θ only. Hence, we restrict ourselves here to the second-order averaged energy equation obtained from the first residual, $\mathcal{R}_\theta(G_0)$ with $G_0 = 1$, as we did for the first order. The result written in terms of the Shkadov scaling is

$$\frac{\mathbb{F} - B\theta}{h} - Pr\delta\left[\frac{1}{2}h\partial_t\mathbb{F} + \partial_t\theta + \frac{3}{8}q\left(\partial_x\mathbb{F} + \mathbb{F}\frac{\partial_x h}{h}\right)\right.$$

$$- \frac{5}{8}\mathbb{F}\partial_x q + \frac{q\partial_x\theta}{h}\right] + \eta\left[2\partial_x\mathbb{F}\partial_x h + \left(\mathbb{F} - \frac{1}{2}B\theta\right)\frac{(\partial_x h)^2}{h}\right.$$

$$+ \frac{\partial_x h\partial_x\theta}{h} + \frac{1}{2}h\partial_{xx}\mathbb{F} + \mathbb{F}\partial_{xx}h + \partial_{xx}\theta\right] = 0, \qquad \text{(E.23)}$$

where the second bracketed expression contains the second-order thermal effects. Equation (E.23) should be coupled with the continuity and the momentum equations (9.43a), (9.43b). This three-equation system constitutes the *regularized model* for the HF case.

Appendix F
Numerical Schemes

F.1 Solving the Orr–Sommerfeld Equation by Continuation

Solving the Orr–Sommerfeld eigenvalue problem is not a straightforward task. Numerical schemes for its solution were proposed as early as 1964 [298] for the problem of a falling film with surfactants (the Orr–Sommerfeld problem for the purely hydrodynamic case was treated for the first time in [10]). The different schemes are based, for example, on the "shooting method" (see, e.g., [67] or [259]), "pseudo-spectral methods" (see, e.g., [44, 202]) or "finite differences." In the last two cases the aim is to discretize the (infinite-dimension) differential eigenvalue problem appropriately and thus convert it to a (finite-dimension) matrix eigenvalue problem, whereas, in the shooting method one looks for the parameter values for which the integration of the equation from one side of the domain (the wall) satisfies the boundary conditions at the other side (the free surface).

Here we present an alternative approach based on *continuation*, which demands a minimum of code writing due to a freely distributed software. The basic idea is the following: Suppose one trivial solution of the problem at hand is known and that this solution is not isolated in the parameter space but lies on a continuous branch of solutions, i.e., a continuous distribution of solutions as a function of a single parameter (a "codim 1 manifold"). One may then construct the whole branch of solutions in small steps, starting from the known trivial solution. An introduction to continuation methods can be found in the monograph by Allgower and Georg [8]. Based on the initial work by Keller [146], accurate continuation algorithms have been developed by Doedel et al. [80, 81] and implemented in the software package AUTO-07P, which can be downloaded from http://indy.cs.concordia.ca. The software is designed as a collection of subroutines that enables the solution of bifurcation problems for ordinary differential equations. It can be installed on most operating systems, including Windows and Linux. For details, we refer the reader to the documentation of the software. Here we provide the basic steps to using the software and how to implement it for the numerical solution of the Orr–Sommerfeld eigenvalue problem.

The user gives a name to the problem to be solved, say xxx, and defines it in the file xxx.f. The numerical constants of each run are specified in separate files `c.xxx.1`, `c.xxx.2` and so on. The computations are initiated by the command `@r`, by typing in a unix shell `@r xxx 1`, for example. The software then compiles the program using the routines defined in `xxx.f`, reads the numerical constants stored in `c.xxx` and the initial solution stored in the file `q.1` produced by a previous run.[1]

The results of the computation are stored using `@sv`. They can be plotted with the software PLAUT run by the command `@p`. Details can be found in the user manual, which contains numerous useful examples, i.e., a variety of dynamical systems, such as reaction-diffusion equations [79].

As an example, let us consider the search for the neutral stability curve of streamwise perturbations in terms of the wavenumber k versus angular frequency ω for an isothermal film falling down an inclined plate. We then need to solve numerically (3.22a) with boundary conditions (3.22c), (3.22g) and (3.22h), where η is given by (3.22f) and M is set to zero.

As the Orr–Sommerfeld system is linear, a constraint/normalization condition on the amplitude must be added. Here we choose the integral constraint

$$\int_0^1 \varphi\,dy = 1/3, \tag{F.1}$$

which is compatible with the normalization of φ_0 in (3.24). In practice, the search for the neutral stability conditions is facilitated by the long-wave nature of the instability. The onset of instability corresponds to a zero wavenumber k and the neutral stability curve emerges from the "trivial solution":

$$\varphi = y^2, \qquad k = 0, \qquad c = 1. \tag{F.2}$$

We therefore proceed in two steps.

Step 1: We start from the trivial solution (F.2) and follow the horizontal axis $\omega = k = 0$ until the critical value of the Reynolds number is detected as a bifurcation point.

Step 2: We restart the computation from $Re = Re_c$ and construct the neutral stability curve. The process is called `stab` and the three necessary files, `stab.f`, `c.stab.1` and `c.stab.2` are in the AUTO source code discussed later.

The first step is implemented with the following sequence of commands:

```
cp c.stab.1 c.stab
@r stab
@sv 1
```

During the computation, AUTO summarizes the results on the screen:

---

[1]The latest distribution AUTO-07P, released in 2007, has a user-friendly command line interface based on the "Python language," as well as a graphical user interface (GUI). Yet, AUTO-07P, like older versions of AUTO, can still be used effectively with the help of the standard Unix commands. Therefore, we leave to the reader the choice of using the Python and GUI interfaces or not and present only the standard Unix procedure to which most current users of AUTO are accustomed.

```
  BR      PT   TY  LAB      PAR(1)                L2-NORM          MAX U(1)
  MAX U(2)        MAX U(3)         MAX U(4)        PAR(2)           PAR(3)
   1       1  EP    1    1.00000E-01    2.42212E+00    1.00000E+00
0.00000E+00    2.00000E+00    0.00000E+00    0.00000E+00    1.00000E+00
   1      15  BP    2    1.19172E+01    2.42212E+00    1.00000E+00
0.00000E+00    2.00000E+00    0.00000E+00    0.00000E+00    1.00000E+00
   1     100  EP    3    9.69172E+01    2.42212E+00    1.00000E+00
0.00000E+00    2.00000E+00    0.00000E+00    0.00000E+00    1.00000E+00
```

The switch from the trivial solution to the neutral stability boundary is enabled by
setting the constant ISW to -1 in c.stab.2. The results are stored in files b.1,
s.1 and d.1. The critical point is detected as a "branch point" (indicated by BP in
the screen outputs) at $Re = 11.91722$, in agreement with the theoretically predicted
value of the critical Reynolds number, $Re_c = 5/6\cot(4\pi/180)$. This point is labeled
as 2 and gives the starting conditions for the second step. Notice that the detection of
the (bifurcation) point 2 is enabled by setting the constant ISP to 2 in the constant
file c.stab.1.

The second step is achieved with the following commands:

```
cp c.stab.2 c.stab
@r stab 1
@sv 2
```

The computation is stopped when Re reaches the value 70 (last row of the constant
file c.stab.2). The results are then stored in files b.2, s.2 and d.2, and can
be visualized with the command @p 2 using the software PLAUT which opens a
Tektronix window. Next, typing the following commands in the PLAUT window

```
ax
1 8
```

select the Reynolds number and the frequency (first and eighth columns of the file
b.2). Finally, the neutral stability curve is plotted and saved in a file 2.fig using
the commands:

```
bd0
sav
2.fig
```

To convert this file into a postscript one, we can use the command @ps 2.fig,
which creates the file 2.fig.ps. The result is displayed in Fig. F.1 for a particular
set of parameters ($\beta = 4°$, $\nu = 2.3\,10^{-6}\mathrm{m}^2\,\mathrm{s}^{-1}$, $\sigma/\rho = 62.6 \times 10^{-6}\ \mathrm{m}^3\,\mathrm{s}^{-2}$) corresponding to an experiment by Liu et al. [170] and to Fig. 7.2. Notice that AUTO
draws the neutral curve for $\omega < 0$. As the curve is symmetric around $\omega = k = 0$, its
part for positive values of ω can be easily recovered through $\omega \to -\omega$, $k \to -k$.

F.1.1 AUTO *Source Code*

The files necessary for the above example on the use of the software AUTO can be
downloaded from extras.springer.com (search for the book by its ISBN, you will

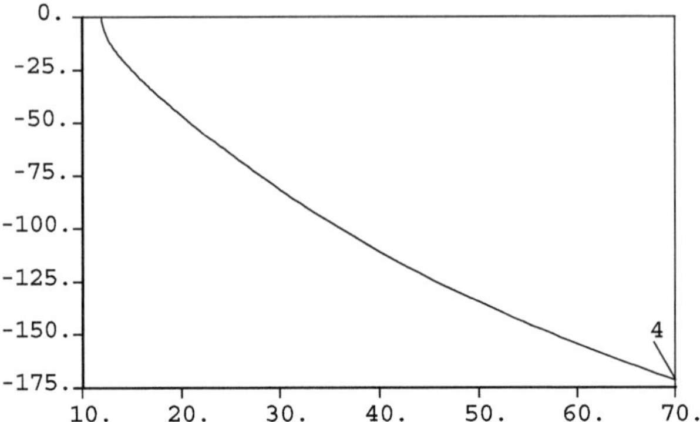

Fig. F.1 Neutral stability curve as plotted by the software `Plaut` in the (Re, ω)-plane. Notice that the second run of the computation, stored in `b.2`, `s.2` and `d.2` follows the first run stored in `b.1`, `s.1` and `d.1`. File `s.1` already contains three solutions labeled 1, 2 and 3, so that the fourth one, stored in `s.1`, is labeled 4 and corresponds to the solution at the end of the computation

then be asked to enter a password, which is given on the copyright page of this print book). They are written in the "old" Fortran style and are therefore compatible with the current release AUTO-07P (and of course with older versions). File `stab.f` contains the Orr–Sommerfeld equation (3.22a) in the format of an autonomous complex dynamical system of dimension 5 (subroutine FUNC), the boundary conditions (3.22c), (3.22g) and (3.22h) (subroutine BCND), the integral condition (F.1), and the trivial solution (F.2) (subroutine STPNT).

There are three "active" continuation parameters in the problem the way it is formulated: Re, k and c, corresponding to the parameters PAR(1), PAR(2) and PAR(3), respectively (subroutine FUNC). There are two more active parameters that are not changed in the runs, Ct and Γ, whose values are set once and for all in STPNT. Since the frequency ω is not a true continuation parameter in the chosen formulation based on the wavenumber k but a combination of the two parameters k and c, the dimensional frequency ω is denoted as PAR(6) in a separate subroutine PVLS that defines specific "solution measures" (non-free parameters). Of course, the value of the dimensionless frequency can be obtained as a combination of k and c, but it is useful to store the dimensional frequency in a user-defined parameter (UZR) in AUTO such that it explicitly appears in the output AUTO files and there is no need for additional post-processing.

F.2 Computational Search for Traveling Wave Solutions and Their Bifurcations

To investigate the behavior of the solutions of the BE, we employ bifurcation analysis using numerical *continuation* techniques [79]. Continuation is a very effective method for determining branches of stationary solutions and their bifurcations, following them in the parameter space using the Newton iteration method. In the context of thin films, continuation was applied in studies of traveling and solitary waves of falling films [52, 227, 228, 239], sliding drops on slightly inclined plates [268, 269], and transverse instabilities of sliding liquid ridges [266].

We seek traveling wave solutions, i.e., stationary solutions of (5.13) in a frame of reference moving downstream at constant speed c. We present the computational methodology for the BE in terms of the Nusselt scaling to illustrate how some of the figures in Chap. 5, prior to the introduction of the Shkadov scaling, are obtained. Introducing $h(x, t) = h(\xi)$ with $\xi = x - ct$, the BE (5.13), truncated at first order for simplicity and dropping ε, can be integrated once to yield

$$-ch + \frac{h^3}{3} - q_0 + \frac{2}{5}Reh^6h' + We\frac{h^3}{3}h''' - Ct\frac{h^3}{3}h' + \frac{h^2}{2}\frac{BMh'}{(1+Bh)^2} = 0, \quad \text{(F.3)}$$

where the primes denote differentiation with respect to ξ; q_0 is the integration constant and represents the flow rate in the moving frame of reference (see (5.51)). Its value is negative because the phase speed c of surface waves is generally higher than the mean velocity of the film. Assuming that no dry spots are possible, i.e., $h \neq 0$, (F.3) can be divided by $-We\,h^3/3$ to get

$$h''' = F[h, h'] = \frac{1}{We}\left[\frac{3}{h^3}(q_0 + ch) - 1 - \frac{6}{5}Reh^3h' + Cth' - \frac{3}{2}\frac{BMh'}{h(1+Bh)^2}\right]. \quad \text{(F.4)}$$

The differential equation (F.4) is recast into a *dynamical system*, as follows:

$$\begin{cases} U_1' = U_2 \\ U_2' = U_3 \\ U_3' = F[U_1, U_2], \end{cases} \quad \text{(F.5)}$$

where $U_1 = h$, $U_2 = h'$ and $U_3 = h''$. The dimension of the dynamical system ($= 3$) is fixed by the third-order surface tension term, which makes the system (F.5) applicable to the majority of the equations for the film thickness discussed in this monograph.

To determine iteratively the periodic solutions of the dynamical system (F.5), we use the continuation and bifurcation tools for ordinary differential equations in the software AUTO-07P. During the computations the periodicity of the solution is enforced, the "phase" of the wave is fixed by $U_1|_{\xi=0} = 1$ (which fixes the origin) and the total volume $\int_0^\lambda U_1\,d\xi = \langle h \rangle_\xi$—with $\lambda = 2\pi/k$—is controlled as specified by the flow condition, open or closed (see Sect. 5.3.1; both boundary conditions have been treated even though for the specific example given in the source code the closed-flow condition is enforced). This amounts to one integral and four boundary conditions, hence the continuation requires three free parameters [146]. By specifying the set of viscous-gravity parameters $\{Re, \Gamma, Ma, Bi, Ct\}$, the remaining free parameters are $\{k, c, q_0\}$. The continuation is started from the neutral mode at criticality corresponding to the *Hopf bifurcation* point with k_c from (5.18a) and

c from (5.16). The starting value of q_0 is fixed by the Nusselt flat film solution $h(\xi) = 1$ such that from (F.3), $q_0 = -2/3$.

Notice that we have avoided specifying the set of Nusselt parameters $\{Re, We, M, B, Ct\}$ (see, e.g., Appendix D.1 and Sect. 4.10). Indeed, the advantage of working with the viscous-gravity parameters is that all these parameters, apart from Re, are independent on the flow rate, which is usually the principal control parameter in experiments.

We now give the necessary steps to compute the results in Figs. 5.2, 5.4 and 5.5, namely how to follow a branch of stationary solutions from the *Hopf bifurcation* point, how to detect and follow a *period-doubling bifurcation* and, finally, how to trace the locus of *saddle-node bifurcation* points in the parameter space.

F.2.1 Hopf Bifurcation

We compute here the γ_2-family of traveling wave solutions of (F.3) using the equation file be.f and the constants files c.be that define the different constants. The command is

```
@r be
```

Once this is executed, one can visualize the results by launching the PLAUT program with the command

```
@p
```

In the PLAUT environment, entering successively the commands

```
ax
1 3
d1bd0
```

gives the γ_2 branch as plotted in Fig. 5.2a for $h_{\max}$ versus k. Now, entering successively

```
2d
a
```

gives solutions for different values of k as specified in the constants file c.be. The labels 2–5 correspond to the wave profiles 1–4 plotted in Fig. 5.2b. Exiting the PLAUT environment using the "quit" command, one can save the γ_2-family with the command

```
@sv g2
```

F.2.2 Period-Doubling Bifurcation

Let us now compute the γ_2-family of solutions for the first harmonic. To do so one has to set $f = 2$, the "harmonic parameter" ($\equiv n$, the number of waves in the wavetrain required to compute the period-doubling bifurcation—see Sect. 5.3.2) in the solution subroutine STPNT() of the file be.f. Once changed and saved, the file must be run with another constant file, c.be.HP2, where the detection of the bifurcation point is enabled (ISP $= 2$), the number of mesh points is doubled (NTST $= 100$), the tolerance parameters (EP*) are slightly decreased and the initial continuation stepsize is adjusted (DS $= -1.e{-}06$). The commands then to run and save are

```
@r be HP2
@sv be
```

AUTO-07P finds several bifurcation points (BP), the first one of which corresponds to the period-doubling bifurcation point for the γ_1-family of traveling waves. We shall then compute the corresponding branch using the third constants file, c.be.PD, where the label of the starting bifurcation point is specified (IRS $= 2$), the branch switching is enabled (ISW $= -1$) and the direction of the continuation is changed (by changing the sign of DS). The commands then to run and save are

```
@r be PD
@sv g1
```

One can finally append the two families of solutions and plot them as follows

```
@ap g1 g12
@ap g2 g12
@p g12
```

In the PLAUT environment, entering as before the commands

```
ax
1 3
d0bd0
```

shows the two families of solution for $h_{\max}$ versus k, while entering

```
ax
1 7
d0bd0
```

shows the two families of solution for c versus k, exactly as presented in Fig. 5.2a. As an exercise, one can try to compute other bifurcating families by changing the value of the BP-starting solution in the constants file c.be.PD, i.e., IRS $= 3$–6, then following the same procedure as above to run, save, append and plot. Once finished and before going to the next section, delete the bifurcation, solution and diagnostic files, respectively b.be, s.be and d.be, with the following command,

```
@dl be
```

and set back to $f = 1$ the harmonic parameter in be.f.

F.2.3 Locus of Saddle-Node Bifurcation Points

The different branches of solutions reported in Fig. 5.4 can be reproduced by tuning the Kapitza number to $\Gamma = 2950$ in the equation file be.f and varying the value of the Reynolds number. Let us trace here the branch for $Re = 3$:

```
@r be
@sv be
```

Because the detection of fold is enabled (ILP $= 1$) in the file c.be, a turning point (see the asterisk in Fig. 5.4) is found (LP) and recorded with the label 6. We shall then track the locus of this turning point in the parameter space. The constants file c.be.TP0—with the turning point as initial solution (IRS $= 6$), the Reynolds number as an additional continuation parameter (PAR(3)), the continuation of fold enabled (ISW $= 2$) and the tolerance parameters (EP*) decreased—is first used to generate starting data

```
@r be TP0
@ap be
```

The fold continuation can then be performed using the constants file c.be.TP starting with the last solution (IRS $= 11$), and where the iteration parameters are increased (ITMX $= 10$, ITNW $= 8$) and the initial continuation stepsize is adjusted (DS $= 0.1$):

```
@r be TP
@ap be
```

Launching the PLAUT environment with @p and typing the commands

```
ax
7 1
d0bd0
```

shows the blow up boundary as plotted in Fig. 5.5 (dashed line).

F.2.4 AUTO Source Code

The Fortran files necessary for the above example on the use of the software AUTO-07P can be downloaded from extras.springer.com (search for the book by its ISBN, you will then be asked to enter a password, which is given on the copyright page of this print book). These are: be.f that contains subroutines FUNC, STPNT, BCND, ICND and all the constants files.

F.3 Time-Dependent Computations Using Finite Differences

We briefly present here the key points of the algorithm implemented to simulate the spatio-temporal evolution of the film based on models of reduced dimensionality, i.e., one evolution equation for h, e.g., the BE (5.12), two coupled evolution equations for h, q, e.g., the Kapitza–Shkadov model (6.13a), (6.13b) and the regularized model (6.1), (6.92), or more than two as, e.g., the three-equation model formulated in [226]. The reader interested in more details of finite difference schemes can refer to one of the many available textbooks, e.g., [220].

In all cases, the equations to be solved can be recast in the form

$$\partial_t \mathbf{H} = \mathcal{L}(\mathbf{H}) + \mathcal{N}(\mathbf{H}), \tag{F.6a}$$

where $\mathbf{H}$ denotes the set of unknowns. Therefore, in the case of the Kapitza–Shkadov model or the more refined two-equation models, $\mathbf{H} = (h, q)$. $\mathcal{L}$ is a linear matrix-differential operator and $\mathcal{N}$ is a nonlinear functional of $\mathbf{H}$ and its spatial derivatives. The spatial evolution of the film is then determined by solving system (F.6a) in a semi-infinite domain with initial conditions

$$\mathbf{H}(x, 0) = \mathbf{H}_0(x) \tag{F.6b}$$

and boundary conditions

$$\mathbf{H}(0, t) = \mathbf{H}_f(t), \qquad \mathbf{H}(\infty, t) = \mathbf{H}_0(\infty) = \text{const.} \tag{F.6c}$$

appropriate for a semi-infinite domain, where $\mathbf{H}_f$ is a function of time corresponding to the forcing at the inlet. Notice that for a semi-infinite domain the boundary condition at infinity will not affect the spatial evolution, provided that it is compatible with the initial condition, more specifically that the limit of the initial condition at infinity coincides with the boundary condition there, as is indeed the case with (F.6a)–(F.6c).

Because boundary conditions (F.6c) are not periodic, a spectral method for the numerical integration of (F.6a)–(F.6c) cannot be used. Instead, a finite-difference scheme is employed.

The computational domain of size L is discretized into a regular grid of N points or "nodes," $x_j = j\Delta x;\ j = 1, 2, \ldots, N$, $\Delta x = L/N$. The discretized variables at the node (x_j, t_n) are denoted by $\mathbf{H}_j^{(n)} = (h_j^{(n)}\ q_j^{(n)}, \ldots)$. Notice that the necessary limitation in size of the actual semi-infinite domain demands the introduction of additional boundary conditions at the downstream limit $x = L$ that do not generate parasitic reflections upstream. We shall return to this point below.

We choose the Crank–Nicholson scheme,

$$\mathbf{H}_j^{(n+1)} - \mathbf{H}_j^{(n)} = \frac{\Delta t}{2}\left(\mathcal{L}(\mathbf{H}_j^{(n+1)}) + \mathcal{N}(\mathbf{H}_j^{(n+1)}) + \mathcal{L}(\mathbf{H}_j^{(n)}) + \mathcal{N}(\mathbf{H}_j^{(n)})\right), \tag{F.7}$$

where $\Delta t = t_{n+1} - t_n$ is the time step. As the solution $\mathbf{H}_j^{(n)}$ is known at time step n, the inversion of (F.7) gives $\mathbf{H}_j^{(n+1)}$. This can be achieved using Newton's method. However, it might not always be easy to implement due to the nature of the nonlinearities $\mathcal{N}(\mathbf{H}_j^{(n+1)})$ and the dimensionality of the equations to be solved. A quasilinearization can be employed instead which gives

$$\mathbf{H}_j^{(n+1)} - \mathbf{H}_j^{(n)} = \frac{\Delta t}{2}\big[\mathcal{L}(\mathbf{H}_j^{(n+1)}) + \mathcal{L}(\mathbf{H}_j^{(n)}) + 2\mathcal{N}(\mathbf{H}_j^{(n)})$$
$$+ \mathcal{N}'(\mathbf{H}_j^{(n)})(\mathbf{H}_j^{(n+1)} - \mathbf{H}_j^{(n)})\big]. \tag{F.8}$$

This scheme substantially simplifies the numerical analysis, though in general the time steps must be smaller than those required for the Newton method. The scheme (F.8) is "consistent" (the solution of the numerical scheme converges to the solution of the partial differential equation (F.6a) as Δt goes to zero) and of "second-order precision" in time (the error associated with the time integration is of $\mathcal{O}(\Delta t^2)$)). Indeed, a Taylor expansion at the node $(x_j, \ t_{n+1/2} = \frac{1}{2}(t_{n+1} + t_n))$ yields

$$\frac{\Delta t}{2}\big[\mathcal{L}(\mathbf{H}_j^{(n+1)}) + \mathcal{L}(\mathbf{H}_j^{(n)}) + 2\mathcal{N}(\mathbf{H}_j^{(n)})$$
$$+ \mathcal{N}'(\mathbf{H}_j^{(n)})(\mathbf{H}_j^{(n+1)} - \mathbf{H}_j^{(n)})\big] - (\mathbf{H}_j^{(n+1)} - \mathbf{H}_j^{(n)})$$
$$- \Delta t\big[\mathcal{L}(\mathbf{H}_j^{(n+1/2)}) + \mathcal{N}(\mathbf{H}_j^{(n+1/2)}) - (\partial_t \mathbf{H})_j^{(n+1/2)}\big] = \mathcal{O}(\Delta t^2). \tag{F.9}$$

The spatial derivatives are approximated by central difference schemes whose precision is again of second-order:

$$(\partial_x \mathbf{H})_j = \frac{1}{2\Delta x}(\mathbf{H}_{j+1} - \mathbf{H}_{j-1}) + \mathcal{O}(\Delta x^2) \tag{F.10a}$$

$$(\partial_{x^2} \mathbf{H})_j = \frac{1}{\Delta x^2}(\mathbf{H}_{j+1} - 2\mathbf{H}_j + \mathbf{H}_{j-1}) + \mathcal{O}(\Delta x^2) \tag{F.10b}$$

$$(\partial_{x^3} \mathbf{H})_j = \frac{1}{2\Delta x^3}(\mathbf{H}_{j+2} - 2\mathbf{H}_{j+1} + 2\mathbf{H}_{j-1} - \mathbf{H}_{j-2}) + \mathcal{O}(\Delta x^2). \tag{F.10c}$$

The quasi-linearized Crank–Nicholson scheme is then formally written as

$$\mathbf{H}^{n+1} = \mathbf{L}^n \mathbf{H}^n, \tag{F.11}$$

where $\mathbf{L}^n$ corresponds to the linear operator $\mathbf{I} - (\Delta t/2)[\mathcal{L} + \mathcal{N}'(\mathbf{H}_j^{(n)})]$ and where $\mathbf{H}^n$ and $\mathbf{H}^{n+1}$ are the unknowns. Thanks to the choice of central differences, $\mathbf{L}^n$ is a diagonally dominant matrix[2] with a nonzero determinant, which then ensures the existence of the inverse of this matrix, which in turn ensures the existence and uniqueness at each time step of the solution to (F.8). $\mathbf{L}^n$ is a "banded matrix," i.e., a sparse matrix whose nonzero entries are confined to a diagonal band whose bandwidth depends on the order of the spatial derivatives. In practice, the numerical

---

[2] A matrix $\mathbf{A} = (a_{ij})$ is said to be "diagonally dominant" if in every row of the matrix the magnitude of the diagonal entry in that row is larger than the sum of the magnitudes of all the other (nondiagonal) entries in that row, i.e., $\forall i \ |a_{ii}| \geq \sum_{j \neq i} |a_{ij}|$. For the properties of diagonally dominant matrices see for example [104].

inversion of (F.11) at each time step can be achieved effectively by direct methods like "LU decomposition," for which, in the case of a banded matrix, the number of required operations is proportional to N and to the bandwidth.

Applying our equations at the first node $j = 1$ demands the values of the film height at two fictitious nodes, $x_0 = 0$ and $x_{-1} = -\Delta x$, upstream of the left end of the computational domain: The finite difference approximation for the third derivative in (F.10c) (the highest derivative for h) involves the nodes $j - 1$ and $j - 2$. For the flow rate, the highest derivative of our equation is a second one, which from (F.10b) involves the node $j - 1$ and hence we need to know the flow rate at the fictitious node $x_0 = 0$. We then impose

$$h_{-1}^{(n)} = h_0^{(n)} = h_f(t_n),$$
(F.12a)

$$q_0^{(n)} = q_f(t_n),$$
(F.12b)

so that nodes h_1 and h_0 are excited simultaneously. Hence, we have a total of three boundary conditions at the inlet.

The treatment of the outlet boundary condition is more subtle. If we had to solve a system of hyperbolic equations (where information travels from left to right) we would not need an outlet boundary condition. But the presence of surface tension makes our equations parabolic and we need boundary conditions at two points. At the same time, we wish to avoid fictitious nodes outside the right end of the computational domain, as we do not have any information about the dynamics there (we wish to keep the "flow" of information from the left to the right). Hence, if $j = N$ is the last node, we could utilize backward finite differences and use only the information at nodes $j = N - 1, N - 2, N - 3, \ldots$. Any boundary condition which would require more than three nodes would introduce more diagonals in the matrix to be inverted (and hence the bandwidth of the matrix to be inverted would be larger) and would therefore increase the number of operations needed at each time step, which in turn might generate spurious reflections of the waves at the endpoint of the domain. A simple and effective way then to deal with the right end of the domain is to impose there a set of linear hyperbolic equations, e.g.,

$$\partial_t \mathbf{H} = v_f \partial_x \mathbf{H},$$
(F.13)

with $v_f > 0$, corresponding to two boundary conditions for h and q at the outlet, which together with (F.12a), (F.12b) means a total of five boundary conditions for our equations (we have a third spatial derivative for h and a second one for q). The wave equation (F.13) is an ad hoc outlet "soft" boundary condition that simulates the wavy behavior of the film, with v_f being a relaxation parameter that is empirically tuned to limit wave reflections. It has a first-order spatial derivative whose discretization in (F.14) requires only the $j = N, N - 1, N - 2$ nodes and hence it does not increase the bandwidth of the matrix to be inverted.

Equation (F.13) ensures that the information is transported downstream and limits wave reflection. It effectively "hyperbolizes" our original system of equations at the end of the domain, ensuring that the information travels from the left to the right. The drawback of using (F.13) is the generation of some numerical errors at the outlet. However, thanks to the convective nature of the primary and secondary instabilities of falling film flows (Sect. 7.1.2), these numerical disturbances cannot invade the numerical domain and therefore affect only a few nodes at an outlet boundary layer. At the last node, the spatial derivative is discretized using an upstream second-order accurate scheme (for consistency with (F.10a)–(F.10c)):

$$(\partial_x \mathbf{H})_N = \frac{1}{2\Delta x}(\mathbf{H}_{N-2} - 4\mathbf{H}_{N-1} + 3\mathbf{H}_N) + \mathcal{O}(\Delta x^2). \tag{F.14}$$

Notice that according to the "Lax–Richtmyer equivalence theorem", for a consistent difference scheme for a linear system of equations, there is equivalence between convergence and stability [220]. For linear systems a stability analysis is therefore sufficient to determine the convergence properties of the scheme. However, in our case, a stability analysis is rather difficult, if not impossible.

As far as the Crank–Nicholson scheme is concerned, Richtmyer and Morton [220] give the following result on its stability. Consider the one-dimensional problem

$$\partial_t u + a(x)\partial_x u = 0, \tag{F.15a}$$

$$u(0, t) = 0, \qquad u(x, 0) = F(x), \tag{F.15b}$$

where $a(x)$ is a positive function. The Crank–Nicholson scheme for (F.15a), (F.15b) is stable and convergent if

$$4K_L \Delta t < 1 \tag{F.16}$$

is satisfied, where K_L is a Lipschitz constant of the function $a(x)$ such that $\forall x, x', \ |a(x) - a(x')| \le K_L|x - x'|$. Since $a(x) = -\partial_t u/\partial_x u \sim -\Delta x/\Delta t$, this result suggests that we impose a condition similar to the "Courant–Friedrichs–Lewy (CFL) condition" [5, 59]. Stated in simple terms, the CFL condition demands that the speed at which the information propagates in the numerical scheme ($\sim \Delta x/\Delta t$) must be larger than the speed of propagation of the physical phenomenon ($\sim K_L$) that is simulated, which in turn implies a constraint on the time step, $\Delta t < K'\Delta x$, where K' is a constant.

Assume for simplicity now that we are dealing with a surface equation so that $\mathbf{H}$ is a scalar, H. For our quasi-linearized Crank–Nicholson scheme (F.11) then let

$$a_j^{(n)} = \Delta t \frac{|(\partial_x H)_j^{(n)}|}{|H_j^{(n)} - H_j^{(n-1)}|}$$

be a measure of the local propagation speed of the physical phenomenon. At each time step, the maximum speed $\max_j |a_j^{(n)}|$ is computed and the time step is adjusted to verify the CFL-like condition,

$$\Delta t < \frac{C\Delta x}{\max_j |a_j^{(n)}|}, \tag{F.17}$$

where C a constant larger than unity.

F.4 Spectral Representation and Aliasing

The time-dependent simulations and the stability analysis of the traveling wave solutions shown in Sects. 8.3 and 8.4 made extensive use of the representation of periodic solutions in Fourier space. For this purpose, an efficient algorithm, the fast Fourier transform (FFT), was utilized. It enables us to go back and forth from the physical space to the Fourier space at a low computational cost. We sketch below the representation of the solutions in the Fourier space and its main limitation related to the *aliasing phenomenon* to be defined soon. Interested readers can consult the book by Press et al. as an introduction to the use of spectral and pseudo-spectral methods [213].

Real periodic solutions can be represented by Fourier series of the form

$$X(\xi) = \hat{\Xi}_0 + \sum_{j=1}^{\infty} \hat{\Xi}_{2j-1} \cos(jk_x\xi) + \hat{\Xi}_{2j} \sin(jk_x\xi), \tag{F.18}$$

where k_x again denotes the streamwise wavenumber. Equation (F.18) can be rewritten in the equivalent form

$$X(\xi) = \frac{1}{L_x} \sum_{j=-\infty}^{\infty} \hat{X}_j \exp(-ijk_x\xi). \tag{F.19}$$

Complex Fourier coefficients $\hat{X}_j$ are defined by means of a continuous Fourier transform,

$$\hat{X}_j = \int_0^{L_x} X(\xi) \exp(ijk_x\xi) \, d\xi, \tag{F.20}$$

where $i = \sqrt{-1}$ denotes the imaginary unit. As X is a real function, the complex coefficients $\hat{X}_j$ satisfy $\hat{X}^\star_{-j} = \hat{X}_j = L_x(\hat{\Xi}_{2j-1} - i\hat{\Xi}_{2j})$, where the star denotes complex conjugation.

An approximation to X can be obtained by truncating the Fourier series (F.18) at N coefficients,

$$X \simeq \hat{\Xi}_0 + \hat{\Xi}_{N-1} \cos\left[(N/2)k_x\xi\right] + \sum_{j=1}^{N/2-1} \hat{\Xi}_{2j-1} \cos(jk_x\xi) + \hat{\Xi}_{2k} \sin(jk_x\xi), \tag{F.21}$$

where N is an even integer; this is equivalent to canceling the coefficients of all frequencies that are not in the interval $[-f_c, f_c]$ with $f_c = N/(2L_x)$, the "Nyquist cut-off frequency." This truncation is acceptable only when the neglected part of the spectrum has a sufficiently small norm, which can be easily controlled by looking at the coefficient $\hat{\Xi}_{N-1}$ of the cut-off frequency.

Whenever the condition $\hat{\Xi}_{N-1} \ll 1$ is satisfied, one can replace the continuous Fourier transform with a discrete one,

$$\hat{X}_j^N = \sum_{p=0}^{N-1} X(\xi_p) \exp(ijk_x\xi_p), \tag{F.22}$$

where the nodes $\xi_p = pL_x/N$ are evenly distributed over a period. We note that $\hat{X}_{N/2}^N = \hat{X}_{-N/2}^N$ are real coefficients. The exponent N here is reserved for the discrete Fourier transform and is introduced to distinguish discrete and continuous Fourier transforms. For periodic signals the two transforms converge to the same result after rescaling:

$$\lim_{N\to\infty} \frac{L_x \hat{X}_j^N}{N} \to \hat{X}_j. \tag{F.23}$$

The bad news about the discrete Fourier transform is that it "confuses" harmonics:

$$\hat{X}_j^N = \sum_{p=0}^{N-1} \frac{1}{L_x} \sum_{q=-\infty}^{\infty} \hat{X}_q \exp(-iqk_x\xi_p) \exp(ijk_x\xi_p)$$

$$= \frac{1}{L_x} \sum_{q=-\infty}^{\infty} \hat{X}_q \sum_{p=0}^{N-1} \exp\left[i2\pi p(j-q)/N\right]$$

$$= \frac{N}{L_x} \sum_{q=-\infty}^{\infty} \hat{X}_{j+qN}. \tag{F.24}$$

As a result, the part of the power spectrum density that does not lie in the frequency range $[-f_c, f_c]$ is moved into that range. This phenomenon is called *aliasing*. Any frequency coefficient outside this range is aliased, that is falsely displaced. Aliasing errors can be limited by ensuring that the norm $\|\hat{X}_{N/2}^N\|$ corresponding to the Nyquist frequency is sufficiently small.

In any algorithm based on pseudo-spectral methods, derivatives are computed in the Fourier space, as a differentiation there corresponds to a simple product,

$$\hat{X'}_j^N = -ijk_x\hat{X}_j^N, \tag{F.25}$$

whereas nonlinearities are computed in the physical space as they correspond to convolutions in the Fourier space. Unfortunately, convolutions in the Fourier space widen the spectrum.

Consider an order σ_{NL} nonlinearity $X^{\sigma_{NL}}$ with σ_{NL} an integer. We have

$$[X(\xi_p)]^{\sigma_{NL}} = \left(\frac{1}{N} \sum_{j=1-N/2}^{N/2} \hat{X}_j^N \exp^{-ijk_x\xi_p} \right)^{\sigma_{NL}}$$

$$= \frac{1}{N^{\sigma_{NL}}} \sum_{j=\sigma_{NL}-\sigma_{NL}N/2}^{\sigma_{NL}N/2} \sum_{D_j^N} \left(\prod_{l=1}^{\sigma_{NL}} \hat{X}_{jl}^N \right) \exp(-ijk_x\xi_p), \tag{F.26}$$

where D_j^N is the set defined by

$$D_j^N = \left\{ (j_1, j_2, \ldots, j_{\sigma_{NL}}) \Big/ \sum_{l=1}^{\sigma_{NL}} j_l = j; \ -N/2 + 1 \le j_l \le N/2 \right\}. \qquad (F.27)$$

Let us assume that σ_{NL} is even (only the details of the proof are modified in the odd case) and let us "aliase" the frequencies outside the interval $[-f_c, \ f_c]$, first by use of the periodicity $\exp(-ijk_x\xi_p) = \exp[-ip(j+N)k_xL_x/N] = \exp[-i(j+N)k_x\xi_p]$,

$$
\begin{aligned}
\left[X(\xi_p)\right]^{\sigma_{NL}} = \frac{1}{N_{NL}^\sigma} \Bigg[&\sum_{j=\sigma_{NL}-\sigma_{NL}N/2}^{-(\sigma_{NL}-1)N/2} \sum_{D_j^N} \left(\prod_{l=1}^{\sigma_{NL}} \hat{X}_{j_l}^N\right) e^{-i(j+\sigma_{NL}N/2)k_x\xi_p} \\
+ &\sum_{j=-(\sigma_{NL}-1)N/2+1}^{-(\sigma_{NL}-2)N/2} \sum_{D_j^N} \left(\prod_{l=1}^{\sigma_{NL}} \hat{X}_{j_l}^N\right) e^{-i(j+(\sigma_{NL}-2)N/2)k_x\xi_p} \\
+ \cdots + &\sum_{j=(\sigma_{NL}-1)N/2+1}^{\sigma_{NL}N/2} \sum_{D_j^N} \left(\prod_{l=1}^{\sigma_{NL}} \hat{X}_{j_l}^N\right) e^{-i(j-\sigma_{NL}N/2)k_x\xi_p} \Bigg], \qquad (F.28)
\end{aligned}
$$

and then through a change of variables,

$$
\begin{aligned}
\left[X(\xi_p)\right]^{\sigma_{NL}} = \frac{1}{N_{NL}^\sigma} \Bigg[&\sum_{j=\sigma_{NL}}^{N/2} \sum_{D_{j-\sigma_{NL}N/2}^N} \left(\prod_{l=1}^{\sigma_{NL}} \hat{X}_{j_l}^N\right) e^{-ijk_x\xi_p} \\
+ &\sum_{j=-N/2+1}^{0} \sum_{D_{j-(\sigma_{NL}-2)N/2}^N} \left(\prod_{l=1}^{\sigma_{NL}} \hat{X}_{j_l}^N\right) e^{-ijk_x\xi_p} \\
+ \cdots + &\sum_{j=-N/2+1}^{N/2} \sum_{D_j^N} \left(\prod_{l=1}^{\sigma_{NL}} \hat{X}_{j_l}^N\right) e^{-ijk_x\xi_p} \\
+ \cdots + &\sum_{j=-N/2+1}^{0} \sum_{D_{j+\sigma_{NL}N/2}^N} \left(\prod_{l=1}^{\sigma_{NL}} \hat{X}_{j_l}^N\right) e^{-ijk_x\xi_p} \Bigg]. \qquad (F.29)
\end{aligned}
$$

In (F.29) the result of the convolution of the frequencies that lie inside the interval $[-f_c, \ f_c]$ is the sum $\sum_{j=-N/2+1}^{N/2} \sum_{D_j^N} (\prod_{l=1}^{\sigma_{NL}} \hat{X}_{j_l}^N) e^{-ijk_x\xi_p}$. All other terms in (F.29) arise from the aliasing of frequencies outside this interval.

The Fourier coefficients of X_{NL}^{σ} for $j < 0$ are thus given by

$$\frac{1}{N}(\hat{X}_{\mathrm{NL}}^{\sigma})_j^N = \frac{1}{N_{\mathrm{NL}}^{\sigma}}\left[\sum_{m\in[0,\,(\sigma_{\mathrm{NL}}-2)/4[\cap\mathbb{N}}\;\sum_{D_{j-(\sigma_{\mathrm{NL}}-2-4m)N/2}^N}\left(\prod_{l=1}^{\sigma_{\mathrm{NL}}}\hat{X}_{jl}^N\right)\right.$$

$$\left. + \sum_{m\in[0,\,\sigma_{\mathrm{NL}}/4[\cap\mathbb{N}}\;\sum_{D_{j+(\sigma_{\mathrm{NL}}-4m)N/2}^N}\left(\prod_{l=1}^{\sigma_{\mathrm{NL}}}\hat{X}_{jl}^N\right) + \sum_{D_j^N}\left(\prod_{l=1}^{\sigma_{\mathrm{NL}}}\hat{X}_{jl}^N\right)\right]. \quad (\text{F.30})$$

Similar expressions are obtained for $j > 0$.

One solution to limit aliasing is to truncate the Fourier spectrum of the function X. Let M be an integer that divides N and $f_{\mathrm{trunc}} = M/(2L_x) = Mf_{\mathrm{c}}/N$, a truncation frequency. If the coefficients of all frequencies outside the range $[f_{\mathrm{trunc}}, f_{\mathrm{trunc}}]$ are set to zero, expression (F.30) is modified by the substitution of the sets D_j^M for D_j^N. It is therefore sufficient to chose M so that the set

$$\left(\bigcup_{m\in[0,\,(\sigma_{\mathrm{NL}}-2)/4[\cap\mathbb{N}} D_{j-(\sigma_{\mathrm{NL}}-2-4m)N/2}^M\right)\cup\left(\bigcup_{m\in[0,\,\sigma_{\mathrm{NL}}/4[\cap\mathbb{N}} D_{j+(\sigma_{\mathrm{NL}}-4m)N/2}^M\right)$$

is empty. A sufficient condition is that the sets $D_{j\pm N}^M$ be empty, which reads

$$j - N < \sum_{l=1}^{\sigma_{\mathrm{NL}}} j_l < j + N; \quad j, j_l \in \{-M/2 + 1, M/2\}, \quad (\text{F.31})$$

or, equivalently,

$$-N < \left(\sum_{l=1}^{\sigma_{\mathrm{NL}}} j_l\right) - j < N; \quad j, j_l \in \{-M/2 + 1, M/2\}. \quad (\text{F.32})$$

Since, $\sigma_{\mathrm{NL}} - (\sigma_{\mathrm{NL}} + 1)M/2 \le (\sum_{l=1}^{\sigma_{\mathrm{NL}}} j_l) - j \le (\sigma_{\mathrm{NL}} + 1)M/2 - 1$, we finally obtain the sufficient condition

$$M \le \frac{2}{\sigma_{\mathrm{NL}} + 1}N. \quad (\text{F.33})$$

For quadratic nonlinearities ($\sigma_{\mathrm{NL}} = 2$), as in the case of the Navier–Stokes equations, condition (F.33) is known as the "two-third rule." Inequality (F.33) is a constraining condition. As the price to pay for the elimination of the cross-stream coordinate y in our modeling strategy is the emergence of high-order nonlinearities, the treatment of the aliasing phenomenon imposes shutting down a significant number of Fourier coefficients.

References

1. Adomeit, P., Renz, U.: Hydrodynamics of three-dimensional waves in laminar falling films. Int. J. Multiph. Flow **26**, 1183–1208 (2000)
2. Adrian, R.J., Meinhart, C.D., Tumkins, A.: Vortex organization in the outer region of the turbulent boundary layer. J. Fluid Mech. **422**, 1–54 (2000)
3. Alekseenko, S.V., Nakoryakov, V.E., Pokusaev, B.G.: Wave Flow in Liquid Films, 3rd edn. Begell House, New York (1994)
4. Alekseenko, S.V., Nakoryakov, V.E., Pokusaev, B.G.: Wave formation on a vertical falling liquid film. AIChE J. **31**, 1446–1460 (1985)
5. Allaire, G., Craig, A.: An Introduction to Mathematical Modelling and Numerical Simulation. Oxford University Press, London (2007)
6. Alleborn, N.: Progress in coating theory. Therm. Sci. **5**(1), 131–152 (2001)
7. Alleborn, N., Durst, F., Lienhart, H.: Use of curtain coating for surface treatment of concrete construction parts. Chem. Ing. Tech. **77**, 84–89 (1996). (In German)
8. Allgower, E.L., Georg, K.: Numerical Continuation Methods. Springer, New York (1990)
9. Amaouche, M., Mehidi, N., Amataousse, N.: An accurate modeling of thin film flows down an incline for inertia dominated regimes. Eur. J. Mech. B, Fluids **24**, 49–70 (2004)
10. Anshus, B.E., Goren, S.L.: A method of getting approximate solutions to the Orr–Sommerfeld equation for flow on a vertical wall. AIChE J. **12**(5), 1004–1008 (1966)
11. Argyriadi, K., Vlachogiannis, M., Bontozoglou, V.: Experimental study of inclined film flow along periodic corrugations: the effect of wall steepness. Phys. Fluids **18**, 012102 (2006)
12. Aris, R.: Vectors, Tensors, and the Basic Equations of Fluid Mechanics. Dover, New York (1962)
13. Bach, P., Villadsen, J.: Simulation of the vertical flow of a thin, wavy film using a finite-element method. Int. J. Heat Mass Transf. **27**, 815–827 (1984)
14. Balmforth, N.J., Liu, J.J.: Roll waves in mud. J. Fluid Mech. **519**, 33–54 (2004)
15. Balmforth, N.J., Ierley, G.R., Spiegel, E.A.: Chaotic pulse trains. SIAM J. Appl. Math. **54**(5), 1291–1334 (1994)
16. Bamforth, J.R., Kalliadasis, S., Merkin, J.H., Scott, S.K.: Modelling flow-distributing oscillations in the CDIMA reaction. Phys. Chem. Chem. Phys. **2**, 4013–4021 (2000)
17. Bejan, A.: Convection Heat Transfer. Wiley-VCH, New York (1995)
18. Bender, C.M., Orszag, S.A.: Advanced Mathematical Methods for Scientists and Engineers. McGraw-Hill, New York (1978)
19. Benjamin, T.B.: Wave formation in laminar flow down an inclined plane. J. Fluid Mech. **2**, 554–574 (1957). Corrigendum in J. Fluid Mech. **3**, 657
20. Benjamin, T.B.: The development of three-dimensional disturbances in an unstable film of liquid flowing down an inclined plane. J. Fluid Mech. **10**, 401–419 (1961)
21. Benney, D.J.: Long waves on liquid films. J. Math. Phys. **45**, 150–155 (1966)

S. Kalliadasis et al., *Falling Liquid Films*, Applied Mathematical Sciences 176,
DOI 10.1007/978-1-84882-367-9, © Springer-Verlag London Limited 2012

22. Bertozzi, A.L., Pugh, M.: Long-wave instabilities and saturation in thin film equations. Commun. Pure Appl. Math. **51**, 625–661 (1998)
23. Bertschy, J.R., Chin, R.W., Abernathy, F.H.: High-strain-rate free-surface boundary layer flows. J. Fluid Mech. **126**, 443–461 (1983)
24. Bestehorn, M., Pototsky, A., Thiele, U.: 3D large scale Marangoni convection in liquid films. Eur. Phys. J. B **33**, 457–467 (2003)
25. Binnie, A.M.: Experiments on the onset of wave formation on a film of water flowing down a vertical plane. J. Fluid Mech. **2**, 551–553 (1957)
26. Bird, R.B., Stewart, W.E., Lightfoot, E.N.: Transport Phenomena. Wiley, New York (1960)
27. Birikh, R.V., Brisman, V.A., Velarde, M.G., Legros, J.C.: Liquid Interfacial Systems. Oscillations and Instability. Dekker, New York (2003)
28. Boag, J.W., Rubinin, P.E., Shoenberg, D.: Kapitza in Cambridge and Moscow. Life and Letters of a Russian Physicist. North-Holland, Amsterdam (1990)
29. Boos, W., Thess, A.: Cascade of structures in long-wavelength Marangoni instability. Phys. Fluids **11**, 1484–1494 (1999)
30. Boussinesq, J.V.: Essai sur la theorie des eaux courantes. Memoires presentes par divers savants a l'Academie des Sciences, Institute de France (Paris) **23**, 1–680 (1877)
31. Brevdo, L., Laure, P., Dias, F., Bridges, T.J.: Linear pulse structure and signaling in a film flow on an inclined plane. J. Fluid Mech. **396**, 37–71 (1999)
32. Briggs, R.J.: Electron-Stream Interaction with Plasmas. MIT Press, Cambridge (1964)
33. Brock, R.R.: Periodic permanent roll waves. J. Hydrol. Eng. **96**(12), 2565–2580 (1970)
34. Bunov, A.V., Demekhin, E.A., Shkadov, V. Ya.: On the non uniqueness of nonlinear waves on a falling film. J. Appl. Math. Mech. **48**(4), 691–696 (1984)
35. Burns, J.C.: Long waves in running water. Proc. Camb. Philol. Soc. **49**, 695–706 (1953)
36. Cachile, M., Schneemilch, M., Hamraoui, A., Cazabat, A.M.: Films driven by surface tension gradients. Adv. Colloid Interface Sci. **96**, 59–74 (2002)
37. Carbone, F., Aubry, N., Liu, J., Gollub, J.P., Lima, R.: Space-time description of the splitting and coalescence of wave fronts in film flows. Physica D **96**, 182–199 (1996)
38. Carr, J.: Applications of Centre Manifold Theory. Springer, New York (1981)
39. Champneys, A.R., Kuznetsov, Y.A.: Numerical detection and continuation of codimension-two homoclinic bifurcations. Int. J. Bifurc. Chaos **4**, 785–822 (1994)
40. Chan, B., Balmforth, N.J., Hosoi, A.E.: Building a better snail: lubrication and adhesive locomotion. Phys. Fluids **17**, 113101 (2005)
41. Chang, H.-C.: Onset of nonlinear waves on falling films. Phys. Fluids A, Fluid Dyn. **1**, 1314–1327 (1989)
42. Chang, H.-C.: Wave evolution on a falling film. Annu. Rev. Fluid Mech. **26**, 103–136 (1994)
43. Chang, H.-C., Demekhin, E.A.: Coherent structures, self similarity, and universal roll-wave coarsening dynamics. Phys. Fluids **12**(9), 2268–2278 (2000)
44. Chang, H.-C., Demekhin, E.A.: Complex Wave Dynamics on Thin Films. D. Möbius and R. Miller. Elsevier, Amsterdam (2002)
45. Chang, H.-C., Demekhin, E., Kalaidin, E.: Interaction dynamics of solitary waves on a falling film. J. Fluid Mech. **294**, 123–154 (1995)
46. Chang, H.-C., Demekhin, E.A., Kalaidin, E.: Simulation of noise-driven wave dynamics on a falling film. AIChE J. **42**, 1553–1568 (1996)
47. Chang, H.-C., Demekhin, E.A., Kalaidin, E.: Generation and suppression of radiation by solitary pulses. SIAM J. Appl. Math. **58**(4), 1246–1277 (1998)
48. Chang, H.-C., Demekhin, E.A., Kopelevitch, D.I.: Construction of stationary waves on a falling film. Comput. Mech. **11**, 313–322 (1993)
49. Chang, H.-C., Demekhin, E.A., Kopelevitch, D.I.: Laminarizing effects of dispersion in an active-dissipative nonlinear medium. Physica D **63**, 299–320 (1993)
50. Chang, H.-C., Demekhin, E.A., Kopelevitch, D.I.: Nonlinear evolution of waves on a vertically falling film. J. Fluid Mech. **250**, 433–480 (1993)
51. Chang, H.-C., Demekhin, E.A., Kopelevitch, D.I.: Local stability theory of solitary pulses in an active medium. Physica D **97**, 353–375 (1996)

52. Chang, H.-C., Cheng, M., Demekhin, E.A., Kopelevitch, D.I.: Secondary and tertiary excitation of three-dimensional patterns on a falling film. J. Fluid Mech. **270**, 251–275 (1994)
53. Cheng, M., Chang, H.-C.: A generalized sideband stability theory via center manifold projection. Phys. Fluids **2**, 1364–1379 (1990)
54. Chicone, C.: Ordinary Differential Equations and Applications. Springer, New York (1999)
55. Christov, C.I., Velarde, M.G.: Dissipative solitons. Physica D **86**, 323–347 (1995)
56. Chu, K.J., Dukler, A.E.: Statistical characteristics of thin, wavy films. Part 2. Studies of the substrate and its wave structure. AIChE J. **20**, 695–706 (1974)
57. Chu, K.J., Dukler, A.E.: Statistical characteristics of thin, wavy films. Part 3. Structure of the large waves and their resistance of gas flow. AIChE J. **21**, 583–593 (1975)
58. Colinet, P., Legros, J.C., Velarde, M.G.: Nonlinear Dynamics of Surface-Tension-Driven Instabilities, p. 512. Wiley-VCH, New York (2001)
59. Courant, R., Friedrichs, K., Lewy, H.: On the partial difference equations of mathematical physics. IBM J. 215–234 (1967)
60. Craster, R.V., Matar, O.K.: On viscous beads flowing down a vertical fibre. J. Fluid Mech. **553**, 85–105 (2006)
61. Craster, R.V., Matar, O.K.: Dynamics and stability of thin liquid films. Rev. Mod. Phys. **81**, 1131–1198 (2009)
62. Cross, M.C., Hohenberg, P.C.: Pattern formation outside of equilibrium. Rev. Mod. Phys. **65**, 851–1112 (1993)
63. Dávalos-Orozco, L.A.: Nonlinear instability of a thin film flowing down a smoothly deformed surface. Phys. Fluids **19**, 074103 (2007)
64. Dávalos-Orozco, L.A., Busse, F.H.: Instability of a thin film flowing on a rotating horizontal or inclined plane. Phys. Rev. E **65**, 026312 (2002)
65. Dávalos-Orozco, L.A., Ruiz-Chavarria, G.: Hydrodynamic stability of a fluid layer flowing down a rotating inclined plane. Phys. Fluids A, Fluid Dyn. **4**(8), 1651–1665 (1992)
66. Dávalos-Orozco, L.A., Ruiz-Chavarria, G.: Hydrodynamic instability of a fluid layer flowing down a rotating cylinder. Phys. Fluids A, Fluid Dyn. **5**(10), 2390–2404 (1993)
67. Davey, A.: A simple numerical method for solving Orr–Sommerfeld problems. Q. J. Mech. Appl. Math. **26**, 401–411 (1973)
68. de Gennes, P.G., Brochard-Wyart, F., Quéré, D.: Capillarity and Wetting Phenomena: Drops, Bubbles, Pearls, Waves. Springer, New York (2004)
69. de Groot, S.R., Mazur, P.: Nonequilibrium Thermodynamics. North-Holland, Amsterdam (1962)
70. Demekhin, E.A., Kaplan, M.A.: Construction of exact numerical solutions of the stationary traveling type for viscous thin films. Izv. Akad. Nauk SSSR, Meh. židk. Gaza **6**, 73–81 (1989)
71. Demekhin, E.A., Shkadov, V.Ya.: Three-dimensional waves in a liquid flowing down a wall. Izv. Akad. Nauk SSSR, Meh. židk. Gaza **5**, 21–27 (1984)
72. Demekhin, E.A., Kaplan, M.A., Shkadov, V.Ya.: Mathematical models of the theory of viscous liquid films. Izv. Akad. Nauk SSSR, Meh. židk. Gaza **6**, 73–81 (1987)
73. Demekhin, E.A., Tokarev, G.Yu., Shkadov, V.Ya.: Hierarchy of bifurcations of space-periodic structures in a nonlinear model of active dissipative media. Physica D **52**, 338–361 (1991)
74. Demekhin, E.A., Tokarev, G.Yu., Shkadov, V.Ya.: Numerical modeling of three-dimensional wave evolution in a falling layer of viscous fluid. Vestn. Mosk. Univ., Ser. 1: Mat., Mekh. **2**, 50–54 (1988)
75. Demekhin, E.A., Kalaidin, E.N., Kalliadasis, S., Vlaskin, S.Yu.: Three-dimensional localized coherent structures of surface turbulence: I. Scenarios of two-dimensional three-dimensional transitions. Phys. Fluids **19**, 114103 (2007)
76. Demekhin, E.A., Kalaidin, E.N., Kalliadasis, S., Vlaskin, S.Yu.: Three-dimensional localized coherent structures of surface turbulence: II. Λ solitons. Phys. Fluids **19**, 114104 (2007)
77. De Wit, A.: Fingering of chemical fronts in porous media. Phys. Rev. Lett. **87**, 054502 (2002)
78. Diez, J.A., Kondic, L., Bertozzi, A.: Global models for moving contact lines. Phys. Rev. E **63**, 011208 (2000)

79. Doedel, E.J.: AUTO07P continuation and bifurcation software for ordinary differential equations. Montreal Concordia University (2008)
80. Doedel, E.J., Keller, H.B., Kernévez, J.P.: Numerical analysis and control of bifurcation problems: (I) Bifurcation in finite dimensions. Int. J. Bifurc. Chaos **1**(3), 493–520 (1991)
81. Doedel, E.J., Keller, H.B., Kernevez, J.P.: Numerical analysis and control of bifurcation problems: (II) Bifurcation in infinite dimensions. Int. J. Bifurc. Chaos **1**(4), 745–772 (1991)
82. Drazin, P.G., Johnson, R.S.: Solitons: an Introduction. Cambridge University Press, London (1989)
83. Drazin, P.G., Reid, W.H.: Hydrodynamic Instability. Cambridge University Press, London (1981)
84. Dressler, R.F.: Mathematical solution of the problem of roll-waves in inclined open channels. Commun. Pure Appl. Math. **2**, 149–194 (1949)
85. Duprat, C., Ruyer-Quil, C., Giorgiutti-Dauphiné, F.: Spatial evolution of a film flowing down a fiber. Phys. Fluids **21**, 042109 (2009)
86. Duprat, C., Ruyer-Quil, C., Kalliadasis, S., Giorgiutti-Dauphiné, F.: Absolute and convective instabilities of a film flowing down a vertical fiber. Phys. Rev. Lett. **98**, 244502 (2007)
87. Duprat, C., Giorgiutti-Dauphiné, F., Tseluiko, D., Saprykin, S., Kalliadasis, S.: Liquid film coating a fiber as a model system for the formation of bound states in active dispersive-dissipative nonlinear media. Phys. Rev. Lett. **103**, 234501 (2009)
88. Dussan, E.B., Davis, S.H.: On the motion of a fluid-fluid interface along a solid surface. J. Fluid Mech. **65**, 71–95 (1974)
89. Dussan, E.B.: On the spreading of liquids on solid surfaces: static and dynamic contact lines. Annu. Rev. Fluid Mech. **11**, 371–400 (1979)
90. Elphick, C., Ierley, G.R., Regev, O., Spiegel, E.A.: Interacting localized structures with Galilean invariance. Phys. Rev. A **44**, 1110–1123 (1991)
91. Émery, H., Brosse, O.: Écoulement d'un film le long d'un plan incliné: développement spatio-temporel des instabilités. PhD thesis, École Polytechnique, Rapport d'option X92 (1995)
92. Finlayson, B.A.: The Method of Weighted Residuals and Variational Principles, with Application in Fluid Mechanics, Heat and Mass Transfer. Academic Press, San Diego (1972)
93. Floquet, G.: Sur les équations différentielles linéaires à coefficients périodiques. Ann. Ec. Norm. Supér. **12**, 47–88 (1883)
94. Floryan, J.M., Davis, S.H., Kelly, R.E.: Instabilities of a liquid film flowing down a slightly inclined plane. Phys. Fluids **30**, 983–989 (1987)
95. Frenkel, A.L.: Nonlinear theory of strongly undulating thin films flowing down vertical cylinders. Europhys. Lett. **18**, 583–588 (1992)
96. Friedman, B.: Principles and Techniques of Applied Mathematics. Wiley, New York (1956)
97. Frisk, D.P., Davis, E.J.: The enhancement of heat transfer by waves in stratified gas–liquid flow. Int. J. Heat Mass Transf. **15**, 1537–1552 (1972)
98. Gallay, T.: A center-stable manifold theorem for differential equations in Banach spaces. Commun. Math. Phys. **152**, 249–268 (1993)
99. Gao, D., Morley, N.B., Dhir, V.: Numerical simulation of wavy falling film flow using VOF method. J. Comput. Phys. **192**, 624–642 (2003)
100. Garazo, A.N., Velarde, M.G.: Dissipative Korteweg–de Vries description of Marangoni–Bénard oscillatory convection. Phys. Fluids **3**, 2295–2300 (1991)
101. Gaspard, P.: Local birth of homoclinic chaos. Physica D **62**, 94–122 (1993)
102. Gjevik, B.: Occurrence of finite-amplitude surface waves on falling liquid films. Phys. Fluids **13**, 1918–1925 (1970)
103. Glendinning, P., Sparrow, C.: Local and global behavior near homoclinic orbits. J. Stat. Phys. **35**, 645–696 (1984)
104. Golub, G.H., Loan, F.V.: Matrix Computations, 3rd edn. John Hopkins University Press, Baltimore (1996)
105. Gondret, P., Rabaud, M.: Shear instability of two-fluid parallel flow in a Hele–Shaw cell. Phys. Fluids **9**(11), 3267–3274 (1997)

106. Goren, S.L., Mani, P.V.S.: Mass transfer through horizontal liquid films in wavy motion. AIChE J. **14**, 57–61 (1968)
107. Goussis, D.A., Kelly, R.E.: Surface wave and thermocapillary instabilities in a liquid film flow. J. Fluid Mech. **223**, 25 (1991). Corrigendum in J. Fluid Mech. **226**, 663
108. Greenberg, M.D.: Foundations of Applied Mathematics. Prentice-Hall, New York (1978)
109. Greenberg, M.D.: Advanced Engineering Mathematics, 2nd edn. Prentice-Hall, New York (1998)
110. Grotberg, J.B.: Pulmonary flow and transport phenomena. Annu. Rev. Fluid Mech. **26**, 529–571 (1994)
111. Guckenheimer, J., Holmes, P.: Nonlinear Oscillations, Dynamical Systems and Bifurcations of Vector Fields. Springer, New York (1983)
112. Häcker, T., Uecker, H.: An integral boundary layer equation for film flow over inclined wavy bottoms. Phys. Fluids **21**, 092105 (2009)
113. Herbert, T.: Secondary instability of boundary layers. Annu. Rev. Fluid Mech. **20**, 487–526 (1988)
114. Hinch, E.J.: A note on the mechanism of the instability at the interface between two shearing fluids. J. Fluid Mech. **144**, 463–465 (1984)
115. Hinch, E.J.: Perturbation Methods. Cambridge University Press, London (1991)
116. Ho, L.W., Patera, A.T.: A Legendre spectral element method for simulation of unsteady incompressible viscous free-surface flows. Comput. Methods Appl. Mech. Eng. **80**, 355–366 (1990)
117. Hocherman, T., Rosenau, P.: On KS-type equations describing the evolution and rupture of a liquid interface. Physica D **67**, 113 (1993)
118. Homsy, G.M.: Model equations for wavy viscous film flow. Lect. Appl. Math. **15**, 191 (1974)
119. Homsy, G.M.: Model equations for wavy viscous film flow, in nonlinear wave motion. In: Newell, A. (Ed.) Lectures in Applied Mathematics, vol. 15, pp. 191–194. American Mathematical Society, Providence (1974)
120. Huerre, P., Monkewitz, P.: Local and global instabilities in spatially developing flows. Annu. Rev. Fluid Mech. **22**, 473–537 (1990)
121. Huerre, P., Rossi, M.: Hydrodynamic instabilities in open flows. In: Godrèche, C., Manneville, P. (Eds.) Hydrodynamic and Nonlinear Instabilities, pp. 81–294. Cambridge University Press, London (1998). Especially §8, 9
122. Huh, C., Scriven, L.E.: Hydrodynamic model of steady movement of solid/liquid/fluid contact line. J. Colloid Interface Sci. **35**, 85–101 (1971)
123. Iooss, G., Joseph, D.D.: Elementary Stability and Bifurcation Theory. Springer, New York (1980)
124. Ito, A., Masunaga, N., Baba, K.: Marangoni effects on wave structure and liquid film breakdown along heated vertical tube. In: Serizawa, A., Fukano, T., Bataille, J. (Eds.) Advances in Multiphase Flow, pp. 255–265. Elsevier, New York (1995)
125. Johnson, R.S.: Shallow water waves on a viscous fluid—the undular bore. Phys. Fluids **15**(10), 1693–1699 (1972)
126. Johnson, R.S.: A Modern Introduction to the Mathematical Theory of Water Waves, 1st edn. Cambridge Texts in Applied Mathematics (No. 19), p. 445. Cambridge University Press, Cambridge (1997)
127. Joo, S.W., Davis, S.H.: Instabilities of three-dimensional viscous falling films. J. Fluid Mech. **242**, 529–547 (1992)
128. Joo, S.W., Davis, S.H., Bankoff, S.G.: Long-wave instabilities of heated falling films: two-dimensional theory of uniform layers. J. Fluid Mech. **230**, 117–146 (1991)
129. Joo, S.W., Davis, S.H., Bankoff, S.G.: A mechanism for rivulet formation in heated falling films. J. Fluid Mech. **321**, 279–298 (1996)
130. Joseph, D.D., Bai, R., Chen, K.P., Renardy, Y.Y.: Core-annular flows. Annu. Rev. Fluid Mech. **29**, 65–90 (1997)
131. Kabov, O.A.: Heat transfer from a small heater to a falling liquid film. Heat Transf. Res. **27**, 221 (1996)

132. Kabov, O.A., Chinnov, E.A.: Heat transfer from a local heat source to a subcooled falling liquid film evaporating in a vapor–gas medium. Russ. J. Eng. Thermophys. **7**, 1–34 (1997)
133. Kabov, O.A., Chinnov, E.A., Legros, J.C.: Three-dimensional deformations in non-uniformly heated falling liquid film at small and moderate Reynolds numbers. In: Proceedings of the Conference on Transport Phenomena with Moving Boundaries, 11–12 October, Berlin, Germany (2003)
134. Kabov, O.A., Marchuk, I.V., Muzykantov, A.V., Legros, J.C., Istasse, E., Dewandel, J.-L.: Regular structures in locally heated falling liquid films. In: Celata, G.P., Di Marco, P., Shah, R.K. (Eds.) 2nd Intern. Symp. on Two-Phase Flow Modelling and Experimentation, May, Pisa (Italy), vol. 2, pp. 1225–1233 (1999)
135. Kalliadasis, S.: Nonlinear instability of a contact line driven by gravity. J. Fluid Mech. **413**, 355–378 (2000)
136. Kalliadasis, S., Chang, H.-C.: Drop formation during coating of vertical fibres. J. Fluid Mech. **261**, 135–168 (1994)
137. Kalliadasis, S., Thiele, U. (Eds.): Thin Films of Soft Matter. Springer-Wien, New York (2007)
138. Kalliadasis, S., Kiyashko, A., Demekhin, E.A.: Marangoni instability of a thin liquid film heated from below by a local heat source. J. Fluid Mech. **475**, 377–408 (2003)
139. Kalliadasis, S., Demekhin, E.A., Ruyer-Quil, C., Velarde, M.G.: Thermocapillary instability and wave formation on a film flowing down a uniformly heated plane. J. Fluid Mech. **492**, 303–338 (2003)
140. Kapitza, P.L.: Wave flow of thin layers of a viscous fluid: I. Free flow. II. Fluid flow in the presence of continuous gas flow and heat transfer. In: ter Haar, D. (Ed.) Collected Papers of P.L. Kapitza (1965), pp. 662–689. Pergamon, Oxford (1948). (Original paper in Russian: Zh. Eksp. Teor. Fiz. **18**, I. 3–18, II. 19–28)
141. Kapitza, P.L., Kapitza, S.P.: Wave flow of thin layers of a viscous fluid: III. Experimental study of undulatory flow conditions. In: ter Haar, D. (Ed.) Collected Papers of P.L. Kapitza (1965), pp. 690–709. Pergamon, Oxford (1949). (Original paper in Russian: Zh. Eksp. Teor. Fiz. **19**, 105–120)
142. Kargupta, K., Sharma, A.: Dewetting of thin films on periodic physically and chemically patterned surfaces. Langmuir **18**, 1893–1903 (2002)
143. Kataoka, D.E., Troian, S.M.: Patterning liquid flow on the microscopic scale. Nature **402**, 794–797 (1999)
144. Kawahara, T.: Formation of saturated solitons in a nonlinear dispersive system with instability and dissipation. Phys. Rev. Lett. **51**, 381–383 (1983)
145. Kawahara, T., Toh, S.: Pulse interactions in an unstable dissipative-dispersive nonlinear system. Phys. Fluids **31**, 2103–2111 (1988)
146. Keller, H.B.: Numerical solution of bifurcation and nonlinear eigenvalue problems. In: Rabinowitz, P.H. (Ed.) Applications of Bifurcation Theory, 3rd edn., pp. 359–384. Academic Press, New York (1977)
147. Kelly, R.E., Goussis, D.A., Lin, S.P., Hsu, F.K.: The mechanism for surface wave instability in film flow down an inclined plane. Phys. Fluids A, Fluid Dyn. **1**, 819–828 (1989)
148. Kistler, S.F., Schweitzer, P.M.: Liquid Film Coating: Scientific Principles and Their Technological Implications. Chapman & Hall, London (1997)
149. Kliakhandler, I.L., Davis, S.H., Bankoff, S.G.: Viscous beads on vertical fibre. J. Fluid Mech. **429**, 381–390 (2001)
150. Kliakhandler, I.L., Porubov, A.V., Velarde, M.G.: Localized finite-amplitude disturbances and selection of solitary waves. Phys. Rev. E **62**(4), 4959–4962 (2000)
151. Kondic, L.: Instabilities in gravity driven flow of thin fluid films. SIAM Rev. **45**(1), 95–115 (2003)
152. Kondic, L., Diez, J.: Flow of thin films on patterned surfaces: controlling the instability. Phys. Rev. E **45**, 045301 (2002)
153. Kondic, L., Palffy-Muhoray, P., Shelley, M.J.: Models of non-Newtonian Hele–Shaw flow. Phys. Rev. E **54**(5), 4536–4539 (1996)

154. Krantz, W.B., Goren, S.L.: Finite-amplitude, long waves on liquid films flowing down a plane. Ind. Eng. Chem. Res. **9**, 107–113 (1970)

155. Krantz, W.B., Goren, S.L.: Stability of thin liquid films flowing down a plane. Ind. Eng. Chem. Res. **10**, 91–101 (1971)

156. Krishnamoorthy, S., Ramaswamy, B.: Spontaneous rupture of thin liquid films due to thermocapillarity: a full-scale direct numerical simulation. Phys. Fluids **7**, 2291 (1995)

157. Lamger, J.S.: Instabilities and pattern formation in crystal growth. Rev. Mod. Phys. **52**(1), 1–28 (1980)

158. Landau, L.D., Lifsitz, E.M.: Fluid Mechanics. Pergamon, Oxford (1959)

159. Leal, L.G.: Laminar Flow and Convective Transport Processes. Butterworth-Heinemann, Stoneham (1992)

160. Leal, L.G.: Fluid Mechanics and Convective Transport Processes. Cambridge University Press, Cambridge (2007)

161. Lee, J.-J., Mei, C.C.: Stationary waves on an inclined sheet of viscous fluid at high Reynolds and moderate Weber numbers. J. Fluid Mech. **307**, 191–229 (1996)

162. Lienhard, J.H. IV, Leinhard, J.H. V: A Heat Transfer Textbook, 3rd edn. Phlogiston Press, Lexington (2004)

163. Levich, V.G.: Physico-Chemical Hydrodynamics. Prentice-Hall, Englewood Cliffs (1962)

164. Lin, S.P.: Finite amplitude side-band stability of a viscous fluid. J. Fluid Mech. **63**, 417–429 (1974)

165. Lin, S.P., Jiang, W.Y.: The mechanism of suppression or enhancement of three-dimensional surface waves in film flow down a vertical plane. Phys. Fluids **14**(2), 4088 (2002)

166. Lin, S.P., Chen, J.N., Woods, D.R.: Suppression of instability in a liquid film flow. Phys. Fluids **8**(12), 3247–3252 (1996)

167. Liu, J., Gollub, J.P.: Onset of spatially chaotic waves on flowing films. Phys. Rev. Lett. **70**, 2289–2292 (1993)

168. Liu, J., Gollub, J.P.: Solitary wave dynamics of film flows. Phys. Fluids **6**, 1702–1712 (1994)

169. Liu, J., Paul, J.D., Gollub, J.P.: Measurements of the primary instabilities of film flows. J. Fluid Mech. **250**, 69–101 (1993)

170. Liu, J., Schneider, J.B., Gollub, J.P.: Three-dimensional instabilities of film flows. Phys. Fluids **7**, 55–67 (1995)

171. Liu, K., Mei, C.C.: Roll waves on a layer of a muddy fluid flowing down a gentle slope—a Bingham model. Phys. Fluids **6**(8), 2577–2589 (1994)

172. Liu, Q.Q., Chen, L., Li, J.C., Singh, V.P.: Roll waves in overland flow. J. Hydrol. Eng. **10**(2), 110–117 (2005)

173. Loglio, G.: Carlo Marangoni and the Laboratory of Physics at the High School "Liceo Classico Dante" in Firenze (Italy). In: Gade, M., Huhnerfuss, H., Korenowski, G.M. (Eds.) Marine Surface Films. Chemical Characteristics, Influence on Air–Sea Interactions and Remote Sensing, pp. 13–15. Springer, Berlin (2006)

174. Lopez, P.G., Miksis, M.J., Bankoff, S.G.: Inertial effects on contact line instability in the coating of a dry inclined plane. Phys. Fluids **9**, 2177–2183 (1997)

175. Malamataris, N.A., Papanastasiou, T.C.: Unsteady free surface flows on truncated domains. Ind. Eng. Chem. Res. **30**, 2211–2219 (1991)

176. Malamataris, N.A., Vlachogiannis, M., Bontozoglou, V.: Solitary waves on inclined films: Flow structure and binary interactions. Phys. Fluids **14**(3), 1082–1094 (2002)

177. Manneville, P.: Dissipative Structures and Weak Turbulence. Academic Press, New York (1990)

178. Marangoni, C.: Sull' Espansione delle Gocce di Un Liquido Galleggiante Sulla Superficie di Altro Liquido. Pavia, Tipographic Fusi (1865)

179. Marangoni, C.: Sul principio della viscosita superficiale dei liquidi stabilito dal Sig. J. Plateau. Part 1. Nuovo Cim., Ser. II **2**(5/6), 238–266 (1872)

180. Marangoni, C.: Sul principio della viscosita superficiale dei liquidi stabilito dal Sig. J. Plateau. Part 2. Nuovo Cim., Ser. II **2**(5/6), 266–273 (1872), p. 268 §14. Abbonacciamento delle onde (pp. 268–270)

181. Meron, E.: Pattern formation in excitable media. Phys. Rep. **218**, 1–66 (1992)
182. Miklavcic, M., Williams, M.: Stability of mean flows over an infinite flat plate. Arch. Ration. Mech. Anal. **80**, 57–69 (1982)
183. Moffatt, H.K.: Viscous and resistive eddies near a sharp corner. J. Fluid Mech. **18**, 1 (1964)
184. Moldavsky, L., Frichman, M., Oron, A.: Dynamics of thin films falling on vertical cylindrical Durfaces subjected to ultrasound forcing. Phys. Rev. E **76**, 045301 (2007)
185. Nakaya, C.: Long waves on a viscous fluid down a vertical wall. Phys. Fluids A, Fluid Dyn. **18**, 1407–1412 (1975)
186. Nakaya, C.: Waves on a viscous fluid down a vertical wall. Phys. Fluids A, Fluid Dyn. **1**, 1143 (1989)
187. Nakoryakov, V.E., Pokusaev, B.G., Alekseenko, S.V., Orlov, V.V.: Instantaneous velocity profile in a wavy liquid film. Inzh.-Fiz Zh. **33**, 399–405 (1977). (In Russian)
188. Nepomnyashchy, A.A.: Three-dimensional spatially periodic motions in liquid films flowing down a vertical plane. Hydrodynam. Perm **7**, 43–52 (1974). (In Russian)
189. Nepomnyashchy, A.A., Velarde, M.G., Colinet, P.: Interfacial Phenomena and Convection. Chapman & Hall/CRC, London (2002)
190. Nguyen, L.T., Balakotaiah, V.: Modeling and experimental studies of wave evolution on free falling viscous films. Phys. Fluids **12**, 2236–2256 (2000)
191. Normand, C., Pomeau, Y., Velarde, M.G.: Convective instability: a physicist's approach. Rev. Mod. Phys. **49**, 581–624 (1977)
192. Nosoko, T., Miyara, A.: The evolution and subsequent dynamics of waves on a vertically falling liquid film. Phys. Fluids **16**(4), 1118–1126 (2004)
193. Nosoko, T., Yoshimura, P.N., Nagata, T., Okawa, K.: Characteristics of two-dimensional waves on a falling liquid film. Chem. Eng. Sci. **51**, 725–732 (1996)
194. Nusselt, W.: Die oberflachenkondensation des wasserdampfes. Z. VDI **50**, 541–546 (1916)
195. Oldeman, B.E., Champneys, A.R., Krauskopf, B.: Homoclinic branch switching: a numerical implementation of Lin's method. Int. J. Bifurc. Chaos **13**, 2977–2999 (2003)
196. Ooshida, T.: Surface equation of falling film flows with moderate Reynolds number and large but finite Weber number. Phys. Fluids **11**, 3247–3269 (1999)
197. Oron, A.: Nonlinear dynamics of three-dimensional long-wave Marangoni instability in thin liquid films. Phys. Fluids **12**, 1633 (2000)
198. Oron, A., Gottlieb, O.: Nonlinear dynamics of temporally excited falling liquid films. Phys. Fluids **14**, 2622–2636 (2002)
199. Oron, A., Gottlieb, O.: Subcritical and supercritical bifurcations of the first- and second-order Benney equations. J. Eng. Math. **50**, 121–140 (2004)
200. Oron, A., Rosenau, P.: Evolution and formation of dispersive-dissipative patterns. Phys. Rev. E **55**, 1267–1270 (1997)
201. Oron, A., Davis, S.H., Bankoff, S.G.: Long-scale evolution of thin liquid films. Rev. Mod. Phys. **69**, 931–980 (1997)
202. Orszag, S.A.: Accurate solution of the Oss–Sommerfeld stability equation. J. Fluid Mech. **50**, 689–703 (1971)
203. Park, C.D., Nosoko, T.: Three-dimensional wave dynamics on a falling film and associated mass transfer. AIChE J. **49**, 2715–2727 (2003)
204. Pascal, J.P.: Linear stability of fluid flow down a porous inclined plane. J. Phys. D, Appl. Phys. **32**, 417–422 (1999)
205. Pascal, J.P.: Instability of power-law fluid flow down a porous incline. J. Non-Newton. Fluid Mech. **133**, 109–120 (2006)
206. Pearson, J.R.A.: On convection cells induced by surface tension. J. Fluid Mech. **4**, 489–500 (1958)
207. Pereira, A., Kalliadasis, S.: Dynamics of a falling film with solutal Marangoni effect. Phys. Rev. E **78**, 036312 (2008)
208. Pereira, A., Trevelyan, P.M.J., Thiele, U., Kalliadasis, S.: Dynamics of a horizontal thin liquid film in the presence of reactive surfactants. Phys. Fluids **19**, 112102 (2007)
209. Pereira, A., Trevelyan, P.M.J., Thiele, U., Kalliadasis, S.: Interfacial hydrodynamic waves driven by chemical reactions. J. Eng. Math. **59**, 207–220 (2007)

210. Petviashvili, V.I., Tsvelodub, O.Yu.: Horse-shoe shaped solitons on an inclined viscous liquid film. Dokl. Akad. Nauk SSSR **238**, 1321–1324 (1978)
211. Pierson, F.W., Whitaker, S.: Some theoretical and experimental observations of the wave structure of falling liquid films. Ind. Eng. Chem. Res. **16**, 401–408 (1977)
212. Pradas, M., Tseluiko, D., Kalliadasis, S.: Rigorous coherent-structure theory for falling liquid films: Viscous dispersion effects on bound-state formation and self-organization. Phys. Fluids **23**, 044104 (2011)
213. Press, W.H., Teukolsky, S.A., Vetterling, W.T., Flannery, B.P.: Numerical Recipes in C—The Art of Scientific Computing, 2nd edn. Cambridge University Press, New York (1992)
214. Prokopiou, T., Cheng, M., Chang, H.-C.: Long waves on inclined films at high Reynolds number. J. Fluid Mech. **222**, 665–691 (1991)
215. de Saint-Venant, A.J.C.: Théorie du mouvement non-permanent des eaux, avec applications aux crues des rivières et à l'introduction des marées dans leur lit. C. R. Acad. Sci. Paris **73**, 147–154 (1871)
216. Pumir, A., Manneville, P., Pomeau, Y.: On solitary waves running down an inclined plane. J. Fluid Mech. **135**, 27–50 (1983)
217. Quéré, D.: Fluid coating on a fiber. Annu. Rev. Fluid Mech. **31**, 347–384 (1999)
218. Ramaswamy, B., Chippada, S., Joo, S.W.: A full-scale numerical study of interfacial instabilities in thin-film flows. J. Fluid Mech. **325**, 163–194 (1996)
219. Ramaswamy, B., Krishnamoorthy, S., Joo, S.W.: Three-dimensional simulation of instabilities and rivulet formation in heated falling films. J. Comput. Phys. **131**, 70–88 (1997)
220. Richtmyer, R.D., Morton, K.W.: Difference Methods for Initial-Value Problems. Interscience, New York (1967)
221. Roberts, A.J.: Low-dimensional models of thin film fluid dynamics. Phys. Lett. A **212**, 63–71 (1996)
222. Roberts, A.J.: Low-dimensional modelling of dynamics via computer algebra. Comput. Phys. Commun. **100**, 215–230 (1997)
223. Roberts, A.J., Li, Z.-Q.: An accurate and comprehensive model of thin fluid flows with inertia on curved substrates. J. Fluid Mech. **553**, 33–73 (2006)
224. Roy, R.V., Roberts, A.J., Simpson, M.E.: A lubrication model of coating flows over a curved substrate in space. J. Fluid Mech. **454**, 235–261 (2002)
225. Ruyer-Quil, C.: Inertial corrections to the Darcy law in a Hele–Shaw cell. C. R. Acad. Sci. Sér. IIb **329**, 337–342 (2001)
226. Ruyer-Quil, C., Manneville, P.: Modeling film flows down inclined planes. Eur. Phys. J. B **6**, 277–292 (1998)
227. Ruyer-Quil, C., Manneville, P.: Improved modeling of flows down inclined planes. Eur. Phys. J. B **15**, 357–369 (2000)
228. Ruyer-Quil, C., Manneville, P.: Further accuracy and convergence results on the modeling of flows down inclined planes by weighted-residual approximations. Phys. Fluids **14**, 170–183 (2002)
229. Ruyer-Quil, C., Manneville, P.: On the speed of solitary waves running down a vertical wall. J. Fluid Mech. **531**, 181–190 (2005)
230. Ruyer-Quil, C., Scheid, B., Kalliadasis, S., Velarde, M.G., Zeytounian, R. Kh.: Thermocapillary long waves in a liquid film flow. Part 1. Low dimensional formulation. J. Fluid Mech. **538**, 199–222 (2005)
231. Ruyer-Quil, C., Treveleyan, P., Giorgiutti-Dauphiné, F., Duprat, C., Kalliadasis, S.: Modelling film flows down a fibre. J. Fluid Mech. **603**, 431–462 (2008)
232. Salamon, T.R., Armstrong, R.C., Brown, R.A.: Traveling waves on vertical films: Numerical analysis using the finite element method. Phys. Fluids **6**, 2202 (1994)
233. Saprykin, S., Demekhin, E.A., Kalliadasis, S.: Self-organization of two-dimensional waves in an active dispersive-dissipative nonlinear medium. Phys. Rev. Lett. **94**, 224101 (2005)
234. Saprykin, S., Demekhin, E.A., Kalliadasis, S.: Two-dimensional wave dynamics in thin films. I. Stationary solitary pulses. Phys. Fluids **17**, 117105 (2005)

235. Saprykin, S., Demekhin, E.A., Kalliadasis, S.: Two-dimensional wave dynamics in thin films. II. Formation of lattices of interacting stationary solitary pulses. Phys. Fluids **17**, 117106 (2005)

236. Saprykin, S., Koopmans, R.J., Kalliadasis, S.: Free-surface thin-film flows over topography: influence of inertia and viscoelasticity. J. Fluid Mech. **578**, 271–293 (2007)

237. Saprykin, S., Trevelyan, P.M.J., Koopmans, R.J., Kalliadasis, S.: Free-surface thin-film flows over uniformly heated topography. Phys. Rev. E **75**, 026306 (2007)

238. Scheid, B., Ruyer-Quil, C., Manneville, P.: Wave patterns in film flows: modelling and three-dimensional waves. J. Fluid Mech. **562**, 183–222 (2006)

239. Scheid, B., Oron, A., Colinet, P., Thiele, U., Legros, J.C.: Nonlinear evolution of non-uniformly heated falling liquid films. Phys. Fluids **14**, 4130–4151 (2002). Erratum: Phys. Fluids **15**, 583 (2003)

240. Scheid, B., Ruyer-Quil, C., Kalliadasis, S., Velarde, M.G., Zeytounian, R. Kh.: Thermocapillary long waves in a liquid film flow. Part 2. Linear stability and nonlinear waves. J. Fluid Mech. **538**, 223–244 (2005)

241. Scheid, B., Ruyer-Quil, C., Thiele, U., Kabov, O.A., Legros, J.C., Colinet, P.: Validity domain of the Benney equation including the Marangoni effect for closed and open flows. J. Fluid Mech. **527**, 303–335 (2005)

242. Scheid, B., Kalliadasis, S., Ruyer-Quil, C., Colinet, P.: Interaction of three-dimensional hydrodynamic and thermocapillary instabilities in film flows. Phys. Rev. E **78**, 066311 (2008)

243. Schlichting, H.: Boundary-Layer Theory, 7th edn. McGraw-Hill, New York (1979)

244. Schmid, P.J., Henningson, D.S.: Stability and Transition in Shear Flows. Springer, Berlin (2001)

245. Scott, S.K.: Chemical Chaos. Oxford University Press, London (1993)

246. Scriven, L.E., Sterling, C.V.: On cellular convection driven surface tension gradients: effects of mean surface tension and surface viscosity. J. Fluid Mech. **19**, 321–340 (1964)

247. Shankar, V., Sharma, A.: Instability of the interface between thin fluid films subjected to electric fields. Langmuir **21**, 3710–3721 (2005)

248. Shkadov, V.Ya.: Wave flow regimes of a thin layer of viscous fluid subject to gravity. Izv. Akad. Nauk SSSR, Meh. židk. Gaza **1**, 43–51 (1967). (English translation in Fluid Dynamics **2**, 29–34 (Faraday Press, New York, 1970))

249. Shkadov, V.Ya.: Solitary waves in a layer of viscous liquid. Izv. Akad. Nauk SSSR, Meh. židk. Gaza **1**, 63–66 (1977)

250. Shkadov, V.Ya., Sisoev, G.M.: Waves induced by instability in falling films of finite thickness. Fluid Dyn. Res. **35**, 357–389 (2004)

251. Shkadov, V.Ya., Velarde, M.G., Shkadova, V.P.: Falling films and the Marangoni effect. Phys. Rev. E **69**, 056310 (2004)

252. Shraiman, B.I.: Diffusive transport in a Rayleigh–Bénard convection cell. Phys. Rev. A **36**, 261–267 (1987)

253. Sisoev, G.M., Shkadov, V.Ya.: A two-parameter manifold of wave solutions to an equation for a falling film of viscous fluid. Dokl. Phys. **44**, 454–459 (1999)

254. Skotheim, J.M., Thiele, U., Scheid, B.: On the instability of a falling film due to localized heating. J. Fluid Mech. **475**, 1–19 (2003)

255. Smith, K.A.: On convective instability induced by surface-tension gradients. J. Fluid Mech. **24**, 401–414 (1966)

256. Smith, M.K.: The mechanism for the long-wave instability in thin liquid films. J. Fluid Mech. **217**, 469–485 (1990)

257. Smith, M.K., Davis, S.H.: Instabilities of dynamic thermocapillary liquid layers. Part 1. Convective instabilities. J. Fluid Mech. **132**, 119–144 (1983)

258. Smith, M.K., Davis, S.H.: Instabilities of dynamic thermocapillary liquid layers. Part 2. Surface-wave instabilities. J. Fluid Mech. **132**, 145–162 (1983)

259. Solorio, F.J., Sen, M.: Linear stability of a cylindrical falling film. J. Fluid Mech. **183**, 365–377 (1987)

260. Spindler, B.: Linear stability of liquid films with interfacial phase change. Int. J. Heat Mass Transf. **25**(2), 161–173 (1982)

261. Sreenivasan, S., Lin, S.P.: Surface tension driven instability of a liquid film flow down a heated incline. Int. J. Heat Mass Transf. **21**, 1517 (1978)
262. Stainthorp, F.P., Allen, J.M.: The development of ripples on the inside of a liquid film moving inside a vertical tube. Trans. Inst. Chem. Eng. **43**, 85–91 (1965)
263. Stuart, J.T.: Nonlinear stability theory. Annu. Rev. Fluid Mech. **3**, 347–370 (1971)
264. Tailby, S.R., Portalski, S.: The hydrodynamics of liquid films flowing on vertical surfaces. Trans. Inst. Chem. Eng. **38**, 324–330 (1960)
265. Thess, A., Orszag, S.A.: Surface-tension-driven Bénard convection at infinite Prandtl number. J. Fluid Mech. **283**, 201–230 (1995)
266. Thiele, U., Knobloch, E.: Front and back instability of a liquid film on a slightly inclined plate. Phys. Fluids **15**, 892–907 (2003)
267. Thiele, U., Goyeau, B., Velarde, M.G.: Stability analysis of thin film flow along a heated porous substrate. Phys. Fluids **21**, 014103 (2009)
268. Thiele, U., Velarde, M.G., Neuffer, K.: Dewetting: film rupture by nucleation in the spinodal regime. Phys. Rev. Lett. **87**(2), 1–4 (2001)
269. Thiele, U., Neuffer, K., Bestehorn, M., Pomeau, Y., Velarde, M.G.: Sliding drops on an inclined plane. Colloids Surf. A **206**, 87–104 (2002)
270. Thomas, H.A.: The propagation of waves in steep prismatic conduits. In: Proc. Hydraulics Conf., pp. 214–229. Univ. of Iowa, Iowa City (1939)
271. Thompson, J.: On certain curious motions observable on the surfaces of wine and other alcoholic liquours. Philos. Mag. **10**, 330–335 (1855)
272. Tiffany, J.M.: The viscosity of human tears. Int. Ophthalmol. **15**, 371–376 (1991)
273. Tihon, J., Serifi, K., Argyiriadi, K., Bontozoglou, V.: Solitary waves on inclined films: their characteristics and the effects on wall shear stress. Exp. Fluids **41**, 79–89 (2006)
274. Trevelyan, P.M.J., Kalliadasis, S.: Dynamics of a reactive falling film at large Péclet numbers. I. Long-wave approximation. Phys. Fluids **16**, 3191–3208 (2004)
275. Trevelyan, P.M.J., Kalliadasis, S.: Dynamics of a reactive falling film at large Péclet numbers. II. Nonlinear waves far from criticality: Integral-boundary-layer approximation. Phys. Fluids **16**, 3209–3226 (2004)
276. Trevelyan, P.M.J., Kalliadasis, S.: Dynamics of a thin-liquid film falling down a heated wall. J. Eng. Math. **50**, 177–208 (2004)
277. Trevelyan, P.M.J., Kalliadasis, S., Merkin, J.H., Scott, S.K.: Dynamics of a vertically falling film in the presence of a first-order chemical reaction. Phys. Fluids **14**, 2402–2121 (2002)
278. Trevelyan, P.M.J., Kalliadasis, S., Merkin, J.H., Scott, S.K.: Mass transport enhancement in regions bounded by rigid walls. J. Eng. Math. **42**, 45–64 (2002)
279. Trevelyan, P.M.J., Scheid, B., Ruyer-Quil, C., Kalliadasis, S.: Heated falling films. J. Fluid Mech. **592**, 295–334 (2007)
280. Trifonov, Yu.Ya.: Steady-state travelling waves on the surface of a viscous liquid film falling down vertucal wires and tubes. AIChE J. **38**, 821–834 (1992)
281. Trifonov, Yu.Ya., Tsvelodub, O.Yu.: Nonlinear waves on the surface of a falling liquid film. Part 1. Waves of the first family and their stability. J. Fluid Mech. **229**, 531–554 (1991)
282. Trifonov, Yu.Ya.: Bifurcations of two-dimensional into three-dimensional wave regimes for a vertically flowing liquid film. Izv. Akad. Nauk SSSR, Meh. židk. Gaza **5**, 109–114 (1989)
283. Tseluiko, D., Kalliadasis, S.: Coherent-structure theory for 3d active dispersive-dissipative nonlinear media (2011, in preparation)
284. Tseluiko, D., Saprykin, S., Kalliadasis, S.: Interaction of solitary pulses in active dispersive-dissipative media. Proc. Est. Acad. Sci. **59**, 139–144 (2010)
285. Tseluiko, D., Blyth, M.G., Papageorgiou, D.T., Vanden-Broeck, J.-M.: Electrified viscous thin film flow over topography. J. Fluid Mech. **597**, 449–475 (2008)
286. Tseluiko, D., Saprykin, S., Duprat, C., Giorgiutti-Dauphiné, F., Kalliadasis, S.: Pulse dynamics in low-Reynolds-number interfacial hydrodynamics: Experiments and theory. Physica D **239**, 2000–2010 (2010)
287. Tsvelodub, O.Yu., Trifonov, Yu.Ya.: Nonlinear waves on the surface of a falling liquid film. Part 2. Bifurcations of the first-family waves and other types of nonlinear waves. J. Fluid Mech. **244**, 149–169 (1992)

288. Usha, R., Uma, B.: Modeling of stationary waves on a thin viscous film down an inclined plane at high Reynolds numbers and moderate Weber numbers using energy integral method. Phys. Fluids **16**(7), 2679–2696 (2004)

289. Van Hook, S.J., Schatz, M.F., Swift, J.B., McCormick, W.D., Swinney, H.L.: Long wavelength surface-tension-driven Bénard convection: experiment and theory. J. Fluid Mech. **345**, 45–78 (1997)

290. Velarde, M.G., Zeytounian, R. Kh. (Eds.): Interfacial Phenomena and the Marangoni Effect. Springer-Wien, New York (2002)

291. Velarde, M.G., Nepomnyashchy, A.A., Hennenberg, M.: Onset of oscillatory interfacial instability and wave motions in Bénard layers. Adv. Appl. Mech. **37**, 167–238 (2000)

292. Verhulst, F.: Nonlinear Differential Equations and Dynamical Systems. Springer, Berlin (1990)

293. Verma, R., Sharma, A., Kargupta, K., Bhaumik, J.: Electric field induced instability and pattern formation in thin liquid films. Langmuir **21**, 3710–3721 (2005)

294. Vlachogiannis, M., Bontozoglou, V.: Observations of solitary wave dynamics of film flows. J. Fluid Mech. **435**, 191–215 (2001)

295. Wasden, F.K., Dukler, A.E.: Insights into the hydrodynamics of free falling wavy films. AIChE J. **35**, 187–195 (1989)

296. Weast, R.W., Selby, S.M.: Handbook of Chemistry and Physics, 63rd edn. CRC, Cleveland (1966)

297. Weinstein, S.J., Ruschak, K.J.: Coating flows. Annu. Rev. Fluid Mech. **36**, 29–53 (2004)

298. Whitaker, S.: Effect of surface active agents on the stability of falling liquid films. Ind. Eng. Chem. Res. **3**, 132–142 (1964)

299. Whitham, G.B.: Linear and Nonlinear Waves. Wiley, New York (1974)

300. Wierschem, A., Linde, H., Velarde, M.G.: Properties of surface wave trains excited by mass transfer through a liquid surface. Phys. Rev. E **64**, 022601 (2001)

301. Wiggins, S.: Introduction to Applied Nonlinear Dynamical Systems and Chaos. Springer, New York (1990)

302. Yang, J., D'Onofrio, A., Kalliadasis, S., De Wit, A.: Rayleigh–Taylor instability of reaction-diffusion acidity fronts. J. Chem. Phys. **117**, 9395–9408 (2002)

303. Yih, C.-S.: Stability of two-dimensional parallel flows for three dimensional disturbances. Quart. Appl. Math. **12**, 434–435 (1955)

304. Yih, C.-S.: Stability of liquid flow down an inclined plane. Phys. Fluids **6**, 321–334 (1963)

305. Yoshimura, P.N., Nosoko, T., Nagata, T.: Enhancement of mass transfer into a falling laminar liquid film by two-dimensional surface waves—some experimental observations and modeling. Chem. Eng. Sci. **51**(8), 1231–1240 (1996)

306. Yu, L.-Q., Wasden, F.K., Dukler, A.E., Balakotaiah, V.: Nonlinear evolution of waves on falling films at high Reynolds numbers. Phys. Fluids **7**, 1886–1902 (1995)

Index

S. Kalliadasis et al., *Falling Liquid Films*, Applied Mathematical Sciences 176,
DOI 10.1007/978-1-84882-367-9, © Springer-Verlag London Limited 2012

MIX
Papier aus verantwortungsvollen Quellen
Paper from responsible sources
FSC® C105338

If you have any concerns about our products,
you can contact us on
ProductSafety@springernature.com

In case Publisher is established outside the EU,
the EU authorized representative is:
**Springer Nature Customer Service Center GmbH
Europaplatz 3, 69115 Heidelberg, Germany**

Printed by Libri Plureos GmbH
in Hamburg, Germany